Climate Obstruction

CLIMATE OBSTRUCTION

A Global Assessment

Edited by J. Timmons Roberts
Carlos R. S. Milani
Jennifer Jacquet
  and
Christian Downie

OXFORD
UNIVERSITY PRESS

Oxford University Press is a department of the University of Oxford.
It furthers the University's objective of excellence in research, scholarship,
and education by publishing worldwide. Oxford is a registered trademark of
Oxford University Press in the UK and in certain other countries.

Published in the United States of America by Oxford University Press
198 Madison Avenue, New York, NY 10016, United States of America.

CIP data is on file at the Library of Congress

ISBN 9780197787151
ISBN 9780197787144 (hbk.)

DOI: 10.1093/oso/9780197787144.001.0001

The manufacturer's authorised representative in the EU for product safety is '
Oxford University Press España S.A., Parque Empresarial San Fernando de Henares,
Avenida de Castilla, 2 – 28830 Madrid (www.oup.es/en or product.safety@oup.com).
OUP España S.A. also acts as importer into Spain of products made by the manufacturer.

CONTENTS

1. Introduction: Understanding Obstruction of Climate Action   *1*
   J. Timmons Roberts, Carlos R. S. Milani, Jennifer Jacquet, and Christian Downie

2. The Global Role of the Oil and Gas Industry in Climate Delay and Denial   *29*
   Lead Authors: Geoff Dembicki, Kristoffer Ekberg, and Kert Davies

   Contributing Authors: Ann-Kristin Bergquist, Ada Nissen, and Stella Levantesi

3. How Coal, Utilities, and Transportation Impede Climate Action   *62*
   Lead Authors: Jen Schneider and Gregory Trencher

   Contributing Authors: Peter K. Bsumek, Christian Downie, Paul K. Gellert, Giulio Mattioli, Jason Monios, Peter Newell, Jennifer Peeples, Joeri Wesseling, Emily Williams, Ryan Wishart, and Ben Youriev

4. The Animal Agriculture Industry's Role in Obstructing Climate Action   *98*
   Lead Authors: Kathrin Lauber and Viveca Morris

   Contributing Authors: Jennifer Jacquet, Peter Li, Ina Möller, Silvia Secchi, Alex Wijeratna, and Melina De Bona

5. Climate Policy Obstruction on the Right and the Far Right   *139*
   Lead Authors: Dieter Plehwe and Justin Farrell

   Contributing Authors: Lucas Araldi, Robert J. Brulle, Jesse Callahan Bryant, William Callison, Kert Davies, Ruth E. McKie, Sotiris Mitralexis, and Alexandru Racu

LIST OF TABLES

LIST OF BOXES

ACKNOWLEDGMENTS

The coeditors would like to extend our thanks to the KR Foundation for supporting this volume. We would also like to express our deep gratitude to our managing editor, Miranda C. Spencer, who has made this volume possible, and who made every sentence read better. The collaboration of the thirty-one lead authors and nearly eighty contributing authors has been simply remarkable. Thanks to August DeVore, Michael Pulgar-Cabrera, Matt Margetta, Robert J. Brulle, and Kim Cobb at the Institute at Brown for Environment and Society for support. We appreciate James Cook, our editor at Oxford University Press, and the able production and editorial team at OUP. It has been an honor to work with you all.

LIST OF CONTRIBUTORS

Lucas Araldi is a PhD candidate in political science at the University of Kassel.

Melissa Aronczyk is a professor of media and politics in the School of Communication and Information at Rutgers University, with faculty affiliations in the Department of Sociology, the Rutgers Climate Institute, and the Eagleton Institute of Politics.

Stefan C. Aykut is a professor of sociology at Hamburg University who studies the societal dynamics of global climate governance and ecological transformations.

Joshua A. Basseches is the Flowerree assistant professor of environmental studies and public policy at Tulane University. His research focuses on US state climate and energy politics and policy.

Larissa Basso is a researcher at the Institute of Advanced Studies at the University of São Paulo.

Salil Benegal is the Joseph B. Board assistant professor of environmental policy at Union College.

Lisa Benjamin is an associate professor at Lewis & Clark Law School. Her research interests include climate impacts and actions of nonstate actors such as energy companies.

Ann-Kristin Bergquist is a professor of economic history in the Department of Economic History at Uppsala University.

Maxwell Boykoff is a professor in and chair of the Environmental Studies Department at the University of Colorado Boulder, and a fellow in its Cooperative Institute for Research in Environmental Sciences. He also leads its Media and Climate Change Observatory.

Rebecca Bromley-Trujillo is an associate professor in the Department of Political Science and research director of the Wason Center for Civic Leadership at Christopher Newport University. Her research explores US state and local climate change policy.

Robert J. Brulle is a visiting professor of environment and society at Brown University and the research codirector of the Climate Social Science Network (CSSN).

Jesse Callahan Bryant is a PhD candidate in sociology at the Yale School of the Environment.

Peter K. Bsumek is a professor in the School of Communication Studies at James Madison University.

William W. Buzbee is a professor of law at Georgetown University, and faculty director of its environmental and policy program.

William Callison is a lecturer in social studies at Harvard University and a member of the Zetkin Collective, an international group researching the political ecologies of the far right.

Lucas G. Christel is a research assistant at the National Council of Scientific and Technical Research and director of the BA in political science program at Universidad Nacional de San Martín.

Travis G. Coan is an associate professor in computational social science at the University of Exeter.

John Cook is a senior research fellow with the Melbourne Centre for Behaviour Change at the University of Melbourne, researching how to use critical thinking to counter misinformation. He coauthored the college textbook *Climate Change: Examining the Facts*.

Kert Davies works on the research team at the Center for Climate Integrity in Washington, DC. Previously, he served as director of the Climate Investigations Center and the Climate Files project, and as research director for Greenpeace USA.

Melina De Bona is a director attorney and adjunct professor at the Earth Rights Research and Action (TERRA) Program at New York University School of Law.

Claire Debucquois is a postdoctoral fellow with the Belgian Fund for Scientific Research.

Laurence L. Delina is an assistant professor of environment and sustainability at the Hong Kong University of Science and Technology.

Geoff Dembicki is an American investigative journalist who focuses on climate justice and oil and gas industry disinformation strategies. He is a regular contributor to *DeSmog* and *The Guardian*.

Kari De Pryck is a lecturer at the Institute for Environmental Sciences at the University of Geneva and affiliated with its Environmental Governance and Territorial Development research hub. She studies the negotiating practices between experts and diplomats within the Intergovernmental Panel on Climate Change.

Natalie Dietrich Jones is a senior research fellow at the Sir Arthur Lewis Institute of Social and Economic Studies, The University of the West Indies (Mona).

Christian Downie is a professor in the School of Regulation and Global Governance at the Australian National University, where he is the director of the Governing Energy Transition (GET) Lab.

Guy Edwards is a doctoral candidate in the Department of International Relations at Sussex University and co-director of the Latin America and the Caribbean Research Network at Sussex.

Kristoffer Ekberg is a historian and associate senior lecturer in human ecology at Lund University whose research focuses on environmental history, particularly business responses to environmental policy and movements.

Danielle Falzon is an assistant professor of sociology at Rutgers University.

Justin Farrell is a professor at Yale University's School of the Environment. He studies culture, with a focus on social class, moral conflict, epistemology, and the environment, blending ethnographic fieldwork with large-scale computational techniques.

M. Omar Faruque is an assistant professor of sociology at the University of New Brunswick. His research interests lie at the intersection of political and environmental sociology with a focus on social movements, resource conflict, environmental governance, and climate politics.

Louise M. Fitzgerald is an assistant professor in political science at Dublin City University in Ireland.

Matías Franchini is an associate professor of international relations at Universidad del Rosario.

Benjamin Franta, a licensed attorney in California, is the founding head of the Climate Litigation Lab and an associate professor of climate litigation at the University of Oxford Sustainable Law Programme.

Francisco Garcia-Gibson is a visiting fellow in LSE government and an adjunct researcher at the National Research Council, Argentina.

Paul K. Gellert is a professor of sociology at the University of Tennessee, Knoxville.

Fergus Green is a lecturer in political theory and public policy in the Department of Political Science at University College London.

Ricardo A. Gutierrez is a researcher at the National Council of Scientific and Technical Research and a professor at the Universidad Nacional de San Martín.

Friederike Hartz is a PhD candidate in geography at the University of Cambridge studying the IPCC.

Noel Healy is a professor of geography and sustainability at Salem State University.

David J. Hess is a professor in the Sociology Department at Vanderbilt University and is the James Thornton Fant chair in Sustainability Studies.

Kathryn Hochstetler is a professor and department head in the Department of International Development at the London School of Economics and Political Science.

Matthew Hornsey is a social psychologist and director of the Business Sustainability Initiative at the University of Queensland.

Hannah Hughes is a senior lecturer in international politics and climate change at Aberystwyth University.

Jennifer Jacquet is a professor in the Department of Environmental Science and Policy at the University of Miami. She also serves as associate research director at the Climate Social Science Network, headquartered at Brown University.

Alyssa Johl is legal director and general counsel for the Center for Climate Integrity.

Tariro Kamuti is a research fellow with the African Wildlife Economy Institute of Stellenbosch University, South Africa.

Andrew B. Kirkpatrick is an associate professor and chair of the political science department at Christopher Newport University.

Aria Kovalovich is a researcher with the US Senate Committee on the Budget.

Arunima Krishna is an associate professor of mass communication, advertising, and public relations at Boston University.

Laura Kuhl is an associate professor of public policy and urban affairs and international affairs at Northeastern University. Her research examines climate adaptation and sustainable transitions.

Myanna Lahsen is a senior researcher at Brazil's National Institute for Space Research (INPE).

Kathrin Lauber is a research fellow at the University of Edinburgh's School of Social and Political Science in the United Kingdom.

Stella Levantesi is a climate journalist and author who writes the "Gaslit" column at *DeSmog*.

Peter Li is an associate professor of East Asian politics, animal policy, and law in China at the University of Houston-Downtown.

Yifei Li is an assistant professor of environmental studies at New York University Shanghai and a global network assistant professor at New York University.

Maria Isabel Santos Lima is a PhD candidate in political science at the Institute for Social and Political Studies of the Rio de Janeiro State University.

Marcela López-Vallejo is a professor of international relations at the Center for North American Studies in the Department of Transpacific Studies at Universidad de Guadalajara.

Cristiana Losekann is an associate professor in the Department of Sociology at Espírito Santo Federal University.

Simone Lucatello is a researcher at Instituto Mora, Mexico City.

Andrew Malmuth is a PhD candidate in sociology at the University of California, Los Angeles.

Giulio Mattioli is a research fellow at the School of Spatial Planning at TU Dortmund University.

Ruth E. McKie is a senior lecturer in quantitative criminology at the School of Applied Social Sciences at De Montfort University. She is a sociologist whose research interests include climate obstruction.

Vinícius Mendes is a postdoctoral student in the Environmental Governance and Politics chair group at Radboud University.

David Michaels is an epidemiologist and professor at the Milken Institute School of Public Health at George Washington University.

Jean Carlos Hochsprung Miguel is an associate professor of science and technology policy at the University of Campinas.

Michael Mikulewicz is an assistant professor of environmental studies at SUNY College of Environmental Science and Forestry.

Carlos R. S. Milani is a professor of international relations at Rio de Janeiro State University's Institute for Social and Political Studies. He coordinates the university's Interdisciplinary Observatory on Climate Change (OIMC).

Meg Mills-Novoa is an assistant professor and director of the Climate Futures Lab at the University of California, Berkeley.

Sotiris Mitralexis is a research fellow at University College London Anthropology and a visiting professor at the Institute for Orthodox Christian Studies, Cambridge.

Elisabeth Möhle is a doctoral fellow in political science at CONICET and Universidad Nacional General San Martín.

Ina Möller is an assistant professor of climate politics in the Environmental Policy Group of Wageningen University.

Jason Monios is a professor of maritime logistics at Kedge Business School.

Bruna Bosi Moreira is a postdoctoral research fellow at the Käte Hamburger Kolleg/Centre for Global Cooperation Research of the University of Duisburg-Essen.

Hanna E. Morris is an assistant professor of climate change communication at the School of the Environment at the University of Toronto.

Viveca Morris is a research scholar at Yale Law School and the executive director of the school's Law, Environment & Animals Program.

Michelle Mycoo is a professor of urban and regional planning in the Department of Geomatics Engineering and Land Management at The University of the West Indies.

Peter Newell is a professor of international relations at the University of Sussex and cofounder and research director of the Rapid Transition Alliance.

Ada Nissen is an associate professor of history in the Department of Archaeology, Conservation, and History at the University of Oslo.

Grace Nosek is a postdoctoral fellow at the University of Toronto Scarborough focused on youth civic engagement and climate action. Her research focuses on corporate tactics to undermine public discourse and democracy.

Oluwaseun J. Oguntuase conducts research with the nonprofit The Anthropocene Research Academy in Nigeria.

Meg Parsons is an associate professor of environment at the University of Auckland.

Jennifer Peeples is a professor and head of the Department of Communication Studies at Utah State University.

Géraldine Pflieger is a professor of urban and environmental policy at the University of Geneva.

Dieter Plehwe is a senior fellow at the Center for Civil Society Research at the Berlin Social Science Center and a lecturer at the University of Kassel's Department of Political Science. He coleads the Climate Social Science Network's Global Census of Climate Obstruction Organizations working group.

Alexandru Racu is an associate professor in the Faculty of Political Science at Hyperion University.

M. Feisal Rahman is an associate fellow in the Living Deltas Research Hub at Newcastle University.

J. Timmons Roberts is the Ittleson professor of environmental and society at Brown University and the executive director of the Climate Social Science Network. His research focuses on social drivers of action on climate change.

Stacy-ann Robinson is an associate professor in the Department of Environmental Sciences at Emory University. She researches the human, social, and policy dimensions of climate change adaptation in Small Island Developing States.

Chris Russill is an associate professor in the School of Journalism and Communication at Carleton University.

E. Lisa F. Schipper is a professor of development geography at the University of Bonn.

Jen Schneider is the associate dean of the College of Innovation and Design and a professor of public policy and administration at Boise State University. Her research addresses challenges in public communication of scientific and environmental controversies.

Silvia Secchi is a professor in the School of Earth, the Environment and Sustainability at the University of Iowa.

Emanuel Semedo is a PhD candidate at the Rio de Janeiro State University and a researcher at the Interdisciplinary Observatory on Climate Change.

Joana Setzer is an assistant professorial research fellow at the Grantham Research Institute on Climate Change and the Environment at the London School of Economics and Political Science. Her main areas of expertise are climate litigation and global environmental governance.

Sonja Solomun is the deputy director of the Center for Media, Technology and Democracy at McGill University.

Dominik A. Stecuła is an assistant professor in the School of Communication at The Ohio State University whose research focuses on the information environment and its effects on society.

Jennie C. Stephens is a climate justice scholar-activist, feminist, and professor of climate justice at the ICARUS Climate Research Centre at the National University of Ireland Maynooth. Her research focuses on power, transformation, and the societal role of universities.

Kimberley Anh Thomas is an environmental justice scholar and associate professor in the Department of Geography, Environment and Urban Studies at Temple University.

Gregory Trencher is an associate professor of energy policy and sustainability transitions at the Graduate School of Global Environmental Studies at Kyoto University. His research focuses on decarbonization strategies in fossil fuel-centric industries and related issues.

Robin Tschoetschel is a postdoctoral researcher at the Department of Social Sciences at Universität Hamburg.

Eduardo Viola is a professor at the Institute of International Relations and a senior researcher of the Brazilian Council for Research (CNPQ). He is the chair of the Brazilian Research Network on the International System in the Anthropocene and Climate Change.

Edward Walker is a professor in and chair of the Department of Sociology at the University of California Los Angeles.

Jonathan R. Walz is an associate professor and chair of the Climate Change and Global Sustainability MA program at the School for International Training-Graduate Institute.

Joeri Wesseling is an assistant professor in innovation studies at the Copernicus Institute of Sustainable Development at Utrecht University.

Alex Wijeratna is an activist, journalist, and senior director at the nonprofit Mighty Earth.

Emily Williams is a postdoctoral scholar with the Sierra Nevada Research Institute at the University of California, Merced.

Ryan Wishart is an assistant professor of sociology at Creighton University and on the faculty of the university's environmental science and sustainability programs.

Adrian Dominik Wójcik is an associate professor of social and conservation psychology at Nicolaus Copernicus University.

Sharon Yadin is a senior lecturer of law and regulation at the Yezreel Valley College School of Public Administration and Public Policy.

Ben Youriev is the Transport Program manager at the think tank InfluenceMap.

CHAPTER 1

Introduction

Understanding Obstruction of Climate Action

J. TIMMONS ROBERTS, CARLOS R. S. MILANI,
JENNIFER JACQUET, AND CHRISTIAN DOWNIE

INTRODUCTION TO THE VOLUME

An overwhelming body of scientific research has established that dangerous destabilization of Earth's climate is occurring and will worsen if humanity continues with business as usual. This knowledge has been summarized in six major assessments released over four decades by the Intergovernmental Panel on Climate Change (IPCC).[1] Despite widespread agreement on the existence and causes of anthropogenic climate change, societies, governments, and economic agents have failed to respond adequately.

The decades of insufficient action on climate change have often been explained by pointing to various barriers—human nature, uncertainties in the science, technical shortcomings of renewable energy sources, democracy deficits, neoliberalism's emphasis on short-term economic logic, and deep religious and other cultural beliefs. The authors and editors of *Climate Obstruction: A Global Assessment* do not dispute the plausibility of these factors. However, based on the accumulating evidence documented here, we offer a different argument: while the major social transformation required to avert dangerous climate change was never going to be easy, it has been made exponentially more difficult by various forms of obstruction. Efforts to address climate change have been unsuccessful, we believe, because of the work to obstruct them and our failure to understand and overcome this obstruction.

This assessment is a project of the Climate Social Science Network (CSSN), an international collaboration of 800 scholars headquartered at Brown University that fosters research and scholarly debate on climate obstruction. With

J. Timmons Roberts et al., *Introduction*. In: *Climate Obstruction*. Edited by: J. Timmons Roberts, Carlos R. S. Milani, Jennifer Jacquet, and Christian Downie, Oxford University Press. © Oxford University Press (2025).
DOI: 10.1093/oso/9780197787144.003.0001

thirteen chapters developed by a global team of 110 scholars led by thirty-one lead authors, the volume summarizes the current state of knowledge on who is impeding action on climate change, and how. While it recounts the history of that obstruction, we focus primarily on the period between 2015, when a global climate treaty, the Paris Agreement, was negotiated, and 2024.

The assessment reviews obstruction efforts by the fossil fuel industries (Chapters 2 and 3), as well as by utilities (Chapter 3), agribusiness (Chapter 4), media and public relations (PR) firms (Chapter 6), and organizations on the political far right (Chapter 5). Research teams also explored the role of the news and social media in disseminating climate disinformation and misinformation (Chapter 6) and how these narratives potentially affected public opinion on climate action (Chapter 7). Other teams examined how obstruction of climate action occurs across the Global South (Chapter 8), at the subnational level (Chapter 9), at the United Nations and in the IPCC (Chapter 10), and in the context of climate adaptation (Chapter 11). The final chapters analyze the surge in regulatory and litigation efforts and civil society movements around the world to curb climate obstruction (Chapters 12 and 13). All chapters underwent peer review and reviews by the volume's four coeditors. As with the IPCC reports, a free download of the assessment is available online.

DEFINING OBSTRUCTION

We define climate obstruction as intentional actions and efforts to slow or block policies on climate change that are commensurate with the current scientific consensus of what is necessary to avoid dangerous human-caused interference with the climate system.[2] That scientific consensus is summarized and brought up to date in the assessments by the IPCC.

Climate obstruction has existed since at least 1980, when the American Petroleum Institute (API) began publicly downplaying the emerging threat of global warming.[3] Research into climate obstruction began in the early 1990s, as civil society groups in the United States, such as the Sierra Club[4] and Ozone Action,[5] helped journalists expose the links between coalitions of fossil fuel companies and contrarian scientists who downplayed the unfolding realities of climate change.[6]

The first scholarly articles on climate obstruction—on contrarian scientists,[7] conservative think tanks[8,9,] and the media[10]—were published in the late 1990s and early 2000s. In the 2010s, interest in the topic rose[11,12,13] and soon the research coalesced to form a picture: a complex network of organizations supporting climate obstruction (Figure 1.1). A 2015 series of investigative pieces by *Inside Climate News* brought to light the extent to which fossil fuel giant Exxon knew about the science of climate change in the 1970s and 1980s,

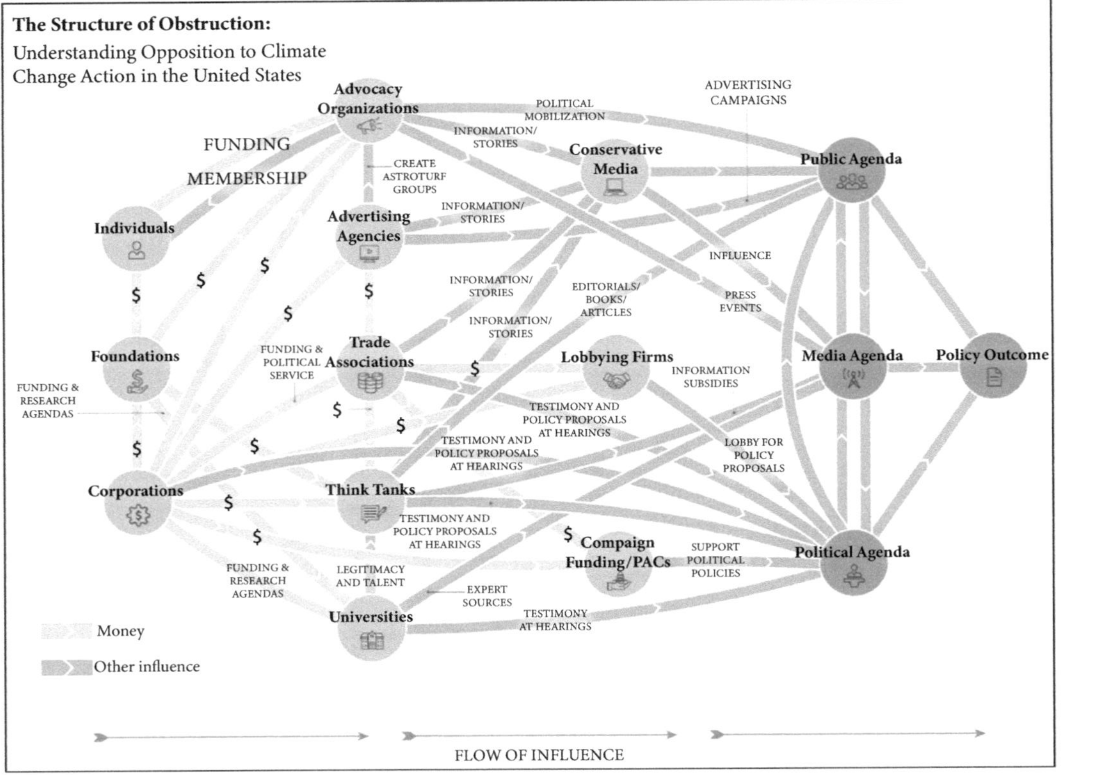

Figure 1.1:

The structure of climate obstruction can be visualized as a complex network of organizations and individuals fighting action on climate change.

Source: R. J. Brulle, "The Structure of Obstruction: Understanding Opposition to Climate Change Action in the United States," CSSN Primer 2021, https://cssn.org/wp-content/uploads/2021/04/CSSN-Briefing_-Obstruction-2.pdf.

but suppressed it and moved aggressively to block climate-related policies.[14] More than twenty-five years of scholarly attention, as well as a range of investigations by civil-society groups, social movements, and journalists, inform this global assessment, the first attempt to summarize what we know—and what we still need to find out—about obstruction of climate action around the world.

The network of organizations and individuals fighting action on climate change is substantial (for one representation, see Figure 1.1).[15] Fossil fuel and allied corporations and wealthy individuals use various channels to seek and achieve policy outcomes that favor their fortunes: media ownership and influence, lobbying firms, campaign donations and organizations, advocacy groups in national capitals and across countries, trade organizations, think tanks, and funding funneled to researchers and centers at strategically selected universities. This funding flows through family or corporate foundations as well as through "donor advised funds," which hide the funders' identities from public view. These influence groups target key politicians and other decision-makers with the goals of instilling doubt about the need to act on climate and creating uncertainty about the ability of renewable energy and other climate solutions to meet society's needs. By raising these doubts, network members hope to achieve delays in rule-making, weak regulations, and postponement of the transition away from their products and practices.

This volume makes clear that types and methods of climate obstruction vary depending on their context: whether we are examining the Global North or South, electoral democracies or autocracies, pluralist or corporatist interest-group systems, low-income territories where extractive industries dominate the economy or high-income neighborhoods enjoying full and well-maintained infrastructure and diverse employment options (Chapters 8 and 10). Similarly, obstruction is starkly different when undertaken by poor and powerless actors than when implemented by wealthy and powerful ones (Chapter 11). The boundaries of what counts as obstruction, then, will be highly dependent upon *who* is taking the action, *where* the decision is taken and *how* it is justified.

We recognize the difficulty in proving that climate obstruction efforts are intentional. Motivations can be complex and may be driven by genuine beliefs, structural constraints, different visions for addressing climate change equitably, and the sometimes-unclear benefits of acting on climate right now. For example, it can be difficult to delineate whether false claims are intentional *dis*information or unintentionally passed-on *mis*information, whether a given policy aims primarily to reduce poverty in developing countries or to support the short-term interests of fossil fuel or agribusiness conglomerates. Motivations can also be difficult to trace through large corporate structures and their vast networks of agents, particularly decades after the fact. Nevertheless, evidence of intentional obstruction has emerged for some actors. For example, certain companies were aware that climate change existed and was dangerous—and in some cases even participated in scientific research on the subject. Yet they chose to challenge the reality of climate change to avoid regulations that would affect their business operations.

Some obstruction is more consequential than others: not all participants in climate obstruction are equally responsible for causing global warming. Attribution science is a fast-growing field seeking to document how much of present and future climate impacts are the result of various actors' behavior.[16] In legal cases, scientific evidence can provide support for plaintiffs' damage claims from those who demonstrably contributed to the harm they experienced.[17,18] However, attribution science must wrestle with many issues when determining what and whom to hold culpable, such as which gasses to count, from which sources, and during what time frame.[19,20,21] Attribution of responsibility for climate obstruction presents a similar set of nascent questions: How is responsibility distributed across the known network of climate obstruction organizations?[22,23]

To address that complexity, some researchers have proposed four arguments that investor-owned fossil fuel producers have a unique liability for anthropogenic climate change.[24] First, these companies produced a large share of the products whose use contributed substantially to climate change. Second, they continued their business operations even after the danger of doing so was scientifically established. Third, they intend to continue expanding the production and use of fossil fuels, even now that the science is clear that these fuels have to be phased out rapidly. Finally, they have done all of this while working to obstruct action on climate change. So while this assessment widens its lens to capture the many other actors involved in blocking climate action nationally and globally, we recognize that fossil fuel producers are one of the most significant generators of climate obstruction.

In each chapter of this volume, the authors describe and analyze the range of actors involved in climate obstruction and the contexts in which it occurs; we summarize them in the following ten key findings.

TEN KEY FINDINGS

1. Climate obstruction is not just about denial

Since the Paris Agreement in 2015, public denial of the fact that climate change exists, is human-caused, and is an existential threat has become far less prevalent.[25] Rather, most actors involved in obstructing climate action today spend less time on denying the scientific reality of climate change and more time on creating doubt about the severity of climate change, challenging the viability of certain climate solutions and policies,[26] or "greenwashing" their products or businesses by claiming that they are sustainable.[27] Typologies have been developed for such "discourses of climate delay," categorizing the different narratives used to argue that climate action is not needed, is someone else's responsibility, should rely on future technology, or is impossible.[28] Since the 1980s, some industries have also spread the idea that climate *action* is dangerous, challenging the feasibility of a modern society not centered on the use of fossil fuels. As one example, the Global Climate Coalition has predicted the demise of national economies and widespread suffering if climate policies (such as adopting alternative forms of energy) are enacted (Chapters 2, 3, and 6).[29,30]

Recently, challenges to climate solutions have escalated (Chapters 6 and 7).[31] While industry-funded or -aligned groups now often claim to be in favor of climate action, they continue to oppose this or that particular policy or project. Researchers have called these tactics "emphasizing the downsides," "impossible expectations," and "policy perfectionism."[32,33] Some are calling this approach simply "delayism."[34] Delayism is playing out in subnational (Chapter 9), national, and international contexts (Chapter 10). For example, conservative think tanks have modeled the projected economic benefits and costs of climate policies, typically claiming vast imbalances. Still, reporters quote such businesses and their associations, and decision-makers seek their input, because they provide "authoritative voices" on the likely impacts of climate action—despite the fact that such analyses routinely assail *any* government intervention in the economy.[35]

The "information and influence campaigns"[36] aiming to shape policy outcomes are complex, and advancing these types of misinformation is only one part of such efforts. Firms also undertake PR efforts to associate themselves with "climate solutions" and "the good life."[37] While they hold little faith in renewables, their belief in the power of other technologies to solve the climate crisis is widespread (Chapters 2, 3, and 7). Such "technological optimism" takes many forms. Fossil fuel corporations promote technological solutions, including carbon capture and sequestration, hydrogen fuel, and algae biofuels, which some argue allow for continued extraction and use of their products because the public is under the impression their business model and main product are changing (Chapter 2). The economic and technical viability of these solutions are hotly contested, however, as are their risks and moral and equity dimensions. The existence of these potential "technofixes" have allowed the fossil fuel industry to position itself as an important player in solving the climate crisis, and in the twenty-first century that it is better to be seen as a partner, not an enemy, of climate action (Chapters 2 and 7). Thus fossil industry delegates at the Conferences of the Parties of the UN Framework Convention on Climate Change (UNFCCC) have routinely portrayed their companies as "part of the solution" (Chapter 10) and made net-zero emissions pledges (Chapters 2, 3 and 4). Meanwhile, the fossil fuel, transport, and animal-food industries also plan to increase production.

Finally, the media for transmitting climate misinformation of all types—where people get their information and from whom—have changed since 2015 (Chapter 6). Climate misinformation affects individuals' climate beliefs and, increasingly, relies on the spread of conspiracy theories (Chapter 7).[38] Addressing the various discourses of climate delay will require both enhancing individual resistance to climate change misinformation and disinformation and instituting systemic changes in information systems to make it more difficult to disseminate falsehoods (Chapters 6 and 7).

2. Big oil is a key player in obstruction, but only one of many

Most of the early literature on climate obstruction focused on the oil and gas industry and the role corporations such as ExxonMobil, Chevron, Total, and Shell have played to slow or block action on climate change. But we now have evidence that as climate science was developing in the 1970s and 1980s, some major oil companies including Exxon also participated in research efforts that documented the likely future trajectory of global warming if fossil fuel use continued unabated.[39] Under President Lee Raymond, the company later discontinued that research and shifted its position from undertaking to attacking climate science (Chapter 2).

While the oil and gas industry led the way on climate obstruction,[40] this assessment shows that numerous other industries—coal companies, utilities, car manufacturers, and meat and dairy producers—joined climate obstruction efforts or deployed their own. Supply chains (both upstream and downstream) create shared interests among corporations across sectors and industries.[41] For example, the aviation, maritime, trucking and railroad industries are all heavy consumers of oil and gas fuels, and a major part of the latter three industries' business is transporting their products. Utilities are still dependent mainly on coal and gas to power their grids; some manufacturing and, increasingly, computing require high volumes of electricity from these grids (Chapter 3). The meat and dairy industries have also engaged in climate obstruction; this sector is uniquely emissions-intensive compared with other agricultural industries due to methane emissions from livestock and deforestation to feed them (Chapter 4). It is not only the largest greenhouse gas emitters that obstruct climate action, but also the businesses in their supply chains.

One manifestation of this interdependency is economy-wide coalitions of business interests that collectively oppose climate policies. In the United States, the US Chamber of Commerce and the National Association of Manufacturers have a history of pooling resources, sharing information, and undertaking joint political activities such as lobbying campaigns to block attempts to legislate a nationwide emissions-trading program.[42,43] Business coalitions such as the European Roundtable of Industrialists (Chapter 3) and the Indonesian Chamber of Commerce and Industry (Chapter 8) have also been involved in weakening or obstructing climate legislation and directives. Industries with strong interests in avoiding regulation have often pushed these associations to extreme anti-climate positions, even when many other member firms might prefer ambitious climate action.

The preferences, strategies, and tactics of the variety of actors engaged in climate obstruction are by no means uniform. For example, in the utility industry, while many firms remain dependent on fossil fuels to generate electricity and therefore tend to oppose climate policies, others hold a more diverse portfolio of generation assets or purchase most of their electricity from renewable energy sources, and hence take different positions (Chapters 3 and 9).[44,45,46] In US states where utilities are permitted to own electricity-generating power plants, those utilities are more likely to oppose climate and renewable-energy legislation than in those where they are not (Chapter 9).[47] Among automakers, there are notable divisions between firms that have embraced manufacturing electric vehicles and those who continue to focus on producing gas- and diesel-engine cars; the latter have lobbied heavily against attempts to strengthen emissions and fuel-efficiency standards (Chapter 3).

3. Trade associations are key agents of climate obstruction

Trade associations, which represent entire industries, and business associations (which represent national or local business interests) allow corporations to "hide in the crowd" when engaging in obstructive actions; this reduces their risk of reputational damage (Chapter 3). Firms in the fossil fuel industries use trade associations as a "command center" for political campaigns to pool resources and coordinate political activities. The API has been at the forefront of climate obstruction for decades, coordinating the oil majors' lobbying and advertising campaigns and launching front groups such as Energy Citizens, to delay US efforts to curb the oil and gas industry's emissions (Chapter 2). API maintains lobbyists in nearly all states, monitoring legislation and routinely meeting with members of Congress and state agency officials to oppose unwanted legislation and programs. The same is true of trade associations in the transport, utilities, coal, and meat and dairy industries. For example, the International Air Transport Association (IATA), which represents airlines, has opposed national climate regulations such as the inclusion of aviation in emissions-trading schemes and international passenger levies for climate damages. Similarly, the Federation of Electric Power Companies in Japan continues to promote coal and downplay the need to transition to renewable energy (Chapter 3).

In the transport sector, the European Automobile Manufacturers' Association, the Japanese Automobile Manufacturers Association, and the Federal Chamber of Automotive Industries in Australia opposed fuel-efficiency standards and a managed transition off internal-combustion engines. Meanwhile, the automakers were able to present their companies as climate leaders (Chapters 3 and 13). Agricultural trade associations have also been involved in climate obstruction: the American Farm Bureau Federation has denied the science of climate change since the early 1990s, and the National Cattlemen's Beef Association has hired university experts to challenge the link between cattle farming and climate change (Chapter 4). Overall, these are vast organizations. In the United States, about ninety trade associations working on climate issues had total revenues of $25.6 billion between 2008 and 2018, with 13% of that, or $3.4 billion, spent directly on political activities.[48]

4. Networks of think tanks are spreading obstruction globally

A group of think tanks—public policy institutes that perform research and advocacy—have promoted disinformation and misinformation about climate science and contested climate policies via reports and presentations to legislatures, news and social media, and other channels.[49,50,51] They present quasi-academic positions while shifting the framing of climate change from a science-based issue to a matter of individual liberty and economic freedom (Chapter 5). They have also funded research reports and conferences to challenge the IPCC consensus on anthropogenic climate change[52] and have networked transnationally to increase their capacity to promote or disrupt specific political discourses on the topic.[53] Their institutional structure is shaped by the predominance of men, especially in the case of European think tanks, illustrating a documented alignment between masculinity and climate contrarianism.[54] There is also a strong convergence between the far-right, neoliberal thought advanced by these think tanks and resistance to climate action (Chapter 5).[55]

Many of the most influential think tanks are based in the United States, but the dominance of think tanks has spread globally, thanks in part to the Atlas Network, a global network of such organizations that advocates free-market (libertarian) and neoliberal policies (Chapter 5). The Heartland Institute in the United States, the Free Market Foundation in South Africa, and the Global Warming Policy Foundation in the United Kingdom all have worked to challenge climate policies and programs (Chapter 5). Similar ideas and practices have also spread to think tanks in the Global South, such as Andes Libre in Peru and the Instituto Liberdade in Brazil (Chapter 8). Funding flows around the world to support these think tanks, which share information, tactics, events, connections, experts, board memberships, and legal support.[56]

5. PR firms enable corporate efforts to obstruct climate action

Public relations and strategic communications firms help companies and trade associations conduct their climate obstruction efforts.[57,58,59] These companies and allied firms collect intelligence on opponents (such as environmental nongovernmental organizations, or NGOs), conduct industry-friendly research, design advertising campaigns, launch front groups, and undertake opinion polling, among other activities (Chapter 6). In the United States, PR firms such as Edelman, the DCI Group, and the Hawthorn Group have run campaigns such as "Advanced Energy for Life" in support of the coal industry (Chapter 3). In Ireland, the PR firm Red Flag Consulting worked for the meat and dairy industry to promote positive messages about animal products (Chapter 4). The growing role of PR firms in blocking climate action has led to their being named in public-interest lawsuits (Chapter 12),[60] being criticized by the United Nations secretary-general,[61] and to the passage of new laws banning advertisements promoting fossil fuels.[62]

PR firms' climate obstruction efforts are facilitated by the characteristics of the global media system. For example, many media companies rely on an advertiser-based business model, which includes revenue streams from corporations that obstruct climate action (Chapter 6). Between 2020 and 2022, Google reportedly received more than $23.7 million from oil majors such as ExxonMobil and BP to purchase online advertising.[63] Meta, the owner of Facebook and Instagram, permitted fossil fuel companies to purchase misleading advertisements despite the assessments of independent fact-checkers that the spots were inaccurate (Chapter 6).[64] PR firms design and place such ads for their clients as part of complex information and influence campaigns to burnish the companies' image.[65] In addition, advertising revenue may also inhibit negative coverage of fossil fuel advertisers. Workers in the PR industry have begun to organize against these practices, targeting the PR firms that have run obstructive campaigns for the highest-polluting clients (Chapters 6 and 13).

6. Some governments have taken leading roles in obstructing climate action

Government interests often align with those of their major industries, including fossil fuel–extracting firms, leading governments to take positions to protect those industries (Chapters 8, 9, and 10). This stance is particularly obvious in the case of state-owned oil and gas companies and when national and local governments are heavily dependent upon one main polluting industry as an income source. Petro-state Saudi Arabia's obstruction of climate action at the UN climate negotiations over the last two decades is a clear example (Chapter 10) but hardly unique: more than two dozen countries have included oil- industry employees in their official delegations to the UNFCCC, and others allow the industry to influence their national negotiating positions (Chapter 10). Many of these governments also seek to support their fossil fuel industries by using public finance to secure export markets. For example, China and Japan have promoted coal-fired power throughout Eurasia by providing public subsidies to countries that import their technology, such as Indonesia, India, and Vietnam (Chapter 8); they have a history of deflecting criticism of such investments at the UN meetings.

In addition, agribusiness and the meat-production industry in Brazil have built strong political connections with government agencies at federal and state levels (Chapters 4 and 8). In South Africa, coal mining is the historical backbone of that nation's economy and social order; in the case of some developed countries (e.g., the Netherlands) and developing ones, such as Venezuela, Iran, and Mexico, the key industry is oil (Chapter 8). Therefore, a just transition away from fossil fuels will require rethinking the central role of the state in promoting new development models and creating the means to make these new models socially and environmentally sustainable in an era of climate change.

Governments in heavily industrialized countries with more diverse economies have also persistently taken obstructive action on climate. The United States, for example, supported and then withdrew from the Paris Agreement, again creating political whiplash in the global effort to address climate change. While numerous governments have obstructed action in the UN negotiations at one time or another, such moves have typically received limited attention despite their substantial implications. For example, at COP 27 in 2022, New Zealand sought to weaken methane accounting rules (Chapter 4); Uruguay has aligned with Argentina and Brazil (the ABU Coalition) when its livestock interests were at stake. Understanding and overcoming climate obstruction thus requires attention to national political economies, and those are often built on local ones: subnational provinces and states that are heavily dependent on fossil fuel extraction often seek to undermine national climate policy, as in the case of Canada and the United States (Chapter 9).

7. Obstruction occurs within multilateral governance structures

The UNFCCC and the IPCC are global institutions that play key roles in establishing priorities on the climate agenda by organizing major international conferences at which multinational negotiations take place and by highlighting areas for future action by governments, corporations, and societies. They are also arenas for establishing and enforcing norms and values such as, for example, climate justice, fair governance, and suitable finance, but they have been unable to respond rapidly and effectively to the climate emergency (Chapter 10).[66,67] Currently, national sovereignty (often defined in opposition to responsibility) is the bedrock of the UN system, and thus global governance institutions lack adequate enforcement and compliance mechanisms. Similarly, international development banks and wealthy "donor nations" inside and outside the UN system have failed to assist lesser developed countries with climate change adaptation (Chapter 11). World maritime and aviation governance institutions face many of the same issues (Chapter 3).

Multilateral forums' tendency to slow and obstruct climate action make them an easy target for polluting companies and trade organizations and allow these groups to play a double-game. They can point to a lack of international action as a reason for slowing action in their own countries. At the same time, the fossil fuel industry, nation-states, and other economic actors employ a variety of diplomatic, procedural, and substantial strategies to obstruct progress in international venues. They have secured a lack of political will to exclude the oil industry, agribusiness, and other obstructionist actors from participating in UN decision-making, a change some campaigning groups have unsuccessfully advocated (Chapter 10).[68]

8. National and local contexts shape obstruction, particularly in the Global South

Climate obstruction looks different depending on national and local contexts. Consider the Global South, which encompasses a group of more than 130 countries outside the core of the global economy, with different trajectories in social and state formation and economic and social development (Chapter 8)[69,70,71]. The Global South's greenhouse gas emissions are also heavily concentrated: India and China alone account for some 60% of the region's emissions, while 120 other Global South countries account for only 22%.[72]

Leaders from the Global South have argued that the Global North—due to its historical responsibility for causing the most emissions—must be more proactive in supporting developing countries in dealing with climate change. Delivering the financing required to decarbonize and adapt to climate change in the Global South is a topic still being negotiated and obstructed within the UNFCCC (Chapter 10).[73] But the Global South can also be an agent of climate obstruction. Political representatives from the Global South often use development priorities as a reason to sideline decarbonization goals, frequently claiming that they have a "right to pollute" as part of their "right to development," and failing to question their own development models or to consider helping the most vulnerable groups in their countries (Chapter 8). Moreover, dominant ideologies can also feed into overt and extreme climate obstruction efforts, including religious influence in government and/or right-wing legacies—be they neoliberalism, conservatism, or farther right. Such trends can be seen in Brazil under former president Jair Bolsonaro, the United States under President Donald Trump, Hungary under Prime Minister Viktor Orban, or Argentina under President Javier Milei (Chapters 5, 7, and 8).[74]

The configuration of organized interests engaged in climate politics shapes how obstruction occurs not just at the national level but also the local, subnational, and international levels. For example, studies have shown that distributive conflicts among energy interests, such as between coal and gas, are vital to explaining climate policy outcomes.[75,76,77,78] How those industries are structured can also shape how obstruction unfolds. For example, in the United States, many utilities hold a monopoly on providing electricity to the city or region in which they are based; therefore, ratepayers may unwittingly be funding the industry's lobbying and advertising efforts to delay and block climate policies (Chapters 3 and 9). Similarly, in countries where media industries are highly concentrated, with a few firms or families owning the majority of news outlets, collective goals including climate action can be undermined (Chapter 6).

9. Obstruction starts at the top but requires social acceptance

As noted earlier, industries such as oil, gas, and coal, utilities, car manufacturing, and meat and dairy production—as well as their trade associations and PR firms—fund and implement climate obstruction from the top down. However, for those efforts to be effective, their rhetoric must be accepted and adopted by the media, political parties, politicians, and individual citizens. For example, a study of more than 1,700 press releases about climate change found that US newspapers covered messages from business coalitions and trade associations that opposed climate policy at double the rate they covered pro-climate policy positions from other types of organizations (Chapters 6 and 7).[79] In addition, most media platforms have few, if any, effective policies in place to prevent the spread of climate misinformation (Chapter 6). Obstructive forces have been able to use social media to exponentially speed the spread of such misinformation among the public.

Individuals are constantly exposed to climate obstruction content from a variety of outlets. Scholars have identified the major categories of anti-climate claims: (1) global warming isn't real, (2) it isn't human-caused, (3) climate impacts aren't bad, (4) climate solutions won't work, and (5) experts are unreliable (Chapter 7). More recently, studies show, attacks on the scientists themselves and on renewable energy and other climate solutions have replaced the first three claims.[80] Whether people are more or less receptive to such rhetoric depends on various social and psychological factors.[81] For example, early research in this area documented how industrial companies targeted Republicans with influence campaigns that helped make climate change a polarized issue in the United States (Chapter 6).[82,83,84] Addressing such climate change misinformation requires systemic changes that make it more difficult for misinformation to spread and that will ensure individuals are able to discern misinformation from fact.

10. There are efforts to address climate obstruction through regulation, litigation, and social movements

Some countries and communities are working to address climate obstruction in formal and informal ways. Options have included direct government regulation and legislation, which attempt to restrict net-zero claims and pledges made by corporations and financial firms, limit corporate influence over climate policy, and prohibit greenwashing. To do so, governments have used a variety of tools including shaming tactics and other forms of mandated disclosure (Chapter 12). They can also limit or prohibit the "revolving door" through which industries co-opt regulators by hiring them, restrict lobbyists' access to key staff and legislators, and limit company or large-donor contributions to political campaigns (Chapter 12).

About 1,500 climate-related lawsuits have been filed around the world; a subset of these suits target climate obstruction, in particular the historical and continuing deception of consumers (Chapter 12). The plaintiffs include individuals (grandmothers in Switzerland, teenagers in Oregon, Indigenous and quilombola groups in Brazil, and many others), subnational actors including city and state governments, federal justices, and major climate organizations. Lawsuits targeting climate obstruction have been brought against fossil fuel producers, including Exxon, Shell, BP, Peabody, and Total, but have recently expanded to other firms. Lawsuits have implicated trade associations, such as the API (Chapter 12) and consulting firms, such as McKinsey & Company, which an Oregon municipality sued along with fossil fuel company trade groups.

Lawsuits have also expanded beyond fossil fuel related cases. In 2021, for example, three NGOs won the first climate lawsuit in Denmark against the meat producer Danish Crown, which had claimed that its pork was "climate friendly."[85] In 2024, the New York State attorney general filed a lawsuit against beef producer JBS USA Food Co., alleging the company had misled consumers about its net-zero commitments (Chapters 4 and 12).

In the Global South, Brazil has one of the highest numbers of climate cases, including 2023 litigation in which the federal environmental protection agency, IBAMA, sued agribusiness leader Dirceu Kruger for climate damage caused by illegal deforestation in the Amazon.[86]

Climate-focused NGOs are finding ways to challenge obstruction. These efforts include social exposure (or "naming and shaming") to shower negative publicity on obstruction activities, lobbying to advocate prevention of obstruction, and campaigns and protests that mobilize citizens concerned about blatant climate obstruction and deception. For example, BankTrack.org monitors the financial activities of private-sector commercial banks around the globe and feeds this information to campaigners who name and shame banks in hope of strengthening their policies on investments that affect the climate (Chapter 13). The record of wins from regulatory, litigative, protest, and shaming campaigns against obstruction is still modest, but may be aided by the broadening understanding of the problem and growing efforts to address disinformation.

Here we synthesize research drawn from our thirteen chapters on the categories of practitioners who are working to counter climate obstruction and highlight who and what their strategies have targeted. For example, some strategies aim directly at fossil firms, others at elected officials, the judiciary branch, or the media. While many of these strategies have proven effective in some countries, such as legal strategies in the United States and greenwashing legislation in the European Union, assessing their effectiveness is notoriously difficult. What follows is a description of current practices and, where available, their records of success.

Policymakers and Lawyers

Political leaders and government officials have used regulation and legislation, litigation, government investigations, public shaming, and diplomatic initiatives to counter obstruction by individuals, civil society groups, business associations, advocacy groups, and governments. Regulation and legislation are critical mechanisms because of their unique power to force change by requiring emissions reductions, increasing transparency and access to information, prohibiting deceptive practices, and promoting accountability (Chapter 12). For example, in some jurisdictions, policymakers seek to regulate corporate miscommunication and fraud in order to better protect investors and consumers by limiting opportunities for businesses to engage in greenwashing and other deceptive activities.

Public and private lawyers have fought climate obstruction by filing lawsuits against firms involved in climate obstruction, either directly (as in the case of state attorneys general), or indirectly (on behalf of plaintiffs such as NGOs). Other lawyers have represented NGOs, academics, and journalists in obstruction-related defamation cases, in both preventative and defensive roles.

More recently, policymakers have initiated government investigations to hold corporations to account and facilitate future litigation. For example, a probe by the Philippines Commission on Human Rights in 2022 found that the carbon majors "engaged in wilful [*sic*] obfuscation and obstruction to prevent meaningful climate action" (Chapter 12).[87] And finally, international coalitions of government representatives have resisted state-backed climate obstruction and promoted government action to limit greenhouse gas emissions. In the international climate negotiations, for example, the Alliance of Small Island States (AOSIS) has for many years deployed the scientific evidence to counter attempts by larger countries, such as the United States and Saudi Arabia, to delay climate action at the UN negotiations (Chapter 10).

NGOs have exposed climate obstruction via research, media monitoring, and debunking misinformation. These efforts have been characterized as "public inoculation" because they can weaken the impact of misinformation campaigns by drawing explicit attention to the actors behind obstructionist efforts, their financial support, and the sources and nature of their misleading claims.[88] For example, Climate Action Against Disinformation (CAAD), a coalition of NGOs, has tracked misleading narratives in the media, highlighting the role that major technology platforms such as Facebook, Instagram, TikTok, and X play in the spread of climate disinformation. Part of their success lies in the "rapid attribution" of misleading content and coordinated efforts to stop its spread (Chapters 6 and 13).

NGOs also name and shame states, firms, and corporate executives involved in climate obstruction (Chapter 13). For example, since 1999, Climate Action Network International and its national chapters have sought to embarrass governments that delay progress at the international climate negotiations by bestowing them the "Fossil of the Day" award in the hope of shaming them into better climate policies (Chapters 10 and 13). The campaigns by the NGOs Clean Creatives, Check My Ads, and Sleeping Giants have challenged companies to remove misleading climate-related advertising from their sites and news platforms.

NGOs also use direct lobbying to counter obstruction, such as by targeting policymakers to rebut disinformation and highlight misleading claims and by offering "carrots and sticks" to offset obstructionist lobbying and political campaign donations (Chapter 13). Given the frequent mismatch in resources between the carbon majors and environmental organizations,[89] NGOs often lobby in coalitions. Coalition work in support of the Inflation Reduction Act (IRA) in 2022, for example, was instrumental in the passage of US President Joe Biden's flagship climate legislation; it included providing detailed reports rebutting claims from business groups about the economic impact of climate action.

Finally, activists sometimes employ strategies operating outside the confines of existing institutions and norms, such as campaigns and protests that involve rallies, civil disobedience, and site occupations.[90] Many of these activities are designed to stigmatize and delegitimize the fossil fuel industry and its enablers. For example, in 2013, student activists in the United States founded UnKoch My Campus to counter the influence of the Koch family, which was using its donations to universities to influence college curricula, research agendas, and the hiring and firing of faculty across numerous campuses (Chapter 13).

Journalists

Journalists, like NGOs, have played a significant role in exposing climate obstruction and debunking misinformation. For example, investigative outlets such as *Inside Climate News* and *DeSmog* have helped to expose misinformation and obstruction campaigns in the oil, gas, and coal industries, among many others. Their work has included revealing the role that "influencers" hired by PR firms play to promote their oil and gas clients (Chapter 6). Other news organizations, such as *The Guardian*, have run series examining the political tactics of fossil fuel corporations (from whom they have stopped accepting advertising; Chapter 13). Early efforts by journalists to expose climate obstruction have developed into coordinated actions to fight its spread. Groups such as Climate Feedback, comprising science journalists, have emerged to fact-check climate news stories and help readers evaluate the sources of these stories (Chapter 6). As discussed, debunking misleading claims in this way is considered vital to inoculating the public and making climate misinformation harder to spread.

Scholars

Academics, including scientists, also aid the efforts of policymakers, NGOs, and journalists to expose and counter climate obstruction, and the results have prompted legislative investigations, legal proceedings, newspaper reports, and social media campaigns. CSSN,[91] which includes 800 scholars worldwide, encourages peer-reviewed research into obstruction. Some academics work closely with journalists from organizations such as Climate Feedback to expose and debunk misinformation. University researchers have also participated in institutional campaigns to convince universities to disclose funding from the fossil fuel industry, and more recently to "disassociate" from firms and industries that advance climate misinformation.[92] While many of these campaigns have been led by students, professors have also pressured their home institutions to act. For example, campaigns by faculty at Brown University and the University of California, San Diego in the United States and the University Cambridge in the United Kingdom, among others, have sought to dissuade the schools from engaging with companies that promote climate disinformation by, for example, refusing their research funding (Chapter 13).

Students

Campuses around the world now face growing pressure to mitigate the risks from taking industry funding, including its influence over research agendas and findings. Indeed, a new international student-led movement is pushing universities to divest or disassociate from fossil fuel corporations.[93] The student-led Fossil Free Research, now renamed Campus Climate Network, is conducting research on corporate influence at a growing number of universities, with a combination of public records requests, archival research, and interviews with university officials. Some law students are trying to address the role their own profession has played in climate obstruction. Law Students for Climate Accountability, for example, releases an annual report card grading major law firms according to the number of fossil fuel clients they represent, and asks law students and young professionals to sign a pledge "to refuse to work for a law firm that represents fossil fuel industry clients" or to "hold my firm accountable for its role in perpetuating climate change."[94]

Businesses

Some climate-concerned business actors have worked to counterbalance the efforts of emissions-intensive incumbent industries to block and delay climate policies. Energy industries such as solar and wind companies and electric-vehicle manufacturers, for example, are building their lobbying capacity to turn the tide of fossil fuel company lobbying. The effort includes creating trade associations such as the Solar Energy Industries Association in the United States, the Smart Energy Council in Australia, the Brazilian Association for the Generation of Clean Energy, and the Asia Wind Energy Association. In the auto industries across North America, Europe, and parts of Asia, firms that could benefit from policies that encourage the use of electric vehicles (EVs) are mobilizing their lobbyists (e.g., Australia's Electric Vehicle Council) to rebut claims from others in the industry that stand to gain from continued demand for internal-combustion engine vehicles (Chapters 3 and 13). Business actors also exert influence outside of their industries by lobbying and pressuring their trade groups or peak business organizations, such as chambers of commerce. Business's employees have also made private and public appeals to their firms to stop supporting climate obstruction, including a 2015 walkout by executives at the global PR firm Edelman Communications[95] and the launch of Clean Creatives, a group of PR professionals that calls on individuals and organizations to refuse to work with the fossil fuel industry (Chapter 13).

DIRECTIONS FOR FUTURE RESEARCH

Although we have attempted to make this volume as comprehensive as possible, many arenas where climate obstruction occurs were omitted due to lack of data. Moving forward, researchers should consider investigating the roles of the following institutions.

Finance

More research is needed on the role of financial institutions, law firms, and large consulting firms in slowing or blocking action on climate change. For example, some NGOs (such as the UK-based Reclaim Finance) have uncovered investments supporting climate denial and called for the insurance industry to fight anti-climate action lawsuits by some US state attorneys general.[96] InfluenceMap, the Center for Climate Integrity, and the Climate Investigations Center, among others have tracked corporate influence on climate politics and funding flows dedicated to obstructing policy. The resources made available by these organizations substantially improve access to information about climate denial and obstructionist actors.

Academia

We also need to better understand the role of universities in climate obstruction, and how and why corporations and trade associations are funding academic research and other initiatives at universities and research centers. Fossil fuel funding and other corporate support of universities have a long history (Chapter 2). This arrangement is problematic because such funding can generate biased research results and distort research agendas.[97,98] For example, a 1998 API plan provided funding to scholars whose research supported the industry's objectives.[99] Internal documents from the oil and gas industry have revealed that such research funding is part of a long-term strategy to keep these companies informed of emerging developments, to influence the mainstream scientific community, and, sometimes, to co-opt and silence an important set of potential critics—all with the goal of slowing efforts to regulate fossil fuels. Documents released by the US House Committee on Oversight and Reform in 2022 and 2023 showed that firms such as BP view relationships with universities and academics as part of an effort to shape policy and public opinion.[100] In some cases, industries have used universities more as platforms to influence public understanding and politics than venues to perform research, as has been found with the animal agriculture industry and at least two university centers.[101] For example, the Danish Centre for Food and Agriculture at the University of Aarhus was forced to withdraw a 2019 report on the climate impacts of beef after it became clear that the meat company Danish Crown and an industry association had helped to write it.[102] This is an emerging area of understanding.

Religion

While some religious groups have activated their national and transnational networks to promote support for climate action, both personal and political, others have engaged in or endorsed climate obstruction on religious grounds. Information on religion and climate obstruction is more abundant in the United States, but efforts to spread denial of anthropogenic climate change has been documented among Orthodox believers, parties, and movements in Eastern Europe including Russia and Romania (Chapter 5). In Latin America, some case studies have been published on the connections between the obstructionist Bolsonaro administration and many religious networks including Orthodox Catholics and Pentecostal evangelicals.[103,104] However, research on religion and climate obstruction is still in its infancy, and could benefit in particular from studies of how religions justify or deny action on climate change, and how religious networks connect with political actors and economic operators to obstruct climate policies both nationally and transnationally.

Labor

Very little is known about the role of labor unions in blocking or delaying climate policies. This gap may reflect the fact that in some regions, such as Europe, Southern Africa, and Latin America, their influence takes place behind closed doors in tripartite negotiations with government leaders and businesses. There is evidence that unions representing coal workers, for example, have worked to obstruct policies targeted at reducing coal emissions in Australia, the United States, and South Africa (Chapter 3).[105] Yet we still lack knowledge about their role in other countries and sectors, including in industries with a large union presence such as vehicle manufacturing, where it is likely that union mobilization could be important to determining the fate of climate policies. Further research on unions in these industries will be important, especially as the labor movement begins to shape policy discussions around what counts as a "just transition" and who should pay for the impacts of climate change.[106]

Militaries

There is also little information available on the role of the military in driving national postures on climate action.[107] Although we know that the world's armed forces are major greenhouse gas emitters, their releases are often unaccounted for due to secrecy and the difficulty of attributing responsibility in complex military operations. This research area is still poorly defined:

militaries have largely avoided pressure to reduce their impact, even as climate-driven disasters accelerate and threaten their facilities and operations.[108,109] There have been anecdotal reports of suppression of climate discourse and planning by the US military, but its obstructive actions have varied.[110] The UK defense department is one of the well-known exceptions, although there are still serious concerns that the official data are neither consistent nor complete enough to adequately guide policymaking in this field.[111,112] The IPCC and UNFCCC have also kept silent on military emissions or obstruction, due to political pressure and a lack of expertise and published data.

Influence Campaigns

The full extent of information and influence campaigns to oppose climate action, and how they are organized and executed, are not well documented.[113] Spreading misinformation is just one small part of those efforts, while direct political spending, lobbying, and PR campaigns to associate one's industry with "climate solutions" and "the good life" are more significant elements.[114] Strategic communications (PR) and law firms have assumed key roles in these broad campaigns, but many of these players remain poorly documented in the journalistic and scholarly literature. Whistleblowers and investigators are needed to help scholars and others understand how these firms funnel money, threaten lawsuits, and engage with other firms to create "astroturf" organizations, infiltrate climate organizations, and defuse pressure for change.

A FUTURE WITHOUT CLIMATE OBSTRUCTION

In the nearly four decades since the establishment of the IPCC wherein scientists raised the alarm about global warming, societies have developed the science, the technology, and the policy tools to address the problem. But the road forward to adequately addressing climate change has remained blocked by a sophisticated and well-funded network of actors claiming that the road is not viable, and that the status quo must be maintained.

What would it mean to end the various forms of climate obstruction documented in this volume? Is it possible? As noted, the research on climate obstruction in some areas is thin, and rigorous analyses of the impact of these various strategies to counter obstruction are lacking. Yet there can be no one strategy or collection of strategies that are likely to succeed in addressing obstructive efforts, which this volume shows vary across institutions, jurisdictions, and time. Publicly held corporations, for example, may be more vulnerable than private ones to being exposed for their climate obstruction, due to a need to protect their reputation, their shareholders, and other investors.

Similarly, in countries with institutions that curtail political freedoms, it is more difficult for climate activists to resist obstructive actions. Campaigning in China, Russia, Saudi Arabia, or Venezuela will look very different from campaigning in Chile, New Zealand, South Africa, or the United States. What is urgently needed, then, are studies that not only assess the effects of different strategies in different contexts, but also their use in combination, over time, and across different stages of the policy process.

While some products and modes of production are simply incompatible with a habitable Earth, viable alternatives exist to address human needs. Better support for industries that aid the transition away from carbon-emissions-intensive lifestyles and cooperation with communities to promote low-carbon livelihoods are urgently needed. However, more science and better technology, additional funding, better policy, and more effective communication will not solve the climate crisis by themselves. Regardless of the weight of scientific knowledge and the logic of making this transition, the balance of evidence presented in this assessment makes one thing clear: there is, and will be, opposition. The success of future climate action will be determined to a substantial degree by the level of understanding and ability to overcome that opposition.

NOTES

1. The six assessments and the full set of other special reports are available at https://www.ipcc.ch/reports/.
2. R. J. Brulle, J. T. Roberts, and M. C. Spencer (eds.), *Climate Obstruction across Europe* (Oxford University Press, 2024), 6.
3. B. Franta, "Early Oil Industry Disinformation on Global Warming," *Environmental Politics* 30, no. 4 (2021): 663–668.
4. M. Wald, "Pro-Coal Ad Campaign Disputes Warming Idea," *New York Times*, July 8, 1991, Section D, page 2.
5. Ozone Action, *Ties That Blind I: Case Studies on Corporate Influence on Climate Change Policy* (Ozone Action, 1996).
6. J. Jacquet, "Guilt and Shame in U.S. Climate Change Communication," *Oxford Research Encyclopedia of Climate Science*, https://oxfordre.com/climatescience/display/10.1093/acrefore/9780190228620.001.0001/acrefore-9780190228620-e-575/ (accessed April 22, 2025).
7. M. Lahsen, "The Detection and Attribution of Conspiracies: The Controversy Over Chapter 8," in *Paranoia within Reason: A Casebook on Conspiracy as Explanation*, ed. G. Marcus (University of Chicago Press, 1999), 111–136.
8. A. M. McCright and R. E. Dunlap, "Challenging Global Warming as a Social Problem: An Analysis of the Conservative Movement's Counter-Claims," *Social Problems* 47 (2000): 499–522.
9. A. M. McCright and R. E. Dunlap, "Defeating Kyoto: The Conservative Movement's Impact on U.S. Climate Change Policy," *Social Problems* 50 (2003): 348–373.
10. J. Dispensa and R. Brulle, "Media's Social Construction of Environmental Issues: Focus on Global Warming—a Comparative Study," *International Journal of Sociology and Social Policy* 23, no. 10 (2003): 74–105.

11. N. Oreskes and E. M. Conway, *Merchants of Doubt: How a Handful of Scientists Obscured the Truth on Issues from Tobacco Smoke to Global Warming* (Bloomsbury Publishing, 2011).

12. M. Lahsen, "Digging Deeper Into the Why: Cultural Dimensions of Climate Change Skepticism Among Scientists," in *Climate Cultures: Anthropological Perspectives on Climate Change*, ed. Jessica Barnes and Michael R. Dove (Yale University Press, 2015).

13. Jane Mayer, *Dark Money: The Hidden History of the Billionaires Behind the Rise of the Radical Right* (Anchor, 2016).

14. N. Banerjee, L. Song, and D. Hasenmyer, "Exxon's Own Research Confirmed Fossil Fuels' Role in Global Warming Decades Ago," InsideClimateNews, 2015. The full series is available at https://insideclimatenews.org/project/exxon-the-road-not-taken/.

15. R. J. Brulle, "The Structure of Obstruction: Understanding Opposition to Climate Change Action in the United States," CSSN Primer, 2021, https://cssn.org/wp-content/uploads/2021/04/CSSN-Briefing_-Obstruction-2.pdf.

16. D. Jamieson, "Responsibility and Climate Change," *Global Justice: Theory Practice Rhetoric* 8 (2015): 23–42

17. M. Burger, W. Jessica, and R. Horton, "The Law and Science of Climate Change Attribution," *Columbia Journal of Environmental Law* 45 (2020): 57.

18. S. Marjanac and L. Patton, "Extreme Weather Event Attribution Science and Climate Change Litigation: An Essential Step in the Causal Chain?," *Journal of Energy & Natural Resources Law* 36, no. 3 (2018): 265–298.

19. E.g., E. A. Page, "Distributing the Burdens of Climate Change," *Environmental Politics* 17, no. 4 (2008): 556–575

20. H. Shue, "Global Environment and International Inequality," *International Affairs* 75 (1999): 531–545.

21. See for example the webtool codeveloped by EcoEquity.org and the Stockholm Environment Institute, at https://www.sei.org/tools/climate-equity-reference-calculator/ .

22. Burger et al., "The Law and Science of Climate Change Attribution."

23. Marjanac and Patton, "Extreme Weather Event Attribution Science and Climate Change Litigation."

24. P. Frumhoff, R. Heede, and N. Oreskes, "The Climate Responsibilities of Industrial Carbon Producers," *Climatic Change* 132, no. 2 (2015): 157–171.

25. T. G. Coan, C. Boussalis, J. Cook, and M. O. Nanko, "Computer-Assisted Classification of Contrarian Claims About Climate Change," *Scientific Reports* 11, no. 1 (2021): 22320.

26. Ibid.

27. N. Healy, J. Stephens, and S. Malin, "Embodied Energy Injustices: Unveiling and Politicizing the Transboundary Harms of Fossil Fuel Extractivism and Fossil Fuel Supply Chains," *Energy Research & Social Science* 48 (2019): 219–234.

28. W. F. Lamb, G. Mattioli, S. Levi, J. T. Roberts, S. Capstick, F. Creutzig, and J. K. Steinberger, "Discourses of Climate Delay," *Global Sustainability* 3 (2020): e17.

29. R. J. Brulle, "Advocating Inaction: A Historical Analysis of the Global Climate Coalition," *Environmental Politics* 32, no. 2 (2023): 185–206.

30. Oreskes and Conway, *Merchants of Doubt*.

31. Coan et al., "Computer-Assisted Classification of Contrarian Claims About Climate Change."

32. Lamb et al., "Discourses of Climate Delay."

33. J. Cook, "A History of FLICC: The 5 Techniques of Science Denial," blog, 2020, https://skepticalscience.com/history-flicc-5-techniques-science-denial.html.

34. H. Araújo, "From Climate Change Denial to Delayism: Oil Firms Send Academics Into the Fray," October 10, 2022, https://english.elpais.com/science-tech/2022-10-10/from-climate-change-denial-to-delayism-oil-firms-send-academics-into-the-fray.html.

35. R. Wetts, "In Climate News, Statements from Large Businesses and Opponents of Climate Action Receive Heightened Visibility," *Proceedings of the National Academy of Sciences* 117, no. 32 (2020): 19054–19060.

36. J. Manheim, *Strategy in Information and Influence Campaigns* (Routledge, 2011).

37. R. Brulle, personal communication to T. Roberts, April 2024.

38. C. Rojas, F. Algra-Maschio, M. Andrejevic, T. Coan, J. Cook, and Y.-F. Li, "Augmented CARDS: A Machine Learning Approach to Identifying Triggers of Climate Change Misinformation on Twitter," *Nature Communications Earth & Environment*, 2024, https://arxiv.org/abs/2404.15673.

39. N. Banerjee, L. Song, and D. Hasemyer, "Exxon's Own Research Confirmed Fossil Fuels' Role in Global Warming Decades Ago," September 16, 2015, *Inside Climate News*, https://insideclimatenews.org/news/16092015/exxons-own-research-confirmed-fossil-fuels-role-in-global-warming/.

40. Franta, "Early Oil Industry Disinformation on Global Warming."

41. J. Cory, M. Lerner, and I. Osgood, "Supply Chain Linkages and the Extended Carbon Coalition," *American Journal of Political Science* 65, no. 1 (2021): 69–87.

42. C. Downie, *Business Battles in the US Energy Sector: Lessons for a Clean Energy Transition* (Routledge, 2019).

43. R. Brulle and C. Downie, "Following the Money: Trade Associations, Political Activity and Climate Change," *Climatic Change* 175, no. 3 (2022): 11.

44. C. Downie, "Fighting for King Coal's Crown: Business Actors in the US Coal and Utility Industries," *Global Environmental Politics* 17, no. 1 (2017): 21–39.

45. N. Kelsey, "Industry Type and Environmental Policy: Industry Characteristics Shape the Potential for Policymaking Success in Energy and the Environment," *Business and Politics* 20, no. 4 (2018): 615–642.

46. L. C. Stokes, *Short Circuiting Policy: Interest Groups and the Battle Over Clean Energy and Climate Policy in the American States* (Oxford University Press, 2020).

47. G. Hall, T. Culhane, and J. T. Roberts, "Climate Coalitions and Anti-Coalitions: Lobbying Across State Legislatures in the United States," *Energy Research & Social Science* 113 (2024): 103562.

48. R. Brulle and C. Downie, "Following the Money: Trade Associations, Political Activity and Climate Change," *Climatic Change* 175, no. 3 (2022): 11.

49. R. J. Brulle, "Institutionalizing Delay: Foundation Funding and the Creation of US Climate Change Counter-Movement Organizations," *Climatic Change* 122 (2014): 681–694.

50. R. J. Brulle, G. Hall, L. Loy, and K. Schell-Smith, "Obstructing Action: Foundation Funding and US Climate Change Counter-Movement Organizations," *Climatic Change* 166 (2021): 1–7.

51. S. Beder, *Global Spin: The Corporate Assault on Environmentalism* (Chelsea Green Publishers, 1997).

52. E. McKewon, "Talking Points Ammo: The Use of Neoliberal Think Tank Fantasy Themes to Delegitimise Scientific Knowledge of Climate Change in Australian Newspapers," *Journalism Studies* 13, no. 2 (2012): 277–297.

53. G. Goldenberg, "Secret Funding Helped Build Vast Network of Climate Denial Think Tanks," February 14, 2013, *The Guardian*, https://www.theguardian.com/environment/2013/feb/14/funding-climate-change-denial-thinktanks-network.

54. N. Almiron, J. A. Moreno, and J. Farrell, "Climate Change Contrarian Think Tanks in Europe: A Network Analysis," *Public Understanding of Science* 32, no. 3 (2023): 268–283.

55. N. Almiron and J. Xifra, *Climate Change Denial and Public Relations: Strategic Communication and Interest Groups in Climate Inaction* (Routledge, 2020).

56. N. Graham, "Think Tanks and Climate Obstruction: Atlas Affiliates in Canada," *Canadian Review of Sociology/Revue canadienne de sociologie* 61, no. 2 (2024): 110–130.

57. M. Aronczyk and M. Espinoza, *A Strategic Nature: Public Relations and the Politics of American Environmentalism* (Oxford University Press, 2022).

58. Sharon Beder, *Global Spin: The Corporate Assault on Environmentalism* (Scribe Publications, 2002).

59. Robert J. Brulle and Carter Werthman, "The Role of Public Relations Firms in Climate Change Politics," *Climatic Change* 169, no. 1 (2021): article 8.

60. D. Drugmand, "Oregon County Sues Fossil Fuel Entities and Enablers for Contributing to Deadly 2021 Heatwave," *Desmog.* June 22, 2023, https://www.desmog.com/2023/06/22/oregon-county-sues-fossil-fuel-entities-and-enablers-for-contributing-to-deadly-2021-heatwave/.

61. "Secretary-General's Special Address on Climate Action 'A Moment of Truth,'" United Nations, https://www.un.org/sg/en/content/sg/statement/2024-06-05/secretary-generals-special-address-climate-action-moment-of-truth-delivered (accessed October 2, 2024).

62. E.g., I. Kaminski, "The Hague Becomes World's First City to Pass Law Banning Fossil Fuel-Related Ads," *The Guardian*, September 14, 2024.

63. Center for Countering Digital Hate, "New Research Shows Google Makes Millions by Greenwashing Big Oil Ahead of COP27," 2022, https://counterhate.com/blog/google-makes-millions-by-greenwashing-big-oil-ahead-of-cop-27/.

64. Scott Waldman, "Climate Denial Spreads on Facebook as Scientists Face Restrictions," *Scientific American*, July 6, 2020.

65. J. Manheim, *Strategy in Information and Influence Campaigns* (Routledge, 2011).

66. R. Kinley, M. Z. Cutajar, Y. de Boer, and C. Figueres, "Beyond Good Intentions, to Urgent Action: Former UNFCCC Leaders Take Stock of Thirty Years of International Climate Change Negotiations," *Climate Policy* 21, no. 5 (2021): 593–603.

67. T. Sommerer, H. Agné, F. Zelli, and B. Bes, *Global Legitimacy Crises: Decline and Revival in Multilateral Governance* (Oxford University Press, 2022), 240.

68. See, e.g., Global Center for Good Governance in Tobacco Control (GGTC), List of Observers Found to Have Links with Tobacco Companies at INC-4, https://ggtc.world/knowledge/partnerships-lobbying-and-corruption/list-of-observers-found-to-have-links-with-tobacco-companies-at-inc-4 (accessed March 24, 2025).

69. D. Allès and E. Brun, "Mobilizing the South: Pluralizing and Complexifying Multilateralism," in *Crisis of Multilateralism? Challenges and Resilience*, ed. A. Guilbaud, F. Petiteville, and F. Ramel (Palgrave Macmillan, 2023), 177–196.

70. N. Dados and R. Connell, "The Global South," *Contexts* 11, no. 1 (2012): 12–13.

71. K. Gray and B. K. Gills, "South-South Cooperation and the Rise of the Global South," *Third World Quarterly* 37, no. 4 (2016): 557–574.

72. H. Fuhr, "The Rise of the Global South and the Rise in Carbon Emissions," *Third World Quarterly* 42, no. 11 (2021): 2724–2746.

73. D. Ciplet, D. Falzon, I. Uri, S. A. Robinson, R. Weikmans, and J. T. Roberts, "The Unequal Geographies of Climate Finance: Climate Injustice and Dependency in the World System," *Political Geography* 99 (2022): 102769.

74. C. Milani, J. Pinto, and A. Facini, "As relações entre autoritarismo, desenvolvimento predatório e obstrução climática no Brasil: uma análise do governo Bolsonaro," in *Política externa, lideranças autoritárias e ultraconservadorismo*, ed. Rubens de S. Duarte & Carlos Milani (APPRIS, 2024), 275–308.

75. R. Brulle, "Denialism: Organized Opposition to Climate Change Action in the United States," in *Handbook of U.S. Environmental Policy*, ed. David M. Konisky, (Edward Elgar Publishing, 2020), 328–341.

76. T. Culhane, G. Hall, and J. T. Roberts, "Who Delays Climate Action? Interest Groups and Coalitions in State Legislative Struggles in the United States," *Energy Research & Social Science* 79 (2021): 102114.

77. C. Downie, *Business Battles in the US Energy Sector: Lessons for a Clean Energy Transition* (Routledge, 2019).

78. K. Hochstetler, *Political Economies of Energy Transition: Wind and Solar Power in Brazil and South Africa* (Cambridge University Press, 2020).

79. Rachel Wetts, "In Climate News, Statements from Large Businesses and Opponents of Climate Action Receive Heightened Visibility," *Proceedings of the National Academy of Sciences* 117 (2020): 19054–19060.

80. Coan et al., "Computer-Assisted Classification of Contrarian Claims About Climate Change."

81. K. M. Norgaard, *Living in Denial: Climate Change, Emotions, and Everyday Life* (MIT Press, 2011).

82. S. Bader, *Global Spin: The Corporate Assault on Environmentalism* (Sage Publications, 1998).

83. R. E. Dunlap, and A. M. McCright, "A Widening Gap: Republican and Democratic Views on Climate Change," *Environment: Science and Policy for Sustainable Development* 50, no. 5 (2008): 26–35.

84. A. M. McCright and R. E. Dunlap, "The Politicization of Climate Change and Polarization in the American Public's Views of Global Warming, 2001–2010," *Sociological Quarterly* 52, no. 2 (2011): 155–194.

85. A. Niranjan, "Danish Firm's 'Climate-Controlled Pork' Claim Misleading, Court Rules," *The Guardian*, March 1, 2024.

86. D. A. Moreira (ed.), *Panorama da litigância climática no Brasil: relatório de 2024* (PUC-Rio 2024). Information on this case is also available at https://climatecasechart.com/non-us-case/ibama-v-dirceu-kruger-illegal-deforestation-in-the-amazon-and-climate-damage/.

87. However the impact of this decision remains uncertain.

88. J. Farrell, K. McConnell, and R. Brulle, "Evidence-Based Strategies to Combat Scientific Misinformation," *Nature Climate Change* 9, no. 3 (2019): 191–195.

89. Robert J. Brulle, "The Climate Lobby: A Sectoral Analysis of Lobbying Spending on Climate Change in the USA, 2000 to 2016," *Climatic Change* 149 (2018): 289–303.

90. D. R. Fisher and S. Nasrin, "Climate Activism and Its Effects," *Wiley Interdisciplinary Reviews: Climate Change* 12, no. 1 (2021): e683.

91. Climate Social Science Network at Brown University, https://cssn.org/about-us/.

92. S. Hiltner, E. Eaton, N. Healy, A. Scerri, J. C. Stephens, and G. Supran, "Fossil Fuel Industry Influence in Higher Education: A Review and a Research Agenda," *Wiley Interdisciplinary Reviews: Climate Change* (2024): e904.

93. Ibid.

94. Law Students for Climate Accountability, https://www.ls4ca.org/take-action (accessed March 24, 2025).

95. March Gunther, "Edelman Loses Executives and Clients Over Climate Change Stance," *The Guardian*, July 7, 2015, https://www.theguardian.com/sustainable-business/2015/jul/07/pr-edelman-climate-change-lost-executives-clients.

96. Reclaim Finance, "In Response to Climate Deniers, Insurers Must Step Up Climate Action," June 7, 2023, https://reclaimfinance.org/site/en/2023/06/07/in-response-to-climate-deniers-insurers-must-step-up-climate-action/.

97. D. Almond, X. Du, and A. Papp, "Favourability Towards Natural Gas Relates to Funding Source of University Energy Centres," *Nature Climate Change* 1–7 (2022): 1122–1128.

98. Hiltner et al., "Fossil Fuel Industry Influence in Higher Education."

99. Center for Climate Integrity, "1998 American Petroleum Institute Global Climate Science Communications Team Action Plan," https://www.documentcloud.org/documents/2840903-1998-API-Global-Climate-Science-Communications/?mode=document#document/p4/a367394.

100. Committee on Oversight and Accountability, "Oversight Committee Releases New Documents Showing Big Oil's Greenwashing Campaign and Failure to Reduce Emissions," Press Release, December 9, 2022, https://oversightdemocrats.house.gov/news/press-releases/oversight-committee-releases-new-documents-showing-big-oil-s-greenwashing.

101. Viveca Morris and Jennifer Jacquet, "The Animal Agriculture Industry, US Universities, and the Obstruction of Climate Understanding and Policy," *Climatic Change* 177, (2024): article 41.

102. H. Stevnhøj, "The Beef Report Case in Brief," Aarhus University, 2019, https://newsroom.au.dk/en/news/show/artikel/koedsagen-faa-overblikket-her/.

103. Jean C. H. Miguel, "A meada do negacionismo climático e o impedimento da governamentalização ambiental no Brasil," *Revista Sociedade e Estado* 37, no. 1 (2022): 293–315.

104. R. W. Santos, "Reconfigurações do Ecossistema Religioso diante da Crise Climática Globa," Cadernos do OIMC, 2024, https://obsinterclima.eco.br/wp-content/uploads/2024/05/Caderno-11-2024-3.pdf.

105. Kathryn Hochstetler, *Political Economies of Energy Transition: Wind and Solar Power in Brazil and South Africa* (Cambridge University Press, 2020).

106. K. Connolly, "'Two Worlds Colliding': Berlin Transport Workers and Climate Activists Unite Over Rights," *The Guardian*, March 1, 2024.

107. N. C. Crawford, *The Pentagon, Climate Change, and War: Charting the Rise and Fall of US Military Emissions* (MIT Press, 2022).

108. Estimates of emissions associated with the military industry range between 1% and 5% of global emissions, which is equivalent to aviation and shipping sectors (approximately 2% each). Nevertheless, due to States' decisions in the field of defense, often rooted in national security priorities and secrecy, the armed forces are largely spared from reporting their share of emissions. R. Bun, G. Marland, T. Oda, L. See, E. Puliafito, Z. Nahorski, Z., . . . Mathias Jonas, Vasyl Kovalyshyn, Iolanda Ialongo, Orysia Yashchun, and Z. Romanchuk, "Tracking Unaccounted

Greenhouse Gas Emissions Due to the War in Ukraine Since 2022," *Science of the Total Environment* 914 (2024): 169879.

109. M. A. Rajaeifar, O. Belcher, S. Parkinson, B. Neimark, D. Weir, K. Ashworth, R. Larbi. . . and O. Heidrich, "Decarbonize the Military—Mandate Emissions Reporting," *Nature* 611, no. 7934 (2022): 29–32.

110. One author reports that "Yet this author was told personally at the National Intelligence University in 2005 that officers there had been "ordered not to talk about that subject." M. M. Andregg, "Climate Change and U.S. National Security," 2020, https://conservancy.umn.edu/items/1439b7f1-dfd4-4929-9b60-7427da2d881c.

111. L. Cottrell, D. Jalili, and D. Burbridge, "The OSCE and Military Emissions: Next Steps and Mitigation of Greenhouse Gases," in *Climate.Changes.Security: Navigating Climate Change and Security Challenges in the OSCE Region*, 2nd Edition, ed. H. Lampalzer and G. Hainzl (Schriftenreihe der Landesverteidigungsakademiep, 2024), 45–70.

112. S. Parkinson, *Comparing Official UK Statistics for Military Greenhouse Gas Emissions*, Scientists for Global Responsibility, 2022, https://coilink.org/20.500.12592/q66r1z on 10 Oct 2024. COI: 20.500.12592/q66r1z.

113. Manheim, *Strategy in Information and Influence Campaigns*.

114. R. Brulle, personal communication to T. Roberts, 2024.

CHAPTER 2

The Global Role of the Oil and Gas Industry in Climate Delay and Denial

LEAD AUTHORS: GEOFF DEMBICKI, KRISTOFFER EKBERG, AND KERT DAVIES

CONTRIBUTING AUTHORS: ANN-KRISTIN BERGQUIST, ADA NISSEN, AND STELLA LEVANTESI

INTRODUCTION: THE NET-ZERO OBSTRUCTION ERA

In the years leading up to and following the signing of the 2015 Paris Accords, something significant changed within the oil and gas sector—at least, on the surface. One major company after another began declaring themselves climate leaders and loudly announcing net-zero targets, a process by which their vast greenhouse gas (GHG) output would either be reduced to zero or neutralized through reductions elsewhere. Companies had for years tried to "greenwash" their images, but this was of another magnitude entirely. "The world's carbon budget is finite and running out fast; we need a rapid transition to net zero," declared BP's CEO Bernard Looney in 2020, promising that the UK oil and gas giant would become a net zero company within three decades. "This will certainly be a challenge, but also a tremendous opportunity."[1] Similar statements emerged from oil companies across the world, from Royal Dutch Shell and Canada's Suncor to Chevron and Saudi Aramco.[2]

We now know, due to the 2024 release of thousands of internal oil and gas company documents obtained through a joint investigation by Democrats in the US House and Senate, that these pledges were more about prolonging the survival of fossil fuel production than fundamentally shifting the industry's business model. In private corporate communications, top oil and gas executives appeared to state this explicitly. Supporting the goals of the Paris Agreement, as well as US involvement in it, was "the most obvious course of

Geoff Dembicki et al., *The Global Role of the Oil and Gas Industry in Climate Delay and Denial*. In: *Climate Obstruction*. Edited by: J. Timmons Roberts, Carlos R. S. Milani, Jennifer Jacquet, and Christian Downie, Oxford University Press. © Oxford University Press (2025). DOI: 10.1093/oso/9780197787144.003.0002 action to take," BP's global head of Sustainability and Climate Policy wrote in a 2017 email thread made public by the investigation. "All the benefits and few of the risks. That was really why the Paris Agreement was designed the way it was—to enable flexible transition from one political regime to the next. No one is committed to anything, other than to stay in the game."[3]

The Paris Agreement and these cynical reactions to it came during an era of growing international consensus in favor of climate action. By 2021, more than 120 countries including leading global emitters such as China, the United States and the member nations of the European Union had committed in some form to reaching net zero by mid-century.

More than 1,500 companies had also adopted organizational net-zero targets.[4] The new aggressive-sounding pledges from oil and gas companies showed the increasing sophistication of the industry's obstruction tactics. Despite all their talk of net-zero and energy transition, a wide body of academic research has not found compelling evidence for an industrywide shift away from fossil fuels during the period following the Paris Agreement. An NGO study in May 2024 showed that six of eight major European and US companies are planning for increased oil and gas production.[5] The plans to increase fossil fuel production are incompatible with scientific calls to keep more than half of known reserves of oil and gas underground—this carbon must never enter the atmosphere—if we are to have a greater than 50% chance of keeping global averaged warming below 1.5 degrees C.[6] The clean energy announcements from the oil and gas industry are out of line with their actions and investments.[7]

The investigation by the US Senate Budget Committee that reviewed thousands of pages of internal documents subpoenaed from ExxonMobil, Chevron, BP, Shell, and the American Petroleum Institute (API) found evidence that companies used deceptive advertising and "made public pledges to support the Paris Agreement and to achieve net-zero emissions while internally recognizing that those goals were outside of their current business plans."[8]

The oil and gas industry commitments to net zero are a pernicious form of *climate obstruction*, herein defined as efforts with the intent to slow or block policies or actions on climate change that are commensurate with the current scientific consensus of what is necessary to avoid dangerous anthropogenic interference with the climate system. Based on an exhaustive survey of academic literature—primarily within the fields of history, sociology, and political science, as well as investigative journalism about the industry—current oil and gas tactics against climate action are a continuation of efforts going back almost seventy years. These tactics have been developed as a response to geopolitics, shifts in the international market for oil, growing public awareness of climate change, and environmental regulation proposed and implemented by governments. The goal remains to protect the market for oil and gas while blocking or delaying the global transition to lower-carbon energy alternatives.[9]

Building on a previous meta-analysis of industry communication,[10] we characterize obstruction by three distinct but interconnected tactics. The first is *denial*, or emphasizing scientific uncertainty. Denial occurs when the industry publicly questions the research of climate scientists and the work of the Intergovernmental Panel on Climate Change (IPCC) in order to hinder a transition. The second tactic, *delay*, or the use of socioeconomic arguments to stall regulations, is a strategy that emerged in tandem with the creation of think tanks and the rise of neoliberal hegemony since the 1970s. Here, industry lobbies against regulations and legislation by appealing to perceived costs for society or citizens at large. The third tactic is industry *co-optation* of the public policy process, or appearing to take the climate crisis seriously while intervening in governing bodies to weaken, neutralize, or undermine potential policy and technological solutions. Much literature assumes a chronological evolution of these tactics moving from denial toward the others. While the intensity with which specific arguments and tactics have been used certainly has changed over time, this does not mean that efforts to delay and co-opt the policy arena are exclusive to the post-Paris period. Rather, our focus on the longer history of obstruction in this chapter demonstrates how all three tactics—denial, delay, and co-optation—have been used by the oil and gas industry throughout decades of environmental and climate governance.

The Oil and Gas Industry

The combustion of oil and gas is one of the most consequential contributors to climate change. But it is important to acknowledge that these fossil fuels are very different from other polluting and hazardous substances such as F-gases, asbestos, or lead. The existing global economic order has for more than a century been built on the inexpensive and powerful energy provided by oil and gas. The industry has also created energy path-dependencies—examples include car-dependent urban design and national highway systems, which disadvantage environmentally safer alternatives.[11] This is not merely a technological dilemma. The development of fossil fuels has shaped global capitalism, democracy, and modernity, and has been the cause of colonization, war and international power struggles.[12] Oil and gas industry leaders and political supporters often argue that without continually expanding global production, the developed world would neither be able to maintain current living standards nor economic growth and stability.[13]

Despite the oil and gas industry's vast global reach and influence, it is dominated by a relatively small handful of private firms (investor-owned companies, or IOCs) and state-owned companies (nationally owned companies, or NOCs). These companies in turn are responsible for a large majority of the greenhouse gas emissions released by humankind over the past century and a half.

To define the industry for the purposes of this chapter, we are drawing from data on the most polluting fossil fuel companies. Fewer than 100 fossil fuel producers have contributed—through the production, transportation, and sale of their products—63% of all CO_2 and methane emissions released between 1854 and 2010. These so-called carbon majors—a list of ninety companies that also includes cement producers, which aren't a focus of this book—have released half of the world's cumulative industrial emissions since 1986 alone.[14]

An updated list from 2023 of the top twenty cumulative fossil fuel climate polluters included the publicly traded multinational oil and gas corporations Chevron, ExxonMobil, BP, Shell, and ConocoPhillips, as well as the state-owned oil companies Saudi Aramco (Saudi Arabia), Gazprom (Russia), National Iranian Oil Company (Iran), Pemex (Mexico), and Petroleos de Venezuela (Venezuela). When we refer to the oil and gas industry in this paper, we are referring to these and other top producers.[15]

Within the oil and gas industry there is ongoing debate about how to respond to the political and economic pressures of climate change, and we acknowledge that individual companies differ widely in how they craft and implement political approaches and obstruction strategies. As the recent trove of oil company documents subpoenaed by federal US Democrats shows, the European oil majors BP and Shell have focused in recent years on publicly touting their commitments to a net-zero future while expanding gas production. The US company Exxon has made weaker sustainability promises, and has more aggressively monitored and sought to discredit climate activists and journalists.[16] Meanwhile, Saudi Arabia has focused on locking developing countries in Africa into long-term oil and gas infrastructure.[17] Comparing the effectiveness of these company-specific approaches to obstruction, along with the opportunities that internal industry disagreements potentially create for climate action proponents, could be a useful area for further research. But despite differences in tactics and approach, the world's top oil and gas climate polluters share a common goal of expanding production for as long as possible.

Here we focus on the industry as a whole, while drawing on strategies of specific producers to illustrate global obstruction trends. Part of the reasoning for our approach has to do with the international market dynamics of oil and gas production and the difficulties of parsing responsibility between industry actors. For instance, crude oil extracted by national oil producer Saudi Aramco might be purchased by ExxonMobil and then imported, refined, and sold as gasoline in the United States. The business model of one company (along with the associated greenhouse gas emissions) is thus inextricably tied to the business model of others.[18]

Another reason to consider the industry as a whole is because of the evidence that oil and gas companies work together to obstruct climate regulations. Throughout the 1990s, companies including Exxon and Mobil, Shell, and BP partnered with other US industrial polluters to disseminate climate change denial through the Global Climate Coalition.[19] The API has also been a critical actor in opposing state and federal climate laws in the United States. State-owned oil and gas companies in the Global South advance their interests through the Organization of the Petroleum Exporting Countries (OPEC).[20] Corporations and national oil and gas companies alike pursue common goals of entrenching global fossil fuel expansion at each year's international climate negotiations, the UN Conference of the Parties (COP). As the recent US Congressional Joint Staff Report investigating the oil and gas industry noted: "Fossil fuel companies collaborate with one another and industry groups to develop pro-fossil fuel disinformation, downplay the industry's responsibility for climate change, and obstruct meaningful climate action."[21]

There is a tension between the seemingly global reach of oil companies and the political context in which they operate. State-owned companies tend to, for obvious reasons, be more dependent on national political context. The actions of transnational investor-owned producers indicate a greater capacity and ambition to influence and obstruct climate policy through global networks.[22] Partly for this reason, this chapter, as well as the studies it relies on, focus primarily on the Global North. For more details on the actions of fossil fuel obstruction in the Global South, see Chapter 8.[23] The focus on the Global North, and especially the United States, is also motivated by the fact that majors in that region "continuously exhibit defensive attitudes to renewables investment and the need to shift from fossil fuels," and that stalling action in major developed world economies can be an effective means of blocking progress globally.[24]

While the scope of this chapter is limited to the general obstruction tactics of a global industry, we do acknowledge the importance of scholarship investigating the specific nuances of localized political struggles and alliances between industry, policymakers, and regulators. Researchers have argued that "the conditions for changing oil companies' climate strategies are likely to be located in the political context rather than in the companies themselves."[25] This context includes intercompany and sectorial competition as climate change governance will force asset re-evaluation within industries.[26] Such contexts, alliances, and networks are further explored in Chapters 5, 9, 10, and 11.

We also acknowledge that oil and gas companies depend on a wide range of actors and organizations outside the industry to assist in enabling its climate obstruction. These include international business groups, banks, chambers of commerce, law firms and think tanks. Crucially, the oil and gas industry depends heavily on public relations and advertising to influence public perception in favor of its business model, a fact that United Nations Secretary-General Antonio Guterres called attention to during a speech in 2024 when he advocated for a global ban on fossil fuel ads and urged PR firms "to stop acting as enablers to planetary destruction."[27] See Chapters 6 and 7 for more.

OBSTRUCTION STRATEGIES

The oil and gas industry has deployed three distinct but interlocked obstruction tactics: undermining or denying the growing scientific consensus on the urgency and dangers of climate change; utilizing socioeconomic scare tactics to delay action by governments; and co-opting or trying to shape the environmental discourse and policies on terms favorable to the industry.

Science Denial: Amplifying Uncertainty to Delay Policy

In the early years of the climate policy debate, which we define roughly as the late 1980s through the early 2000s, the lead obstruction tactic deployed by fossil fuel corporate interests was to downplay or deny the scientific consensus, casting doubt in order to forestall action in the policy arena. This tactic continues intermittently in various forms to this day.

One recent example took place during the international climate negotiations in Dubai in 2023, when COP28 president Sultan Al Jaber stated during a panel discussion that there is "no science" supporting the idea that phasing out fossil fuels is essential for restricting global warming to 1.5 degrees C, the level agreed by scientists is a dangerous threshold. Al Jaber, who at the time was also the chief executive of the United Arab Emirates' state oil company, Adnoc, faced backlash for the statement, which one scientist called "verging on climate denial."[28] A large body of scientific literature—including a landmark 2021 report from the International Energy Agency[29]—has demonstrated that managing a rapid decline of fossil fuel production is crucial for stabilizing global temperatures.[30]

While Al Jaber's comments were widely publicized and debated, denial about the required pace of transition off of fossil fuels is so commonplace within the oil and gas industry that it is often ignored. During the Dubai talks, IPCC chair and former professor of energy Jim Skea noted that while a full phaseout of oil and gas might not be required to achieve 1.5 degrees C, global oil use by 2050 needs to be reduced by 60% and gas use must fall by 45%.[31] Yet when researchers compared the production output of 142 coal, oil, and gas companies against necessary fossil fuel reductions, they determined that from 2014 to 2020, 63% of the oil companies and 70% of the gas companies in their study exceeded the production limits required for global temperature stabilization. If the companies continued producing at these levels, the oil industry would overshoot its cumulative production budget by 42% by 2050, while the gas industry would overshoot its budget by 53%, vastly increasing the likelihood that humankind would fail to meet the Paris climate targets.[32] A 2011 study showed that keeping the Earth's carbon budget within safe limits would require leaving 80% of proven oil gas and coal reserves in the ground. These reserves are "on the books" of oil and gas companies.[33] Some social scientists consider the industry's continued insistence on expansion of production a tacit rejection of science and refer to it as "a subtle form of climate change denialism."[34]

Internal communications from oil and gas producers explicitly anticipate that the world will fail to meet climate targets. One Exxon official wrote in a 2018 email that "we don't yet see the world even approaching a 2 deg C pathway . . . let alone a 1.5 deg C pathway," adding that achieving either of these "would require unprecedented changes to the global energy system and economy, at a rate and scale never before demonstrated."[35] Yet publicly, ExxonMobil and all other major producers have announced plans to reach net-zero emissions by 2050. This disconnect between publicly stated climate ambition and business plans that expand global oil and gas production serves a similar goal as outright denial. The process by which companies portray themselves as climate leaders, particularly through promotion of unproven or nonexistent technologies without concrete plans to shrink their extractive operations, has deflected public attention away from the industry's disproportionately large contribution to the emergency.[36]

Industry-Led Climate Research

Although the particulars of industry-led climate denial have shifted since the signing of the Paris climate agreement, the roots stretch back many decades. A growing body of academic literature has traced the origins of this form of climate obstruction to the late 1950s and early 1960s. That era was when oil companies in the United States, in line with the most advanced science of the time, were first alerted by reputable sources that "increased carbon dioxide concentrations in the atmosphere could be expected to produce global warming, and that fossil fuels were in fact causing atmospheric carbon dioxide to build up."[37] The API commissioned a report about climate in 1968, one of the first known instances of the industry privately studying the link between fossil fuels and climate change, explaining that "temperature changes are almost certain to occur by the year 2000 and these could bring about climatic changes."[38]

By the early 1970s, major companies in the United States and Europe were conducting increasingly sophisticated research on the impacts of their products on the atmosphere and the climate.[39] Exxon, as well its Canadian subsidiary Imperial Oil, was leading efforts among industry to monitor climate science and policy efforts.[40] Subsequent analysis has demonstrated that Exxon's early climate research and temperature modeling was sophisticated and precise.[41] The French oil major Total in 1971 published an article in the company magazine *Total Information* entitled "Atmospheric pollution and climate," which attributed a steady increase of carbon dioxide in the atmosphere to "consumption of coal and oil." The publication described this trend as "quite worrying" and stated that "it is not impossible, according to some, to foresee at least a partial melting of the polar ice caps, which would certainly result in significant sea level rise. The catastrophic consequences are easy to imagine."[42]

Royal Dutch Shell, as it was then called, was also researching the consequences and impacts of GHGs. In the late 1960s, Shell commissioned British scientist James Lovelock to study pollution from fossil fuels, part of a broader effort to take seriously the threat of global warming.[43] Shell then donated £10,000 in 1972 to help establish the Climatic Research Unit at the University of East Anglia, meaning that the company was at the forefront of a worldwide scientific effort to measure and understand global temperature rise. A confidential internal Shell research committee formed in the early 1980s concluded that "by the time the global warming becomes detectable it could be too late to take effective countermeasures to reduce the effects or even to stabilize the situation."[44] Studies on climate change were also being commissioned and published by Italian oil major Eni. A recently surfaced study Eni commissioned between 1969 and 1970 from its Isvet research center made clear that the company was aware that rising fossil fuel use, if left unchecked, could lead to "catastrophic" impacts on the global climate.

A separate 1978 report produced by the Eni-created Tecneco company contained additional warnings of the climate risks posed by fossil fuels and included a projection of atmospheric CO_2 levels. "It is assumed that with the increasing consumption of fossil fuels, which began with the industrial revolution, the CO_2 concentration will reach 375–400 [parts per million or ppm] in the year 2000," stated the report.[45] This projection, as well as others created by leading oil and gas companies, would prove broadly accurate in the decades to come.

Pivoting to Denial

Despite this early knowledge, the oil and gas industry began publicly denying its contribution to climate change starting in the late 1980s. One turning point was the June 1988 testimony to the US Congress of scientist James Hansen of the National Aeronautics and Space Administration, who explained that he had "99% confidence" humans were altering the atmosphere by releasing GHGs, a statement that made front page of the *New York Times*. The same year, the IPCC was established with the task of providing scientific evidence on climate change to guide policy. The IPCC's findings, accompanied by rapidly growing public awareness, inspired the creation of a regulatory framework for addressing global emissions under the United Nations umbrella.[46]

In Europe, oil and gas companies challenged the fixed emission cuts suggested at the Toronto Conference on the Changing Atmosphere, a 1988 event held the same week as Hansen's testimony. These oil companies opposed a European-community-wide tax on carbon emissions, which was proposed by environmental commissioner Carlo Ripa di Meana from the Italian Greens.[47] The response by US companies to early international climate efforts was also aggressively obstructive.[48] Across Europe and North America, major oil and gas producers led some of the first communications campaigns questioning the need for climate action and created a broad alliance of anti-regulatory business interests, including manufacturing firms and coal-burning electric utilities (see Chapters 3 and 4).

The Global Climate Coalition (GCC) was formed in 1988 with a clear goal of monitoring and undermining climate science in order to block and delay US government efforts to restrict fossil fuels or shift economies away from those fuels entirely. An early GCC bulletin, disseminated to mainstream media organizations, claimed that "many scientists have growing doubts about the accuracy" of climate projections which were the rationale for "government action to restrict the burning of fossil fuels."[49] By the early to mid-1990s, the GCC had amplified the voices of a roster of academics such as Dr. Patrick Michaels at the University of Virginia and Dr. Richard Lindzen of the Massachusetts Institute of Technology (MIT), who each publicly questioned the scientific consensus on climate change and were frequently quoted in major media outlets.[50]

The GCC knew that the climate skepticism it was pitching to the media was not scientifically sound. A seventeen-page draft "primer" prepared for coalition members in 1995 noted that "the contrarian theories raise interesting questions about our total understanding of climate processes, but they do not offer convincing arguments against the conventional model of greenhouse gas emission-induced climate change."[51] That rebuttal was later dropped from the final version of the primer after pressure from the coalition's operating committee.

A similar dynamic seems to have played out within ExxonMobil. In 2017, researchers demonstrated that roughly 80% of internal documents and peer-reviewed papers produced by Exxon/Mobil/ExxonMobil from the 1970s until the 2010s acknowledged that global temperature rise is caused by human activities including burning fossil fuels. Yet most of Exxon's paid advertorials in mainstream media (81%) expressed doubt that human-induced climate change is real.[52] Meanwhile, the libertarian think tank the Cato Institute, backed by the oil and gas company Koch Industries, held one of the first climate denial conferences in 1991.[53] A pamphlet from the event claimed there is "very little evidence at all" for catastrophic climate projections.[54]

By some measures these denial campaigns appear to have been highly successful. In 1989, when climate change was first coming to the general public's attention, nearly 65% of Americans said they worried a "great deal" or a "fair amount" about it, according to a global study of public opinion conducted by several UK universities. Only eight years later, that number had decreased to 50%. "Through the 1990s, at a critical point when the fossil fuel usage needed to be brought under control, public concern about the risks and causes of climate change waned," reads a separate academic study from that period.[55]

International policymaking continued to threaten oil and gas operations, including and especially the 1997 Kyoto conference (UNFCCC COP3). If negotiations there were successful in creating a compulsory international treaty to deal with climate change, they would set in motion massive economic shifts away from fossil fuels. As a result of oil and gas companies emphasizing uncertainty about the science, politicians in the US were convinced that no action was needed, which effectively stalled negotiations and the ratification of the Kyoto Protocol by the US Senate.[56] The GCC gained allies around the world including Saudi Arabia, which effectively lobbied against the IPCC and the Kyoto Protocol on grounds of scientific uncertainty.[57] The United States signed the Kyoto Protocol in 1997, but the Senate refused to ratify it after an industry pressure campaign.

The utility of science denial during this period was unevenly distributed around the world, depending on national interests and public opinion. It was less effective and less prevalent in Europe. Russia, a country that stood to benefit financially from Kyoto under an international emissions-trading scheme (because its emissions had fallen due to economic collapse) was less concerned at this early stage with denial and obstruction.[58]

Despite divisions between oil-producing nations and within the industry itself, oil and gas companies were successful in using scientific denial to undermine the Kyoto Protocol. After becoming US President in 2001, George W. Bush immediately pulled the United States, the world's second largest national emitter of greenhouse gases, out of the Kyoto agreement. Pressure from the GCC appears to have played a significant role in that decision. A 2001 US State Department meeting briefing explained: "POTUS [President of the United States] rejected Kyoto in part based on input from you [the Global Climate Coalition]."[59]

The lead-up to Kyoto may have been the peak of outright climate science denial as an obstruction tactic. The GCC disbanded in 2001, and though individual companies including ExxonMobil[60] and Koch Industries[61] continued to finance climate denial efforts, the industry as a whole, as well as individual companies, began publicly accepting the basic reality and causes of climate change over the coming years. But to this day, rhetorical acknowledgment of the scientific consensus continues to contrast with growth plans that deny the urgent need to phase out fossil fuels. This softer form of denial is regularly presented with little controversy to shareholders and regulators.

When researchers analyzed the corporate behavior of ten of the major publicly traded oil and gas companies between 2005 and 2019 based on coded corporate-earnings calls, that is, the quarterly communications between firms and major investors, there was no evidence that any of these companies were planning to shift their core business away from fossil fuels.[62] For example, a 2021 Climate Action Framework from the API declared a goal of "the continued promotion of natural gas in a carbon constrained economy." A Shell public relations plan explained that "Shell has no immediate plans to move to a net-zero emissions portfolio over our investment horizon of 10–20 years," even as the company publicly stated that it wanted to help the world shift away from fossil fuels.[63]

Tactical Delay: Using Economic Arguments to Maintain the Status Quo

The use of economic scare tactics as part of lobbying efforts to delay or weaken policy action has occurred with science denial from early on. Whether through advertising campaigns, widely disseminated reports from economic consulting firms or other third parties, or internal communications with policymakers, the oil and gas industry has lobbied legislators with claims that climate policy initiatives will wreck local, state, or national economies. Such claims often conflate the industry's private business interests with the economic well-being of the public.

Contemporary examples of economic arguments to undermine climate action are numerous. In the lead-up to the Biden administration proposing spending hundreds of billions of dollars in tax credits and other incentives on climate initiatives through the $3.5 trillion Build Back Better plan, oil and gas organizations blitzed social media with ads stating that fossil fuels are essential to maintaining Americans' quality of life.[64] Though the administration ultimately passed a major infrastructure bill called the Inflation Reduction Act, spending on climate was reduced due to obstruction from oil and gas-friendly policymakers like Democratic Senator Joe Manchin. In the European Union, where regulators are implementing an ambitious program of infrastructure spending known as the Green Deal, an oil and gas executive recommended resisting a rapid transition to renewables with the narrative that the industry "ensures security of supply and supports the world with affordable energy that helps to foster economic growth as well as welfare for billions of people in the developing countries."[65] Such messages aim to catalyze public opposition to climate action as oil and gas producers heavily lobby policymakers against the implementation of legislation.

Opposing environmental action with altruistic-sounding appeals to the economic greater good was a tactic developed during the same decades the oil and gas industry was first studying climate change. From the late 1970s onward, industry leaders sought to influence public perceptions about regulation, framing the administrative state as damaging to free enterprise and economic growth. Those leaders included Charles Koch, CEO of Koch Industries, who stated in 1978 that "we need to attack government regulation for wreaking havoc on those it is allegedly designed to help." Koch further called on like-minded businessmen to "be involved in politics and political action—from local tax revolts to campaigns for Congress and the presidency."[66] (For more on populist and far-right campaigns to obstruct climate action see Chapter 5.)

The tactic of encouraging business to fight ideological battles against government intervention in society was part of a larger trend. Several studies have shown that starting in the 1970s, the oil and gas industry intensified its opposition to environmental regulation and environmentalism.[67] Fighting against state efforts to regulate the industry's air and water pollution helped consolidate conservative voices against burgeoning conservation movements and accompanied broader neoliberal attacks on a Keynesian planned economy. Over time, these attacks helped transform climate change mitigation and environmental regulation at the elite level into an issue of price mechanisms and consumer demand.[68] A core element of this tactic has been to portray environmental movements as opposed to basic welfare, comprising a privileged class indifferent to the hardships of the ordinary worker. Positioning environmentalists as being against the working class has helped create divisions in the environmental movement and also aided in the criminalization of protests, thereby protecting fossil fuel investments.[69]

As companies backed climate science denial efforts during the 1990s, they also hired economic consultants to analyze proposed climate policies. Economic consultants helped the oil and gas industry to frame potential solutions to rising emissions on terms favorable to their business interests.[70] One example comes from a 1993 document produced by Exxon's Canadian subsidiary Imperial Oil. The company hired the consulting firm DRI/McGraw-Hill to look at the potential economic impacts of a national tax on carbon. The firm concluded that a carbon tax could result in "stabilization" of Canada's emissions without a major impact on the national economy due to "enormous amounts of additional revenue" generated by the tax. However, the firm warned that the tax could be harmful to Canada's oil and gas industry and cause specific losses of CAD $940 million for Imperial. The company, along with Exxon, drew from that report and the credibility of the consulting firm to create talking points framing the policy in the worst possible light. Targeting government and media, company leaders warned that a carbon tax would hurt a "precarious economy and international competitiveness."[71]

Leveraging expert studies to make dire economic claims about climate action became common across the oil and gas industry.[72] Industry-hired economists disseminated research that inflated potential costs of environmental regulations while ignoring policy benefits. These economic arguments "played a key role in undermining numerous major climate policy initiatives in the US over a span of decades, including carbon pricing and participation in international climate agreements."[73] The API, for example, commissioned the economic consulting firm Charles River Associates to produce a 1991 report concluding that effective climate action would be destructive to the US Gross National Product. The report's conclusions, which were based on shaky or scant evidence according to a later analysis, were then reported in major media outlets such as *USA Today* and the *New York Times*.[74]

Similarly, a 1996 background paper released by the GCC declared that "U.S. living standards and lifestyles would be seriously damaged by many of the greenhouse gas abatement proposals currently under consideration, especially those that would stabilize or reduce carbon emissions for taxing fossil fuels."[75] The coalition's members reiterated that message in prominent advertisements. "Let's face it: the science of climate change is too uncertain to mandate a plan of action that could plunge economies into turmoil," read 1997 ads paid for by Mobil in the *Washington Post* and the *New York Times*.[76] Economic arguments proved rhetorically useful for opposing stricter government regulation of industry and more forceful climate action and descend from a longer history of anti-environmentalist arguments.[77]

The oil and gas industry in the process helped to create narratives that became conventional wisdom about climate change, such as the idea that regulating GHGs is bad for economic growth. Such advertisements worked to shift responsibility for global warming away from the fossil fuel industry and onto consumers. The ads also claimed "that climate change was a 'risk,' rather than a reality, that renewable energy is unreliable, and that the fossil fuel industry offered meaningful leadership on climate change."[78] The oil and gas industry made these economic claims to great effect in opposing climate policies introduced during the late 2000s and into the 2010s.

This economic rhetoric coincided with a rapid political reorientation around climate politics globally. Following major cultural moments for global warming, such as Hurricane Katrina in 2005 and the release of Al Gore's 2006 documentary *An Inconvenient Truth*, public and elite concern about the climate emergency intensified, including among influential conservatives.[79] Even Exxon seemed to acknowledge the new reality. Its CEO, Rex Tillerson, said during a 2007 speech at the global energy conference CERAWeek that "the risks to society and ecosystems from climate change could prove to be significant."[80] Climate policy in the European Union advanced quickly during the 2000s with development of the EU Emissions Trading System (EU ETS) as a response to the Kyoto Protocol.

Given US President George W. Bush administration's refusal to address climate change amid growing public concern about the issue, the incoming administration of Barack Obama in 2009 attempted to use a Democratic majority in Congress to implement the first comprehensive climate legislation in American history. The prospects initially looked good for the passage of the American Clean Energy and Security Act (known as the Waxman-Markey bill); it would have established a system of cap-and-trade in GHG emissions and potentially spurred a global market for emissions trading. The bill passed in the House of Representatives, but soon came under heavy attack, especially by oil interests. The API paid for national ads raising economic fears about Waxman-Markey, with one 2009 advertisement in the *Washington Post* stating "If you like $4 gasoline, you'll love the House Climate Bill."[81] The API also launched its "Energy Citizens" initiative, an "astroturf" campaign with the look and feel of grassroots support later revealed to have been designed by Edelman Public Relations.[82] Libertarian think tanks financed in part by oil and gas interests effectively branded the policy as "cap and tax."[83] The Tea Party, a far-right social movement that received financial backing from fossil fuel billionaires the Koch brothers, made stopping Waxman-Markey a priority.[84]

Lobbying pressure was intense and focused on the purported negative economic impacts of the legislation. One study found that companies that expected to suffer financially under Waxman-Markey spent more than $700 billion lobbying against the bill, radically reducing the possibility of its passing.[85] From 2000 to 2016, the fossil fuel industry spent more than $370 million lobbying the US government, its spending peaking in the years leading up to Waxman-Markey. Combined with additional spending from utilities and transportation companies (see Chapter 3) as well as meat and dairy companies[86] (see also Chapter 4), this tsunami of political cash engulfed the lobbying efforts of environmental groups and renewable energy companies in favor of the legislation.[87] ExxonMobil alone spent $33.3 million on lobbying from January 2009 to June 2010, according to an analysis from the organization the Center for American Progress Action, followed closely behind by ConocoPhillips ($30 million), Chevron ($27.8 million), BP ($19.3 million), Koch Industries ($16.2 million), and Shell ($16.6 million).[88]

Meanwhile, top investor-owned oil and gas companies were engaged in their own lobbying efforts in Europe. A report from a group of NGOs calculated that between 2010 and 2018, BP, Shell, Chevron, ExxonMobil, and Total as well as their industry groups spent at least €251.3 million lobbying EU institutions. The sector employed more than two hundred lobbyists in Brussels during this period, and the efforts seemed to have paid off. Report authors noted that climate and renewables targets were weakened from their original versions. "Big polluters like Shell, BP and their lobby groups have delayed, weakened, and sabotaged EU action on the climate emergency thanks to their hefty lobby spending. A cool quarter of a billion over the last decade buys a lot of access and influence in Brussels," Pascoe Sabido, a researcher at Corporate Europe Observatory, told *The Guardian*.[89]

Despite mounting scientific evidence that climate change will have catastrophic impacts for much of the world's population, oil and gas companies to this day portray their opposition to climate policy as being the only sensible and rational economic approach. At the COP28 talks in Dubai in 2023, for example, Saudi Arabia led efforts to oppose language calling for a fossil fuel phaseout in the final agreement. "It would be unacceptable that politically motivated campaigns put our people's prosperity and future at risk," Haitham Al-Ghais, the secretary general of OPEC, said during the talks.[90] Meanwhile, Saudi Arabia had been leading an ambitious global investment plan to create new demand for oil and gas in Africa. In public materials touting the plan, it talked altruistically about "removing barriers" to energy in poor countries and promoting "sustainability."[91]

This type of discourse fits the pattern shown above, in which the oil and gas industry frames opposition to more environmentally sound technologies and policies as benefiting society overall. However, one study articulated the narrow interests being served by the industry's climate obstruction by analyzing several years of company annual reports from BP, Chevron, ExxonMobil, and Shell to calculate how much senior executives and board members were being compensated for continued fossil fuel production. In 2018 alone, top executives at Shell and Chevron personally earned more than $20 million, with ExxonMobil's CEO not far behind. The authors came to an obvious but underappreciated conclusion: "Slowing down the low-carbon transition is profitable for the boards of these Carbon Majors."[92]

Co-Opting the Policy Arena: Positioning the Industry as Part of the Solution

The industry's positioning of itself as part of the solution to climate change could be seen as the most evolved of its strategies—or as the industry's last resort. With global concern about the crisis growing following the 2015 Paris talks, along with ambitious new government legislation to address it, the oil and gas industry has increasingly inserted itself into the policymaking process. The goal is to be accepted as a legitimate actor in solving climate change. Companies aim to be seen as a partner, not an enemy. As part of this process, they advance expensive technological fixes for climate change that give the appearance of taking the crisis seriously. This advocacy allows them to slow climate action and continue production, all while portraying themselves as part of the solution.

The strategy of solutionism has become visible at COP climate talks. In 2022, more than 600 fossil fuel industry delegates attended the international negotiations in Egypt, up 25% from the year before.[93] Yet that record was exceeded the following year in Dubai, when at least 2,456 lobbyists associated with the oil, gas, and coal industries attended.[94] Among them were representatives of a trade organization called the Pathways Alliance, representing top Canadian oil-sands producers. "We're going there in a very constructive way to say, 'We're here, we're a big source of emissions and we're going to be a big part of the solution,'" the organization's president said at the time.[95] Critics had a much different interpretation. To them, the fossil fuel presence at COP28 was an industry effort to co-opt and weaken the international policy response to climate change.

Though being part of the solution is an obstruction tactic that has been used by oil and gas companies for decades, its use has increased in the net-zero era. Although voluntary, the Paris Agreement set an expectation for countries to halt global temperature rise at 2 degrees C above preindustrial levels, with an aspirational commitment to aim for the even more aggressive stabilization goal of 1.5 degrees. This helped catalyze an effort among scientists to calculate specific limits on the amount of carbon that could be added to the atmosphere. These rough calculations became a target, leading many countries to pledge a goal of net zero starting with Sweden in 2017. By 2022, countries as well as companies representing more than 90% of the global economy had made net-zero pledges.[96] Global energy-related carbon dioxide emissions reached a new record high in 2023.[97]

Addressing climate change in any substantial way constitutes an existential dilemma for oil and gas companies because even modest international efforts to bend the global emissions curve through regulations or pro-climate industrial policies are likely to affect the industry's business model.[98] But the post-Paris net-zero era has been accompanied by revelations about the private internal climate knowledge of oil and gas companies—what they knew and when about the atmospheric impacts of the products they were selling. Starting in 2015, journalists with *Inside Climate News* and the *Los Angeles Times*/Columbia University began publishing corporate documents retrieved from industry archives and other sources, revealing, as noted earlier, that Exxon and other oil producers had privately studied the threat of climate change and their industry's contribution to it as early as the 1970s, but then led efforts throughout the 1990s and 2000s to mislead the public and policymakers about climate science and attack climate policies with economic fearmongering.[99]

Revelations about what these companies knew combined with attribution research on their emissions led to new thinking about responsibility, including legally. Lawsuits have now been filed against major privately owned oil and gas producers including ExxonMobil, BP, Shell, and ConocoPhillips.[100] These cases utilize a diversity of legal approaches and strategies to hold the industry legally liable for knowing its products were causing global harm, then deceiving the public and policymakers about the certainty and urgency of climate change to stall or defeat policies to stem the crisis (see Chapter 12).[101]

The oil and gas industry found itself facing new climate regulations, adopted by countries in response to the Paris Agreement, at a time when two of its historically favored obstruction tactics—scientific denial and economic scare tactics—were being subjected to intense legal scrutiny. These fossil fuel companies also continued to face immense financial pressure to increase oil and gas reserves and production and meet their obligations to shareholders.[102] We argue that the combination of global climate action and litigation following the Paris Agreement was the impetus for the industry's rapid embrace of the language and optics of climate leadership, which began with European oil majors such as Shell, BP, and Total making voluntary net-zero pledges and soon spread throughout the industry, even among stakeholders historically resistant to any kind of climate action including ExxonMobil and the API. This approach—publicly stating concern and even occasionally accepting responsibility for their products causing climate change—could be described as a way of co-opting the climate solutions discourse in order to forestall legislation and maintain the oil and gas business model.

Oil and gas companies attempt to legitimize their image as climate leaders by supporting solutions that reduce operational emissions while still allowing the industry to expand. The oil and gas industry has, for instance, advocated carbon capture and storage or sequestration (CCS), which involves capturing carbon dioxide and burying it underground. There have been dozens of pilot projects and years of discussion and debate about the technical difficulties and prohibitively high costs of this technology. But the majority of the most-advanced CCS projects have underperformed worldwide and in one internal corporate document obtained by US Congressional investigators, Exxon predicted that "global scale is limited" for the deployment of CCS by 2050.[103] Even if the industry was successful in deploying CCS, the companies' expressed intent is to use it to address Scope 1 emissions, which would abate only a small percentage of oil and gas emissions, given that a company's Scope 3 emissions, the emissions from actually burning fossil fuels produced in the form of gasoline or other consumer products, represent 80% to 95% of total carbon emissions from oil and gas.[104] A strategy to address Scope 3 emissions is absent from the net-zero plans of companies like Chevron and Exxon.[105] Instead, the industry paradoxically says it intends to make the production process for oil and gas more climate-friendly as it grows the overall global supply of fossil fuels. "We need to keep in mind that this is about reducing emissions and not reducing production," a Canadian oil industry organization stated in 2022 about carbon-capture technology and net zero.[106]

The focus on operational emissions rather than the big picture—global consumption of fossil fuels and polluting energy systems—has been at the core of international climate governance since the 1990s.[107] With alarms about climate change getting louder every year, fossil fuel–producing nations still plan for continued or even increased oil and gas supply.[108]

The contradiction between climate solutions that lower emissions from oil and gas production while allowing consumption of fossil fuels to increase is apparent in Norway. The energy company Equinor is by far Norway's most important state-controlled enterprise. Today Equinor attempts to reduce its production emissions by using and developing CCS technology and by electrifying its drilling platforms (similar to how meat and dairy producers focus on mitigating emissions from the energy they use rather than the methane they produce or the land-use changes caused by their products; see Chapter 4). Focusing on electrifying fossil fuel production is in line with Norwegian government policy, which for decades has struggled with reaching climate goals without jeopardizing the country's biggest industry and emissions source— the oil industry. Equinor aims for nearly zero emissions from its oil and gas production on the Norwegian continental shelf by 2050. However, the effect of carbon capture on global emissions is uncertain. This strategy postpones change as it allows for discussions about phasing out oil and gas entirely to be marginalized, while emissions-trading systems, electrification, CCS, and voluntary emissions- reduction agreements take center stage.[109]

Indeed, despite their talk of compliance with climate imperatives, top oil and gas producers lobbied heavily against the Biden administration's IRA and the European Union's Green Deal, two of the most significant pieces of climate legislation ever proposed in major world economies. And in the wake of Russia's Ukraine invasion in 2022, along with sanctions cutting off Russian oil and gas from global markets, other major oil companies and their investors, flush with the biggest profits in their corporate history, began retracting climate promises they had made only several years earlier. BP, which earned $27.7 billion in 2022, announced it would scale back its earlier pledge to reduce oil and gas production and instead increase its annual investment in the sector by $1 billion annually. "The conversation three or four years ago was somewhat singular around cleaner energy, lower-carbon energy," its former CEO Bernard Looney explained.[110] Shell during this same period weakened and in some cases abandoned entirely its previous investments in renewable energy projects while earning record profits of $40 billion.[111] Leading Canadian oil sands producers, meanwhile, announced plans to keep expanding their oil exports to Asia and other foreign markets even as they continued to brand themselves as net-zero leaders.[112]

To better understand the tactic of claiming to be a climate leader while increasing production of the fossil fuels responsible for climate change, it is useful to look at history. As we discussed earlier in this chapter, during the 1970s oil and gas companies were beginning privately to study the impacts of their business model on the global climate and other natural systems. This advanced knowledge of environmental issues allowed companies to shape society's response on terms beneficial to the industry. Companies then engaged in proactive strategies "aim[ed] at securing a position from which they will be able to take part in the policy design process and orient it in directions that suit them—e.g. more market instruments, less command-and-control regulations."[113]

Leading oil producers in 1974 created the International Petroleum Industry Environmental Conservation Association (IPIECA), which promoted the narrative of the industry as an innovative environmental leader. Oil executives also exerted influence on the formation of institutions such as the United Nations Environment Programme (UNEP), helping to shape contemporary thinking about global environmental problems through a pro-economic development framework, dismissing or rejecting the idea of constraining business to reduce pollution.[114]

As a way to circumvent the boundaries of the 1970s Limits to Growth debate and preferences for command-and-control regulation, the International Chamber of Commerce (ICC) worked closely with the UNEP throughout the 1970s and the 1980s. Oil companies including Shell were key players within the ICC and helped construct an alternative regulatory approach in global environmental governance that was based on industry self-regulation and voluntary commitments, based on new concepts such as sustainable development and eco-efficiency. In 1984, UNEP co-organized the World Industry Conference on Environmental Management (WICEM) with the ICC, held at Versailles, Paris. Oil companies, including Exxon, Gulf Oil, and Shell, both sponsored and participated in WICEM along with a range of other multinational corporations.[115]

One of the key outcomes of the conference was a definition of sustainable development as "development that can be maintained indefinitely without damaging the environment—or threatening development itself." These were some of the ideas that the Brundtland Commission embraced in its seminal final report *Our Common Future*, which was a major success for industry.[116]

The transnational community of "green" business networks that emerged in the 1980s had one point in common: their perception of sustainable development based on self-regulation and voluntary codes.[117] The notion that the oil and gas industry should frame itself as an environmental leader took a backseat during the 1990s. Motivated by calls for strong government action on climate change at the same time oil prices were plummeting, as we noted earlier, companies engaged in a more aggressive campaign of scientific denial and economic scare-mongering to oppose climate regulations. But the soft tactic of delay and co-optation began to gain more prominence within the industry following the Kyoto climate talks in 1997 and the subsequent emergence of the cap-and-trade system ETS in Europe. With such a system, regulators put a cap on the total emissions allowed and let industry actors buy and sell emission quotas, effectively leaving industry to decide on how to lower emissions rather than subsidizing specific technologies.

Many oil and gas companies sought to portray themselves as part of the solution within this framework, increasingly embracing the concept of corporate social responsibility that was being widely adopted by corporations from the mid-1990s through the 2000s. In industry communications, this kind of "responsible business" approach to climate change rose to dominance in the 2010s.[118] In the Australian context, the focus on industrial innovation and solutions comprised almost 35% of the communication from industry between 2008 and 2019.[119]

An emphasis on industry-derived false solutions is now a key part of the obstruction strategy in our present era. Researchers have pointed to the growing prominence of "natural" (methane) gas in climate debates, whereby gas producers use the language of "transition" to tout gas as a "bridge fuel," as one prominent example.[120] The industry's argument goes that burning methane emits less greenhouse gases than coal and thus is an important step toward widespread deployment of renewables. However, emerging science and more accurate leak detection have shown that when one accounts for losses of methane along the natural-gas drilling and supply chain, the additional climate forcing of this methane negates any advantage over coal.[121] A BP public relations campaign in 2024 used the slogan "It's and, not or," indicating that continued oil and gas production goes hand in hand with developing energy alternatives.[122] But executive-level documents from BP raised the concern that "gas locks in future emissions above a level consistent with 2 degrees," while additional internal documents show that the company was warned by Princeton University researchers that new global supplies of shale gas could accelerate catastrophic levels of warming.[123]

Co-opting the public policymaking process allows the industry to present itself as a committed leader that needs to be at the negotiating table while prolonging the production of fuels whose combustion continues to accelerate the crisis the negotiations are attempting to solve. Net-zero pledges reliant on carbon offsets, direct air capture, CCS, or even geoengineering from oil companies shift attention away from mitigation here and now.[124] Executives have become increasingly candid about the lifeline that industry-favored solutions such as carbon capture technologies can provide for fossil fuels during the global energy transition. "We believe that our direct capture technology is going to be the technology that helps to preserve our industry over time," the CEO of the US oil company Occidental said in 2023. "This gives our industry a license to continue to operate for the 60, 70, 80 years that I think it's going to be very much needed."[125] Rather than address humankind's dangerous reliance on oil and gas, industry's co-optation of the policymaking process via technologies such as carbon capture make that reliance on planet-warming fuels even stronger.

CONCLUSION

On the surface it may seem like we are in a different era for climate action following the Paris Agreement. Most major oil companies across the planet now loudly tout their ambition of helping the world achieve net-zero emissions by 2050, a requirement for avoiding global catastrophe. Yet the transitionary rhetoric from oil companies is not matched by the necessary actions. While companies have developed and touted projects to develop renewables or address environmental pollution,[126] the results have been marginal. The ultimate impact of these activities has been to legitimize oil and gas operations as well as keeping open the possibility for continued extraction.[127] Production of the fuels at the center of our climate crisis continues to rise.

The current net-zero pledges and solutions put forward by industry are the latest phase in a process of climate obstruction going back at least to the 1970s. During those decades, the oil and gas industry's tactics have shifted depending on the political and economic context. The industry has alternately pushed denial of the basic science on climate change, used economic scare tactics to justify aggressive lobbying against solutions, and attempted to co-opt the policymaking process on terms favorable to its business model. Though each of these tactics has risen and fallen in prominence at various times, we argue that all three are currently in play.

While outright science denial from oil and gas companies has subsided since its peak in the 1990s, it has since 2015 morphed into a subtler and less public form of denial through fossil fuel expansion plans that seemingly ignore the increasingly dire warnings of climate breakdown.

As we have shown, companies are ignoring and/or obstructing global efforts to curb emissions to match the goal of the IPCC guidance defining 1.5 degrees C of average warming as "dangerous" climate change. Similarly, the oil and gas industry continues to frame climate action as destructive to the economy and people's livelihoods in its attack on national legislation in the US and European Union. All the while, producers push false solutions such as carbon capture and natural gas that merely prolong our reliance on fossil fuels.

Even though oil and gas producers are currently portraying themselves as climate leaders, they could shift to cruder or more aggressive obstruction tactics as the need arises. Companies, for example, may return to explicitly backing and funding populist social movements questioning strong climate action—a topic for future research to track. If far-right and authoritarian politics continue to gain hold, this trend may herald a new form of denial and obstruction that is not passive or opaque but blatantly refuses to comply with the scientific warnings about climate change.[128] A potential return to more overt anti-science politics and the impacts it could have on the oil and gas industry's tactics and rhetoric, could be a valuable research area.

As we have seen with Vladimir Putin's 2021 invasion of Ukraine, which catalyzed a global discussion about Europe's dependance on Russian gas supplies and helped spur a boom in liquefied natural gas export projects worldwide, the oil and gas industry will continue to exploit fears about energy security to avoid any infringements on its core business model. More research is therefore needed on how climate mitigation efforts and fossil fuel phase out can actually improve geopolitical energy security across the world.

There is a need to shift our attention to the supply of fossil fuels. For too long, climate politics in the international arena has been focused on the emissions from production of fossil fuels. The topic of significantly reducing or phasing out fossil fuel supply has been absent from both reports and international governance. Few countries plan for a decline in supply despite ambitious national climate targets. But at the same time, the oil and gas industry has never been more precarious or vulnerable. Companies face hundreds of lawsuits in dozens of jurisdictions over their role in denying climate science and obstructing action (see Chapter 12) at a time when economical alternatives to oil are accelerating and national and subnational governments are adopting major climate legislation. It may be easier to create an obstruction campaign to undermine an emissions target than it is to undermine the simple and urgent message to "keep it in the ground."

NOTES

1. "BP Sets Ambition for Net Zero by 2050, Fundamentally Changing Organisation to Deliver," BP, February 12, 2020, https://www.bp.com/en/global/corporate/news-and-insights/press-releases/bernard-looney-announces-new-ambition-for-bp.html.

2. Jillian Ambrose, "Shell Unveils Plans to Become Net-Zero Carbon Company by 2050," April 16, 2020, *The Guardian*, https://www.theguardian.com/business/2020/apr/16/shell-unveils-plans-to-become-net-zero-carbon-company-by-2050; "Getting to Net-Zero," Suncor, July 28, 2021, https://www.suncor.com/en-ca/news-and-stories/our-stories/getting-to-net-zero; "Chevron Sets Net Zero Aspiration and New GHG Intensity Target," Chevron, October 12, 2021, https://www.chevron.com/newsroom/2021/q4/chevron-sets-net-zero-aspiration-and-new-ghg-intensity-target; "Aramco Expands Climate Goals, Stating Ambition to Reach Operational Net-Zero Emissions by 2050," Aramco, October 23, 2021, https://www.aramco.com/en/news-media/news/2021/ambition-to-reach-operational-net-zero-emissions-by-2050.

3. House Committee on Oversight and Accountability Democrats and Senate Committee on the Budget, Joint Staff Report, "Denial, Disinformation and Doublespeak: Big Oil's Evolving Efforts to Avoid Accountability for Climate Change," May 2024, https://www.budget.senate.gov/imo/media/doc/fossil_fuel_report1.pdf.

4. Sam Fankhauser, Stephen M. Smith, Myles Allen, Kaya Axelsson, Thomas Hale, Cameron Hepburn, J. Michael Kendall, Radhika Khosla, Javier Lezaun, Eli Mitchell-Larson, Michael Obersteiner, Lavanya Rajamani, Rosalind Rickaby,

Nathalie Seddon, and Thom Wetzer, "The Meaning of Net Zero and How to Get It Right," *Nature Climate Change* 12, no. 1 (2022): 15–21.

5. Oil Change International, "Big Oil Reality Check: Aligned in Failure," May 2024, https://www.oilchange.org/wp-content/uploads/2024/05/big_oil_reality_check_may_21_2024.pdf; see also Jessica Green, Jennifer Hadden, Thomas Hale, and Paasha Mahdavi, "Transition, Hedge, or Resist? Understanding Political and Economic Behavior toward Decarbonization in the Oil and Gas Industry," Review of International Political Economy 29, no. 6 (2022): 2036–2063.

6. Dan Welsby, James Price, Steve Pye and Paul Ekins, "Unextractable Fossil Fuels in a 1.5 °C World," *Nature* 597 (2021): 230–234. See also *Net Zero Roadmap: A Global Pathway to Keep the 1.5 °C Goal in Reach*, IEA, 2023, https://www.iea.org/reports/net-zero-roadmap-a-global-pathway-to-keep-the-15-0c-goal-in-reach, Licence: CC BY 4.0.

7. Mei Li, Gregory Trencher, and Jusen Asuka, "The Clean Energy Claims of BP, Chevron, ExxonMobil and Shell: A Mismatch Between Discourse, Actions and Investments," *PLOS One* 17, no. 2 (2022): e0263596; Gregory Trencher, Matheiu Blondeel, and Jusen Asuka, "Do All Roads Lead to Paris?: Comparing Pathways to Net-Zero by BP, Shell, Chevron and ExxonMobil," *Climatic Change* 176, no. 7 (2023): Article 83.

8. House Committee on Oversight and Accountability Democrats, "Denial, Disinformation and Doublespeak."

9. Benjamin Franta, "Big Carbon's Strategic Response to Global Warming, 1950–2020" (PhD diss., Stanford University, 2022); Kristoffer Ekberg, Bernhard Forchtner, Martin Hultman, and Kirsti M Jylhä, *Climate Obstruction: How Denial, Delay and Inaction Are Heating the Planet* (Taylor & Francis, 2022).

10. Inga Schlichting, "Strategic Framing of Climate Change by Industry Actors: A Meta-Analysis," *Environmental Communication* 7, no. 4 (2013): 493–511.

11. See for example Gregory C. Unruh, "Understanding Carbon Lock-In," *Energy Policy* 28, no. 12 (2000): 817–830; Matthew T. Huber, *Lifeblood: Oil, Freedom, and the Forces of Capital* (Minneapolis: University of Minnesota Press, 2013); Robert Groß, Odinn Melstedt, and Nocolas Chachereau, "Creating the Conditions for Western European Petroculture: The European Refinery Expansion Program and the Transition from Coal to Oil, 1948–1955," *Journal of Energy History* 10 (2023): 1–22.

12. Timothy Mitchell, *Carbon Democracy: Political Power in the Age of Oil* (Verso, 2011); Andreas Folkers, "Fossil Modernity: The Materiality of Acceleration, Slow Violence, and Ecological Futures," *Time and Society* 30, no. 2 (2021): 223–246; Matthias Schmelzer and Melissa Büttner, "Fossil Mentalities: How Fossil Fuels Have Shaped Social Imaginaries," *Geoforum* 150 (2024): 103981; Anna Åberg, Kristoffer Ekberg, and Susanna Lidström, "Pervasive Petrocultures: Histories, Ideas and Practices of Fossil Fuels," *Journal of Energy History/Revue d'histoire de l'énergie* 10 (2023), https://energyhistory.eu/en/node/366.

13. See for example the social justice frame in William F. Lamb, Giulio Mattioli, Sebastian Levi, J. Timmons Roberts, Stuart Capstick, Felix Creutzig, Jan C. Minx, Finn Müller-Hansen, Trevor Culhane, and Julia K. Steinberger, "Discourses of Climate Delay," *Global Sustainability* 3 (2020), https://doi.org/10.1017/sus.2020.13.

14. Richard Heede, "Tracing Anthropogenic Carbon Dioxide and Methane Emissions to Fossil Fuel and Cement Producers, 1854–2010," *Climatic Change* 122, no. 1 (2014): 229–241.

15. Marco Grasso and Richard Heede, "Time to Pay the Piper: Fossil Fuel Companies' Reparations for Climate Damages," *One Earth* 6, no. 5 (2023): 459–463.

16. House Committee on Oversight and Accountability Democrats, "Denial, Disinformation and Doublespeak."

17. Damian Carrington, "Revealed: Saudi Arabia's Grand Plan to 'Hook' Poor Countries on Oil," *The Guardian*, November 27, 2023, https://www.theguardian.com/environment/2023/nov/27/revealed-saudi-arabia-plan-poor-countries-oil

18. Peter C. Frumhoff, Richard Heede, and Naomi Oreskes, "The Climate Responsibilities of Industrial Carbon Producers," *Climatic Change* 132, no. 2 (2015): 157–171; Marco Grasso, "Towards a Broader Climate Ethics: Confronting the Oil Industry with Morally Relevant Facts," *Energy Research and Social Science* 62 (2020): 101383; Marco Grasso, "Oily Politics: A Critical Assessment of the Oil and Gas Industry's Contribution to Climate Change," *Energy Research & Social Science* 50 (2019): 106–115; Marco Grasso, *From Big Oil to Big Green. Holding the Oil Industry to Account for the Climate Crisis* (MIT Press, 2022); Marco Grasso, "Fossil Fuel Companies' Duty of Reparation: Why the Industry Must Concur to Foot the Climate Bill," *Globalizations* 21, no. 4 (2023): 1–18. This responsibility for a larger share of emissions was already acknowledged by oil companies themselves, such as Shell in their 1988 confidential report on global warming.

19. Robert J. Brulle, "Networks of Opposition: A Structural Analysis of U.S. Climate Change Countermovement Coalitions 1989–2015," *Sociological Inquiry* 91, no. 3 (2020): 603–624; Robert J. Brulle, "Advocating Inaction: A Historical Analysis of the Global Climate Coalition," *Environmental Politics* 32 (2022): 185–206.

20. Guiliano Garavini, *The Rise and Fall of OPEC in the Twentieth Century* (Oxford University Press, 2019).

21. House Committee on Oversight and Accountability Democrats, "Denial, Disinformation and Doublespeak."

22. Sonya Ahamed, Gillian L. Galford, Bindu Panikkar, Donna Rizzo, and Jennie C. Stephens, "Carbon Collusion: Cooperation, Competition, and Climate Obstruction in the Global Oil and Gas Extraction Network," *Energy Policy* 190 (2024): 114103.

23. Guy Edwards, Paul K. Gellert, Omar Faruque, Kathryn Hochstetler, Pamela D. Elwee, Prakash Kashwan, Ruth E. McKie, Carlos Milani, Timmons Roberts, and Jonathan Waltz, "Climate Obstruction in the Global South: Future Research Trajectories," *PLOS Climate* 2, no. 7 (2023), https://doi.org/10.1371/journal.pclm.0000241; Laura Kuhl, Jennie C. Stephens, Carlos Arriaga Serrano, Marla Perez-Lugo, Cacilio Ortiz Garcia, and Ryan Ellis, "Fossil Fuel Interests in Puerto Rico: Perceptions of Incumbent Power and Discourses of Delay, *Energy Research & Social Science* 111 (2024): 82024, https://doi.org/10.1016/j.erss.2024.103467; David M. Hughes, *Energy without Conscience: Oil, Climate Change, and Complicity* (Duke University Press, 2017).

24. Li et al., "The Clean Energy Claims of BP, Chevron, ExxonMobil and Shell."

25. Jon Birger Skjærseth and Tora Skodvin, "Climate Change and the Oil Industry: Common Problems, Different Strategies," *Global Environmental Politics* 1, no. 4 (2001): 43–64. For a discussion on different strategies, see Leticia Canal Vieira, Mariolina Longo, and Matteo Mura, "From Carbon Dependence to Renewables: The European Oil Majors' Strategies to Face Climate Change," *Business Strategy and the Environment* 32, no. 4 (2023): 1248–1259.

26. Jeff D. Colgan, Jessica F. Green, and Thomas N. Hale, "Asset Revaluation and the Existential Politics of Climate Change," *International Organization* 75 no. 2 (2021): 586–610. Compare with the division of fossil capital in primitive accumulation of fossil capital and fossil capital in general in Andreas Malm and the

Zetkin Collective, *White Skin Black Fuel, on the Dangers of Fossil Fascism* (Verso, 2021); see also Yue Guo, Michael Bradshaw, Chang Wang, and Mathieu Blondeel, "Globalization and Decarbonization: Changing Strategies of Global Oil and Gas Companies," *WIREs Climate Change* 14, no. 6 (2023): e849; Irja Vormedal and Jonas Meckling, "How Foes Become Allies: The Shifting Role of Business in Climate Politics," *Policy Sciences* 57 (2023): 101–124; Julian Gregory and Frank W. Geels, "Unfolding Low-Carbon Reorientation in a Declining Industry: A Contextual Analysis of Changing Company Strategies in UK Oil Refining (1990–2023)," *Energy Research & Social Science* 107 (2024): 103345; Julien Chevallier, Stéphane Goutte, Qiang Ji, and Khaled Guesmi, "Green Finance and the Restructuring of the Oil-Gas-Coal Business Model Under Carbon Asset Stranding Constraints," *Energy Policy* 149 (2021): 112055; Krista Halttunen, "'We Don't Want to Be the Bad Guys': Exploring the Future of International Oil Companies in a 'Well Below 2 Degrees' World" (PhD diss., University of London, 2023); Christian Bogmans, Andrea Pescatori, and Ervin Prifti, "The Impact of Climate Policy on Oil and Gas Investment: Evidence from Firm-Level Data," Working Paper, International Monetary Fund, 2023; Guri Bang and Bård Lahn, "From Oil as Welfare to Oil as Risk? Norwegian Petroleum Resource Governance and Climate Policy," *Climate Policy* 20 no. 8 (2020): 997–1009; Wim Carton and Andreas Malm, *Overshoot* (Verso, 2024).

27. Geoff Dembicki, Ellen Ormesher, and T. J. Jordan, "UN Chief Calls for Ban on Fossil Fuel Advertising," June 5, 2024, https://www.desmog.com/2024/06/05/un-chief-calls-for-ban-on-fossil-fuel-advertising/.

28. Damian Carrington and Ben Stockton, "Cop28 President Says There Is 'No Science' Behind Demands for Phase-Out of Fossil Fuels," December 3, 2023, https://www.theguardian.com/environment/2023/dec/03/back-into-caves-cop28-president-dismisses-phase-out-of-fossil-fuels.

29. IEA, "Net Zero by 2050—A Roadmap for the Global Energy Sector," May 2021, https://www.iea.org/reports/net-zero-by-2050.

30. Cleo Verkuijl, Georgia Piggot, Michael Lazarus, Harro van Asselt, and Peter Erickson, "Aligning Fossil Fuel Production with the Paris Agreement Insights for the UNFCCC Talanoa Dialogue," Stockholm Environment Institute, 2018, http://www.jstor.org/stable/resrep17207.

31. Jim Skea, "On Committing to Keeping 1.5 Degrees Within reach," December 2023, https://www.youtube.com/watch?v=_rVrpWl5BoQ&ab_channel=COP28UAE.

32. Saphira Rekker, Guangwu Chen, Richard Heede, Matthew C. Ives, Belinda Wade, and Chris Greig, "Evaluating Fossil Fuel Companies' Alignment with 1.5 °C Climate Pathways." *Nature Climate Change* 13, no. 9 (2023): 927–934.

33. "Unburnable Carbon: Are the World's Financial Markets Carrying a Carbon Bubble?," Carbon Tracker, July 2011, https://carbontracker.org/reports/carbon-bubble/.

34. Matthew Megura and Ryan. Gunderson, "Better Poison Is the Cure? Critically Examining Fossil Fuel Companies, Climate Change Framing, and Corporate Sustainability Reports," *Energy Research and Social Science* 85 (2022), https://doi.org/10.1016/j.erss.2021.102388.

35. Joint Staff Report, "Denial, Disinformation, and Doublespeak: Big Oil's Evolving Efforts to Avoid Accountability for Climate Change," April 2024, https://oversightdemocrats.house.gov/sites/evo-subsites/democrats-oversight.house.gov/files/evo-media-document/2024-04-30.COA%20Democrats%20-%20Fossil%20Fuel%20Report.pdf.

36. Sylvia Jaworska, "Change But No Climate Change: Discourses of Climate Change in Corporate Social Responsibility Reporting in the Oil Industry," *International Journal of Business Communication* 55, no. 2 (2018): 194–219. See also Wim Carton, "Carbon Unicorns and Fossil Futures. Whose Emission Reduction Pathways Is the Ipcc Performing?," in *Has It Come to This?: The Promises and Perils of Geoengineering on the Brink*, ed. J. P. Sapinski, Holly Jean Buck, and Andreas Malm (Rutgers University Press, 2020); Wim Carton, "'Fixing' Climate Change by Mortgaging the Future: Negative Emissions, Spatiotemporal Fixes, and the Political Economy of Delay," *Antipode* 51, no. 3 (2019): 750–769; Robert Watt, "The Fantasy of Carbon Offsetting," *Environmental Politics* 30, no. 7 (2021): 1069–1088; Dinah Rajak, "Waiting for a Deus Ex Machina: 'Sustainable Extractives' in a 2°C World," *Critique of Anthropology* 40, no. 4 (2020): 471–489.

37. Benjamin Franta, "Big Carbon's Strategic Response," 19; Benjamin Franta, "Early Oil Industry Knowledge of Co2 and Global Warming," *Nature Climate Change* 8, no. 12 (2018): 1024–1025.

38. CIEL, *Smoke and Fumes, The Legal and Evidentiary Basis for Holding Big Oil Accountable for the Climate Crisis*, 2017, https://www.ciel.org/wp-content/uploads/2017/11/Smoke-Fumes-FINAL.pdf, 11–12; E. Robinson and R. C. Robbins, "Sources, Abundance, and Fate of Gaseous Atmospheric Pollutants: Final Report and Supplement" (Stanford Research Institute, 1968).

39. CIEL, *Smoke and Fumes*.

40. Neela Banerjee, Lisa Song, and David Hasemyer, "Exxon's Own Research Confirmed Fossil Fuels' Role in Global Warming Decades Ago," *Inside Climate News*, September 16, 2015, https://insideclimatenews.org/news/16092015/exxons-own-research-confirmed-fossil-fuels-role-in-global-warming/.

41. Geoffrey Supran, Stefan Rahmstorf, and Naomi Oreskes, "Assessing Exxonmobil's Global Warming Projections," *Science* 379, no. 6628 (2023), https://doi:10.1126/science.abk0063. See also Geoffrey Supran and Naomi Oreskes, "Assessing ExxonMobil's Climate Change Communications (1977–2014)," *Environmental Research Letters* 12, no. 8 (2017), https://doi:10.1088/1748-9326/aa815f.

42. Cristophe Bonneuil, Pierre-Louis Choquet, and Benjamin Franta, "Early Warnings and Emerging Accountability: Total's Responses to Global Warming, 1971–2021," *Global Environmental Change* 71 (2021), https://doi.org/10.1016/j.gloenvcha.2021.102386.

43. Leah Aronowsky, "Gas Guzzling Gaia, or: A Prehistory of Climate Change Denialism," *Critical Inquiry* 47, no. 2 (2021): 306–327; Matthew Green, "Lost Decade: How Shell Downplayed Early Warnings Over Climate Change," *Desmog*, March 31, 2023, https://www.desmog.com/2023/03/31/lost-decade-how-shell-downplayed-early-warnings-over-climate-change/.

44. Shell, Greenhouse Effect Working Group, "Greenhouse Effect," 1988, https://www.documentcloud.org/documents/4411090-Document3#document/p4/a415539. See also Jenny Andersson, "Ghost in a Shell: The Scenario Tool and the World Making of Royal Dutch Shell," *Business History Review* 94, no. 4 (2020): 729–751.

45. Stella Levantesi, "Italian Oil Firm Eni Faces Lawsuit Alleging Early Knowledge of Climate Crisis," May 2023, https://www.theguardian.com/environment/2023/may/09/italian-oil-firm-eni-lawsuit-alleging-early-knowledge-climate-crisis. See also Marco Grasso, Stella Levantesi, and Serena Beqja, "Climate Obstruction in Italy: From Outright Denial to Widespread Climate Delay," in *Climate Obstruction Across Europe*, ed. Robert J. Brulle, J. Timmons Roberts, and Miranda Spencer (Oxford University Press, 2024), 268–293.

46. Ruth Morgan, *Climate Change and International History: Negotiating Science, Global Change, and Environmental Justice* (Bloomsbury, 2024).

47. Christophe Bonneuil and Fabienne Jouty, "Undermining Climate Action: European Oil Companies' Fight Against the European Carbon Tax Project, 1989–1992," Working Paper, *Capitalism and the Environment from Stockholm to COP 28: New Historical Research on the Role of Business and Labor*, University of Lausanne, November 9–10, 2023; Christophe Bonneuil, "Genèse et abandon d'une politique climatique: France, 1988–1992," *20 & 21. Revue d'histoire* 159, no. 3 (2023): 79–95; Freddie Daley, Peter Newell, Ruth McKie, and James Painter, "Climate Obstruction in the United Kingdom: Charting the Resistance to Climate Action," in *Climate Obstruction Across Europe*, ed. Robert J. Brulle, J. Timmons Roberts, and Miranda C. Spencer (Oxford University Press, 2024), 26–56.

48. David L. Levy, "Business and the Evolution of the Climate Regime: The Dynamics of Corporate Strategies," in *The Business of Global Environmental Governance*, ed. David L. Levy and Peter Newell (MIT Press, 2005), 73–104, 81.

49. Geoff Dembicki, *The Petroleum Papers: Inside the Far-Right Conspiracy to Cover Up Climate Change* (Graystone Books, 2022), 91.

50. Robert J. Brulle, "Advocating Inaction: A Historical Analysis of the Global Climate Coalition," *Environmental Politics* (2022): 1–22; see also Naomi Oreskes and Erik M. Conway, *Merchants of Doubt: How a Handful of Scientists Obscured the Truth on Issues from Tobacco Smoke to Global Warming* (Bloomsbury, 2010).

51. Andrew C. Revkin, "Industry Ignored Its Scientists on Climate," *New York Times*, April 23, 2009, https://www.nytimes.com/2009/04/24/science/earth/24deny.html; GCC, "Predicting Future Climate Change: A Primer," 1995, https://www.climatefiles.com/denial-groups/global-climate-coalition-collection/global-climate-coalition-draft-primer/, 16.

52. Supran and Oreskes, "Assessing ExxonMobil's Climate Change Communications."

53. Geoff Dembicki, "How Koch Industries, Fake Scientists, and Rush Limbaugh Invented Climate Denial," *Vice*, October 14, 2022, https://www.vice.com/en/article/n7zqxm/how-koch-industries-fake-scientists-and-rush-limbaugh-invented-climate-denial.

54. Cato Institute, "Global Environmental Crisis, Science or Politics?" June 1991, https://www.documentcloud.org/documents/4618065-1991-CATO-Climate-Denial-Conference#document/p3/a442309 https://www.climatefiles.com/deniers/patrick-michaels-collection/1991-cato-climate-denial-conference-flyer-schedule/.

55. Stuart Capstick, Lorraine Whitmarsh, Wouter Poortinga, Nick Pidgeon, and Paul Upham, "International Trends in Public Perceptions of Climate Change Over the Past Quarter Century," *WIREs Climate Change* 6 (2015), https://doi:10.1002/wcc.321.

56. See for example the Byrd-Hagel resolution, https://www.congress.gov/105/bills/sres98/BILLS-105sres98ats.pdf; Eric Pooley, *The Climate War: True Believers, Power Brokers, and the Fight to Save the Earth*, (Grand Central, 2010); PBS Frontline, *The Power of Big Oil*, Episode 1, 2022, https://www.pbs.org/wgbh/frontline/documentary/the-power-of-big-oil/.

57. Joanna Depledge, "Striving for No: Saudi Arabia in the Climate Change Regime," *Global Environmental Politics* 8, no. 4 (2008): 9–35.

58. Alexander Gusev, "Evolution of Russian Climate Policy: from the Kyoto Protocol to the Paris Agreement," *L'Europe en Formation* 2, no. 380 (2016): 39–52; Teresa Ashe and Marianna Poberezhskaya, "Russian Climate Scepticism: An Understudied Case," *Climatic Change* 172, no. 3 (2022): 41; Marianna Poberezhskaya and

Ellie Martus, "Climate Obstruction in Russia: Surviving a Resource- Dependent Economy, an Authoritarian Regime, and a Disappearing Civil Society," in *Climate Obstruction Across Europe*, ed. Robert J. Brulle, J. Timmons Roberts, and Miranda C. Spencer (Oxford University Press, 2024), 214–242.

59. United States Department of State, "Briefing Memorandum, no 200113080," June 20, 2001, acquired via Freedom of Information Act request by Greenpeace, https://www.documentcloud.org/documents/4407192-Global-Climate-Coalition-Meeting-2001.

60. Greenpeace, "Exxon's Climate Denial History: A Timeline," https://www.greenpeace.org/usa/fighting-climate-chaos/exxon-and-the-oil-industry-knew-about-climate-crisis/exxons-climate-denial-history-a-timeline/ (accessed January 16, 2024).

61. Greenpeace, "Koch Industries: Secretly Funding the Climate Denial Machine," https://www.greenpeace.org/usa/fighting-climate-chaos/climate-deniers/koch-industries/ (accessed January 16, 2024).

62. Jessica Green, Jennifer Hadden, Thomas Hale, and Paasha Mahdavi, "Transition, Hedge, or Resist? Understanding Political and Economic Behavior toward Decarbonization in the Oil and Gas Industry," *Review of International Political Economy* 29, no. 6 (2022): 2036–2063; Paasha Mahdavi, Jessica Green, Jennifer Hadden, and Thomas Hale, "Using Earnings Calls to Understand the Political Behavior of Major Polluters," *Global Environmental Politics* 22, no. 1 (2022): 159–174.

63. Rachel Frazin, "Internal Documents Cast Doubt on Big Oil Climate Promises," *The Hill*, September 14, 2022, https://thehill.com/policy/energy-environment/3643691-internal-documents-cast-doubt-on-big-oil-climate-promises/.

64. Influence Map, "Climate Change and Digital Advertising," August 2021, https://content.influencemap.org//site/data/000/822/InfluenceMap_ClimateChange&DigitalAdvertisingReport_August2021.pdf.

65. Ludwig Moehring, "The EU Green Deal—or: European Oil & Gas Industry to Make Its Mark," https://www.iogp.org/blog/opinions/the-eu-green-deal-or-european-oil-gas-industry-to-make-its-mark/ (accessed January 16, 2024).

66. Charles Koch, "The Business Community: Resisting Regulation," *Libertarian review* 7, no. 7 (1978): 30–34.

67. Alex Boynton, "Confronting the Environmental Crisis? Anti-Environmentalism and the Transformation of Conservative Thought in the 1970s," (PhD diss., University of Kansas, 2015).

68. Caleb Wellum. *Energizing Neoliberalism* (Johns Hopkins University Press, 2023); Niklas Olsen and Rasmus Skov Andersen, "Shielding the Market from the Masses: The Origins of Libertarian Anti-environmentalism in the 1960s and 1970s," *Journal of Modern European History* 20, no. 3 (2022): 304–310; Troy Vettese, "Limits and Cornucupianism: A History of Neo-Liberal Environmental Thought 1920–1970" (PhD diss., New York University, 2019).

69. See Grace Nosek, "The Fossil Fuel Industry's Push to Target Climate Protesters in the U.S.," *Pace Environmental Law (PELR) Review* 38, no. 1 (2021): 53–108; Amy Westerwelt and Geoff Dembicki, "Meet the Shadowy Global Network Vilifying Climate Protesters," *The New Republic*, September 12, 2023; Irwin, Randi, Vanessa Bowden, Daniel Nyberg, and Christopher Wright, "Making Green Extreme: Defending Fossil Fuel Hegemony through Citizen Exclusion," *Citizenship Studies* 26, no. 1 (2022): 73–89; Katrin Arning and Martina Ziefle, "Defenders of Diesel: Anti-Decarbonisation Efforts and the Pro-Diesel Protest Movement in Germany," *Energy Research & Social Science* 63 (2020): 101410; Caroline Sassan, Priyanka

Mahat, Melissa Aronczyk, and Robert J. Brulle, "Energy Citizens 'Just Like You'? Public Relations Campaigning by the Climate Change Counter-Movement," *Environmental Communication* 17, no. 7 (2023): 794–810.

70. Benjamin Franta, "Weaponizing Economics: Big Oil, Economic Consultants, and Climate Policy Delay," *Environmental Politics* 31, no. 4 (2021): 555–575.

71. Geoff Dembicki, "Exxon Could Have Helped Stop Climate Change 30 Years Ago, 'Proprietary' Docs Show," September 20, 2022, https://www.desmog.com/2022/09/20/exxon-imperial-oil-climate-change-proprietary-docs/.

72. See also William Carroll, Nicolas Graham, Michael K. Lang, Zoë Yunker, and Kevin D. McCartney, "The Corporate Elite and the Architecture of Climate Change Denial: A Network Analysis of Carbon Capital's Reach into Civil Society," *Canadian Review of Sociology* 55, no. 3 (2018): 425–450; for historical precedents see Melissa Aronczyk and Maria I. Espinoza, *A Strategic Nature: Public Relations and the Politics of American Environmentalism* (Oxford University Press, 2022). More on this in Chapter 6.

73. Franta, "Weaponizing Economics."

74. Ibid.

75. GCC, "Economic and Lifestyle Impacts From Proposed Greenhouse Gas Emission Restrictions," Briefing Paper, 1996, https://www.climatefiles.com/denial-groups/global-climate-coalition-collection/1996-economic-impacts-from-proposed-emission-restrictions/.

76. Dembicki, *The Petroleum Papers*, 96.

77. See for example Alex Boynton, "Formulating an Anti-Environmental Opposition: Neoconservative Intellectuals During the Environmental Decade," *The Sixties* 8, no. 1 (2015): 1–26; Kristoffer Ekberg and Victor Pressfeldt, "A Road to Denial: Climate Change and Neoliberal Thought in Sweden, 1988–2000," *Contemporary European History* 31, no. 4 (2022): 627–644; Vettese, "Limits and Cornucupianism."

78. Geoffrey Supran and Naomi Oreskes, "Rhetoric and Frame Analysis of ExxonMobil's Climate Change Communications," *One Earth* 4, no. 5 (2021): 696–719.

79. Elisabeth Bumiller and John M Broder, "McCain Differs With Bush on Climate Change," May 13 2008, https://www.nytimes.com/2008/05/13/us/politics/12cnd-mccain.html.

80. Rex Tillersson, "The State of the Energy Industry: Strengths, Realities, and Solutions," CERAWeek opening adress, February 13, 2007, https://www.climatefiles.com/exxonmobil/2007-speech-exxonmobil-ceo-rex-tillerson-ceraweek/.

81. Jane van Ryan, "$4 Gasoline," API, June 24, 2009, https://www.api.org/news-policy-and-issues/blog/2009/06/24/4-gasoline.

82. Yale Climate Connections, "Greenpeace Releases Oil Industry's 'Energy Citizen' E-mail on P.R. Rallies," August 18, 2009, https://yaleclimateconnections.org/2009/08/energy-citizen/.

83. John M. Broder, "'Cap and Trade' Loses Its Standing as Energy Policy of Choice," March 25, 2010, https://www.nytimes.com/2010/03/26/science/earth/26climate.html.

84. In fact, researchers have, rather than astro-turf, referred to such mobilizations as subsidized publics. See Jacob McLean, "United They Roll? How Canadian Fossil Capital Subsidizes the Far-Right," in *Political Ecologies of the Far Right—Fanning the Flames*, ed. Irma Kinga Allen, Kristoffer Ekberg, Ståle Holgersson, and Andreas Malm (Manchester University Press, 2024), 80–98.

85. Kyle C. Meng and Ashwin Rode, "The Social Cost of Lobbying Over Climate Policy," *Nature Climate Change* 9 (2019): 472–476.

86. Oliver Lazarus, Sonali McDermid, and Jennifer Jacquet, "The Climate Responsibilities of Industrial Meat and Dairy Producers," *Climatic Change* 165, no. 30 (2021), https://doi.org/10.1007/s10584-021-03047-7.

87. Robert J. Brulle, "The Climate Lobby: A Sectoral Analysis of Lobbying Spending on Climate Change in the USA, 2000 to 2016," *Climatic Change* 149, no. 3 (2018): 289–303.

88. Daniel J. Weiss, Rebecca Lefton, and Susan Lyon, "Dirty Money Oil Companies and Special Interests Spend Millions to Oppose Climate Legislation," *Capaction*, September 27, 2010, https://www.americanprogressaction.org/article/dirty-money/.

89. Sandra Laville, "Fossil Fuel Big Five 'Spent €251m Lobbying EU' Since 2010," *The Guardian*, October 24, 2019, https://www.theguardian.com/business/2019/oct/24/fossil-fuel-big-five-spent-251m-lobbying-european-union-2010-climate-crisis.

90. "Saudi Arabia Is Trying to Block a Global Deal to End Fossil Fuels, Negotiators Say," *New York Times*, December 10, 2023, https://www.nytimes.com/2023/12/10/climate/saudi-arabia-cop28-fossil-fuels.html.

91. Damian Carrington, "Revealed: Saudi Arabia's Grand Plan to 'Hook' Poor Countries on Oil," *The Guardian*, November 27, 2023, https://www.theguardian.com/environment/2023/nov/27/revealed-saudi-arabia-plan-poor-countries-oil.

92. Dario Kenner and Richard Heede, "White Knights, or Horsemen of the Apocalypse? Prospects for Big Oil to Align Emissions with a 1.5 °C Pathway," *Energy Research & Social Science* 79 (2021): 102049.

93. Ruth Michaelson, "'Explosion' in Number of Fossil Fuel Lobbyists at COP27 Climate Summit," November 10, 2022, https://www.theguardian.com/environment/2022/nov/10/big-rise-in-number-of-fossil-fuel-lobbyists-at-cop27-climate-summit.

94. "Record Number of Fossil Fuel Lobbyists Granted Access to COP28 Climate Talks," press release, December 5, 2023, https://www.globalwitness.org/en/press-releases/record-number-fossil-fuel-lobbyists-granted-access-cop28-climate-talks/.

95. Amanda Stephenson, "Canadian Oil and Gas Execs to Talk Up Emissions Reduction Plans at COP28," November 29, 2023, https://www.cbc.ca/news/canada/calgary/oilsands-executives-cop-28-carbon-capture-1.7044163.

96. "An Affordable, Reliable, Competitive Path to Net Zero," McKinsey, November 30, 2023, https://www.mckinsey.com/capabilities/sustainability/our-insights/an-affordable-reliable-competitive-path-to-net-zero.

97. International Energy Agency, "CO2 Emissions in 2023: A New Record High, but Is There Light at the End of the Tunnel?" March 3, 2024, https://www.iea.org/reports/co2-emissions-in-2023/executive-summary.

98. Marten Boon, "A Climate of Change? The Oil Industry and Decarbonization in Historical Perspective," *Business History Review* 93, no. 1 (2019): 101–125.

99. Sara Jerving, Katie Jennings, Masako Melissa Hirsch, and Susanne Rust, "Special Report: What Exxon Knew About Global Warming's Impact on the Arctic," October 10, 2015, https://www.latimes.com/business/la-na-adv-exxon-arctic-20151011-story.html.

100. See for example the investigations into Exxon in New York, in Justin Gillis and Clifford Krauss, "Exxon Mobil Investigated for Possible Climate Change Lies by New York Attorney General," *New York Times*, November 6, 2015, https://www.nytimes.com/2015/11/06/science/exxon-mobil-under-investigation-in-new-york-over-climate-statements.html; and Massachusetts, "Attorney General's

Office Exxon Investigation," https://www.mass.gov/lists/attorney-generals-office-exxon-investigation#:~:text=On%20April%2019%2C%202016%2C%20Attorney,Massachusetts%20consumer%20protection%20statute%2C%20M.G.L (accessed January 16, 2024).

101. For an updated list of these see https://climateintegrity.org/cases (accessed March 26, 2025).

102. Krista Halttunen, Raphael Slade, and Iain Staffell, "'We Don't Want to Be the Bad Guys': Oil Industry's Sensemaking of the Sustainability Transition Paradox," *Energy Research & Social Science* 92 (2022): 102800; Pete Tillotson, Raphael Slade, Iain Staffell, and Krista Halttunen, "Deactivating Climate Activism? The Seven Strategies Oil and Gas Majors Use to Counter Rising Shareholder Action," *Energy Research & Social Science* 103 (2023): 103190.

103. Michael Buchsbaum and Edward Donnelly, "Fossil Fuel Companies Made Bold Promises to Capture Carbon. Here's What Actually Happened," September 25, 2023, https://www.desmog.com/2023/09/25/fossil-fuel-companies-made-bold-promises-to-capture-carbon-heres-what-actually-happened/; and Cartie Werthman, "Despite Advertising Carbon Capture, ExxonMobil Saw Marginal Role for It in Fighting Climate Change," May 21, 2024, *DeSmog*, https://www.desmog.com/2024/05/21/despite-advertising-carbon-capture-exxonmobil-saw-marginal-role-for-it-in-fighting-climate-change-shell/.

104. "How Will Oil and Gas Companies Get to Scope 3 Net Zero?," Wood MacKenzie, October 2022, https://www.woodmac.com/press-releases/few-oil-and-gas-companies-commit-to-scope-3-net-zero-emissions-as-significant-challenges-remain/.

105. Li et al., "The Clean Energy Claims of BP, Chevron, ExxonMobil and Shell."

106. Geoff Dembicki, "Going 'Net-Zero' Will Boost Oil Sands Production, Industry Tells Canadian Government," March 6, 2023, *DeSmog*, https://www.desmog.com/2023/03/06/net-zero-to-boost-oil-sands-production-pathways-alliance-tells-canadian-government/.

107. Stefan C. Aykut and Monica Castro, "The End of Fossil Fuels? Understanding the Partial Climatisation of Global Energy Debates," in *Globalising the Climate*, ed. Stefan Aykut, Jean Foyer, and Edouard Morena (Routledge, 2017), 173–193.

108. Amy Janzwood and Kathryn Harrison, "The Political Economy of Fossil Fuel Production in the Post-Paris Era: Critically Evaluating Nationally Determined Contributions," *Energy Research & Social Science* 102 (2023), https://doi.org/10.1016/j.erss.2023.103095.

109. For a more in-depth analysis of this process see Ada Nissen, "A Greener Shade of Black? Statoil, the Norwegian Government and Climate Change,1990–2005," *Scandinavian Journal of History* 46 no. 3 (2021): 408–429; see also Robert Duffy, "'The Cleanest Oil Fields in the World': Narratives Encouraging Oil Development in California," *Energy Research & Social Science* 97 (2023): 102998.

110. Stanley Reed, "BP, in a Reversal, Says It Will Produce More Oil and Gas," *The Guardian*, February 7, 2023, https://www.nytimes.com/2023/02/07/business/bp-oil-gas-profits.html.

111. Ron Buosso, "Exclusive: Shell Pivots Back to Oil to Win Over Investors," *Reuters*, June 9 2023, https://www.reuters.com/business/energy/shell-pivots-back-oil-win-over-investors-sources-2023-06-09/.

112. Geoff Dembicki, "Going 'Net-Zero' Will Boost Oil Sands Production, Industry Tells Canadian Government," *DeSmog*, March 6, 2023, https://www.desmog.com/2023/03/06/net-zero-to-boost-oil-sands-production-pathways-alliance-tells-canadian-government/.

113. Sybille van den Hove, Marc Le Menestrel, and Henri-Claude de de Bettignies, "The Oil Industry and Climate Change: Strategies and Ethical Dilemmas," *Climate Policy* 2, no. 1 (2002): 3–18.

114. Ben Huf, Glenda Sluga, and Sabine Selchow, "Business and the Planetary History of International Environmental Governance in the 1970s," *Contemporary European History* 31, no. 4 (2022): 553–569; see also Ann-Kristin Bergquist and Thomas David, "Business In (Action): The International Chamber of Commerce and Climate Change from Stockholm to Rio," *Uppsala Papers in Economic History* 3 (2024): 1–26.

115. Ann-Kristin Bergquist and Thomas David, "Beyond Planetary Limits! The International Chamber of Commerce, the United Nations, and the Invention of Sustainable Development," *Business History Review* 97, no. 3 (2023): 1–31.

116. Bergquist and David, "Beyond Planetary Limits!"; Peter Bright, "Report from Bergen," *Science and Cristian Belief* 3 no. 1 (1991): 19–24.

117. Steven B. Hellem and Amy M. Goldman, "The Global Environmental Management Initiative (GEMI): A Case for Corporate Leadership; Two Decades of Environmental, Health and Safety (EHS) and Sustainability Progress" (GEMI, 2009), 1–5, www.gemi.org

118. Inga Schlichting, "Strategic Framing of Climate Change by Industry Actors."

119. Cristopher Wright, Daniel Nyberg, and Vanessa Bowden, "Beyond the Discourse of Denial: The Reproduction of Fossil Fuel Hegemony in Australia, *Energy Research & Social Science* 77 (2021): 102094.

120. Irja Vormedal, Lars H. Gulbrandsen, and Jon Birger Skjærseth, "Big Oil and Climate Regulation: Business as Usual or a Changing Business?," *Global Environmental Politics* 20, no. 4 (2020): 143–166; Yutong Si, Dipa Desai, Diana Bozhilova, Sheila Puffer, and Jennie C. Stephens, "Fossil Fuel Companies' Climate Communication Strategies: Industry Messaging on Renewables and Natural Gas," *Energy Research & Social Science* 98 (2023): 103028.

121. RMI, "Reality Check: Natural Gas's True Climate Risk," July 13, 2023, https://rmi.org/reality-check-natural-gas-true-climate-risk/.

122. BP America, @bp_america, post on x/twitter, "We are one of," January 19, 2024, https://twitter.com/bp_America/status/1748420110488912154.

123. Geoff Dembicki, "BP Was Warned Gas-Driven Climate Change Could Cause 'Unprecedented Famine,'" May 3, 2024, *DeSmog*, https://www.desmog.com/2024/05/03/bp-was-warned-gas-driven-climate-change-could-cause-unprecedented-famine-us-congressional-investigation/.

124. Ellen Palm, Joachim Peter Tilsted, Valentin Vogl, and Alexandra Nikoleris, "Imagining Circular Carbon: A Mitigation (Deterrence) Strategy for the Petrochemical Industry," *Environmental Science & Policy* 151 (2024), https://doi.org/10.1016/j.envsci.2023.103640; Wim Carton, Inge-Merete Hougaard, Nils Markusson, and Jens Friis Lund, "Is Carbon Removal Delaying Emission Reductions?," *WIREs Climate Change* 14, no. 4 (2023), https://doi.org/10.1002/wcc.826; Ryan Gunderson, Diana Stuart, and Brian Petersen, "The Political Economy of Geoengineering as Plan B: Technological Rationality, Moral Hazard, and New Technology," *New Political Economy* 24 no. 5 (2019): 696–715; Carton and Malm, *Overshoot*.

125. Ben LeFebvre, "Oil Industry Sees a Vibe Shift on Climate Tech," *Politico*, March 8, 2023, https://www.politico.com/news/2023/03/08/oil-industry-shift-climate-tech-00085853; Benoît Morenne, "Occedental Plans to Suck Carbon from the Air So It Can Keep Pumping Oil," April 10, 2023, *Wall Street Journal*. https://www.

wsj.com/articles/occidental-plans-to-suck-carbon-from-the-airso-it-can-keep-pumping-oil-2990c5a.

126. Odinn Melstedt, "Between Pushback and Collaboration: The International Oil Industry and Environmental Regulation and Governance, 1970s–1980s," Working paper for Capitalism and the Environment from Stockholm to COP 28: New Historical Research on the Role of Business and Labor, University of Lausanne, November 9–10, 2023.

127. Jenny Andersson, "Ghost in a Shell: The Scenario Tool and the World Making of Royal Dutch Shell," *Business History Review* 94, no. 4 (2020): 729–751.

128. Cara Daggett, "Petro-Masculinity and the Politics of Climate Refusal," *Autonomy*, March 1, 2022, https://autonomy.work/portfolio/petro-masculinity-climate-refusal/; see also Tadzio Müller, "Beyond Ethics and Rationality: Welcome to the Assholocene," *Friedliche Sabotage*, 2023, https://steadyhq.com/de/friedlichesabotage/posts/a1c14c1a-feb9-42ec-9678-358be77f8d99.

CHAPTER 3

How Coal, Utilities, and Transportation Impede Climate Action

LEAD AUTHORS: JEN SCHNEIDER
AND GREGORY TRENCHER
CONTRIBUTING AUTHORS: PETER K. BSUMEK,
CHRISTIAN DOWNIE, PAUL K. GELLERT,
GIULIO MATTIOLI, JASON MONIOS, PETER NEWELL,
JENNIFER PEEPLES, JOERI WESSELING, EMILY
WILLIAMS, RYAN WISHART, AND BEN YOURIEV

INTRODUCTION: THE ROLE OF COAL, TRANSPORTATION, AND UTILITIES IN THE CLIMATE CRISIS

Globally, coal, transportation and utilities have generated an outsized footprint in the climate crisis. Coal currently accounts for 29% of global primary energy supply. Not only is it responsible for more than 40% of annual CO_2 emissions, but it has also contributed to around one-third of planetary warming to date.[1] Though coal is also consumed by the steel and cement industries, the vast majority of it is used for power generation. Transportation, meanwhile, accounts for approximately one-third of end-use emissions worldwide, with road vehicles, maritime operations, and aviation all contributing a major share.[2] The utilities sector, too, has contributed significantly to global emissions. Worldwide, electricity and heat production accounted for 44% of fossil fuel emissions in 2021, with coal plants representing 73% of this figure.[3] In the United States alone, also in 2021, the American electric utility industry was responsible for approximately 4.5% of global CO_2 emissions.[4]

Thus, climate action consistent with objectives of the 2015 Paris Agreement and other multilateral and national plans requires a radical reduction in fossil fuel use and a transition to less polluting forms of energy. This, however,

Jen Schneider et al., *How Coal, Utilities, and Transportation Impede Climate Action*. In: *Climate Obstruction*.
Edited by: J. Timmons Roberts, Carlos R. S. Milani, Jennifer Jacquet, and Christian Downie,
Oxford University Press. © Oxford University Press (2025). DOI: 10.1093/oso/9780197787144.003.0003

represents a profound challenge to these enterprises, which have accordingly become major players on the field of climate obstruction.

Interconnected Sectors, Similar Challenges and Obstruction Strategies

We consider coal, transportation, and utilities together because, as the research synthesized in this chapter reveals, not only do these industries share business linkages, but they also have also leveraged common strategies to delay or obstruct effective climate action. The particular forms of obstruction these sectors have employed reflect their interdependence, their reliance on fossil fuels, the difficulty of decarbonizing, and the financial losses that decarbonization would likely incur.

The transportation sector, for example, is tightly connected with the carbon majors due to its dependence on oil and gas for fuel, and also faces some of the same business challenges. In the United States, China, and elsewhere, rail companies rely heavily on revenue from transporting coal; as coal's fortunes turn, so do those of the railroad industry.[5,6] Similarly, depending on location, utilities still rely on fossil fuels including coal to power their grids. Together, coal mining, transportation, and electric utilities form a massive profit-making consortium that depends heavily on fossil fuels and on each other for both current profitability and future survival.

Moreover, coal, transportation, and utilities are prone to financial and technological lock-in, hampering the transition away from fossil fuels. This challenge is most notable for the coal industry, because coal mines typically possess long lifespans and require several decades to exhaust reserves and to recover sunken financial capital.[7] This lock-in is exacerbated by the young age of coal-fired power plants[8] and the intimately linked and mutually dependent relationships between political actors and the coal industry.

Yet as recognition of the threat of climate change continues to rise, coal has also become the most precarious fossil fuel, as it can be replaced relatively cheaply and easily for power generation with lower-carbon energy sources including renewables (or with other fossil fuels such as methane gas). The transition away from coal means that reserves must be left in the ground, thereby creating the risk of stranded assets.

The transportation sector also suffers from lock-in. Like large-scale automobile factories, aircraft and ships are costly to build while needing to operate over several decades to recoup investments.[9] Both the aviation and shipping industries also remain heavily fossil fuel–dependent. For example, in 2022, only 0.1% of all jet fuel used globally came from sustainable aviation fuels (SAF),[10] which include nonpetroleum feedstocks such as biomass. The International Air Transport Association (IATA) has repeatedly missed its self-set targets on sustainable fuels uptake.

The Intergovernmental Panel on Climate Change (IPCC) describes aviation as "hard-to-decarbonize."[11] Besides its fossil fuel dependence, this status reflects the limited decarbonization potential of emerging technologies such as hydrogen, biofuels, and synthetic fuels due to their low availability and high costs.

Shipping, unlike aviation, has numerous technological and operational options to reduce its emissions, with strong though seldom utilized near-term opportunities for rapid decarbonization including deploying low- or zero-carbon fuel sources (e.g., ammonia, hydrogen), e-fuels (synthetic methanol produced from renewable electricity, CO_2, and hydrogen), increased efficiency measures, electrification, and slow-steaming.[12] However, as we will discuss below, the maritime industry has been as slow to decarbonize as the other industries and sectors because of political and organizational challenges.

In short, all industries and sectors examined here are seeing slow or uneven rates of decarbonization. For example, the idea that that we are witnessing a global "death of coal," as some environmental groups have argued, only makes sense when applied to Global North countries, such as the United States and in Western Europe. In contrast, in many Global South countries' dependence on coal remains strong or, in some regions, is even growing.[13] Many of these countries face technical and economic barriers to decarbonization due to the higher upfront costs of renewable energy.[14]

We see this same unevenness when examining the electric utilities industry. In the aggregate, the industry has reduced its GHG emissions by around 36% since 2006.[15] Furthermore, around half of American utilities have either pledged to be carbon free or are at least subject to state orders to reach 100% carbon-free electricity by 2050.[16] Yet many utilities still rely heavily on fossil fuels. Moreover, the industry is not on track to meet the targets it has set, with some of the largest electricity producers continuing to produce the highest emissions intensity in generation.[17,18] Meanwhile, a few specific utilities are disproportionately large emitters. It is thus not surprising that the electric utilities that are the slowest to decarbonize in the United States tend to be those that are the most engaged in climate obstruction.[19]

Our Approach

To deepen our understanding of how the coal industry, electric utilities, and the transportation sector have obstructed meaningful climate action, this chapter outlines three typologies that illustrate their relationships and obstructive activities across various industrial, sectoral, and political scales: (1) the actors and organizations involved; (2) their methods of engagement in direct political action, including policymaking; and (3) the various rhetorical and discursive tactics they use to articulate their economic and social value in hopes of maintaining the status quo. As we will illustrate, similar obstructive actors, political strategies, and communication tactics have been employed across all three sectors, and while climate science denial may have been these industries' primary strategy prior to the Paris Agreement, the tactics they use to block, weaken, or delay decarbonization efforts have continued to evolve and diversify.

Despite the commonalities in obstruction strategies across industries and sectors, however, there is no single, overarching story to tell about the decarbonization of coal, utilities, or transportation worldwide. This chapter demonstrates that these different sectors may share similar approaches to obstruction, but the complexity of the systems they operate within make it difficult to make sweeping claims about how fast decarbonization is occurring, or how effective obstruction efforts are writ large. We do know that obstruction is happening across scales and is facilitated by industry participation in regulatory bodies. But the details of what occurs in particular sectors matter, and below we provide examples that illustrate the complexity at play.

ACTORS, ORGANIZATIONS, AND COALITIONS

Table 3.1 shows the key actors, organizations, and coalitions engaged in climate obstruction efforts in these industries. We begin this section by identifying the individual actors (industrialists, politicians, oligarchs, and firms) who wield the power to drive or block energy transitions.[20] We then examine trade associations, front groups, public relations firms, and unions, which play an important role obstructing climate action at both the national and the international levels.[21] By functioning as a sole representative of their industry, these organizations are able to augment the perceived legitimacy of that industry and increase their "social license to operate."[22,23,24] Moreover, their collective structure allows individual companies to engage indirectly in obstructionist activities, minimizing the risk of reputational damage they may face when acting alone. In turn, these actors and organizations often form or participate in regional and global coalitions designed to advance the interests of industry and slow climate action.[25] Furthermore, as we discuss below, such corporate actors are thus able to participate in global rulemaking mechanisms, including those headed by the United Nations (UN).

Elite Individual Industrialists, Politicians, and Corporations

Elite actors such as CEOs or owners of large, private coal companies have exerted significant political influence in many jurisdictions across the Global North and South.[26] In Indonesia, for example, both high-level politicians and

Table 3.1 EXAMPLES OF VARIOUS ACTORS, ORGANIZATIONS, AND COALITIONS OPERATING IN VARIOUS INDUSTRIES/SECTORS

	Coal	Utilities	Automotive	Maritime	Aviation
Elite individual industrialists, politicians, and firms	Robert Murray, Donald Trump, Members of Indonesia's Parliament, Oligarchs in the Philippines, Peabody Coal (US)	Southern Company, American Electric Power (US), Perusahaan Listrik Negara (Indonesia), Adani (India), RWE (Germany)	Original Equipment Makers (OEMs) such as Toyota and BMW Group, parts suppliers	Nation-states such as Argentina or Brazil	—
Trade associations, unions, front groups, and PR firms	American Coal Council, Minerals Council of Australia	Edison Electric Institute, American Coalition for Clean Coal Electricity (US), Federation of Electric Power Companies (Japan)	European Automotive Manufacturers' Association, Japanese Automobile Manufacturers Association	International Chamber of Shipping, Baltic and International Maritime Council (BIMCO)	International Air Transport Association (IATA)
Regional/global regulatory and governance bodies	National Coal Association (US), World Coal Association	Public utility commissions, Federal Energy Regulatory Commission (US), regional transmission organizations	European Commission expert groups	International Maritime Organization (IMO)	International Civil Aviation Organization (ICAO)

business leaders possess financial stakes in coal mines, resulting in vested interests and energy policies that favor coal power.[27,28,29] In the Philippines, preferential support for coal power has been traced to a handful of powerful oligarchs.[30,31] The interests of coal-industry elites have also had a profound impact on politics in the United States. For example, in 2016, Robert Murray, the CEO of coal mining company Murray Energy, used political donations and lobbying to influence presidential candidate Donald Trump,[32,33] who promised to end the "war on coal."[34]

Although such high-profile industrialists and politicians garner extensive media attention, it is more common for large corporations to "flatten" their public image so that individual companies, their public relations firms and industry groups, and coal-mining communities become indistinguishable. One image-management strategy US coal companies frequently use is to support local organizations such as arts and community groups with admirable missions. Many of these organizations appear to be grassroots, but paid by corporate actors to lobby on their behalf while hiding this objective under a guise of neutrality.[35] These "astroturf" organizations play an important role in disseminating the coal industry's messages to counter environmental concerns about coal[36] by making their interests seem synonymous with those of coal-mining communities, seeking to blend in and become one with the communities they serve or exploit. Such techniques have allowed coal companies, such as Peabody Energy, to become household names.

Similarly, in the transportation sector, many corporate actors participate in climate obstruction, including automakers (i.e., original equipment makers, or OEMs) and parts suppliers. The big OEMs play an important role in blocking and diluting regulations that directly impair their marketing and innovation strategies.[37] Toyota and BMW Group, despite possessing electric vehicle portfolios, have actively lobbied against policymaker efforts to strengthen emissions and fuel-efficiency standards in Washington and Europe.[38,39] Though parts suppliers have generally played a much smaller role in climate obstruction than OEMs and their associations,[40] these companies are important members of cross-sector coalitions that include trade associations and front groups, especially as they face decreasing demand for the traditional components used in internal combustion engines.

Trade Associations, Front Groups, PR Firms, and Unions

In the coal industry, trade associations—nonprofit alliances established to represent and advance the interests of specific industries—were particularly active in climate obstruction in the lead-up to the Paris Agreement. Notable examples in the United States include the American Coal Council and the American Coalition for Clean Coal Electricity (ACCCE). Individual firms use such associations to coordinate and advance their common political interests by, for example, lobbying or donating to politicians.[41] Coal industry actors have also pursued obstruction by creating and/or funding "front groups," typically nonprofit entities that advocate agendas of the private sector but disguise their activities as citizenry-led efforts (a strategy frequently considered synonymous with "astroturfing," mentioned earlier).[42] In the United States, prominent examples include the National Mining Association and the West Virginia Coal Association's "Faces of Coal" campaign.[43] Such front groups tend to be created by public relations and strategic communications firms including Edelman, the DCI Group, and the Hawthorn Group; prominent campaigns have included "Advanced Energy for Life"[44] (launched for Peabody Energy) and "America's Power" (for ACCCE).[45] Rail companies and electric utilities in the United States have been long-term partners with coal mining companies in supporting such efforts.[46] ACCCE (later renamed America's Power), was formed when the Center for Energy and Economic Development (CEED) and Americans for Balanced Energy Choices (ABEC) joined forces, and played a pivotal role from 2008 until the early 2020s in blocking federal environmental regulation of coal, particularly then-President Obama's Clean Power Plan.

Unions that represent coal workers have also obstructed climate action, most frequently by seeking to block efforts to close coal mines and power stations.[47] In the United States, coal unions have used narratives such as a "war on coal" to deflect blame for decreased employment opportunities in the industry—which are due mainly to corporate decisions to mechanize the extraction process—onto environmental regulations. This blame-shifting has aimed to block federal climate action and align support for Republican candidates locally and nationally.[48,49,50] Meanwhile, states responding to industry demands have rolled back laws for mine worker safety and environmental protection and subverted enforcement of federal law to protect coal company profits and slow the industry's decline.[51] Researchers have described a similar situation in Australia, where some unions have pressed consecutive governments not to introduce policies that might threaten carbon-intensive industries including coal mining.[52,53]

Trade associations in the utilities industry have also engaged in obstructionist efforts (Brulle, 2021; Dunlap and Brulle, 2020; Stokes, 2020; Williams et al., 2022).[54,55,56,57] In Japan, the Federation of Electric Power Companies, the trade association of the largest regional power utilities, has adopted a strong pro-coal stance, downplaying the need to move away from fossil power or rapidly increase renewables.[58,59] In the United States, an analysis of more than two thousand organizations that oppose climate action (the so-called climate change countermovement) showed that organizations representing electric utilities and the coal industry were the most numerous and influential.[60] Through its trade organizations, including the Edison Electric Institute (EEI) for investor-owned utilities and National Rural Electric

Cooperative Association for cooperative utilities, the US utilities industry is highly active in coordinating and engaging with the political process.[61] Much obstructionism has occurred through utilities' dues-paying membership in the EEI and in the more controversial Utility Air Resources Group (UARG).[62] For example, UARG repeatedly filed suit and used public comment opportunities to block climate policy. All the while, however, it was funded largely in secret via ratepayer funds, and was only disbanded once legislative hearings revealed this practice.[63]

In addition, electric utilities were involved in three of the most well-known climate denial front groups: the Global Climate Coalition, the Information Council on the Environment, and the Greening Earth Society.[64,65] The Global Climate Coalition, active between 1989 and 2002, was established to obstruct climate policy in the United States and to discredit the activities of the United Nations Framework Convention on Climate Change (UNFCCC) and IPCC by rejecting climate science and propagating disinformation[66] (see Chapter 10). Notably, most of its members were from electric utility organizations and the coal industry.[67] The Information Council on the Environment was created to "reposition global warming as theory (not fact)," and was composed solely of electric utility organizations[68,69] and the EPRI—the research arm of the electric utility industry. Today, EPRI focuses on championing the benefits, but not the drawbacks, of carbon capture and storage technologies.[70]

OEMs and parts suppliers in road transportation also rely heavily on industry associations to lobby against climate regulations. Notable examples include the US-based Auto Alliance (AAM), the international Truck and Engine Manufacturers Association, the European Automobile Manufacturers' Association, and the Japanese Automobile Manufacturers Association (JAMA).[71] Like those of the other industries examined, these associations allow individual companies to hide in the crowd when engaging in obstructive actions, thereby reducing the risk of incurring reputational damage.[72,73,74] As a further risk- aversion strategy, individual transportation-related companies tend to make public statements that are more or less supportive of government policy while hiding their concurrent support for the obstructive actions of their affiliated industry association.[75] Other parts of the transportation industry employ similar dynamics on a broader scale, as discussed next.

International, Regional, and Subnational Coalitions and Governance Bodies

Maritime corporations enjoy strategic relationships with their trade associations, largely because of the inner workings of the International Maritime Organization (IMO), a UN committee charged with oversight of the shipping industry.[76] First, the IMO allows member states to be represented by private industry representatives. Second, trade associations, such as the

International Chamber of Shipping (ICS), the Baltic and International Maritime Council (BIMCO), and the World Shipping Council, which all oppose any binding climate regulation,[77] exert a strong influence there. ICS, representing national shipowner associations and 80% of the world's merchant fleet, dispatches large numbers of delegates to IMO negotiations, including at UNFCCC meetings. Consequently, individual shipping lines are spared from having to engage directly in obstructionist tactics because associations such as ICS do it for them.[78,79] Furthermore, a small number of players dominate maritime debates.[80] Recent research has found that only 20% of the industry-focused delegations attending meetings submit 90% of the documents.[81] Consequently, industry delegations are able to control the debate agenda, further promoting the interests of developed countries, large flag states, major shipbuilding nations, and trade organizations.[82]

The aviation industry also follows this pattern. For example, the International Air Transport Association (IATA) has consistently promoted the global CORSIA scheme, a market-based mechanism requiring airlines to purchase carbon offsets or sustainable aviation fuels to compensate for GHG emissions above a 2019–2020 baseline. By advocating for CORSIA, IATA has worked to avoid the introduction of regulations and taxes on companies' operations.[83] At the same time, the association has lobbied against national and regional climate policies, including a proposed kerosene tax in the European Union. InfluenceMap has also traced the UN International Civil Aviation Organization's (ICAO's) influence in promoting the CORSIA scheme.[84] It found that ICAO decision-making on CORSIA had consistently reflected IATA's demands, including those that weakened CORSIA during the COVID-19 crisis to change the baseline year for measuring GHG emissions, significantly weakening the policy.

There are numerous other examples of close collaboration between corporations, trade agencies, and regulators at wide scales. Coal interests have organized themselves into global coalitions, such as the World Coal Institute (now called FutureCoal) and also coordinate with relevant global and national bodies charged with promoting industrial interests.[85] Along with transportation actors, coal actors participate in the European Roundtable of Industrialists and Business Europe, among others, and also work through organizations such as the National Coal Association in the United States. These coalitions magnify the efforts of individual companies to delay the transition away from coal and oil. For example, transportation actors in Europe have leveraged membership in such bodies to promote road construction.[86]

Utilities also leverage these close connections at multiple levels. For example, in the United States, the electric utility industry is a conglomeration of investor-owned, state, municipal, cooperative, and federal entities who produce and distribute electricity. Investor-owned utilities in particular have proven to be highly effective at protecting the interests of certain corporate actors.[87] These utilities frequently enjoy a natural monopoly when providing electricity production in the regions in which they operate, though exceptions do exist in states where deregulation has occurred.[88] Because ratepayers in monopolized regions have no choice but to pay for electricity and gas from their local utility, in many cases they are effectively forced to fund utilities' climate denial and anti–climate legislation lobbying.

Corporations, trade associations, front groups, and international governance bodies do not always present a united front on climate obstruction. Rather, their interests can diverge, as reflected with divisions on policy preferences appearing both within and between types of actors. Box 3.1 illustrates these divisions using the example of road transportation in the European Union.

Box 3.1: FAULT LINES AMONG ACTORS: THE COMPLEX CASE OF ROAD TRANSPORTATION IN THE EUROPEAN UNION

Analyses[89,90,91,92] of the political struggle around CO_2 emission standards for vehicles and policies to phase out internal combustion engines in Europe over the past decade highlight multiple fault lines dividing actors in the road transportation industry, including those between:

- automakers specializing in larger, more powerful and therefore more difficult to electrify vehicles in the luxury segment and those building smaller cars. The former struggle more than the latter to comply with emission standards and are thus more likely to lobby against regulation.
- automakers supporting policies to accelerate the shift to electric vehicles (from France, Spain, and Italy) and laggards (largely from Germany, including Volkswagen, BMW, and Daimler), with the latter found to be more active in lobbying against climate policy.
- parts supplier companies and their trade groups. A rift has opened between the world's largest auto-parts manufacturer (Bosch), which is more positive toward an EV transition, and the EU's primary industry association (the European Association of Automotive Suppliers, known as CLEPA), which remains opposed.

These divisions among manufacturers sometimes overlap with national divisions (e.g., German automakers vs. Italian and French automakers), but can also play out within countries—for example, in Germany, BMW and Daimler have held different policy positions than Volkswagen.[93,94] Recent reports have noted an increasing fragmentation of the actors in

Continued

the European context, as major players such as Volkswagen and Bosch have softened their opposition to regulation aimed at accelerating the transition to EVs.[95,96,97] This fragmentation makes it easier for governments to achieve the measures required for climate mitigation.

Furthermore, automotive trade unions are often aligned with industry when it comes to climate policy, although their position can be complex.[98,99,100,101] Worker unions including IG Metall have engaged in climate obstruction when climate action risks job losses, but have been more supportive of green job policies they believe will support employment.[102,103] These trends also apply to utilities and shipping companies, which are often characterized by frontrunners that have actively pursued decarbonization, and laggards that prefer to continue incumbent technologies.[104,105,106] These divisions provide opportunities for policymakers, as we discuss later.

DIRECT POLITICAL OBSTRUCTION STRATEGIES AND MECHANISMS

In this section, we identify five common strategies that the coal industry, electric utilities, and the transportation sector have used to shape policymaking and influence political systems, thereby weakening, delaying, or blocking government climate action. Table 3.2 provides a list of examples of the distinct strategies deployed by different industries and sectors. In practice, these strategies and mechanisms are frequently used in parallel, so synergies and overlap do occur.

Participation in Rulemaking and Policymaking

Industry actors actively participate in rulemaking and policymaking processes to block or weaken climate policy, as illustrated in the earlier example of the (now defunct) Global Climate Coalition, which participated in UNFCCC meetings to advocate a voluntary approach to climate mitigation while disseminating views that rejected the scientific consensus on climate change.[107,108] Although it is both logical and common for policymakers to seek expert knowledge from industry when formulating policies, the involvement of private-sector actors can disproportionally amplify the voice of industry interests in such policy discussions, decelerating the momentum for ambitious climate action.[109] Another concern is that close and/or frequent contact with policymakers increases opportunities for insider lobbying, in which industry

Table 3.2 DIRECT POLITICAL OBSTRUCTION STRATEGIES AND MECHANISMS ACROSS INDUSTRIES/SECTORS

Strategy or mechanism	Industry/ Sector	Examples
Participation in rulemaking and policymaking	Shipping, aviation, road transport, utilities	Sourcing state delegates for UN agencies such as the IMO and ICAO from nations' transportation, aviation, and maritime ministries, which have weaker climate interests and stronger industry ties than climate ministries.
Lobbying	Aviation, road transport	Lobbying globally and nationally against policies such as short flight bans and carbon taxes Lobbying in EU by road transportation sector to slow policies on vehicle electrification
Revolving doors	Coal, road transport, utilities	In Australia, placing coal industry representatives into powerful political positions, affecting climate policy across successive national governments
Political suppression	Coal	Suppressing or criminalizing dissent and protest using government powers, frequently supported by industry through public relations campaigns and direct political contributions
Litigation	Road transport, utilities	Launching lawsuits against policymakers, e.g., against the California Air Resources Board's attempts to introduce regulations more stringent than federal requirements. German-based energy company RWE's attempted suit against the Dutch government's law that sought to ban coal burning for power generation by 2030.

actors enjoy a privileged status as insiders due to their inclusion in formal and informal consultations with policymakers.[110] Not only is insider lobbying less transparent and harder to track than outsider lobbying, it provides another pathway for corporate or state capture to occur.[111]

Coal industry actors continue to participate in international rulemaking and policymaking with the aim of legitimating themselves as critical players in tackling climate change and thus influencing outcomes. Indeed, fossil fuel interests were reportedly behind the dilution of the wording in the final report from the 2021 COP26 meeting in Glasgow, where references to a phase-out of coal and fossil fuel subsidies were changed to "phase down".[112] As another example, the COP24 meeting held two years earlier in Katowice was strongly influenced by the attendance of numerous coal-industry interests, due largely to its official sponsorship by Poland's leading coal companies, such as Jastrzębska Spółka Węglowa (JSW).[113]

In the road transportation sector in Europe, several studies have shown how business actors populate the membership of the expert groups that draft or provide key input on policy strategies, proposals, and legislation as a means of delaying the transition to electric vehicles and/or weakening climate regulations.[114,115,116,117,118,119] Similar strategies have been identified in other key auto markets such as the United States, Japan, and Australia.[120,121]

EU regulation was updated in 2016 to prevent corporate dominance and conflicts of interest, and to increase transparency. However, this change has had limited success in removing industry conflicts of interest from expert groups.[122] For example, the expert group on motor vehicles, of which 78% of members represented corporate interests, included companies that had been caught up in the 2015 "Dieselgate" scandal and criticized in a European Parliament inquiry.[123] The car industry's obstruction of government regulation also surfaced in Dieselgate (which involved German automakers installing software in their diesel models to circumvent US emissions standards) although it primarily concerned air pollution rather than climate change. But the inquiry highlighted the excessive representation of car-industry interests in expert groups consulted by government policymakers as a key factor in the scandal as well as in the weak governance of the auto industry.

Regulatory dynamics in aviation and shipping are similar, with industry members often over-represented. As discussed, unlike at the UNFCCC, where state delegations typically consist of staff from climate and foreign ministries, state delegates negotiating on climate issues in IMO and ICAO delegations typically derive from transportation, aviation, and maritime ministries.[124] Participation in these activities has allowed the aviation and maritime industries to leverage their support for weak, top-down and international climate regulation at their respective UN agencies to strategically oppose more ambitious and bottom-up climate policies from the regional or national level.[125] More specifically, the maritime industry has sought to delay climate action by manipulating formal policymaking processes via the IMO. Proceedings and deliberations over regulation in the IMO are characterized by lengthy technical discussions and analysis over small details and measures.[126] Not only does this tactic provide a mechanism of delay, but it also results in ongoing technical discussions leading to conservative or weakened emissions targets. Initially, the IMO set a 50% reduction goal for 2050 compared with 2008, far short of complete decarbonization. Moreover, by choosing 2008 as the benchmark for peak emissions (i.e., one year prior to the onset of the global financial crisis) the organization considerably weakened the industry's decarbonization ambitions. In July

2023, the IMO finally set a new target that is much closer to net zero by 2050, though the mechanisms for achieving that goal remain opaque.

Lobbying

Lobbying is prevalent across the industries. In the United States, the coal industry's lobbyists countered attempts by the Obama administration to limit emissions from coal-fired electricity through the Waxman-Markey bill and the Clean Power Plan.[127] Electric utilities did the same,[128] spending more than $500 million over the last two decades.[129] More recently, from 2008 to 2018, researchers[130] found that trade associations in the utility industry spent $44 million on direct lobbying, $44 million on grants, $18.6 million on advertising and promotion, and $24 million on political contributions. This volume of spending makes electric utilities one of the most politically active industries on climate change. In the United States, contract lobbyists working in house for state electric utilities are reported to be regarded by state policymakers as more influential than those from other industries.[131] Moreover, the lobbying activities of utilities have often succeeded in obstructing policies aimed at accelerating decarbonization and the shift to renewable energy. Consider California's Pacific Gas & Electric, which fought electric metering for rooftop solar, and First Energy, which led the successful campaign to repeal Ohio's clean energy standards, replacing them with a bailout for coal and nuclear power.[132]

The transportation sector has also used lobbying to block regulatory action. In the European Union, decades of effort to decarbonize the automotive industry have finally borne fruit with the adoption of a rule in 2022 to ban the sale of new fossil fuel–based vehicles with internal combustion engines by 2035. But this rule to phase out internal combustion vehicles was delayed by several years due to the participation of automakers in regulatory and political processes and their lobbying to achieve concessions from the European Union. These obstructive actions succeeded in considerably weakening the rule, by securing:[133,134,135,136]

- A voluntary agreement rather than binding regulations with manufacturers;
- Less stringent targets than originally proposed;
- A phase-in of decarbonization measures, an absence of interim targets, and the postponement of implementation of the targets, allowing more time than originally planned for manufacturers to comply with regulations; and
- The introduction of several loopholes and the absence of penalties for manufacturers that do not meet targets for electric car sales.

Taken together, these successful lobbying results delayed the introduction of climate regulations in road transportation, increasing emissions in the sector and making it more difficult to achieve climate mitigation targets. Researchers have argued that "this has enabled vehicle manufacturers to make sure that any new stipulations are so toothless that they are spared any serious efforts beyond pursuing the technological innovations that they were already working on."[137]

Revolving Doors

The revolving door is a phenomenon whereby personnel move back and forth between industry and government positions, facilitating lobbying and in some cases regulatory capture.[138,139,140] Our analysis shows that multiple industries benefit from this mechanism.

For example, coal mining in Australia is marked by close ties between political leaders and representatives of industry. Individuals linked to coal interests have subsequently been appointed to powerful political positions, successfully influencing climate policy across successive national governments.[141,142,143] In road transportation, actors have gained insider knowledge about policy-making and developed personal ties with policymakers through revolving doors.[144] This phenomenon is particularly well documented for Germany, where revolving-door dynamics are common and where summits and conferences are organized to facilitate interactions between government and industry, but shut out NGOs and unions.[145,146,147,148] The utilities and aviation industries have also been criticized for revolving-door practices.[149,150,151] For example, in July 2021, ICAO hired the former director of aviation and environment at IATA to become that UN agency's director of legal affairs and external relations. This person had reportedly been part of IATA's previous advocacy efforts to weaken ICAO's CORSIA offsetting scheme in 2020.[152]

Political Suppression

Political actors and government bodies, often serving the interests of industry, regularly seek to limit climate activism by banning protests, arresting activists, censoring information, and restricting public access to policymaking debates.[153] For example, evidence suggests that the coal industry, which relies on land acquisition for extraction, transportation, and the construction of large-scale power stations, has suppressed opposition from civil society organizations and frontline communities facing the negative effects of mining and climate change because it serves as a means of obstructing and delaying the energy transition.[154] In the Philippines, Indonesia, and Colombia, members of civil society organizations have experienced human right violations including arrests, intimidation—and even abduction and murder—for resistance to coal extraction.[155,156,157] In such contexts, even when one is not menaced directly, the specter of political suppression tends to discourage activists from pursuing confrontational tactics, resulting in weaker societal opposition to climate obstruction activities.[158] This form of silencing reduces pressure on government officials, weakening the imperative to introduce policies to transition away from fossil fuels.[159] Furthermore, human rights violations aimed at preserving coal-industry interests are typically concentrated in the Global South, where they lack visibility to downstream countries in the Global North that import fossil fuels.[160]

Litigation

Litigation has become a powerful weapon for many companies to slow the pace and ambition of climate action.[161] Successful suits to prevent or weaken climate policies have been particularly visible in the automotive industry. In the United States, the California Air Resources Board's attempts to introduce regulations more stringent than federal requirements have triggered various lawsuits by automobile and truck associations.[162] A notable case in 2022 focused on California's proposed nitrous oxide emissions-reduction policy, which targeted heavy-duty vehicles.[163] Legal strategies of delay have also emerged in the coal and utility industries. For example, the German-based utility RWE tried to sue the Dutch government for €1.4 billion in compensation for a law that sought to ban coal power by 2030. Subnational governments have also sued national governments to protect their fossil fuel industries. For example, in 2015 the US state of West Virginia sued the federal government to block the implementation of the Clean Power Plan, President Barack Obama's signature climate policy, which aimed to accelerate the shift from coal to renewable energy.[164,165] Seven years later, the US Supreme Court ruling on the case led to greater restrictions on the Environmental Protection Agency's ability to regulate CO_2 emissions.

RHETORICAL AND DISCURSIVE TACTICS

Coal, transportation, and utilities also use strategic communication tactics to hinder action on climate change.[166,167,168,169] Their rhetorical and discursive tactics are circulated via the news media, social media, policy position papers, government testimony, and information campaigns. Leading up to the Paris Agreement, and in the years since, tactics have evolved from outright denial about the causes and impacts of climate change to what can be called "discursive resistance."[170] This technique consists of deploying

Table 3.3 RHETORICAL AND NARRATIVE OBSTRUCTION TACTICS

	Definition
Rhetorical obstruction tactics	
Climate denial	Casting doubt on the basic scientific facts or impacts of climate change
Corporate ventriloquism	Transmitting messages through other entities such as community and/or astroturf (self-created) organizations to construct and animate an alternative ethos, voice, or identity that advances their interests
Narratives	
Apocalyptic consequences for industry	Fearmongering about the imminent demise of a particular industry, economic system, or political system if climate regulation succeeds, and the catastrophic ramifications associated with that loss
Technological optimism and fossil fuel solutionism	Painting a positive and future-oriented image of fossil fuels by emphasizing the ability for new technologies (e.g., carbon capture, ammonia co-firing) to contribute to decarbonization, thereby reducing the imperative to shift away from fossil fuels
The right to develop and energy poverty	Positioning the use of fossil fuels as a core right for emerging economies and a means to attain economic development and overcome energy poverty

narratives that emphasize, for example, the costs of action, particularly inflated estimates of the socioeconomic impacts of climate action or the threat of carbon leakage when emissions-intensive industries move to other regions or countries with less stringent laws. Another discursive tactic involves propagating hyped expectations about the potential of decarbonization technologies such as carbon capture and sequestration (CCS) to ensure a future for coal.[171] These tactics are summarized in Table 3.3 and described briefly in the next sections.

Climate Denial

Climate denial activities across the fossil fuel sector have been explored in great depth (see Chapter 2).[172] In the United States, the electric utility industry knew about climate change in the 1960s and 1970s, but then spread doubt and denial to prevent action.[173,174] By the 1990s, in full coordination with groups representing the oil and coal industries, utilities also increasingly embraced outright climate denial, including founding front groups to spread climate disinformation. Examples include the already mentioned PR campaign headed by the Edison Electric Institute (EEI) and Southern Company in 1991, which aimed to "reposition global warming as theory (not fact)." These disinformation efforts continued over several decades. As recently as 2017, the then-CEO of Southern Company, Thomas Fanning, who also served as EEI's chair, declared during a television interview with CNBC that CO_2 is "certainly not" the cause of planetary warming, and that climate change has "been happening for millennia."[175] In this way, the utility industry has mirrored coal industry efforts by leveraging denial as an obstruction tactic.[176]

The transportation sector has also participated in climate denial. In the United States, Chrysler, Ford, and General Motors became members and supported the activities of the previously discussed Global Climate Coalition.[177] As a more recent example of contesting the science, Toyota has reacted to efforts by the Japanese state to accelerate the shift to EVs by claiming as recently as 2023 that battery-electric drivetrains offer limited decarbonization benefits in Japan due to the electricity sector's dependance on fossil fuel generation.[178,179]

Corporate Ventriloquism

Corporate ventriloquism is a rhetorical process by which companies transmit messages through other entities, usually of their own creation, to construct and animate an alternative ethos, voice, or identity that advances their interests.[180] Some companies have used corporate ventriloquism to advance their agendas through support for community organizations.[181] Companies may create front groups and information campaigns, but also leverage corporate social responsibility strategies that invest in community development. Such strategies have been used for decades as a means for industry actors to underscore their legitimacy and to shore up the perception that they are positive social contributors to the environments in which they operate.[182,183,184] Research has shown how extractive industries such as coal have invested heavily in local communities to sow solidarity and support for the industry.[185] Material investments such as parks, community centers, sponsored events (concerts, car shows, etc.), and museums are not only a means of winning support through material investment in communities (a form of philanthropy or bribery), but also of symbolically "branding" the community and of making it more difficult for the community to oppose industry goals. Similarly, investor-owned utilities sponsor sports stadiums and museums—branding efforts that double as philanthropy—in their efforts to effectively blur the line between public and private interests.

In the case of road transportation, this blurring between public and private interests typically happens at the state level. Research[186] draws on the concept of "state dependence"[187] to characterize the "privileged position of the automotive industry within capitalist states due to governments' dependence on the jobs, growth, and state revenue it provides." This dependence means that in practice state actors might act in the interests of the industry even in the absence of outright lobbying.[188,189] This pattern can be even more pronounced in countries characterized by coordinated or corporatist state–business relationships, such as Germany.[190,191,192]

Fearmongering About the Consequences of Climate Action

All industries and sectors examined share the tactic of fearmongering about potential negative economic and societal impacts of climate action. This strategy creates a "jobs-versus-environment" dilemma to garner support for the former.[193] The US coal industry has used diverse forms of this argument to mobilize support for its own survival. One narrative built around a theme of industrial apocalypse argues that coal is "too big to fail," and if it fails, the entire economy would be dragged along. This narrative has been used to argue for extending the life of existing coal plants through subsidies to ensure grid stability and reliability.[194,195]

Coal actors in both the Global North and the Global South have also adopted the language of a "just transition"—originally used by anti-coal actors—to argue for slow and nondisruptive phase-out timelines and to foster fear about the socioeconomic consequences of a forced contraction of coal extraction.[196] For example, in South Africa, the Congress of South African Trade Unions called early plant closures by the utility Eskom "a hostile act of provocation directed at workers and their unions," and called for Eskom and the government to suspend the closures until "a just transition-solution is arrived at by all affected stakeholders."[197]

Similarly, the auto industry lobby argues that the economic well-being of car companies translates directly into national and economic prosperity.[198,199] OEMs and trade unions similarly fearmonger and obstruct based on claims about job losses.[200,201,202,203] It is well documented that carmakers sometimes use exaggerated forecasts of reduced employment in the sector as a bargaining tool with policymakers.[204,205] These arguments have been used by the car industry to oppose regulations to phase out internal combustion engines in the European Union and Japan.[206,207]

Technological Over-Optimism and Fossil Fuel Solutionism

Several of the industries examined employ discourses that paint an optimistic picture about the ability for niche or unproven technologies to decarbonize the use of fossil fuels,[208] thereby downplaying the need to transition away from incumbent energy systems. For example, utilities in Japan have promoted the potential for coal co-firing with ammonia, an unproven technology, rather than planning a managed phaseout of their fleet of coal-fired power plants.[209] A more established example of technological optimism is CCS.[210] Not only do the companies championing CCS and ammonia co-firing try to extend the life of existing power plants, but they also seek to deploy these technologies.[211] In this way, nontransformative technologies that perpetuate dependence on fossil fuels are framed as "clean" and as a solution to the climate problem. In the US, for example, a 2004 leaked memo from a coal front group president to the CEO of Peabody Energy stated: "Our belief is that, on climate change like other issues, you must be for something rather than against everything. The combination of carbon sequestration and technology is what we preach and we are looking for more members in the choir."[212] Indeed, after 2000, electric utility-affiliated organizations shifted toward promoting CCS and "clean" coal when seeking to delay climate action.[213] This is despite the fact that according to the industry itself, CCS remains prohibitively expensive and difficult to deploy at scale. Although this technology has largely fallen out of coal industry discourse in the United States and Europe, in some Asian regions including Japan, CCS is still promoted by electric utilities as a means of ensuring the continued operation of coal-fired power plants.[214]

The shipping industry also uses this delay tactic, promoting incremental efficiency improvements of existing fossil fuel technologies. A 2011 amendment to the International Convention for the Prevention of Pollution from Ships (MARPOL) created the Energy Efficiency Design Index (EEDI) and Ship Energy Efficiency Management Plan (SEEMP). The former requires certain new vessels to be designed to meet rising energy-efficiency standards, while the latter requires all vessels to create and follow energy efficiency plans that adhere to certain criteria. However, the EEDI has major deficiencies that limit its efficacy in reducing emissions. For example, it applies only to new ships and its targets can easily be met by reducing the speed of travel.[215,216] While it is desirable for each generation of ships to be more efficient than the last and use less fuel per ton or mile, efficiency alone cannot lead to decarbonization—not least because continued growth in shipping leads to increased emissions, even if the vessels are more efficient. The same is true for aviation.

The Right to Develop and Energy Poverty

Many actors in the surveyed industries have linked narratives that stress a "right to develop" in the Global South and/or the need to overcome "energy poverty" with agendas to continue the use of coal and various fossil fuel technologies (see Chapter 6). These discourses have been promulgated by exporting nations in both the Global North and the Global South, where many countries are still engaged in the process of building out their energy systems and developing their economies.[217] For example, in India, China, Indonesia, and South Africa, industry and government actors have stressed their right to develop when defending their continued use of coal in the power sector.[218] Many actors in the Global North that are providing technology, finance, or fuel to these countries also have a financial interest in preserving coal dependence. In pursuing this interest, they claim that coal is a critical means of overcoming energy poverty by ensuring widely available and cheap electricity, thereby co-opting and building upon right to develop discourses. For example, the World Coal Association claims that "coal has a fundamental role in providing access to base load electricity" and that "energy at affordable prices [is] critical to global development."[219] In the United States, the front group Advanced Energy for Life (funded by Peabody Energy) claimed in its PR campaign that energy poverty was "the world's *number one* human and environmental crisis" (emphasis added), and that coal was the solution.[220] Meanwhile, the coal industries in the United States and Australia have also framed their exported coal as necessary for solving energy poverty.[221,222,223,224]

The auto making industry has used the same narrative. A foundation associated with the British motoring association, the Royal Automobile Club (RAC), opposed a fuel tax increase by arguing that 80% of the British population was in "transport poverty."[225] Yet this estimate was based on a flawed indicator, with more accurate estimates finding that only 9% of British households were vulnerable to fuel price increases.[226] At the European level, automotive actors seeking to deflect support for electric vehicles have also argued that e-fuels, their preferred decarbonization solution, would help ensure "affordable mobility." This is despite evidence showing that synthetic e-fuels (derived from renewable electricity, CO_2, and hydrogen) are a more expensive decarbonization solution than electrification.[227]

POSSIBLE RESPONSES TO OBSTRUCTION

A variety of responses at both international and national levels have emerged to counter the obstruction strategies and tactics used by these industries and sectors.

International Responses

At the international level, states have begun to build coalitions to phase out fossil fuels. These include the Powering Past Coal Alliance (launched by Canada and the United Kingdom) and the Beyond Oil and Gas Alliance (launched by

Quebec, Denmark, Costa Rica, France, and other nations and provinces). More recently, in 2024, member nations of the Group of Seven (G7) industrialized countries committed to phasing out unabated coal power by 2035. In 2021, fourteen countries (including France, Germany, and the United States) challenged the IMO's current low level of ambition and called for a target of full decarbonization of the shipping sector by 2050, including the establishment of intermediate targets for each decade.[228] Such clubs are an essential means of injecting ambition and speed into international governance frameworks, where progress is often stalled by the need for consensus and vulnerability to industry lobbying.

Many multilateral responses have sought to restrict international financing for fossil fuel projects, especially coal. States have also used such initiatives to address carbon lock-in. For example, the Asian Development Bank's Energy Transition Mechanism aims to catalyze the early retirement of coal-fired power plants by providing public and private financing to developing Asian countries. One recipient, Indonesia, which has deepened its reliance on coal while developing its power sector in recent years, was promised up to $20 billion over three to five years via the Just Energy Transition Partnership (JETP), established in 2022 and led by the United States and Japan. While such multilateral efforts are critical for lowering the economic hurdles to transitioning away from coal-fired power in the Global South, some enactments of the idea of "paying a country to transition" have flaws in their design. For example, though promised $20 billion in payments, Indonesia will only see some $160 million, with the rest provided as loans and grants, deepening the country's debt burden to rich countries.[229]

The challenge for policymakers and civil society organizations is thus to ensure that these multilateral mechanisms are used as intended and, where possible, strengthened. Besides, rather than simply outsourcing the task to the Global South, it is imperative that the Global North lead by example in the transition away from fossil fuels. This task requires strong leadership from high-ambition countries, and for non–state actors such as NGOs to use their influence to hold laggards to account, not least by shaming states and associated initiatives that fail to drive the phase-out of coal and other fossil fuels (see Chapter 13).

National Responses

At the national level, this review of obstructionist activities by actors in coal, electric utilities, and transportation suggests at least three responses.

First, there are clear divisions between firms in these industries. Policymakers could leverage these disagreements to weaken the obstructive effect of the united front of opposition that trade associations tend to create.[230] Previous studies have shown that these divisions are especially prevalent in the utilities, car, and aviation industries.[231,232,233,234,235,236,237] As Box 3.1 shows, these fault lines are prevalent across the road transportation industry in the European Union, where some companies are more enthusiastically embracing EVs while others are resisting or delaying electrification efforts.

Second, policymakers, civil society actors, and researchers should be aware of and facilitate the conditions that can convert actors from climate obstructionists to climate pioneers.[238] In the case of the utility and car industries, important conditions include ensuring the widespread availability and affordability of key technologies, such as renewable power, batteries, and charging infrastructure. Government policies can play an important role in incentivizing the production and use of these technologies and creating economic incentives that reward early-adopter firms.[239]

Unions have historically been powerful actors in climate obstruction.[240] Yet they tend to lack cohesion, since they are afraid of job losses from sustainability transitions, while simultaneously supporting the prospect of gaining "green" job benefits in the mid-term.[241,242] This dichotomy leads to conflicting political positions, suggesting that addressing concerns about job losses from the energy transition could also weaken organized labor's obstructive stance.[243,244]

Third, increasing public and investor pressures for disclosure of lobbying activities and involvement in rulemaking negotiations are prompting positive changes (see Chapter 13). For example, some auto manufacturers are already reluctant to lobby individually on politically sensitive climate policies.[245] This trend may be partly attributable to the backlash following the Dieselgate scandal.[246,247] Furthermore, as mentioned, EU regulations were updated in 2016 to reduce corporate dominance and conflicts of interests and to increase transparency.[248] However, the effects of these rule changes is still unclear, and car makers continue to exploit their industry associations to impede climate policies.[249] Thus confronting these kinds of loopholes and industry behavior remains essential.[250]

Progress can be seen in the aviation industry as well. In 2022, the United States responded to public criticism by releasing the policy papers it submitted to the UN's ICAO for the first time. It also supported making all documents related to key decisions by Committee on Aviation Environmental Protection (CAEP) meetings publicly available in advance of the meetings, aligning itself with IMO and UNFCCC practices. Other reforms are still needed, including ending the practice of requiring CAEP delegates to sign nondisclosure agreements, removing external restrictions on observers' ability to speak freely about ICAO, and granting free media access to the aviation industry's climate negotiation meetings.

CONCLUSION

Several important findings emerge from the three typologies we have explicated in this chapter. First, as Table 3.1 illustrates, actors, organizations, and states frequently coordinate across scales to implement obstruction strategies. By masking the actions of individual organizations, this strategy affords these actors plausible deniability when others hold them accountable for climate obstruction activities. Second, there is significant involvement by industry players in governance bodies, both informal and formal. This access grants them an outsized influence on climate policymaking compared with other important social and political players, such as NGOs, think tanks, and researchers.

Our second typology (shown in Table 3.2) describes and analyzes five direct political strategies employed across the industries and sectors examined. The analysis suggests that given revolving-door effects and existing regulatory structures, it can be difficult to distinguish between state and corporate interests, resulting in an underrepresentation of civil society interests in climate-related policymaking. Additionally, we found that regulatory institutions in many geographies and at various scales have been populated, influenced, or captured to varying degrees by corporate interests. We expect, however, that the efficacy of direct political engagement as a climate obstruction strategy will vary depending on the industry and the differing governance cultures of countries in the Global North and South. These factors likely include varying levels of stakeholder pressure and the extent to which industries or various companies' roles in climate obstruction has been brought into the public spotlight. Thus, while our chapter has highlighted the similarities in tactics used across the industries and sectors examined in this chapter, obstruction activities also vary by timing, national context, and political culture. An important future research task is to move beyond descriptive analyses toward explanatory studies that deepen understanding of how obstruction tactics vary across regions and industries and the conditions that influence this variation.

Finally, our third typology (shown in Table 3.3) clearly demonstrates that the industries and sectors examined share common communication tactics. This finding likely reflects their interdependence as much as the ways in which rhetorical tactics tend to circulate via industry and front-group activities. We also observed that tactics have evolved from outright climate denial to a more multifaceted and agile discursive resistance, particularly the use of so-called common sense arguments for maintaining fossil fuels that present a business case for their continued use or frame them as part of the solution to climate change. These industries have also shown skill in mimicking or co-opting social-movement narratives, such as by claiming that fossil fuels will help achieve energy justice objectives or using the "just transition" argument against a rapid phase-out of fossil fuels (see Chapter 6).

Despite the robust evidence supporting these conclusions, there are specific knowledge gaps. First, we need to identify and analyze obstruction activities in other industries, such as steel and rail, that are closely connected to the coal industry and transportation sector.[251] However, future work should recognize that the rail industry's dependence on coal does not automatically mean that it uniformly opposes the energy transition in all geographic contexts. Some actors in the rail industry, such as those in India, are eager to reduce their reliance on coal transportation for revenues.[252] In road transportation, an important under-researched topic is the links between the industry's obstruction activities, public (mis)information about the environmental performance of decarbonization technologies such as EVs, and their public acceptance.

Second, little academic attention has been paid to the *transnational* obstacles to transitioning away from coal. As researchers[253] have argued, "choices are not confined to national borders, as the availability of foreign finance and technical capacity may affect whether new coal plants are built." Most of the new capacity recently built or still under construction in Asian countries including Vietnam, Indonesia, Bangladesh, and India has involved financial institutions and manufacturers based outside these countries, especially in China, Japan, South Korea, Europe, and the United States.[254,255] Because coal is generally framed as the form of power best suited to rapidly emerging economies, the role of these transnational networks requires further investigation. While scholarship on the obstruction activities of the coal industry in Global South countries is limited (see Chapter 8), there exists even less understanding of how other industries such as road transportation, shipping, and aviation might be obstructing climate action outside the Global North.

Finally, it is clear that climate negotiations continue to happen behind closed doors, involving delegations that merge corporate and state interests. As a result, there is little data on how decisions are made. Furthermore, civil society interests, such as NGOs, are frequently under-represented in these negotiations. In the shipping industry, for example, until organizations such as the IMO adopt more formal decision-making processes that allow for transparency and accountability, such as voting, it will be difficult to closely analyze these processes.

To avoid crossing dangerous levels of planetary warming, society must accelerate the decarbonization of electricity generation and all forms of transportation while simultaneously phasing out coal combustion.[256] As documented in this chapter, confronting the ongoing obstruction activities of these industries may be as pivotal to global decarbonization efforts as the development and deployment of renewable technologies.

NOTES

1. International Energy Agency, "Global Energy & CO2 Status Report 2019," 2019, https://www.iea.org/reports/global-energy-co2-status-report-2019/emissions.
2. International Energy Agency, "Tracking Clean Energy Progress 2023: Transport," 2023, https://www.iea.org/energy-system/transport#tracking.
3. International Energy Agency, "Greenhouse Gas Emissions from Energy Data Explorer," 2023, https://www.iea.org/data-and-statistics/datatools/greenhouse-gas-emissions-from-energy-data-explorer.
4. U.S. Environmental Protection Agency, "Greenhouse Gas Inventory Data Explorer," 2024, https://cfpub.epa.gov/ghgdata/inventoryexplorer/#electric powerindustry/entiresector/allgas/category/all.
5. Z. Shao, H. He, S. Mao, S. Liang, S. Liu, X. Tan, M. Gao, and Y. Xin, "Toward Greener and More Sustainable Freight Systems: Comparing Freight Strategies in the United States and China," 2022, https://theicct.org/wpcontent/uploads/2022/01/China-US-freight_final.pdf.
6. Association of American Railroads, "Railroads and Coal," 2019, https://www.aar.org/wp-content/uploads/2018/05/AAR-Railroads-Coal.pdf.
7. C. Hauenstein, "Stranded Assets and Early Closures in Global Coal Mining Under 1.5oC, *Environmental Research Letters* 18, no. 2 (2023): 024021.
8. A. Edianto, G. Trencher, N. Manych, and K. Matsubae, "Forecasting Coal Power Plant Retirement Ages and Lock-In with Random Forest Regression," *Patterns* 4, no. 7 (2023): 100776.
9. P. Erickson, S. Kartha, M. Lazarus, and K. Tempest, "Assessing Carbon Lock-In," *Environmental Research Letters* 10, no. 8 (2015): 084023.
10. H. Mistry, "Sustainable Aviation Fuel," presentation, 2023, IATA, https://www.iata.org/en/iata-repository/pressroom/presentations/saf-gmd2023/.
11. IPCC, "AR6 Climate Change 2022: Mitigation of Climate Change, the Working Group III Contribution," 2022, www.ipcc.ch/report/sixthassessment-report-working-group-3/.
12. J. Wesseling and N. Meijerhof, "Towards a Mission-Oriented Innovation Systems (MIS) Approach, Application for Dutch Sustainable Maritime Shipping," *PLOS Sustainability and Transformation* 2, no. 8 (2023): e0000075.
13. J. Steckel and M. Jakob, "To End Coal, Adapt to Regional Realities," *Nature* 607 (2022: 29–31.
14. K. S. Gallagher, R. Bhandary, E. Narassimhan, and Q. T. Nguyen, "Banking on Coal? Drivers of Demand for Chinese Overseas Investments in Coal in Bangladesh, India, Indonesia and Vietnam," *Energy Research & Social Science* 71 (2021): 101827.
15. EPA, "Greenhouse Gas Inventory Data Explorer."
16. Lawrence Berkeley National Laboratory, "U.S. Power Sector is Halfway to Zero Carbon Emissions," 2021, https://emp.lbl.gov/news/uspower-sector-halfway-zero-carbon-emissions.
17. A. M. Galli Robertson and M. B. Collins, "Super Emitters in the United States Coal-Fired Electric Utility Industry: Comparing Disproportionate Emissions Across Facilities and Parent Companies," *Environmental Sociology* 5, no. 1 (2019): 70–81.
18. C. Fogler and N. Ver Beek, "The Dirty Truth About Utility Climate Pledges," 2023, https://coal.sierraclub.org/sites/nat-coal/files/dirty_truth_report_2023.pdf?utm_source=sierraclub&utm_medium=web&utm_id=dirty-truth&utm_content=page.

19. E. L. Williams, S. A. Bartone, E. K. Swanson, and L. C. Stokes, "The American Electric Utility Industry's Role in Promoting Climate Denial, Doubt, and Delay," *Environmental Research Letters* 17, no. 9 (2022): 094026.

20. B. K. Sovacool and M.-C. Brisbois, "Elite Power in Low-Carbon Transitions: A Critical and Interdisciplinary Review," *Energy Research & Social Science* 57, (2019): 101242.

21. R. Brulle and C. Downie, "Following the Money: Trade Associations, Political Activity and Climate Change," *Climatic Change* 175, no. 3 (2022): 11.

22. C. L. Toenshoff, "Hiding in the Crowd: Corporate Climate Lobbying Under Investor and Consumer Pressure, in Department of Political Science" (PhD diss., Stanford University, 2023).

23. G. Curran, "Social Licence, Corporate Social Responsibility and Coal Seam Gas: Framing the New Political Dynamics of Contestation," *Energy Policy* 101 (2017): 427–435.

24. M. A. Latapí Agudelo, L. Johannsdottir, and B. Davidsdottir, "Drivers That Motivate Energy Companies to Be Responsible: A Systematic Literature Review of Corporate Social Responsibility in the Energy Sector," *Journal of Cleaner Production* 247 (2020): 119094.

25. R. J. Brulle, "Networks of Opposition: A Structural Analysis of U.S. Climate Change Countermovement Coalitions 1989–2015," *Sociological Inquiry* 91, no. 3 (2021): 603–624.

26. R. D. Eller, "The Coal Barons of the Appalachian South, 1880–1930," *Appalachian Journal* 4, no. 3/4 (1977): 195–207.

27. J. A. Ordonez, M. Jakob, J. C. Steckel, and A. Fünfgeld, "Coal, Power and Coal-Powered Politics in Indonesia," *Environmental Science & Policy* 123 (2021): 44–57.

28. Greenpeace, JATAM, and Indonesian Corruption Watch, "Shedding Light on Political Corruption in Indonesia's Coal Mining Sector," 2018, https://www.greenpeace.org/static/planet4-indonesia-stateless/2018/12/727d7a2dcoalruption-english-web.pdf.

29. I. Anugrah, "Land Control, Coal Resource Exploitation and Democratic Decline in Indonesia," *TRaNS: Trans -Regional and -National Studies of Southeast Asia* 11, no. 2 (2023): 195–213.

30. N. Manych and M. Jakob, "Why Coal? The Political Economy of the Electricity Sector in the Philippines," *Energy for Sustainable Development* 62 (2021): 113–125.

31. P. G. Saculsan and A. Mori, "Why Developing Countries Go Through an Unsustainable Energy Transition Pathway? The Case of the Philippines from a Political Economic Perspective," *Journal of Sustainability Research* 2, no. 2 (2020): e200012.

32. L. Friedman, "How a Coal Baron's Wish List Became President Trump's To-Do List," *New York Times*, January 9, 2018, https://www.nytimes.com/2018/01/09/climate/coalmurray-trump-memo.html.

33. P. Taddonio, "Bob Murray Helped Shape Trump's Energy Policies. Now, His Coal Company Is Facing Bankruptcy," PBS, 2019, https://www.pbs.org/wgbh/frontline/article/bob-murray-helped-shape-trumps-energypolicies-now-his-coal-company-is-facing-bankruptcy/.

34. V. Volcovici, "Coal Baron Murray, a Trump Ally Who Fought Environmental Regulation, Dead at 80," *Reuters*, October 27, 2020, https://www.reuters.com/article/world/coal-baron-murray-a-trump-ally-who-fought-environmental-regulation-dead-at-80-idUSKBN27B22E.

35. T. P. Lyon and J. W. Maxwell, "Astroturf: Interest Group Lobbying and Corporate Strategy," *Journal of Economics & Management Strategy* 13, no. 4 (2004): 561–597.

36. P. K. Bsumek, J. Schneider, S. Schwarze, and J. Peeples, "Corporate Ventriloquism: Corporate Advocacy, the Coal Industry, and the Appropriation of Voice," in *Voice and Environmental Communication*, eds. J. Peeples and S. Depoe (Palgrave Macmillan, 2014), 21–43.

37. Influence Map, "The Automotive Sector and Climate Change: Assessing Automakers' Climate Strategies Against a 1.5°C Aligned Transition, 2022, https://influencemap.org/report/The-Automotive-Sector-and-Climate-Change-18218.

38. H. Tabuchi, "Toyota Led on Clean Cars. Now Critics Say It Works to Delay Them," *New York Times*, July 25, 2021, https://www.nytimes.com/2021/07/25/climate/toyotaelectric-hydrogen.html.

39. Influence Map, "The Automotive Sector and Climate Change."

40. Corporate Europe Observatory, "Two Years After Dieselgate: Car Industry Still Drives Berlin and Brussels," 2017, https://www.corporateeurope.org/en/powerlobbies/2017/09/two-years-after-dieselgate-car-industry-still-drives-berlin-and-brussels.

41. Brulle and Downie, "Following the Money."

42. R. E. Dunlap and A. M. McCright, "Organized Climate Change Denial," in *The Oxford Handbook of Climate Change and Society*, ed. J. S. Dryzek, R. B. Norgaard, and D. Schlosberg (Oxford University Press, 2011), 144–160.

43. Bsumek et al., "Corporate Ventriloquism."

44. M. Gaworecki, "Peabody Energy Goes on Offense with New PR Campaign Designed to Sell Same Old Dirty Coal," *DeSmog*, November 11, 2014, https://www.desmog.com/2014/11/11/peabody-energy-goes-offense-new-pr-campaigndesigned-sell-same-old-dirty-coal/.

45. J. Smyth, "America's Power Loses More Member Companies, Deletes Member List," Energy and Policy, 2023, https://energyandpolicy.org/americas-power-loses-moremember-companies-deletes-member-list/.

46. Brulle, "Networks of Opposition."

47. M. Mildenberger, *Carbon Captured: How Business and Labor Control Climate Politics* (MIT Press, 2020).

48. A. Bodenhamer, "King Coal: A Study of Mountaintop Removal, Public Discourse, and Power in Appalachia," *Society & Natural Resources* 29, no. 10 (2016): 1139–1153.

49. P. G. Lewin, "'Coal is Not Just a Job, It's a Way of Life': The Cultural Politics of Coal Production in Central Appalachia," *Social Problems* 66, no. 1 (2017): 51–68.

50. S. Ryerson, "Precarious Politics: Friends of Coal, the UMWA, and the Affective Terrain of Energy Identification," *American Quarterly* 72 (2020): 719–747.

51. R. Wishart and P. Greenberg, "Coal and Environmental Inequality," in *Handbook on Inequality and the Environment*, ed. M. Long, M. Lynch, and P. Stretesky (Edward Elgar Publishing, 2023), 506–530.

52. C. Wright, R. Irwin, D. Nyberg, and V. Bowden, "'We're in the Coal Business': Maintaining Fossil Fuel Hegemony in the Face of Climate Change," *Journal of Industrial Relations* 64, no. 4 (2022): 544–563.

53. H. A. Baer, "The Nexus of the Coal Industry and the State in Australia: Historical Dimensions and Contemporary Challenges," *Energy Policy* 99 (2016): 194–202.

54. Brulle, "Networks of Opposition."

55. R. E. Dunlap and R. J. Brulle, "Sources and Amplifiers of Climate Change Denial," in *Research Handbook on Communicating Climate Change*, ed. D. C. Holmes and L. M. Richardson (Edward Elgar Publishing, 2020), 49–61.

56. L. C. Stokes, *Short Circuiting Policy: Interest Groups and the Battle Over Clean Energy and Climate Policy in the American States* (Oxford University Press, 2020).

57. Williams et al., "The American Electric Utility Industry's Role in Promoting Climate Denial."

58. G. Trencher, N. Healy, K. Hasegawa, and J. Asuka, "Discursive Resistance to Phasing Out Coal-Fired Electricity: Narratives in Japan's Coal Regime," *Energy Policy* 132 (2019): 782–796.

59. D. Tanner, "Japan's Energy Choices Towards 2020: How Its Shift to Coal Is Misaligned with the Strategic Interests of Japan Inc.," Influence Map, 2018, https://influencemap.org/site/data/000/302/Japan_Report_October_2017.pdf.

60. Brulle, "Networks of Opposition."

61. Brulle and Downie, "Following the Money."

62. Williams et al., "The American Electric Utility Industry's Role in Promoting Climate Denial."

63. Ibid.

64. R. J. Brulle, "Advocating Inaction: A Historical Analysis of the Global Climate Coalition," *Environmental Politics* 32, no. 2 (2023): 185–206.

65. Williams et al., "The American Electric Utility Industry's Role in Promoting Climate Denial."

66. Brulle, "Networks of Opposition."

67. Brulle, "Advocating Inaction."

68. Brulle, "Networks of Opposition."

69. Williams et al., "The American Electric Utility Industry's Role in Promoting Climate Denial."

70. A. Abdulla, R. Hanna, K. R. Schell, O. Babacan, and D. G. Victor, "Explaining Successful and Failed Investments in U.S. Carbon Capture and Storage Using Empirical and Expert Assessments," *Environmental Research Letters* 16, no. 1 (2021): 014036.

71. Influence Map, "The Automotive Sector and Climate Change."

72. Brulle and Downie, "Following the Money."

73. Toenshoff, "Hiding in the Crowd."

74. J. Wesseling, J. Farla, D. Sperling, and M. Hekkert, "Car Manufacturers' Changing Political Strategies on the ZEV Mandate: Transportation Research Part D," *Transport and Environment* 33 (2014): 196–209.

75. Influence Map, "The Automotive Sector and Climate Change."

76. Influence Map, "Corporate Capture of the International Maritime Organization: How the Shipping Sector Lobbies to Stay Out of the Paris Agreement," 2017, https://influencemap.org/report/Corporate-capture-of-the-IMO-902bf81c05a0591c551f965020623fda.

77. Ibid.

78. Ibid.

79. H. N. Psaraftis, "Decarbonization of Maritime Transport: To Be or Not to Be?," *Maritime Economics & Logistics* 21, no. 3 (2019): 353–371.

80. Ibid.

81. R. Baumler, M. C. Arce, and A. Pazaver, "Quantification of Influence and Interest at IMO in Maritime Safety and Human Element Matters," *Marine Policy* 133 (2021): 104746.

82. P. Cariou and L. M. Randrianarisoa, "Stakeholders' Participation at the IMO Marine Environmental Protection Committee," *Marine Policy* 149 (2023): 105506.

83. IATA, "Offsetting CO2 Emissions with CORSIA," 2024, https://www.iata.org/en/programs/environment/corsia/.

84. Influence Map, "Corporate Capture and the UN International Civil Aviation Organization," 2022, https://influencemap.org/report/Corporate-Captureof-the-UN-International-Civil-Aviation-Organization-19779.

85. P. Newell and M. Paterson, "A Climate for Business: Global Warming, the State and Capital," *Review of International Political Economy* 5, no. 4 (1998): 679–703.

86. B. Van Apeldoorn, "Transnational Class Agency and European Governance: The Case of the European Round Table of Industrialists," *New Political Economy* 5, no. 2 (2000): 157–181.

87. J. A. Basseches, "Who Pays for Environmental Policy? Business Power and the Design of State-Level Climate Policies," *Politics & Society* 52, no. 3 (2023): 00323292231195184, https://doi:10.1177/00323292231195184.

88. Stokes, *Short Circuiting Policy*.

89. Influence Map, "The Automotive Sector and Climate Change."

90. Influence Map, "European Automotive Suppliers & EU Climate Policy," 2022, https://influencemap.org/report/European-Automotive-Suppliers-EU-Climate-Policy-17388.

91. Schwedes et al., "Lobbying im Verkehr."

92. Haas and Sander, "The European Car Lobby."

93. Influence Map, "The Automotive Sector and Climate Change."

94. Richter and Haas, "Greening the Car?"

95. Influence Map, "German Automakers And Climate Policy: An InfluenceMap Report," 2021, https://influencemap.org/report/German-Automakers-And-Climate-Policy-a3edf15c64b2e258c29f83beb93337f6.

96. Influence Map, "European Automotive Suppliers & EU Climate Policy."

97. Influence Map, "The Automotive Sector and Climate Change."

98. Haas, "Cracks in the Gearbox of Car Hegemony."

99. Keil and Kreinin, "Slowing the Treadmill for a Good Life for All?"

100. A. K. Keil and J. K. Steinberger, "Cars, Capitalism and Ecological Crises: Understanding Systemic Barriers to a Sustainability Transition in the German Car Industry, *New Political Economy* 29, no. 1 (2023): 1–21.

101. Pichler et al., "Beyond the Jobs-Versus-Environment Dilemma?"

102. Allan and Robinson, "Working Towards a Green Job?"

103. Richter and Haas, "Greening the Car?"

104. Basseches, "Who Pays for Environmental Policy?"

105. Influence Map, "Corporate Capture of the International Maritime Organization."

106. Williams et al., "The American Electric Utility Industry's Role in Promoting Climate Denial."

107. Brulle, "Advocating Inaction."

108. B. Franta, "Weaponizing Economics: Big Oil, Economic Consultants, and Climate Policy Delay," *Environmental Politics* 31, no. 4 (2022): 555–575.

109. G. Trencher, A. Rinscheid, M. Duygan, N. Truong, and J. Asuka, "Revisiting Carbon Lock-In in Energy Systems: Explaining the Perpetuation of Coal Power in Japan," *Energy Research & Social Science* 69 (2020): 101770, https://doi.org/10.1016/j.erss.2020.101770.

110. W. Grant, "Pressure Politics: The Changing World of Pressure Groups," *Parliamentary Affairs* 57, no. 2 (2004): 408–419.

111. A. Lucas, "Investigating Networks of Corporate Influence on Government Decision-Making: The Case of Australia's Climate Change and Energy Policies," *Energy Research & Social Science* 81 (2021): 102271, https://doi.org/10.1016/j.erss.2021.102271.

112. F. Harvey, D. Carrington, and A. Morton, "Second Cop26 Draft Text: Coal Phaseout Remains in But Some Language Softened," *The Guardian*, November 2021, https://www.theguardian.com/environment/2021/nov/12/second-cop26-draft-criticisedfor-weakened-language-on-fossil-fuels.

113. C. Farand, "Polish Coal Company Announced as First Sponsor of UN Climate Talks in Katowice," *DeSmog*, November 27, 2018, www.desmogblog.com/2018/11/27/polishcoalcompany-announced-first-sponsor-un-climate-talks-katowice.

114. E. L. Williams, S. A. Bartone, E. K. Swanson, and L. C. Stokes, "The American Electric Utility Industry's Role in Promoting Climate Denial, Doubt, and Delay," *Environmental Research Letters* 17, no. 9 (2022), DOI 10.1088/1748-9326/ac8ab3.

115. Influence Map, "The Automotive Sector and Climate Change."

116. Influence Map, "European Automotive Suppliers & EU Climate Policy."

117. Influence Map, "German Automakers and Climate Policy."

118. J. Jackson, "(Re)coordinating the German Political Economy: E-mobility and the Verkehrswende," *German Politics* 33, no. 4 (2023): 1–23.

119. I. Richter and T. Haas, "Greening the Car? Conflict Dynamics within the German Platform for Electric Mobility," *Sustainability* 12, no. 19 (2020): 8043.

120. Influence Map, "The Automotive Sector and Climate Change."

121. Influence Map, "The FCAI and Australian Climate Policy: The Automotive Industry's Playbook to Weaken Australian Fuel Economy (CO2) Standards," 2023, https://influencemap.org/briefing/The-FCAI-and-Australian-Climate-Policy-22253#:~:text=Set%20a%20weak%20emissions%20reduction,SUVs%20and%20light%20commercial%20vehicles.

122. Corporate Europe Observatory, "Corporate Interests Continue to Dominate Key Expert Groups, New Rules, Little Progress," 2017, https://corporateeurope.org/en/expertgroups/2017/02/corporate-interests-continue-dominate-key-expert-groups.

123. J. Gieseke and G. Gerbrandy, "Report on the Inquiry Into Emission Measurements in the Automotive Sector (2016/2215(INI)) by the Committee of Inquiry into Emission Measurements in the Automotive Sector," European Parliament, 2017, https://www.europarl.europa.eu/doceo/document/A-8-2017-0049_EN.pdf.

124. J. Earsom, "It's Not as Simple as Copy/Paste: The EU's Remobilisation of the High Ambition Coalition in International Climate Governance," *International Environmental Agreements* 23 (2023): 27–42.

125. Influence Map, "Airlines and European Climate Policy," 2022, https://influencemap.org/report/Airlines-and-European-Climate-Policy-19388.

126. C. Hendriksen, "Inside the Blue Box: Explaining Industry Influence in the International Maritime Organization" (PhD diss., Copenhagen Business School, 2020).

127. C. Downie, "Fighting for King Coal's Crown: Business Actors in the US Coal and Utility Industries," *Global Environmental Politics* 17, no. 1 (2017): 21–39.

128. Basseches, "Who Pays for Environmental Policy?"

129. Stokes, *Short Circuiting Policy*.

130. Brulle and Downie, "Following the Money."

131. Basseches, "Who Pays for Environmental Policy?"

132. Stokes, *Short Circuiting Policy*.

133. S. Goulding-Carroll, "Lawmakers Back EU-Wide Ban on New Fossil Fuel Cars from 2035, Despite Strong Lobbying," Euractiv, 2022 https://www.euractiv.com/section/transport/news/lawmakers-back-eu-wide-ban-on-newfossil-fuel-cars-from-2035-despite-strong-lobbying/.

134. Haas and Sander, "The European Car Lobby."

135. Richter and Haas, "Greening the Car? "

136. O. Schwedes, B. Sternkopf, and F. Nowak, "Lobbying im Verkehr—ein vernachlässigtes Praxisfeld," *Umweltpsychologie* 19 (2015): 146–168.

137. Haas and Sander, "The European Car Lobby."

138. J. Heern, "Who's Controlling Our Energy Future? Industry and Environmental Representation on United States Public Utility Commissions," *Energy Research & Social Science* 101 (2023): 103091, https://doi.org/10.1016/j.erss.2023.103091.

139. E. Moe, "Vested Interests, Energy Efficiency and Renewables in Japan," *Energy Policy* 40 (2012): 260–273.

140. Lucas, "Investigating Networks of Corporate Influence on Government Decision-Making."

141. Baer, "The Nexus of the Coal Industry and the State in Australia."

142. C. Wright, D. Nyberg, and V. Bowden, "Beyond the Discourse of Denial: The Reproduction of Fossil Fuel Hegemony in Australia," *Energy Research & Social Science* 77 (2021): 102094, https://doi.org/10.1016/j.erss.2021.102094.

143. Wright et al., "We're in the Coal Business."

144. S. Wheaton, "Driving Through the Revolving Door Before Emissions Vote," *Politico*, 2023 https://www.politico.eu/newsletter/politico-euinfluence/driving-through-the-revolving-door-before-emissions-vote-2/.

145. Corporate Europe Observatory, "Two Years After Dieselgate."

146. Jackson, "(Re)coordinating the German Political Economy."

147. J. Posaner, F. Müller, and G. Hervey, "Too Big to Fail: Why Germany's Autostate Keeps Winning," *Politico*, 2017, https://www.politico.eu/interactive/germany-autostate-winning-too-big-to-fail/.

148. Greenpeace, "Schwarzbuch Autolobby," 2016, https://www.greenpeace.de/sites/default/files/publications/s01841_web_greenpeace_schwarzbuch_autolobby_04_16.pdf.

149. Heern, "Who's Controlling Our Energy Future?"

150. Moe, "Vested Interests, Energy Efficiency and Renewables in Japan."

151. S. Roth, "The Revolving Door at Public Utilities Commissions? It's Alive and Well," *Los Angeles Times*, June 8, 2023, https://www.latimes.com/environment/newsletter/2023-06-08/the-revolving-door-at-public-utilities-commissions-its-alive-and-well-boiling-point.

152. Influence Map, "Corporate Capture and the UN International Civil Aviation Organization."

153. International Center for Not-for-Profit Law and Law, E.C.f.N.-f.-P, "Closing Civic Space for Climate Activists," 2020, https://www.icnl.org/wpcontent/uploads/Climate-Change-and-Civic-Space-Briefer-vf.pdf.

154. Ibid.

155. Anugrah, "Land Control, Coal Resource Exploitation and Democratic Decline in Indonesia."

156. L. L. Delina, "Topographies of Coal Mining Dissent: Power, Politics, and Protests in Southern Philippines," *World Development* 137 (2021): 105194, https://doi.org/10.1016/j.worlddev.2020.105194.

157. N. Healy, J. C. Stephens, and S. A. Malin, "Embodied Energy Injustices: Unveiling and Politicizing the Transboundary Harms of Fossil Fuel Extractivism and Fossil Fuel Supply Chains," *Energy Research & Social Science* 48 (2019): 219–234.

158. A. M. Mudhoffir, "The Limits of Civil Society Activism in Indonesia: The Case of the Weakening of the KPK," *Critical Asian Studies* 55, no. 1 (2023): 62–82.

159. X. Gao, M. Davidson, J. Busby, C. Shearer, and J. Eisenman, "The Challenges of Coal Phaseout: Coal Plant Development and Foreign Finance in Indonesia and Vietnam," *Global Environmental Politics* 21, no. 4 (2021): 110–133.

160. Healy et al., "Embodied Energy Injustices."

161. K. Tienhaara, R. Thrasher, B. A. Simmons, and K. P. Gallagher, "Investor-State Disputes Threaten the Global Green Energy Transition," *Science* 376, no. 6594 (2022): 701–703.

162. Wesseling et al., "Car Manufacturers' Changing Political Strategies on the ZEV Mandate."

163. Influence Map, "US Heavy-Duty Transport & Climate Change: How Heavy-Duty Manufacturers Have Lobbied to Weaken US Climate Policy While Publicly Promoting a Zero-Emission Vehicle Transition," 2022, https://influencemap.org/report/US-Heavy-Duty-Transport-Climate-Change-20434.

164. C. Davenport, "Court Gives Obama a Climate Change Win," *New York Times*, June 10, 2015, https://www.nytimes.com/2015/06/10/us/coal-epa-clean-power-plan.html.

165. B. Magill, "The Suit Against the Clean Power Plan, Explained," Climate Central, 2016, https://www.climatecentral.org/news/the-suit-against-the-clean-powerplan-explained-20234.

166. C. Strambo, A. C. González Espinosa, A. J. Puertas Velasco, and L. M. Mateus Molano, "Contention Strikes Back? The Discursive, Instrumental and Institutional Tactics Implemented by Coal Sector Incumbents in Colombia," *Energy Research & Social Science* 59 (2020): 101280, https://doi.org/10.1016/j.erss.2019.101280.

167. G. Curran, "Coal, Climate and Change: The Narrative Drivers of Australia's Coal Economy," *Energy Research & Social Science* 74 (2021): 101955, https://doi.org/10.1016/j.erss.2021.101955.

168. I. Richter and K. Smith Stegen, "A Choreography of Delay: The Response of German Auto Incumbents to Environmental Policy," *Environmental Innovation and Societal Transitions* 45 (2022): 1–13.

169. A. K. Keil and H. Kreinin, "Slowing the Treadmill for a Good Life for All? German Trade Union Narratives and Social-Ecological Transformation," *Journal of Industrial Relations* 64, no. 4 (2022): 564–584.

170. Trencher et al., "Discursive Resistance to Phasing Out Coal-Fired Electricity."

171. G. Trencher, Y. Okubo, and A. Mori, "Phasing Out Carbon Not Coal? Identifying Coal Lock-In Sources in Japan's Power Utilities," *Climate Policy* 24, no. 6 (2024): 766–784.

172. G. Supran and N. Oreskes, "Rhetoric and Frame Analysis of ExxonMobil's Climate Change Communications," *One Earth* 4, no. 5 (2021): 696–719.

173. D. Anderson, M. Kasper, and D. Pomerantz, "Utilities Knew: Documenting Electric Utilities' Early Knowledge and Ongoing Deception on Climate Change From 1968–2017," *Energy and Policy*, 2017, https://energyandpolicy.org/utilities-knew-about-climatechange/.

174. Williams et al., "The American Electric Utility Industry's Role in Promoting Climate Denial."

175. Anderson et al., "Utilities Knew."

176. B. Miller Gaither and T. K. Gaither, "Marketplace Advocacy by the U.S. Fossil Fuel Industries: Issues of Representation and Environmental Discourse," *Mass Communication and Society* 19, no. 5 (2016): 585–603.

177. Dunlap and McCright, "Organized Climate Change Denial."

178. K. Wada and M. Inoue, "Japan and the Global Transition to Zero Emission Vehicles, The Climate Group, 2022, https://www.theclimategroup.org/ourwork/resources/japan-and-global-transition-zero-emission-vehicles.

179. Toyota, "Strength in Diversity: Scientist Explains Japan's Unique Road to Decarbonization," 2023, *Toyota Times*, https://toyotatimes.jp/en/toyota_news/1024_1.html#anchorTitles.

180. J. Schneider, S. Schwarze, P. Bsumek, and J. Peeples, *Under Pressure: Coal Industry Rhetoric and Neoliberalism* (Palgrave Macmillan, 2016).

181. Ibid.

182. R. Scott and E. Bennett, "Branding Resources: Extractive Communities, Industrial Brandscapes and Themed Environments," *Work, Employment and Society* 29, no. 2 (2015): 278–294.

183. J. Prno and D. Scott Slocombe, "Exploring the Origins of 'Social License to Operate' in the Mining Sector: Perspectives from Governance and Sustainability Theories," *Resources Policy* 37, no. 3 (2012): 346–357.

184. K. Deonandan, J. Schoenfeld, A. Salim, and M. Bourassa, "Social License to Operate (SLO): Private Governance and Barriers to Community Engagement," *The Extractive Industries and Society* 17 (2024): 101404, https://doi.org/10.1016/j.exis.2024.101404.

185. S. E. Bell and R. York, "Community Economic Identity: The Coal Industry and Ideology Construction in West Virginia," *Rural Sociology* 75, no. 1 (2010): 111–143.

186. G. Mattioli, C. Roberts, J. K. Steinberger, and A. Brown, "The Political Economy of Car Dependence: A Systems of Provision Approach," *Energy Research & Social Science* 66 (2020):101486.

187. S. Luger, *Corporate Power, American Democracy, and the Automobile Industry* (Cambridge University Press, 2000).

188. T. Haas, "Cracks in the Gearbox of Car Hegemony: Struggles Over the German Verkehrswende Between Stability and Change," *Mobilities* 15, no. 6 (2020): 810–827.

189. Haas and Sander, "The European Car Lobby."

190. J. Meckling and J. Nahm, "When Do States Disrupt Industries? Electric Cars and the Politics of Innovation," *Review of International Political Economy* 25, no. 4 (2018): 505–529.

191. J. Herberg, T. Haas, D. Oppold, and D. von Schneidemesser, D., "A Collaborative Transformation beyond Coal and Cars? Co-Creation and Corporatism in the German Energy and Mobility Transitions," *Sustainability* 12, no. 8 (2020): 3278.

192. Jackson, "(Re)coordinating the German Political Economy."

193. M. Pichler, N. Krenmayr, D. Maneka, U. Brand, H. Högelsberger, and M. Wissen, "Beyond the Jobs-Versus-Environment Dilemma? Contested Social-Ecological Transformations in the Automotive Industry," *Energy Research & Social Science* 79 (2021): 102180.

194. J. Lu and G., Nemet., "Market-Led Decline amidst Intense Politicization: Coal in the United States," in *The Political Economy of Coal: Obstacles to Clean Energy Transitions*, ed. M. Jakob and J. Steckel (Routledge, 2022), 96–113.

195. A. M. Galli Robertson, "Privileged Accounts in the Debate Over Coal-Fired Power in the United States," *Society & Natural Resources* 34, no. 2 (2021): 188–207.

196. M. Duygan, K. Aya, P. Temocin, and G. Trencher, "A Tale of Two Coal Regimes: An Actororiented Analysis of Destabilisation and Maintenance of Coal Regimes in Germany and Japan," *Energy Research and Social Science* 105 (2023), https://105.doi:10.1016/j.erss.2023.103297.

197. T. Spencer, M. Colombier, O. Sartor, A. Garg, V. Tiwari, J. Burton, T. Caetano, F. Green, F. Teng, and J. Wiseman, "The 1.5°C Target and Coal Sector Transition: At the Limits of Societal Feasibility," *Climate Policy* 18, no. 3 (2018): 335–351.

198. Haas and Sander, "The European Car Lobby."

199. Mattioli et al., "The Political Economy of Car Dependence."

200. K. Allan and J. Robinson, "Working Towards a Green Job? Autoworkers, Climate Change and the Role of Collective Identity in Union Renewal," *Journal of Industrial Relations* 64, no. 4 (2022): 585–607..

201. Pichler et al., "Beyond the Jobs-Versus-Environment Dilemma?"

202. Richter and Haas, "Greening the Car? "

203. Richter and Smith Stegen, "A Choreography of Delay."

204. Influence Map, "European Automotive Suppliers & EU Climate Policy."

205. Mattioli et al., "The Political Economy of Car Dependence."

206. Influence Map, "European Automotive Suppliers & EU Climate Policy."

207. Influence Map, "Japanese Industry Groups and Climate Policy," 2020, https:// influencemap.org/presentation/Japanese-Industry-Groups-and-Climate-Policy- 899704d005cb96359cc5b5e2a9b18a84.

208. S. Capstick, F. Creutzig, T. Culhane, W. F. Lamb, S. Levi, G. Mattioli, J. C. Minx, F. Müller-Hansen, J. R. Roberts, S. Capstick, F. Creutzig, J. C. Minx, F. Müller- Hansen, T. Culhane, and J. K. Steinberger, "Discourses of Climate Delay," *Global Sustainability* 3 (2020): e17.

209. Trencher et al., "Phasing Out Carbon Not Coal? "

210. Trencher et al., "Discursive Resistance to Phasing Out Coal-Fired Electricity."

211. I. Kikuma, M. Annex, T. Brandily, D. Kang, and M. Tengler, "Japan's Costly Ammonia Coal Co-Firing Strategy," BNEF, 2022, https://about.bnef.com/blog/ japans-ammonia-coal-co-firing-strategy-a-costly-approach-todecarbonization- renewables-present-more-economic-alternative/.

212. Schneider et al., *Under Pressure.*

213. Williams et al., "The American Electric Utility Industry's Role in Promoting Climate Denial."

214. Trencher et al., "Phasing Out Carbon Not Coal?"

215. P. Gilbert and A. Bows, "Exploring the Scope for Complementary Sub-Global Policy to Mitigate CO2 from Shipping," *Energy Policy* 50 (2012): 613–622.

216. Psaraftis, "Decarbonization of Maritime Transport."

217. Steckel and Jakob, "To End Coal, Adapt to Regional Realities."

218. N. Ohlendorf, M. Jakob, and J. C. Steckel, "The Political Economy of Coal Phase- Out: Exploring the Actors, Objectives, and Contextual Factors Shaping Policies in Eight Major Coal Countries," *Energy Research & Social Science* 90 (2022): 102590, https://doi.org/10.1016/j.erss.2022.102590.

219. World Coal Association, "Coal's Contribution," 2023, https://www.worldcoal.org/ coal-facts/coals-contribution/.

220. Gaworecki, "Peabody Energy Goes on Offense with New PR Campaign."

221. Curran, "Coal, Climate and Change."

222. A. Lucas, "Fossil Networks and Dirty Power: The Politics of Decarbonisation in Australia," in *Handbook of Anti-Environmentalism*, ed. D. Tindall, M. Stoddart, and R. E. Dunlap (Edward Elgar Publishing, 2022), 192–215.

223. Wright et al., "We're in the Coal Business."

224. Wright et al., "Beyond the Discourse of Denial."

225. Royal Automobile Club Foundation, "21 Million UK Households in Transport Poverty," 2012, https://www.racfoundation.org/media-centre/transport- poverty.

226. G. Mattioli, Z. Wadud, and K. Lucas, "Vulnerability to Fuel Price Increases in the UK: A Household Level Analysis: Transportation Research Part A," *Policy and Practice* 113 (2018): 227–242.

227. Goulding-Carroll, "Lawmakers Back EU-Wide Ban on New Fossil Fuel Cars from 2035."

228. J. Monios and G. Wilmsmeier, "Maritime Governance after COVID-19: How Responses to Market Developments and Environmental Challenges Lead Towards Degrowth," *Maritime Economics & Logistics* 24, no. 4 (2022): 699–722.

229. N. Ferris, "Don't Write Off the Just Energy Transition Partnership Just Yet," Energy Monitor, 2023, https://www.energymonitor.ai/just-transition/dont-write-off-thejetp-just-energy-transition-partnership-just-yet/?cf-view.

230. C. Downie, *Business Battles in the U.S. Energy Sector* (Routledge, 2019).

231. Basseches, "Who Pays for Environmental Policy? "

232. Downie, "Fighting for King Coal's Crown."

233. J. Eun, M. Lee, and Y. H. Jung, "Green Product Portfolio and Environmental Lobbying," *Business and Politics* 25, no. 3 (2023): 195–214.

234. Influence Map, "European Automotive Suppliers & EU Climate Policy."

235. Influence Map, "Airlines and European Climate Policy."

236. Toenshoff, "Hiding in the Crowd."

237. Wesseling et al., "Car Manufacturers' Changing Political Strategies on the ZEV Mandate."

238. I. Vormedal and J. Meckling, "How Foes Become Allies: The Shifting Role of Business in Climate Politics," *Policy Sciences* 37 (2023): 101–124, https://doi:10.1007/s11077-023-09517-2.

239. Ibid.

240. Allan and Robinson, "Working Towards a Green Job?"

241. Ibid.

242. Richter and Haas, "Greening the Car?"

243. K. Hochstetler, Political Economies of Energy Transition: Wind and Solar Power in Brazil and South Africa (Cambridge University Press, 2020).

244. Pichler et al., "Beyond the Jobs-Versus-Environment Dilemma?"

245. Wesseling et al., "Car Manufacturers' Changing Political Strategies on the ZEV Mandate."

246. Corporate Europe Observatory, "Two Years After Dieselgate."

247. Influence Map, "Japanese Industry Groups and Climate Policy."

248. Corporate Europe Observatory, "Corporate Interests Continue to Dominate Key Expert Groups."

249. Influence Map, "Automakers and Climate Policy Advocacy: A Global Analysis," 2024, https://influencemap.org/report/Automakers-and-Climate-Policy-Advocacy-A-Global-Analysis-27906.

250. Toenshoff, "Hiding in the Crowd."

251. Brulle, "Networks of Opposition."

252. L. Montrone, N. Ohlendorf, and R. Chandra, "The Political Economy of Coal in India: Evidence from Expert Interviews," *Energy for Sustainable Development* 61 (2021): 230–240.

253. Gao et al., "The Challenges of Coal Phaseout."

254. M. R. Davidson, X. Gao, J. Busby, C. Shearer, and J. Eisenman, "Hard to Say Goodbye: South Korea, Japan, and China as the Last Lenders for Coal," *Environmental Politics* 32, no. 7 (2023): 1–22.

255. Gallagher et al., "Banking on Coal?"

256. IEA, "Net Zero by 2050: A Roadmap for the Global Energy Sector," 2021, https://www.iea.org/reports/net-zero-by-2050.

CHAPTER 4

The Animal Agriculture Industry's Role in Obstructing Climate Action

LEAD AUTHORS: KATHRIN LAUBER AND VIVECA MORRIS

CONTRIBUTING AUTHORS: JENNIFER JACQUET, PETER LI, INA MÖLLER, SILVIA SECCHI, ALEX WIJERATNA, AND MELINA DE BONA

INTRODUCTION: LIVESTOCK AND THE CLIMATE CRISIS

Collectively, the greenhouse gas (GHG) emissions of the fifteen largest meat and dairy companies exceed those of some carbon majors.[1] Several leading meat and dairy companies have made public commitments to becoming net zero in recent years, including Danone, Danish Crown, Nestlé, Tyson Foods, and JBS, the world's largest meat company.[2] Yet it is unclear how these companies will achieve their climate promises, and the lack of comprehensive data on their emissions and mitigation strategies impedes accountability and independent verification.[3] While the sector is highly heterogenous, ranging from small-scale famers and producers to transnational firms with integrated supply chains, it has become clear that many leading animal agribusiness actors in particular have contributed to efforts to influence the public's understanding of the livestock sector's role in the climate crisis and obstruct policy responses that threaten their ability to profit from business as usual.

Addressing the climate impacts of animal agriculture is critical to prevent catastrophic global heating. Even if fossil fuel use ended immediately, emissions related to food production alone are on course to push global warming beyond 1.5°C above preindustrial levels between 2051 to 2063.[4] Given the limited mitigation potential of technological measures alone, reducing the production and consumption of animal-based foods in high-consuming

Kathrin Lauber et al., *The Animal Agriculture Industry's Role in Obstructing Climate Action*. In: *Climate Obstruction*. Edited by: J. Timmons Roberts, Carlos R. S. Milani, Jennifer Jacquet, and Christian Downie, Oxford University Press. © Oxford University Press (2025). DOI: 10.1093/oso/9780197787144.003.0004 societies is necessary to meaningfully lower emissions.[5] Shifts toward more plant-based diets could contribute a significant proportion of the mitigation required to limit global warming to 2°C while maintaining food security and cobenefitting public health and biodiversity.[6]

Animal-based foods are responsible for an estimated 57% of food production emissions.[7] In 2006, the United Nations (UN) Food and Agriculture Organization (FAO) published *Livestock's Long Shadow*, the first global estimate of animal agriculture's contributions to anthropogenic climate change.[8] Since then, a multitude of studies have documented the significance of the sector's pollution,[9] estimating that it contributes between 11.2% and 19.6% of total global GHG emissions.[10] In the report, the FAO contended that the livestock sector "has such deep and wide-ranging environmental impacts that it should rank as one of the leading focuses for environmental policies" and warned that these environmental harms would worsen without "major corrective" measures.[11] Yet nearly two decades later, measures to effectively reduce livestock-related emissions are rarely at the forefront of climate policy. Animal agribusinesses operate within what the nonprofit Institute for Agriculture and Trade Policy (IATP) has termed an "accountability vacuum."[12] As documented in this chapter, available evidence suggests that industry actors have played key roles in keeping it that way.

Emissions related to animal-based food production derive from multiple sources. Cropland and grazing land that produce livestock feed are responsible for about 21% of the world's food production-based GHG emissions.[13] These emissions result from management activities, such as plowing fields (which reduces carbon storage), applying nitrogen fertilizer (nitrogen not taken up by crops runs off into waterways and gets broken down by microbes in the soil, releasing nitrogen oxide), and burning fossil fuel to run farm equipment.[14] Enteric fermentation from ruminant animals such as cows and sheep is also a leading source, representing about a fifth of total food production emissions, while manure management contributes about 2%.[15] Land-use changes that cause soil disturbance and biomass loss contribute an additional 12% of total food-production emissions.[16] Overall, livestock are estimated to emit about one-third of all human-caused methane emissions[17] and around half of human-caused nitrous oxide emissions.[18] The extensive acreage required to produce livestock and feed—which accounts for more than 80% of all land used for agriculture[19]—also incurs a significant and often uncounted "carbon opportunity cost" given the potential carbon that could be sequestered if ecosystems were restored on land used for livestock production.[20]

The livestock sector's emissions of methane, a highly potent but short-lived GHG, are particularly significant. According to the nonprofits IATP and Changing Markets Foundation, the fifteen largest meat and dairy corporations combined emit roughly 12.8 million tons (MT) of the GHG, representing around

3.4% of global anthropogenic methane emissions and 11.1% of all livestock-related methane.[21] A growing number of studies indicate that shifting to diets with fewer animal products would significantly reduce GHG emissions.[22] Similarly, the Intergovernmental Panel on Climate Change (IPCC)'s 2019 special report on climate change and land concluded that healthy and sustainable diets, underpinned by a focus on just agricultural transitions, present "major opportunities for reducing GHG emissions from food systems."[23]

Of the world's nations, China is the number one emitter of GHGs from meat and dairy production,[24] responsible for about 14% of global animal-based food production GHG emissions,[25] followed by Brazil (11%), the United States (8%), and India (7%).[26] Brazil, the United States, the European Union, Argentina, Canada, Australia, and New Zealand are all major meat and dairy exporters with high per capita meat and dairy consumption.[27] As a group, China and these so-called surplus protein exporters are responsible for the majority of global emissions from meat and dairy production[28] and host the headquarters of most of the world's largest animal agribusinesses.[29] Meat, dairy, and feed production are characterized by growing corporate concentration.[30] Globally, the top four agrochemical and animal pharmaceutical firms are estimated to receive 66% and 58% of sector revenues, respectively.[31] In the United States, 80% of soybean processing, 73% of beef processing, 67% of pork processing, and 54% of chicken processing are controlled by the top four firms.[32] Brazil-based JBS SA is the world's largest animal protein company by a large margin, with a daily slaughter capacity of 16,000 lambs, 75,000 cattle, 115,000 pigs, and 14 million poultry birds.[33] As described in this chapter, many of these companies have successfully leveraged their outsized market power for out-sized political power, which they have used to maintain their social license to operate under business as usual.

Key to maintaining this license has been the industry's promise to reduce emissions, mainly by mitigating energy-related emissions through the use of renewables[34] (similar to oil and gas producers; see Chapter 2) or through technical interventions in production such as anaerobic digesters on manure lagoons to reduce waste, feed additives to decrease methane emitted by rumi-nant animals, and targeted measures aimed at improving animal productiv-ity.[35] Global meat and dairy consumption is, however, projected to rise by 14% and 20%, respectively, by 2030 (compared with 2018–2020 averages)[36] and methane emissions are predicted to reach more than 140 MT annually by 2050 under current policies.[37] Although estimates for the mitigation poten-tial of technical interventions vary significantly across studies—ranging from 4 MT to 42 MT per year[38]—it is evident that such measures are not sufficient to offset projected emissions increases.[39] Furthermore, while meat and dairy companies speak frequently about their interest in using methane-reducing feed additives, *Bloomberg* reported in 2023 that they are not following through on using available products at any significant scale.[40]

In this chapter, we examine the role of obstruction in creating and maintaining this unsustainable status quo. We aim to provide an overview of the evidence on the recent history of climate obstruction related to animal agriculture, followed by a synthesis of the narratives and practices constituting contemporary obstruction within the animal agriculture sector. We conclude the chapter with possible efforts to counteract obstruction and a research agenda.

A MODERN HISTORY OF ANIMAL AGRICULTURE CLIMATE OBSTRUCTION IN TOP-EMITTING REGIONS

The production and consumption of animal-based foods is shaped by a wide range of policies, from subsidies to crop insurance, national dietary guidelines, procurement, trade policy, climate policy, environmental regulations, and more. The net result of these policies is that the prices of industrially produced meat and dairy products in much of the world are artificially low, with their true costs externalized in the form of health and environmental harms.

Policy action at the scale needed to address livestock's impact on climate change remains rare. Efforts to date focus largely on tweaks around the edges of our current industrial food system and we have yet to see sustained policy action focused on transforming diets, setting binding GHG reduction targets for the agriculture sector, requiring comprehensive disclosure of emissions, regulating pollutants from industrial livestock operations, or transforming subsidy programs to cease incentivizing unsustainable levels of meat and dairy production. For example, neither recent US nor EU plans to address methane emissions include direct measures to regulate animal agriculture emissions, a major source of methane in both regions.[41] The Global Methane Pledge, spearheaded by the United States and European Union, commits to "achieve all feasible reductions in the energy and waste sectors" yet for agriculture, merely seeks to mitigate "emissions through technology innovation as well as incentives and partnerships with farmers."[42] An analysis of major US and EU policies from 2014 to 2020 found that public funding for animal-based farming exceeded $44 billion.[43] The few countries that have sought to address livestock emissions through policies such as taxes, the removal of environmentally harmful subsidies, binding emissions-reduction targets, and a shift toward more sustainable diets have met pushback and seen their policies weakened as a result. For example, in 2022, New Zealand—where agriculture accounts for almost half of the country's total emissions—proposed the world's first tax on cow emissions.[44] Following backlash from industry groups, the Labour government initially revised the proposal to keep the levy fixed at a lower rate for five years, but a change of government saw the new center-right coalition scrap the policy altogether before it came into force.[45]

In 2006, the United Nations' FAO published *Livestock's Long Shadow*, a 390-page report that presented the first global estimate of the livestock sector's contributions to climate change and stated the need for measures to hold producers accountable for their environmental damage.

The publication put the livestock industry on the defensive. *Beef Today* described the report as "red meat for the vegetarian activists" and as "UN cover for their pre-existing bias."[46] Industry-funded groups' focus on climate-related messaging appears to have expanded significantly in the years that followed.[47] In 2009, the producer-funded Beef Checkoff program awarded a grant to an academic researcher at the University of California (UC) Davis, Dr. Frank Mitloehner, to assess the FAO's findings.[48] Mitloehner's critique did not focus on the report's empirical evidence, but rather criticized the authors' comparison of the emissions of the livestock sector to those of the transportation sector because the former included a full life-cycle analysis while the latter included only direct emissions.[49] Mitloehner's challenge to *Livestock's Long Shadow* was promoted in press releases, including by UC Davis ("Don't Blame Cows for Climate Change"[50]) and the American Chemical Society ("Eating less meat and dairy products won't have a major impact on global warming"[51]). Popular media outlets reported on Mitloehner's efforts to challenge the UN's findings as if the link between animal agriculture and climate change had been debunked.[52] In the years since, Mitloehner—who, according to his CV, has received more than $5 million in research funding throughout his career from industry groups—has continued to downplay the livestock industry's role in the climate crisis and is quoted regularly by the media as an expert on the climate emissions of livestock.[53]

Following the release of *Livestock's Long Shadow*, meat and dairy corporations and countries with major livestock industries—including Brazil, Argentina, Uruguay, Paraguay, the United States, and Australia—reportedly complained to the FAO.[54] In interviews with *The Guardian*, former FAO officials said they were "censored, sabotaged, undermined and victimized for more than a decade" and that attempts to further illuminate connections between livestock and climate change were discouraged and at times suppressed.[55]

The United States

Animal agribusiness involvement in climate obstruction in the United States dates back to at least the 1990s, when major agriculture industry groups worked hand-in-hand with other highly polluting industries to block policies aimed at reducing emissions. As *Inside Climate News* documented, the American Farm Bureau Federation (AFBF), among the nation's most powerful political lobbying groups, has derailed climate policy in the United States for more than four decades.[56] AFBF is a national tax-exempt nonprofit organization and lobbying group that leads a network of state-level Farm Bureau nonprofit organizations, some of which have affiliated for-profit companies that sell insurance. Many of the organization's 5.9 million members must pay dues as a condition of their insurance policies.[57] AFBF questioned the attribution of extreme weather to anthropogenic climate change as late as 2019,[58] and continued to oppose attempts to regulate or tax GHG emissions in 2023.[59] AFBF also opposes "any tie and/or connection" of climate-focused practices to federal crop insurance programs, as well as efforts to legislate mandatory cap-and-trade provisions and agricultural GHG emissions reporting.[60]

By the mid-1990s, AFBF was an active member of the Global Climate Coalition, an international lobbying group that opposed climate action and contested the science of global warming.[61] The Coalition spearheaded opposition to the 1997 Kyoto Protocol, which committed industrialized nations to limiting and reducing GHG emissions in accordance with set targets.[62] In a 1997 congressional hearing about the Protocol, AFBF's president argued against legally binding GHG emissions caps, stating that the science of climate change is "unclear that we even have a problem."[63]

In 2004, the US Environmental Protection Agency (EPA) launched a $15 million study to collect data to create methodologies to estimate livestock farm emissions, with the purpose of informing regulation of these facilities under federal air-pollution laws.[64] Major livestock industry trade associations funded the study in exchange for an agreement with the EPA, which granted participating farms immunity from civil action by the agency for the duration of the data- collection program, which was supposed to be completed in two years.[65] Nearly 14,000 farms, including 90% of the country's largest livestock farms at the time, received regulatory immunity through this agreement.[66] Ultimately, data was collected at only about two dozen farms, but all continue to receive regulatory immunity.[67] As of April 2023, the EPA had yet to publish any final methodologies resulting from the process.[68] Meanwhile, animal agribusiness groups—including the AFBF, state farm bureaus, the National Pork Producers Council, and the National Cattlemen's Beef Association (NCBA)—have lobbied the country's financial regulatory agency to protect them from new proposed emissions disclosure requirements[69] and backed a federal bill, the Protect

Farmers from the SEC Act, to prohibit the agency from requiring GHG disclosures for agricultural products.[70] Today, animal agriculture's GHG emissions in the United States remain effectively unregulated.

The European Union

With farming and agribusiness representing a powerful political force across Europe, recent EU efforts to shift to more sustainable production and consumption have encountered substantial opposition. A 2023 report by the European Commission's Group of Chief Scientific Advisors, for example, identified "evidence that some meat-industry representative bodies have influenced public discourse in order to counter scientific evidence on the negative impact of meat consumption on health and the climate."[71] The EU's 2020 Farm to Fork strategy promised reforms to align agricultural and food policy with environmental goals, including an acknowledgment that adopting "a more plant-based diet with less red and processed meat . . . will reduce not only risks of life threatening diseases, but also the environmental impact of the food system."[72] Efforts to do so have, however, stalled following opposition not only from some member states, conservative members of the European Parliament, and the agriculture sector, but also the European Commission's Directorate-General for Agriculture and Rural Development (DG AGRI).[73]

DG AGRI is the Directorate-General responsible for the EU Common Agricultural Policy (CAP), initiated in 1957 to ensure secure food supplies and support the livelihoods of European farmers. CAP supports emissions-intensive livestock farming and research has estimated that 82% of EU agricultural subsidies in 2013 were used to produce animal-based foods or animal feed.[74] Despite many reforms, scholars agree that the integration of environmental and climate concerns into the CAP has remained limited ever since. This inertia is due partly to a persisting discourse of agricultural exceptionalism, as well as the extensive access agricultural producers and traders have to decision-makers in the European Commission and Parliament.[75] In line with the primary focus of its mandate on ensuring agricultural production and food security, DG AGRI historically has been closely aligned with agricultural interests.

Copa-Cogeca, among the oldest and most-established EU lobby groups, formed soon after CAP in 1962, through the merger of two groups representing farmers and agri-cooperatives.[76] Together with other sectoral organizations such as the Liaison Centre for the Meat Processing Industry in the European Union and the European Feed Manufacturers Federation, they now form a coalition that responds specifically to growing concerns about Europe's high meat and dairy consumption. Launched in 2019 by eleven industry groups,

European Livestock Voice has been focused on defending the reputation of European meat as environmentally friendly and nutritionally essential.[77] Its *Meat the Facts* campaign website serves as a hub for the dissemination of news, opinion pieces, and resources in support of this goal.[78]

Despite superficial changes to reorient the CAP toward climate and environmental goals, for example through direct financial incentives to farmers,[79] meaningful action to address animal agriculture emissions remains limited. The EU's continued investment in promoting meat and dairy products has been criticized as standing at odds with its rhetoric on sustainable diets.[80] More than half of the 2016–2020 marketing funding under the CAP agricultural promotion program supported marketing for meat and dairy products, according to an analysis by Greenpeace Europe.[81] The Greenpeace report was met with claims that restricting funding for meat and dairy advertising would in fact harm EU environmental goals, as illustrated by European Livestock Voice's claim that it "feeds a kind of populism against the work put in place by the Commission and EU farmers to constantly improve the sustainability of EU agriculture."[82] A proposal to reform the promotion policy, due in early 2022, was blocked, according to an internal DG AGRI memo,[83] exemplifying tensions resulting from growing pressure on the Directorate to enact EU environmental commitments.

Within EU climate policy, agricultural GHG emissions also remain a major blind spot.[84] Provided with an opportunity to strengthen rules on emissions reporting and reduction through the 2024 revision of the EU Industrial Emissions Directive, EU decision-makers shied away from comprehensive action on livestock farming. While the legislative draft proposed to extend the Directive's scope to large cattle farms and lower inclusion thresholds for other livestock farms, the final Directive, negotiated amidst intensive lobbying and pressure from conservative political groups,[85] excludes cattle (pending a 2026 review) and weakens the proposed threshold for inclusion of pig and poultry farms.[86]

China

Over the past four decades, China's meat consumption and production have grown dramatically alongside increasing urbanization and rising income levels.[87] Today, China is not only the world's largest consumer meat market[88] but also the largest livestock producer[89] and importer of animal feed.[90] While the Chinese consume 27% of the world's meat, per capita intake—excluding fish—remains approximately half that of Americans.[91] Chinese demand for animal-based food and animal feed creates significant extraterritorial impacts. In New Zealand, for example, 15% of nitrogen and irrigation water use is attributed to feed used for livestock products exported to China.[92] Similarly, China was the largest importer of Brazilian soy—a major animal feed crop—and responsible for 51% of CO_2 emissions embodied in the country's soy exports between 2010 and 2015.[93]

This enormous growth in production and consumption of animal products is conditioned by the country's "industrial meat regime" that emerged as part of China's late twentieth-century economic reforms: "a strategically managed set of policies, discourses, relations, and resources enacted with the goal of increasing commodity meat production, 'modern' forms of meat consumption, and agribusiness profits."[94] Food security is high on the Chinese government's agenda out of a concern for regime stability, and meat continues to symbolize the nation's progress against hunger.[95]

Chinese domestic meat production is concentrated in a small number of corporations, many of which are supported by the state.[96] These include WH Group Ltd., the world's largest producer of pork since its purchase of US-based Smithfield Foods in 2013, dairy giants China Mengniu Dairy Company Limited, Yili Group, and New Hope Liuhe Co., a leading producer of animal feed, swine, and poultry. As of 2021, five Chinese companies were listed in the top ten feed companies in the world ranked by volume.[97] State-owned COFCO—China Oil and Foodstuffs Corporation—is China's largest agriculture processor, manufacturer, and trader. COFCO plays a major role in the trade of soy and palm oil, two commodities central to the destruction of the world's rainforests.[98] An investigation by Global Witness and the Pulitzer Center found that COFCO was purchasing from deforesters in Brazil and Indonesia while simultaneously receiving sizable "green" loans from major Chinese and European banks.[99] Ironically, COFCO, as a trade agent of the Chinese government, has publicly pledged to pursue sustainable soy and palm oil sourcing.[100]

GHG emissions disclosure among Chinese meat and dairy companies is extremely limited.[101] Although Scope 3 emissions (emissions resulting from activities by assets not directly owned or controlled by the reporting organization, such as from supply chains) account for the majority of the sector's emissions, none of the twelve largest listed animal protein companies in China report them comprehensively or have a target to reduce them.[102] WH Group's Smithfield Foods has committed to become carbon negative by 2030, but includes only its US operations in this pledge.[103] Smithfield Foods' net-zero commitment further relies heavily on converting pig manure into methane gas, which is then sold as "renewable natural gas," a term some critics have called greenwashing.[104]

In its Nationally Determined Contribution submitted ahead of the 2021 UN Climate Change Conference (COP26),[105] China pledged to reach peak emissions by 2030 and noted plans to reduce emissions from food production overall, but did not include plans to reduce animal-based food consumption or production.[106] At times, the Chinese government has recognized the need to reduce meat and dairy consumption: a 2016 semi-official dietary guideline, for instance, recommended a 50% reduction in animal protein intake.[107] Yet the government has continued to support the interests of industrial meat producers, notably by setting ambitious growth targets for the livestock and feed sectors in China's 2021–2025 Five-Year Plan for animal agriculture.[108] Currently, there is little available evidence to assess whether and to what extent climate obstruction within China comes directly from agribusinesses.

Brazil

Brazil's agricultural GHG emissions have grown more than 160% since 1970.[109] As of 2023, Brazil is the world's number-one producer and exporter of soybeans, number-two producer and number-one exporter of beef and chicken, and the world's leading exporter of corn.[110] Its growth in livestock and feed production is considered the main driver of deforestation, land degradation, and forest fires in biodiversity hotspots and carbon sinks including the Amazon rainforest and the Cerrado savanna.[111] Government policy, shaped by close proximity with the agribusiness sector, has led to a continued expansion of livestock and feed production in the Amazon region even after the 2015 Paris Agreement.[112]

Since 2000, Brazil's leading meat-processing companies—most notably JBS, Marfrig, and Minerva[113]—have grown dramatically in terms of both size and concentration of power.[114] This is due in part to the Brazilian National Development Bank's "national champions" policy, which actively supported internationalization by offering subsidized loans and purchasing shares and debentures of selected companies from 2007 to 2013.[115] This government funding heavily benefited large Brazilian meatpacking companies, including JBS and Marfrig.[116]

A pro-agribusiness, cross-party voting bloc within the Brazilian government, Frente Parlamentar da Agropecuária (FPA) has been a central force in promoting the dismantling of environmental protections.[117] As of 2023, more than half of the seats within the Chamber of Deputies and Federal Senate are affiliated with FPA. This ruralist coalition, financed at least in part through food industry conglomerates, is supported by the think tank Instituto Pensar Agro (IPA), which emerged in 2011 as an informal coalition between agricultural representatives from Mato Grosso and some members of Congress.[118]

De Olho nos Ruralistas, a Brazilian agribusiness watchdog, reports that the growth of IPA—especially in the last decade—catalyzed the industry's consolidation of influence and increased its political power. The IPA-FPA coalition provided a space in which Brazil's agricultural elites—including old and new agricultural associations together with domestic and international multinational corporations[119]—could reconcile internal divisions to present a united front at the Congress and Senate.[120] It has done so through an increasing flow of resources and newfound strategic focus, allowing for the swift construction of joint business-parliamentary positions.[121] The IPA-FPA coalition has, for example, opposed the demarcation of Indigenous territories and the creation of conservation units.[122] Moreover, the FPA's submission of Bill PL 3729/2004, as part of a trio of environmentally detrimental legislation referred to as the "destruction package" by Brazilian NGOs and civil society, risks weakening environmental licensing requirements.[123] In addition to these collective efforts, individual companies reportedly benefit from significant access to key officials, facilitated by a revolving door whereby corporate employees transition to government posts and vice-versa.[124]

An analysis of Brazilian newspapers' climate coverage from 2002 to 2010 found that climate change was generally framed as an energy problem, even though the country's energy emissions are relatively small compared with its land use and agriculture emissions. The study found that discussions of meat production in the context of climate change were marginal—in the case of some leading newspapers, less than 0.5% of climate coverage mentioned meat—and often minimized the role of the meat industry in the climate crisis.[125]

POLITICAL AIMS AND STRATEGIES

In the next sections we detail common narratives that represent forms of obstruction in the context of animal agriculture and the practices actors use to disseminate these narratives. We divide practices into those aimed at (1) influencing policy and politics, (2) shaping science and public perceptions, and (3) building supportive coalitions.

Common Narratives

The climate impacts of animal agriculture are scientifically well-established, but the livestock industry continues to contest the scale, severity, and very existence of this problem. A key obstruction strategy is the creation of doubt about and controversy over evidence of the negative impacts of animal agribusiness. Narratives deployed to obstruct action to mitigate animal agriculture's climate impacts have positioned undesired interventions as unscientific, ineffective, or harmful. Notably, shifting the parameters of the debate toward technical interventions—away from measures such as herd size reduction or dietary change—is a powerful, subtle way to maintain the status quo. While some of these narratives include overt denialism, the majority of arguments are presented as seemingly objective scientific critiques or constructive proposals for better, science-based solutions. In the section that follows we summarize key narratives employed to counter progress toward addressing the climate impacts of livestock production[126] and briefly discuss them in light of current evidence.

"Animal Agriculture's Contributions to the Climate Crisis Are Uncertain or Overstated"

This narrative includes claims such as (1) there is no scientific consensus on the climate impacts of livestock, (2) fossil fuels are the real problem and deserve sole blame for climate change, and (3) biogenic methane from livestock is different—and less of a problem—than methane from fossil fuel sources. The IPCC does assign fossil fuel methane emissions slightly higher metric values than biogenic methane emissions (29.8 vs. 27.2 for their respective global warming potential over one hundred years, or GWP100) because the CO_2 produced by the breakdown of fossil methane is considered additional.[127] This difference is, however, relatively minimal. As discussed in the introduction, livestock farming is estimated to contribute 11.2% to 19.6% of total global GHG emissions. With methane being a short-lived but potent GHG, it has an important impact on peak warming and the feasibility of remaining within internationally agreed temperature limits.[128]

"Livestock Production Is Essential and/or Good for the Climate"

The narrative that animal agriculture and its outputs are necessary for people and the planet is used to justify business as usual, undermining attempts to hold producers accountable for their climate impacts. It positions the continuation of animal-based food production at current rates as essential for (1) food security and nutrition, (2) economic growth and farmer livelihoods, and (3) responsible environmental stewardship (by emphasizing that animals sequester carbon on grazed land, benefit soil quality, and/or serve as upcyclers of human food byproducts). Related marketing of meat as "green" or "climate friendly"—such as Brazen Beef, a product from Tyson Foods' Climate-Smart Beef Program[129]—may serve to alleviate consumer concerns about high levels of meat consumption.

Livestock companies are also promoting the message that biogas from industrial livestock farms is an important source of renewable energy.[130] The industry has embraced opportunities to profit related to climate change, such as by selling soil carbon-offset credits, even though questions of how much carbon can be sequestered in soil and for how long are yet to be resolved.[131]

While food systems play an important role in climate mitigation and adaptation, shifts toward more healthful, plant-based diets have been identified as a key option for reducing GHG emissions while maintaining food security.[132]

"Any Climate Impacts of Animal Agriculture Can Be Addressed Through Technical Fixes"

This narrative claim accepts that livestock farming affects the climate to some extent, but suggests those impacts can be mitigated without reducing animal-foods production and consumption through (1) changes in livestock farming methods, and/or (2) technological interventions such as seaweed in cattle feed and anaerobic digesters.[133] Claims that technological interventions will effectively address the climate impacts of livestock are unfoundedly optimistic, largely unproven at scale, and stand at odds with high-level assessments that conclude that such measures alone are not sufficient without demand reduction.[134] Such claims often rely on their lower emissions *intensity*, even where *total* emissions are predicted to increase with higher production levels. Changes to farming practices—for example, switching from grain-finishing feedlot systems to exclusively pasture-based beef production systems—are not only infeasible at current consumption levels, but also unlikely to meaningfully mitigate the climate impacts of livestock production.[135]

"Regulating Livestock Emissions and/or Animal-Source Food Consumption Is Unnecessary and/or Infeasible"

This narrative revolves around claims that undesired interventions will negatively affect (1) farmers' livelihoods, (2) food security and nutrition, and (3) climate and the environment.[136] In this context, agribusiness is positioned as an ally to farmers, governments, and civil society in tackling these issues. The claim that regulation is not necessary often hinges on the argument that agribusinesses are already voluntarily and proactively taking sufficient action to reduce their own emissions, or that technical interventions suffice.[137] However, reducing production and consumption of meat and dairy, particularly in high-consuming, wealthy nations, is considered essential to meet climate goals,[138] and adapting our food system to climate change will require a shift toward lower-impact food production.[139]

"Proponents of Reduced Animal-Source Food Consumption and/or Livestock Emissions Regulation Are Misguided or Extremist"

Focusing on discrediting the messenger, this narrative positions those who support policies to reduce livestock emissions either as (1) unscientific and ideologically driven or (2) unrealistic.[140] Claims that those supporting such measures act from a place of emotion and ideology—as opposed to the science-based approach of animal agribusiness and its allies—serve to undermine trust pre-emptively. Some evidence-based calls to adopt more sustainable diets are undermined using a straw-man argument implying that entire populations would be forced into vegetarianism or veganism, invoking animal-based food consumption as a cultural norm and animal farming as important cultural heritage.

Common Practices for Influencing Policy and Politics

As noted above, actors with an interest in maintaining or expanding industrial livestock production have impeded progress toward a more sustainable food system, including by keeping reduced production and consumption off policy agendas. In national and subnational settings, agribusinesses have lobbied, made campaign donations, and formally participated in policy processes with the aim to counteract progress on climate mitigation. Research using US lobbying reports, for instance, found that between 2000 and 2019, the ten largest US-based meat and dairy companies spent a combined $109 million on lobbying activities and $26 million on donations to federal political candidates.[141] The dairy industry also stood out as one of the most active sectors in lobbying around Canada's Healthy Eating Strategy, a roadmap for government action on more healthful diets.[142] Though empirical investigations of agribusiness's political obstruction remain rare, it is clear that livestock-related measures found in major initiatives such as the US Farm Bill, the EU Industrial Emissions Directive, and New Zealand's proposal to tax cattle emissions have been subjected to pressure from animal agribusiness and allied interests.

Investigative reporting suggests that obstruction has contributed to the striking neglect of livestock's climate impacts, and the mitigation potential of sustainable diets, within global climate and food governance. Following the Paris Agreement, the presence of the livestock sector in global climate governance has grown consistently, with 120 meat and dairy delegates—triple the number from the previous year—counted at the 28th UN Climate Change Conference.[143] The summit, held in 2023, was hailed as the first "food COP," where 134 nations pledged to transform food systems to address, and adapt to, climate change.[144] Livestock industry groups reacted positively to the COP28 outcomes, welcoming the emphasis on production efficiency over reduced consumption.[145] COP28 was accompanied by the launch of an FAO roadmap to address the climate crisis and end hunger.[146] Described as a "music to our ears" by a livestock industry representative,[147] the roadmap emphasized the high GHG emissions associated with animal agriculture but omits from its list of 120 recommendations any interventions to reduce meat production and consumption.[148] An accompanying FAO study has been accused by scientists of using cherry-picked and misrepresented evidence that served to underestimate the mitigation potential of reduced meat consumption.[149] This follows accusations by former FAO staff of the UN agency "censoring and sabotaging their work when it challenged livestock industry positions."[150]

The lack of global ambition on cutting livestock-related emissions and the limitations inherent in inviting the same companies contributing to a problem to develop its solutions were illustrated by the outcome of the 2021 UN Food Systems Summit. The Summit, which aimed to drive food-systems transformation to achieve the organization's Sustainable Development Goals, concluded without a clear message on meat reduction.[151] A draft paper prepared by industry representatives who formed the Summit's "sustainable livestock" cluster promoted intensified production, prompting criticism by scientists and NGOs. Following the ad-hoc addition to the cluster of several independent experts, a compromise solution ultimately saw three position papers published on livestock, only one of which mentioned reduced consumption.[152] More widely, the Summit was seen to reflect a depoliticized approach to food systems transformation that prioritizes corporate-friendly, technological interventions and fails to problematize corporate power in food systems.[153]

Pressure by member states—which, unlike companies, have formal powers in intergovernmental decision-making—is central to efforts within the UN system to obstruct the inclusion of targets and recommendations perceived as threatening to animal agriculture. This influence was illustrated by, for example, reports that messages supporting a shift toward more plant-based diets were removed from the Summary for Policymakers and main report of the IPPC Sixth Assessment Report mitigation working group under pressure from Brazil and Argentina, countries with strong cattle and feed industries and correspondingly influential lobbies.[154]

Common Practices for Shaping Science, Evidence for Policy, and Public Perceptions

Practices aimed at shaping science, the translation of evidence into policy, and public opinion play an instrumental role in the obstruction of efforts to mitigate animal agriculture's climate impacts, serving to undermine the case for action and support alternative narratives.

Influencing the Conduct, Publication, and Interpretation of Science

In response to growing concern about the climate impacts of livestock production, animal agribusiness has increased its sponsorship of research and scholarship. Industry-funded academics have repeatedly challenged or downplayed the scientific evidence establishing the livestock industry's role in the climate crisis (Boxes 4.1 and 4.2).[155] Initial livestock industry funding of individual academics and climate-related research has, in recent years, been followed by much larger investments in researchers and university centers—such as the UC Davis's Clarity and Leadership for Environmental Awareness and Research (CLEAR) Center and AgNext at Colorado State University—whose activities include promoting industry-supported climate solutions and producing policy analyses aligned with industry interests.[156] In 2022, an investigative journalist published documents showing that the CLEAR Center had been formed from an agreement between UC Davis and the Institute for Feed Education & Research (IFEEDER), an arm of the American Feed Industry Association[157] whose members include many major animal agriculture companies.[158] The documents indicated that CLEAR's industry funders "considered its greatest benefit to be its ability—as an apparently independent, academically credible voice—to make a positive case to the wider world about meat and dairy's environmental impact."[159] Although commercial partnerships are common in agricultural research, the risks this poses are illustrated by cases where undue influence is publicly documented. The Danish Centre for Food and Agriculture at the University of Aarhus, for example, withdrew a 2019 report on the climate impacts of beef after it emerged that Danish Crown and an industry association had cowritten it.[160]

The debunking of independent scientific consensus is a powerful strategy to obstruct climate action, if we follow the argument that consensus is a more crucial foundation for policy action than the quantity and quality of available evidence.[161] One way of undermining an emerging consensus, then, is to foster the appearance of an *alternative* consensus, illustrated by the Dublin Declaration of Scientists on the Societal Role of Livestock. Launched in 2022 and re-published in a 2023 special issue of the academic journal *Animal Frontiers*, the Declaration describes the livestock industry as "too precious to society to become the victim of simplification, reductionism, or zealotry."[162] The statement—which was coordinated by agribusiness consultants and reportedly signed by more than one thousand scientists—was endorsed by the EU Commissioner for Agriculture and shared with EU officials in an effort to prevent the adoption of ambitious environmental policies.[163]

Box 4.2: THE USDA-LAND-GRANT COMPLEX

During the first presidential administration of Donald Trump (2017–2021), the US Department of Agriculture (USDA) suppressed the term "climate change" in its communications and research reports.[164]

Continued

This minimization of the science and urgency of climate change to advance a political agenda was not an exception but rather a manifestation of the agency's long-standing role in promoting climate denial, and, more broadly, boosting unsustainable agricultural practices.

In the United States, deep ties between agribusiness, federal agencies, and the research community are strong and well-established.[165] Land-grant universities and their extension arms are tightly linked to the USDA via funding channels, joint appointments, revolving doors, and co-location of labs.[166] This *USDA-land-grant complex* played a key role in the development of the biological, mechanical, and organizational innovations that formed the basis for the dramatic growth in concentrated animal feeding operations (CAFOs).[167] Yet researchers at USDA and land-grant universities who helped develop these technologies paid little attention to their environmental effects.

USDA-administered public funding and conservation farm bill programs continue to be used to subsidize livestock production, perhaps most notably through feed-crop subsidies, supplying livestock producers with feed at a price below the cost of production.[168] The Environmental Quality Incentive Program, the largest USDA-funded conservation program, is required by the US Farm Bill to use 50% of its funding for livestock.[169] However, the NGO Environmental Working Group found that little of the funding from such programs supported practices the USDA considers climate-smart and "some of the practices that received the most funding actually exacerbate the climate crisis."[170] Thus, US federal policies to mitigate agricultural GHG emissions, particularly from livestock, are based on voluntary, subsidized, and often ineffective or counterproductive practices.[171]

A recent example of how public money administered by USDA has been used to obfuscate the climate impacts of livestock while purporting to address them is the funding of $3.1 billion Climate Smart Commodities public–private partnerships.[172] The agency has not released specific and detailed information on the beneficiaries of the funding beyond heavily redacted grant agreements and has refused to fulfill public record requests for information on how the money will be spent to protect "trade secrets and commercial or financial information."[173] Project summaries show that projects involving beef and dairy products alone account for over $1.3 billion in funding and involve companies such as Land O'Lakes, Tyson Foods, Archer-Daniels-Midland Company, and Cargill. The program lacks both a standardized methodology to assess GHG emissions and standardized implementation monitoring.[174]

Such actions are not new. The 2008 Farm Bill included a provision to exempt from Freedom of Information Act requests, and de facto shield from public view and researchers' analysis, georeferenced USDA information regarding farm payments.[175] In 2011, the industry shut down a proposed rule that would have required CAFOs to report basic information to USDA; as a result, there is no national database on CAFO locations, nor the pollution they generate.[176] High-profile USDA supporters from both parties continue to argue that current policies are sustainable and efficient.[177]

Influencing the Use of Science in Decision-Making: Metrics and Measurement

The choice of metrics for attributing responsibility is inherently political. How we measure the climate impacts of meat and dairy products has implications for both how responsibility for climate change is understood and how it is addressed by public policies. In the context of livestock, an alternative metric to the commonly used GWP100—GWP*—has been embraced by animal agribusinesses operating in countries with historically large herd sizes.[178] While GWP100 reports entities' contributions to total global emissions according to their current emissions, GWP* places more emphasis on methane's shorter lifespan and "considers an entity's contribution to be its additions to its own baseline, as measured in a past year."[179] As a result, the use of GWP* can result in proportionally greater animal farming CO_2-equivalent (CO_2e) emissions being attributed to developing countries with historically low emissions but growing herd sizes, while countries with historically high emissions but stable or declining herd sizes benefit.[180]

Influencing Media and Public Perception

Growing attention to the environmental impacts of meat and dairy has been met with public relations campaigns such as Meat and Dairy Facts (Ireland), the European Livestock Voice's Meat the Facts, as well as the US National Cattlemen's Beef Association (NCBA)'s BeefUp Sustainability and Masters of Beef Advocacy programs.[181] In response to the Beyond Beef campaign, which encouraged people to halve their meat consumption, the NCBA responded with a countercampaign and the development of an elementary school curriculum that emphasized cattle ranchers as stewards of the environment.[182] It further questioned "the USDA's commitment to US farmers and ranchers" after the agency's employee newsletter endorsed Meatless Mondays in 2012,[183] calling the effort an "animal rights extremist campaign to ultimately end meat consumption."[184] The USDA subsequently removed the post and said it did not support Meatless Mondays.[185]

Media organizations have played a key role in shaping public perception by amplifying industry messaging.[186] More broadly, a now substantial body of peer-reviewed articles analyzing media coverage of the livestock-climate connection indicates that media coverage has often focused on consumer responsibility and choice rather than corporate responsibility and policy change.[187] In practice, this emphasis means that much coverage of agriculture and climate change is concerned with meat-eating, veganism, or vegetarianism as individual dietary choices, whereas it rarely features livestock corporations' contributions to climate change and the role of government policies—taxes and farm subsidies, for example—in promoting or hindering sustainable production and consumption. An analysis by Faunalytics of one thousand climate-related articles in top US media outlets in 2021 and 2022 found that only 7% mentioned animal agriculture and fewer discussed its contributions to the climate crisis.[188] This paucity is reflected in limited public understanding of livestock's climate impacts.[189] For example, more than one in five polled Americans do not think that meat and dairy production contributes at all to global warming, and more than one in five said they do not know.[190]

Coalition Building and Management

Efforts to obstruct mitigation of the climate impacts of animal agriculture involve a wide range of actors, extending beyond those directly involved in the production or sale of animal-based foods to other commercial actors from adjacent industries including public relations, lobbying, media, finance, and industry-supported nonprofit organizations.

Business and trade associations unite companies across sectors or regions, playing a crucial role in coordinating and representing policy positions. In the context of livestock, such groups include, for example, the European Livestock and Meat Trades Union, the Istituto Salumi Italiani Tutelati, Beef+Lamb New Zealand, and the Global Dairy Platform. In recent years, animal protein groups have formed new industry organizations that focus specifically on climate-related messaging. For example, in 2021, the Meat Institute (previously the North American Meat Institute) and more than a dozen supporting companies and industry groups formed Protein PACT (for People, Animals & Climate of Tomorrow), which promotes "animal agriculture at the center of global solutions."[191]

Farmers occupy a pivotal role within food politics, and disaffected agrarian communities are a driving force behind right-wing populist pressure against policies aimed at reducing agriculture's climate impacts (see Box 4.3). A powerful political force in regions with high agricultural production, some groups with the power to mobilize the political capital of farming, such as Copa-Cogeca, have been a prominent voice in opposition to tougher climate policies.

On October 1, 2019, a caravan of tractors blocked more than a thousand kilometers of highway in the Netherlands as they traveled into the capital, The Hague. Their anger had been sparked by a proposal issued by the Dutch green-liberal party Democrats 66 to halve the national livestock herd to address nitrogen pollution, responsible for significant damage to the climate, ecosystems, and human health.[192] The nitrogen problem had been neglected by governing parties for many years until a ruling by the Dutch Council of State in May 2019 obliged the Dutch government to conform to its own environmental goals and EU biodiversity regulations.[193]

The series of farmer protests that ensued from 2019–2023 represent one of the highest-profile countermovements against European environmental policy. The protesters engaged in tactics such as lighting manure piles on fire,[194] threatening green campaigners and politicians,[195] and blocking roads, supermarket distribution centers, bridges, and harbors.[196] Similar tactics were subsequently adopted by farmers in Belgium, France, Germany, and Poland. This fight against nitrogen regulations, rooted in allegations that agriculture is unfairly and wrongly framed as a driver of climate change,[197] has been described as an attempt to generate enough political backlash to discourage any future political efforts to reduce herd size.

While not all Dutch farmers believe in the need for continued growth and enhanced production of animal agriculture,[198] those who do received much support from animal feed and livestock processing companies. For example, a Dutch newspaper revealed the role of the Netherlands-based feed company ForFarmers, which provided Agractie, one of the leading protest organizations, with financial, logistical, and communications support.[199] The investigative journalism platform Follow the Money reported that a coalition of feed producers and meat and dairy processors contributed to the protests by financing a lobbying platform called "Agri Facts," disguised as an independent fact-checker.[200] Furthermore, key players in the negotiations between protesters and government included the dairy arm of the Dutch agriculture organization LTO Nederland and a newly formed Agriculture Collective chaired by a former board member of two animal feed companies.[201]

In public, the farmers' movement in the Netherlands presented itself as the voice of traditional farmers whose livelihoods are threatened, an image that has garnered public support across the country. Meanwhile, the political negotiations sparked by these protests seemed to focus largely on maintaining industrial animal farming and limiting nitrogen policy to voluntary measures.[202] Initial negotiations between protesters

Continued

and the government resulted in concessions toward industrial meat and dairy production, with the agricultural minister promising that there would be no reduction of herds.[203] Grievances that affect smaller farmers, such as the adaptation of monitoring and reporting measures or the possibility for farmers to choose the most appropriate mix of environmentally friendly practices themselves, were not addressed.[204]

Agribusiness interests often hire public relations and lobbying firms to support their efforts to influence policy and public perception. For example, the Irish public relations firm Red Flag reportedly worked for Meat and Dairy Facts to promote positive messaging around animal products.[205] Although the role of think tanks and front groups in climate obstruction is studied mostly in the context of fossil fuels, emerging research on European think tanks indicates that such groups have also contributed to the creation of ignorance about the connection between livestock and climate change.[206] New climate-focused animal agriculture think tanks and nonprofit groups continue to be formed and promoted by the industry. For example, the US-based Center for Environment and Welfare was launched in 2023 to "educate the public and corporate leaders about animal extremists" and "to properly frame sustainability and environment issues."[207] The center is led by a partner at Berman and Company, a public relations firm with a history of working on behalf of the carbon majors.[208]

As discussed earlier, several high-profile academics and academic centers have served as contrarian voices in public and policymaking spaces, downplaying the livestock sector's climate impacts or undermining proposed policy responses.

While major environmental NGOs invite significant reputational risk if they partner with fossil fuel firms, agribusinesses continue to provide funding for environmental groups and boost their public images by partnering with them. In exchange for vague corporate commitments, such NGOs have played a role in facilitating agribusiness corporations' environmental claims (e.g., WWF's promotion of responsible and sustainable food labeling schemes with low standards[209]). The public-interest credentials of leading NGOs can lend legitimacy to questionable climate commitments and the companies making them, thus contributing to the narrative that the issue is being addressed. For example, environmental disclosure group CDP gave JBS an A- rating on climate, and the Nature Conservancy and Environmental Defense Fund worked with Tyson Foods to legitimize the company's climate-friendly beef claims.[210]

EFFORTS TO COUNTER OBSTRUCTION

The main form of response to animal agribusiness' climate obstruction is exposure: academic research, civil-society research projects, and investigative journalism have together increased public understanding of the nature and scope of animal agriculture-related obstruction efforts. Some of the obstructive narratives and practices documented in this chapter mirror those used by other industries, including oil and gas.[211] In 2020, the news platform *DeSmog*, whose reporting has focused largely on the carbon majors, launched an Agribusiness Database of companies and organizations that use climate change in their messaging, lobby around climate action, and may have ties to climate science denial.[212]

Climate change litigation focused on animal agriculture may emerge as a significant driver of climate mitigation and adaptation efforts in the years ahead.[213] There have been several attempts so far to formally sanction the industry. For example, in 2021, three Danish NGOs filed a suit against Danish Crown, one of the world's largest meat producers, over claims that its pork production is climate friendly, which the NGOs contend violate Denmark's Marketing Act. In response to the lawsuit, Danish Crown announced it would end the use of the "climate-controlled pig" label on pork packaging.[214]

Misleading climate-related claims are at the heart of other formal complaints as well. For example, in 2021, the US-based nonprofit Physicians Committee for Responsible Medicine petitioned the Federal Trade Commission, which regulates US advertising, to prohibit the NCBA from releasing advertisements portraying beef as an environmentally sustainable product.[215] In early 2023, the US-based nonprofit Mighty Earth filed a complaint with the National Advertising Division of the Better Business Bureau over JBS's issuing of sustainability-linked bonds that were tied to its stated goal to achieve net zero by 2040, and in early 2024 the New York Attorney General sued JBS's US subsidiaries for misleading climate-related claims (Box 4.4). In 2022, the Dutch city of Haarlem became the world's first municipality to adopt a ban on meat advertising as part of an effort to reduce GHG emissions.[216] In 2023, the UK-based environmental law charity ClientEarth submitted a complaint under the responsible business conduct guidelines of the Organization for Economic Cooperation and Development (OECD) against US-based Cargill for failing to adequately address its role in soy-driven deforestation and human rights violations in Brazil.[217]

More broadly, transnational agrarian movements that promote food sovereignty—"the right of peoples to healthy and culturally appropriate food produced through sustainable methods and their right to define their own food and agriculture systems"[218]—have emerged as a crucial force in countering the productivity-focused status quo and growing corporate control of global food systems.[219] The concept of food sovereignty was introduced in 1996 by La Via Campesina, a global peasant movement that today continues to represent groups such as small and medium-scale producers, Indigenous people, and agricultural workers, advocating for sustainable, agroecological approaches to farming underpinned by social-justice principles.

Box 4.4: ATTEMPTS TO HOLD JBS ACCOUNTABLE FOR ITS NET-ZERO CLIMATE COMMITMENTS

The world's largest meat company, JBS, issued $3.2 billion in sustainability-linked bonds (SLBs) on US capital markets in 2021. These bonds were tied to the company's stated promise to cut emissions and achieve "net zero greenhouse gas emissions across our entire value chain by 2040."[220] With more than 500 facilities and products sold in 190 countries, JBS is a leading emitter of GHGs. In 2020, the company's emissions exceeded those of Spain.[221]

In January 2023, the NGO Mighty Earth filed a whistleblower complaint at the US Securities and Exchange Commission (SEC), alleging that the company's "net zero by 2040"-based SLBs were misleading and constituted US securities fraud.[222] The complaint alleged that the company's representations concerning GHG emissions reductions were false and misleading and that JBS failed to disclose to investors material information needed to evaluate the truth of its emissions-related claims, including data on the number of animals it slaughters annually and the company's Scope 3 supply-chain emissions, which are estimated to account for an estimated 97% of its emissions overall.[223]

In a 2016 SEC filing, JBS had disclosed detailed animal-slaughter figures by region but omitted these figures in its SEC filings since then, reporting figures only for its processing facilities' slaughter capacities.[224] This omission has allowed JBS to avoid independent verification of its emissions claims and to reject emissions estimates that independent analysts extrapolated from the company's slaughter capacity figures. The *Washington Post* compared this strategy to "an oil company neglecting to include emissions from burning the oil that it sells."[225]

An assessment of JBS's climate plans by the *Corporate Climate Responsibility Monitor 2022* also found no evidence of sufficient GHG emission-reduction plans.[226] In 2023, the US National Advertising Review Board recommended that JBS discontinue using five challenged net-zero claims—including "JBS is committing to be net zero by 2040"—in its advertising because the claims communicate misleading messages and the company does not have a "formulated and vetted plan at present" for achieving its goal.[227] In February 2024, the New York Attorney General sued JBS's US subsidiaries for misleading customers over its climate goals and impacts.[228] The lawsuit stated that JBS "has had no viable plan to meet its commitment to be Net Zero by 2040."[229]

RESEARCH NEEDS AND INFORMATION GAPS

Compared with research on the fossil fuel sector, scholarship examining the nature and significance of climate obstruction related to animal agriculture is still nascent and sparse. To date, investigative journalists at outlets such as *InsideClimate News*, the *New York Times*, *DeSmog*, and *Unearthed* have played an instrumental role in uncovering and analyzing climate deception related to animal agriculture, as have NGOs such as IATP.

As discussed earlier, a significant amount of research on climate change and animal agriculture is conducted at universities that receive agribusiness funding. There is much need for independent environmental science and social science research about the linkages between animal agriculture and the climate crisis produced by research institutions unburdened by financial conflicts of interests and long-standing industry ties. Research needs and opportunities include:

- Investigations of corporate interactions with—and capture of—relevant government agencies and international standard-setting bodies, including efforts to shape the use of metrics. More broadly, research on the role of governments and state-owned enterprises in climate obstruction, including within the UN system and through trade negotiations.
- Investigations of the actions, internal knowledge, and choices of companies, trade associations, and their financial backers related to impeding climate action and supporting the continuation and expansion of emissions-intensive agriculture systems.
- Peer-reviewed quantification, tracking, and attribution of emissions related to industrial agriculture, including for feed companies. Emerging satellite and artificial-intelligence technologies hold the potential to help overcome the lack of comprehensive public GHG databases of dairy and livestock operations and to improve accountability.[230]
- Examinations of the overlap between the animal agriculture climate change countermovement and obstruction in other sectors such as oil and gas, tobacco, and ultra-processed foods, in terms of both strategies and supporting organizations, such as public relations firms and think tanks.

CONCLUSION

Reducing emissions from animal agriculture can contribute significantly to limiting climate change and maintaining a livable planet, whereas failing to act will render reaching current climate goals unfeasible. Effectively and justly reducing the climate impacts of animal agriculture will require dramatic changes in how we produce, consume, regulate, and subsidize livestock products, particularly in high-consuming countries. Although reducing production and consumption of animal-based products in these countries is considered necessary, the limited commitments and policy actions to date have focused predominantly on financial incentives for technical interventions with limited proven mitigation potential. Efforts aimed at effectively reigning in livestock-related emissions have been slowed or obstructed repeatedly by actors with vested financial interests in maintaining or expanding industrial animal agriculture. A better understanding of such obstruction, backed with rigorous, peer-reviewed evidence, will be a crucial step toward better safeguarding of climate and food policy from undue influence.

ACKNOWLEDGMENTS

Thank you to Nathalia Dutra and Michael Cao for research assistance.

NOTES

1. IATP and Changing Markets Foundation, "Emissions Impossible: Methane Edition," November 15, 2022, https://www.iatp.org/emissions-impossible-methane-edition.
2. Danone, "Our Climate Actions," https://www.danone.com/impact/planet/climate-actions.html (accessed September 27, 2023); "Combatting Climate Change," Danish Crown, https://www.danishcrown.com/en-gb/sustainability/accountability-and-key-figures/sustainability-strategy/combating-climate-change/ (accessed September 27, 2023); "Our Road to Net Zero," Nestlé, https://www.nestle.com/sustainability/climate-change/zero-environmental-impact (accessed September 27, 2023); "Tyson Foods Targets 2050 to Achieve Net Zero Greenhouse Gas Emissions," Tyson Foods, June 9, 2021, https://www.tysonfoods.com/news/news-releases/2021/6/tyson-foods-targets-2050-achieve-net-zero-greenhouse-gas-emissions; "JBS Is Committing to Be Net Zero by 2040," JBS, https://jbs.com.br/netzero/en/ (accessed September 27, 2023).
3. IATP and Changing Markets Foundation, "Emissions Impossible: Methane Edition"; "Corporate Climate Responsibility Monitor 2023," New Climate and Market Watch Institute, February 2023, https://newclimate.org/sites/default/files/2023-04/NewClimate_CorporateClimateResponsibilityMonitor2023_Feb23.pdf.
4. Michael A. Clark et al., "Global Food System Emissions Could Preclude Achieving the 1.5° and 2°C Climate Change Targets," *Science* 370 no. 6517 (2020): 705–708;

Catherine Ivanovich, Tianyi Sun, Doria R. Gordon, and Ilissa B. Ocko, "Future Warming from Global Food Consumption," *Nature Climate Change* 13 no. 3 (2023): 297–302.

5. Ivanovich et al., "Future Warming from Global Food Consumption"; Joseph Poore and Thomas Nemecek, "Reducing Food's Environmental Impacts through Producers and Consumers," *Science* 360, 6392 (2018): 987–992; Drew Shindell et al., *Global Methane Assessment: Benefits and Costs of Mitigating Methane Emissions* (United Nations Environment Programme and Climate and Clean Air Coalition, 2021), https://www.unep.org/resources/report/global-methane-assessment-benefits-and-costs-mitigating-methane-emissions; Marco Springmann et al., "Options for Keeping the Food System within Environmental Limits," *Nature* 562, 7728 (2018): 519–525.

6. Walter Willet et al., "Food in the Anthropocene: The EAT–Lancet Commission on Healthy Diets from Sustainable Food Systems," *The Lancet* 393, no. 10170 (2019): 447–492.; Intergovernmental Panel on Cimate Change, IPCC Working Group I Technical Support Unit, *Global Warming of 1.5°C: An IPCC Special Report on the Impacts of Global Warming of 1.5°C Above Pre-industrial Levels and Related Global Greenhouse Gas Emission Pathways, in the Context of Strengthening the Global Response to the Threat of Climate Change, Sustainable Development, and Efforts to Eradicate Poverty* (IPCC, 2018), http://www.ipcc.ch/report/sr15.

7. Xiaoming Xu, Prateek Sharma, Shijie Shu, Tzu-Shun Lin, Philippe Ciais, Francesco N. Tubiello, Pete Smith, Nelson Campbell, and Atul K. Jain, "Global Greenhouse Gas Emissions from Animal-Based Foods Are Twice Those of Plant-Based Foods," *Nature Food* 2, no. 9 (2021): 724–739.

8. Henning Steinfeld, Pierre Gerber, Tom Wassenaar, Vincent Castel, Mauricio Rosales, and Cees de Haan, *Livestock's Long Shadow: Environmental Issues and Options* (Food and Agriculture Organization of the United Nations, 2006), https://www.fao.org/3/a0701e/a0701e00.htm.

9. These studies include: Mario Herrero, Petr Havlík, Hugo Valin, An Notenbaert, Mariana C. Rufino, Philip K. Thornton, Michael Blümmel, Franz Weiss, Delia Grace, and Michael Obersteiner, "Biomass Use, Production, Feed Efficiencies, and Greenhouse Gas Emissions from Global Livestock Systems," *Proceedings of the National Academy of Sciences* 110, 52 (2013): 20888–20893; Gidon Eshel et al., "Land, Irrigation Water, Greenhouse Gas, and Reactive Nitrogen Burdens of Meat, Eggs, and Dairy Production in the United States," *Proceedings of the National Academy of Sciences* 111, 33 (2014): 11996–12001; William J. Ripple et al., "Ruminants, Climate Change and Climate Policy," *Nature Climate Change* 4, 1 (2014): 2–5; Eva Wollenberg et al., "Reducing Emissions from Agriculture to Meet the 2 C Target," *Global Change Biology* 22, no. 12 (2016): 3859–3864; Willet et al., "Food in the Anthropocene."

10. Estimates of animal agriculture's contribution to global GHG emissions range from 11.2% to 19.6%, with significant uncertainty. UN FAO, GLEAM v 3.0 Dashboard, 2022, https://foodandagricultureorganization.shinyapps.io/GLEAMV3_Public/; Xu et al., "Global Greenhouse Gas Emissions";Sophie Kevany, "UN Numbers Say Meat Is Bad for the Climate. The Reality Is Worse," *Vox*, May 27, 2023, https://www.vox.com/future-perfect/23738600/un-fao-meat-dairy-livestock-emissions-methane-climate-change.

11. Steinfeld et al., *Livestock's Long Shadow*, xxiv, 275, 277.

12. IATP and Changing Markets Foundation, "Emissions Impossible: Methane Edition."

13. Xu et al., "Global Greenhouse Gas Emissions."

14. Ibid.; Xiaoming Xu and Atul Jain, "How Much Do Crops Contribute to Emissions?," *GreenBiz*, October 14, 2021, https://www.greenbiz.com/article/how-much-do-crops-contribute-emissions.

15. Xu et al., "Global Greenhouse Gas Emissions."

16. Ibid.

17. Shindell et al., *Global Methane Assessment.*

18. Gerber at al. 2013; Hanqin Tian et al., "A Comprehensive Quantification of Global Nitrous Oxide Sources and Sinks," *Nature* 586, no. 7828 (2020): 248–256;Michael B. Eisen and Patrick O. Brown, "Rapid Global Phaseout of Animal Agriculture Has the Potential to Stabilize Greenhouse Gas Levels for 30 Years and Offset 68 Percent of CO2 Emissions This Century," *PLoS Climate* 1, no. 2 (2022): e0000010, https://doi.org/10.1371/journal.pclm.0000010.

19. Poore and Nemecek, "Reducing Food's Environmental Impacts through Producers and Consumers"; Eisen and Brown, "Rapid Global Phaseout of Animal Agriculture."

20. Matthew Hayek, Helen Harwatt, William J. Ripple, and Nathaniel D. Mueller "The Carbon Opportunity Cost of Animal-Sourced Food Production on Land," *Nature Sustainability* 4, no. 1 (2020): 21–24.

21. IATP and Changing Markets Foundation, "Emissions Impossible: Methane Edition."

22. Springmann et al., "Options for Keeping the Food System within Environmental Limits"; Michael A. Clark, Marco Springmann, Jason Hill, and David Tilman, "Multiple Health and Environmental Impacts of Foods," *Proceedings of the National Academy of Sciences of the United States of America* 116, no. 46 (November 2019): 23357–23362; Ivanovich et al., "Future Warming from Global Food Consumption"; Willett et al., "Food in the Anthropocene."

23. "IPCC Special Report on Climate Change and Land," Intergovernmental Panel on Climate Change, 2019, https://www.ipcc.ch/srccl/, 58.

24. Xu et al., "Global Greenhouse Gas Emissions," Figure 4C.

25. Ibid. China is estimated to be responsible for 8% of total global animal-based food production emissions, which make up 57% of total global food production emissions (8/57 = 14%). The figures for the United States, China, and India were calculated in the same manner.

26. Xu et al., "Global Greenhouse Gas Emissions."

27. "Emissions Impossible: How Big Meat and Dairy are Heating Up the Planet," IATP and GRAIN, July 18, 2018, https://grain.org/article/entries/5976-emissions-impossible-how-big-meat-and-dairy-are-heating-up-the-planet.

28. Ibid.

29. Ibid.

30. Jennifer Clapp, *Food*, 3rd ed. (Wiley, 2020); Philip McMichael, "Political Economy of the Global Food and Agriculture Aystem," in *Rethinking Food and Agriculture*, ed. Amir Kassam and Laila Kassam (Woodhead Publishing, 2020), 53–75.

31. Shefali Sharma, "Companies: Dominating the Market from Farm to Display Case," *Heinrich Böll Stiftung*, September 7, 2021, https://eu.boell.org/en/2021/09/07/companies-dominating-market-farm-display-case.

32. Ibid.

33. Ibid.

34. O. Lazarus, S. McDermid, and J. Jacquet, "The Climate Responsibilities of Industrial Meat and Dairy Producers," *Climatic Change* 165 (2021): 30.

35. Andy Reisinger, Harry Clark, Annette L. Cowie, Jeremy Emmet-Booth, Carlos Gonzalez Fischer, Mario Herrero, Mark Howden, and Sinead Leahy, "How Necessary and Feasible Are Reductions of Methane Emissions from Livestock to Support Stringent Temperature Goals?," *Philosophical Transactions of the Royal Society A* 379 no. 2210 (2021): 20200452, https://doi.org/10.1098/rsta.2020.0452.

36. OECD and FAO, "OECD-FAO Agricultural Outlook 2021–2030," 2021, https://doi.org/10.1787/19428846-en. Numbers based on projected consumption of meat (beef/veal, pig, poultry, sheep) and dairy (milk, butter, cheese).

37. Reisinger et al., "How Necessary and Feasible Are Reductions of Methane Emissions."

38. Shindell et al., *Global Methane Assessment*, 13.

39. Tim Searchinger, Richard Waite, Craig Hanson, Janet Ranganathan, and Emily Matthews, "World Resources Report: Creating a Sustainable Food Future: A Menu of Solutions to Feed Nearly 10 Billion People by 2050" (World Resources Institute, 2019), https://agritrop.cirad.fr/593176/1/WRR_Food_Full_Report_0.pdf.

40. Ben Elgin, "Why Won't Companies Use This Quick Fix to Reduce Cow Methane Emissions?," *Bloomberg*, June 28, 2023, https://www.bloomberg.com/news/features/2023-06-28/this-quick-fix-reduces-methane-emissions-from-cow-burps.

41. Viveca Morris, "Opinion: The Cow-Shaped Hole in Biden's Methane Plan," *Politico*, November 16, 2021, https://www.politico.com/news/agenda/2021/11/16/methane-emissions-cows-agriculture-climate-change-522550; "High Steaks: How Focusing on Agriculture Can Ensure the EU Meets Its Methane-Reduction GAoals," *Changing Markets Foundation*, June 2022, http://changingmarkets.org/wp-content/uploads/2022/06/CM-REPORT-ENG-HIGH-STEAKS.pdf.

42. "Global Methane Pledge," Climate & Clean Air Coalition, https://www.globalmethanepledge.org (accessed September 25, 2023).

43. Simona Vallone and Eric F. Lambin, "Public Policies and Vested Interests Preserve the Animal Farming Status Quo at the Expense of Animal Product Analogs," *One Earth* 6, no. 9 (2023): 1213–1226

44. Rachel Pannett, "How New Zealand Plans to Tackle Climate Change: Taxing Cow Burps," *Washington Post*, February 1, 2023, https://www.washingtonpost.com/climate-solutions/interactive/2023/new-zealand-cows-burps-methane-tax/.

45. Ibid., Craymer, "New Zealand ends plans to price agricultural emissions," *Reuters*, June 11, 2024, https://www.reuters.com/business/environment/new-zealand-ends-plans-price-agricultural-emissions-2024-06-11/

46. Steve Cornett, "Too Many Livestock?," *Beef Today*, February 2007, https://www.proquest.com/magazines/too-many-livestock/docview/204496067/se-2.

47. Morris and Jacquet, "The Animal Agriculture Industry"; Boren, "Revealed"; Hiroko Tabuchi, "He's an Outspoken Defender of Meat: Industry Funds His Research, Files Show," *New York Times*, October 31, 2022, https://www.nytimes.com/2022/10/31/climate/frank-mitloehner-uc-davis.html.

48. Morris and Jacquet, "The Animal Agriculture Industry"; Sylvia Wright, "Don't Blame Cows for Climate Change," University of California Davis, December 7, 2009, https://www.ucdavis.edu/news/don%E2%80%99t-blame-cows-climate-change.

49. Morris and Jacquet, "The Animal Agriculture Industry"; Maurice E. Pitesky, Kimberly R. Stackhouse, and Frank M. Mitloehner, "Clearing the Air:

Livestock's Contribution to Climate Change," *Advances in Agronomy* 103 (2009): 1–40.

50. Wright, "Don't Blame Cows for Climate Change."

51. American Chemical Society, "Eating Less Meat and Dairy Products Won't Have Major Impact on Global Warming," American Chemical Society, March 22, 2010, https://www.acs.org/pressroom/newsreleases/2010/march/eating-less-meat-and-dairy-products-wont-have-major-impact-on-global-warming.html.

52. Vasile Stanescu, "'Cowgate': Meat Eating and Climate Change Denial," in *Climate Change Denial and Public Relations*, ed. Núria Almiron and Jordi Xifra (Taylor and Francis, 2019): 178–194.

53. Morris and Jacquet, "The Animal Agriculture Industry"; Tabuchi, "He's an Outspoken Defender of Meat."

54. Arthur Neslen, "'The Anti-Livestock People Are a Pest': How UN Food Body Played Down Role of Farming in Climate Change," *The Guardian*, October 20, 2023, https://www.theguardian.com/environment/2023/oct/20/the-anti-livestock-people-are-a-pest-how-un-fao-played-down-role-of-farming-in-climate-change.

55. Arthur Neslen, "Ex-Officials at UN Farming Body Say Work on Methane Emissions Was Censored," *The Guardian*, October 20, 2023, https://www.theguardian.com/environment/2023/oct/20/ex-officials-at-un-farming-fao-say-work-on-methane-emissions-was-censored.

56. Neela Banerjee, Neela Banerjee, Georgina Gustin, and John H. Cushman Jr., "The Farm Bureau: Big Oil's Unnoticed Ally Fighting Climate Science and Policy," *Inside Climate News*, December 21, 2018, https://insideclimatenews.org/news/21122018/american-farm-bureau-fossil-fuel-nexus-climate-change-denial-science-agriculture-carbon-policy-opposition/.

57. American Farm Bureau Federation, "It Pays to Be A Farm Bureau Member," https://www.fb.org/about/benefits (accessed March 27, 2025); "Amber Waves of Gain," *Defenders of Wildlife*, April 2000, https://web.archive.org/web/20230217033654/https://defenders.org/sites/default/files/publications/amber_waves_of_gain.pdf.

58. American Farm Bureau Federation, "Global Climate Change," https://www.fb.org/files/climate_2019.pdf (accessed November 17, 2023).

59. American Farm Bureau Federation, "Farm Bureau National Policies 2023," January 10, 2023, 183–184, https://www.idahofb.org/media/01ppihy4/afbf-policy-book-2023-final-pdf.pdf.

60. Ibid., 62, 183–184.

61. Banerjee et al., "The Farm Bureau"; Robert J. Brulle, "Advocating Inaction: A Historical Analysis of the Global Climate Coalition," *Environmental Politics* 32 no. 2 (2023): 185–206.

62. Brulle, "Advocating Inaction."

63. Banerjee et al., "The Farm Bureau"; US Congress, "Global Climate Negotiations: Obligations of Developed and Developing Countries: Hearing Before the Committee on International Relations, House of Representatives, One Hundred Fifth Congress, first session, July 24, 1997," 1998, https://babel.hathitrust.org/cgi/pt?id=pst.000032135077&view=1up&seq=38, 34.

64. Leah Douglas, "The Breathtaking Lack of Oversight for Air Emissions From Animal Farms," *The Nation*, December 20, 2019, https://www.thenation.com/article/archive/air-emissions-environment/.

65. Douglas, 2019; "Environmental Protection Agency: Animal Feeding Operations Consent Agreement and Final Order; Notice," *Federal Register* 70 no. 19 (2005),

https://www.epa.gov/sites/default/files/2016-06/documents/afolagooneem report2012draftappe.pdf.

66. Ibid.

67. Ibid.

68. Madison McVan, "18 Years and Counting: EPA Still Has No Method for Measuring CAFO Air Pollution," *Investigate Midwest*, April 20, 2023, https://investigatemidwest.org/2023/04/20/18-years-and-counting-epa-still-has-no-method-for-measuring-cafo-air-pollution/.

69. Georgina Gustin, "Big Agriculture and the Farm Bureau Help Lead a Charge Against SEC Rules Aimed at Corporate Climate Transparency," *Inside Climate News*, August 22, 2022, https://insideclimatenews.org/news/22082022/big-agriculture-the-farm-bureau-sec-rules/.

70. US Senate Committee on Agriculture, Nutrition, and Forestry, "Boozman, Braun & Lucas Reintroduce Legislation to Protect Family Farmers & Ranchers from Burdensome SEC Climate Rules," February 13, 2023, https://www.agriculture.senate.gov/newsroom/rep/press/release/boozman-braun-and-lucas-reintroduce-legislation-to-protect-family-farmers_ranchers-from-burdensome-sec-climate-rules.

71. DG Research and Innovation, Group of Chief Scientific Advisors, *Towards Sustainable Food Consumption: Promoting Healthy, Affordable and Sustainable Food Consumption Choices* (European Commission, 2023), 11, https://op.europa.eu/en/web/eu-law-and-publications/publication-detail/-/publication/9f582c41-1565-11ee-806b-01aa75ed71a1.

72. "Farm to Fork Strategy: For a Fair, Healthy and Environmentally-Friendly Food System," *European Commission*, 2020, https://food.ec.europa.eu/system/files/2020-05/f2f_action-plan_2020_strategy-info_en.pdf.

73. Eddy Wax, "From Farm to Flop? Political Risks Choke EU's Green Food Plan," *Politico*, January 26, 2023, https://www.politico.eu/article/blocked-and-delayed-political-risks-choke-eus-green-food-plan-farmers/; Andrew Rettman, "How 'Big Meat' Lobbies Brussels to Keep Carnivore Status Quo," *EU Observer*, August 10, 2023, https://euobserver.com/alt-protein/156892.

74. Anniek J. Kortleve, Neela Banerjee, Georgina Gustin, John H. Cushman Jr., "Over 80% of the European Union's Common Agricultural Policy Supports Emissions-Intensive Animal Products," *Nature Food* 5 (2024): 288–292.

75. Gerry Alons, "Environmental Policy Integration in the EU's Common Agricultural Policy: Greening or Greenwashing?," *Journal of European Public Policy* 24 (2017): 1604–1622.

76. "CAP vs Farm to Fork: Will We Pay Billions to Destroy, or to Support Biodiversity, Climate, and Farmers?," *Corporate Europe Observatory*, October 2020, https://corporateeurope.org/sites/default/files/2020-10/CAP_Farm-to-Fork-Final_0.pdf; DeSmog, "Copa-Cogeca," https://www.desmog.com/copa-cogeca/ (accessed May 24, 2024).

77. DeSmog, "European Livestock Voice," https://www.desmog.com/european-livestock-voice/ (accessed May 24, 2024).

78. European Livestock Voice, "Meat the Facts," https://meatthefacts.eu (accessed May 24, 2024).

79. European Court of Auditors, "Greening: A More Complex Income Support Scheme, Not Yet Environmentally Effective," 2017, https://www.eca.europa.eu/Lists/ECADocuments/SR17_21/SR_GREENING_EN.pdf.

80. Joop de Boer and Harry Aiking, "EU Citizen Support for Climate-Friendly Agriculture (Farm) and Dietary Options (Fork) Across the Left-Right Political Spectrum," *Climate Policy* 23, no 4 (2023): 509–521.

81. Sini Eräjää, "Marketing Meat: How EU Promotional Funds Favour Meat and Dairy," *Greenpeace European Unit*, April, 2021, https://www.greenpeace.org/static/planet4-eu-unit-stateless/2021/04/20210408-Greenpeace-report-Marketing-Meat.pdf.

82. European Livestock Voice, "Opinion: What Greenpeace's Latest Report Won't Tell You About the EU Promotion Policy," *Meat the Facts*, April 23, 2021, https://web.archive.org/web/20230729131414/https://meatthefacts.eu/home/activity/beyond-the-headlines/opinion-piece-what-greenpeaces-latest-report-wont-tell-you-about-the-eu-promotion-policy/.

83. Wax, "From Farm to Flop?"

84. Changing Markets Foundation, "High Steaks"; "High Steaks Part 2: Taking Methane from Animal Farming Out of Its Blindspot," Changing Markets Foundation, October 2022, http://changingmarkets.org/wp-content/uploads/2022/10/CM-Report-ENG-Web-High-Stakes-02-October-2022-Final.pdf.

85. Changing Markets Foundation, "High Steaks Part 2"; Maria Simon Arboleas, "'Ciao' Cows: Cattle Excluded from EU's Industrial Emissions Cut Plan," *Euractiv*, November 30, 2023, https://www.euractiv.com/section/agriculture-food/news/ciao-cows-cattle-excluded-from-eus-industrial-emissions-cut-plan/; Antonia Zimmermann, "EU Conservatives Score Big Win on Industrial Emissions Rules," *Politico*, July 11, 2023, https://www.politico.eu/article/eu-conservative-big-win-industrial-emissions-directive/.

86. "The EU Indulges the Largest Industrial Polluters with New Emissions Rules," European Environmental Bureau, November 29, 2023, https://eeb.org/the-eu-indulges-the-largest-industrial-polluters-with-new-emissions-rules/; "Pollution: Deal with Council to Reduce Industrial Emissions," European Parliament, November 29, 2023, https://www.europarl.europa.eu/news/en/press-room/20231127IPR15436/pollution-deal-with-council-to-reduce-industrial-emissions.

87. Anne Grimmelt, Sheng Hong, Roberto Uchoa de Paula, Cherie Zhang, and Jia Zhou, "For Love of Meat: Five Trends in China That Meat Executives Must Grasp," McKinsey & Company, February 2023, https://www.mckinsey.com/industries/consumer-packaged-goods/our-insights/for-love-of-meat-five-trends-in-china-that-meat-executives-must-grasp.

88. Ibid.

89. Zhaohai Bai, Zhaohai Bai, Wenqi Ma, Lin Ma, Gerard L. Velthof, Zhibiao Wei, Petr Havlík, Oene, Michael R. F. Lee, and Fusuo Zhang, "China's Livestock Transition: Driving Forces, Impacts, and Consequences," *Science Advances* 4, no. 7 (2018): eaar8534.

90. United States Department of Agriculture Foreign Agricultural Service (USDAFAS), "2022/23 China Soybean Imports Raised to Record High," August 2023, https://apps.fas.usda.gov/psdonline/circulars/oilseeds.pdf.

91. Grimmelt et al., "For Love of Meat."

92. Hao Zhao, Jinfeng Chang, Petr Havlík, Michiel van Dijk, Hugo Valin, Charlotte Janssens, Lin Ma, Zhaohai Bai, Mario Herrero, Pete Smith, and Michael Obersteiner, "China's Future Food Demand and Its Implications for Trade and Environment," *Nature Sustainability* 4, 12 (2021): 1042–1051.

93. Chris Arsenault, "China and EU Appetite for Soy Drives Brazilian Deforestation, Climate Change: Study," *Monga Bay*, June 2020, https://news.mongabay.com/2020/06/china-and-eu-appetite-for-soy-drives-brazilian-deforestation-climate-change-study/; Neus Escobar, E. Jorge Tizado, Erasmus K.H.J. zu Ermgassen, Pernilla Löfgren, Jan Börner, and Javier Godar, "Spatially-Explicit Footprints of Agricultural Commodities: Mapping Carbon Emissions Embodied in Brazil's Soy Exports," *Global Environmental Change: Human and Policy Dimensions* 62 (2020): 102067, https://doi.org/10.1016/j.gloenvcha.2020.102067.

94. Mindi Schneider, "Wasting the Rural: Meat, Manure, and the Politics of Agro-Industrialization in Contemporary China," *Geoforum; Journal of Physical, Human, and Regional Geosciences* 78, (2017): 89–97; see also Peter Li, "Exponential Growth, Animal Welfare, Environmental and Food Safety Impact: The Case of China's Livestock Production," *Journal of Agriculture and Environmental Ethics* 22, 3 (2009): 230–231.

95. Han Yang, "The Philosophical, Historical and Policy Basis of China's Food Security Strategy," *Reform* 1 (2022), https://www.drc.gov.cn/DocView.aspx?chnid=379&leafid=1338&docid=2904915.

96. IATP and GRAIN, "Emissions Impossible"; Schneider, "Wasting the Rural."

97. Shefali Sharma, "The Need for Feed: China's Demand for Industrialized Meat and Its Impacts," IATP, February 2014, https://www.iatp.org/sites/default/files/2017-05/2017_05_03_FeedReport_f_web_0.pdf; Jackie Roembke, "Top Feed Companies: 144 Global Producers Rank in 2022," *Feed Strategy*, September 2022, https://www.feedstrategy.com/business-markets/feed-production-by-region/article/15443042/top-feed-companies-144-global-producers-rank-in-2022.

98. Global Witness, "Why Should Global Financiers Worry about Agribusiness COFCO's Deforestation Links?," *Global Witness*, September 5, 2023, https://www.globalwitness.org/en/campaigns/forests/why-should-global-financiers-worry-about-agribusiness-cofcos-deforestation-links/.

99. Ibid.

100. COFCO, "Our Policies," https://www.cofcointernational.com/sustainability/our-policies/ (accessed March 27, 2025).

101. "Transforming Animal Agriculture in China," FAIRR, September 2021, https://www.fairr.org/resources/reports/transforming-animal-agriculture.

102. "Sustainable Protein Transformation in China," FAIRR, October 2022, https://www.fairr.org/resources/reports/sustainable-protein-transformation-in-china.

103. Smithfield Foods, Inc., "Smithfield Foods to Become Carbon Negative by 2030," PR Newswire, September 2020, https://www.prnewswire.com/news-releases/smithfield-foods-to-become-carbon-negative-by-2030-301124087.html; Shefali Sharma and Ben Lilliston, "From Net Zero to Greenwash—Global Meat and Dairy Companies," IATP, October 2021, https://www.iatp.org/net-zero-greenwash-global-meat-and-dairy-companies.

104. Smithfield Foods, Inc., 2020. Critics argue that the company's biogas capture projects fail to address water and air pollution from factory farms, and ignore leakage concerns from methane transport and storage; see, e.g., Melba Newsome, "Turning Hog Waste Into Biogas: Green Solution or Greenwashing?" *Yale Environment 360*, https://e360.yale.edu/features/turning-hog-waste-into-biogas-green-solution-or-greenwashing (accessed March 27, 2025).

105. Ministry of Ecology and Environment (China), *China's Achievements, New Goals and New Measures for Nationally Determined Contributions*, October 2021, https://disasterlaw.ifrc.org/media/3644.

106. Lazarus et al., "The Climate Responsibilities of Industrial Meat and Dairy Producers."

107. For details of the guidelines, see Xiaodong Wang, "Ministry Tweaks Eating Guidelines," *China Daily*, May 14, 2016, http://www.chinadaily.com.cn/china/2016-05/14/content_25269469.htm; for positive reactions to the reported guidelines, see "China Aims to Cut Meat Consumption in Half," *AWI Quarterly*, Fall 2016, https://awionline.org/awi-quarterly/2016-fall/china-aims-cut-meat-consumption-half.

108. "Livestock Production in the 14th Five-Year Plan Era: Building Two Trillion-yuan Sectors," *China Food News*, December 28, 2021, http://www.news.cn/food/20211228/3f33ce29a5c6447c860f6067cbe550a5/c.html.

109. Shefali Sharma and Sergio Schlesinger, "The Rise of Big Meat: Brazil's Extractive Industry," IATP, November 30, 2017, 31, https://www.iatp.org/sites/default/files/2017-11/2017_11_30_RiseBigMeat_f.pdf.

110. Information from USDA Foreign Agricultural Service, https://apps.fas.usda.gov/psdonline/app/index.html#/app/topCountriesByCommodity (accessed September 25, 2023).

111. Sharma and Schlesinger, "The Rise of Big Meat," 17–23.; Alencar et al., "Desafios e Oportunidades Para Redução das Emissões de Metano no Brasil";Marcos H. Costa et al., "The Physical Hydroclimate System of the Amazon," *Science Panel for the Amazon* (May 2022), https://www.theamazonwewant.org/wp-content/uploads/2022/05/Chapter-15-Bound-May-11.pdf, 3; Alceu Luís Castilho, Bernardo Fialho, Bruno Stankevicius Bassi, Eduardo Luiz Damiani Goyos Carlini, Hugo Souza, Katarina Moraes, Luma Ribeiro Prado, Nanci Pittelkow, and Natália Freire Bellentani ., "Invaders: Who Are the Businessmen and Corporations Whose Farms Overlap Indigenous Lands in Brazil," *Agribusiness Watch*, April 19, 2023, https://deolhonosruralistas.com.br/wp-content/uploads/2023/05/Invaders_ENG-2023.pdf; "Derstroying [*sic*] Native Vegetation to Produce Beef and Soy Is the Main Driver Affecting Biodiversity in the Cerrado and Amazon," *World Wildlife Foundation*, December 2021, https://shorturl.at/rvzO2; Eduardo Eiji Maeda, Eduardo Eiji Maeda, Luiz E. O. C. Aragão, Jessica C. A. Baker, Luiz Carlos Balbino, Yhasmin Mendes de Moura, Antônio Donato Nobre, Matheus Henrique Nunes, Celso H. L. Silva Junior, and Júlio César dos Reis, "Land Use Still Matters After Deforestation," *Communications Earth & Environment* 4, no. 29 (2023), https://www.nature.com/articles/s43247-023-00692-x.

112. Eder Johnson de Area Leão Pereira, Luiz Carlos de Santana Ribeiro, Lúcio Flávio da Silva Freitas, and Hernane Borges de Barros Pereira, "Brazilian Policy and Agribusiness Damage the Amazon Rainforest," *Land Use Policy* 92 (2020): 104491, https://doi.org/10.1016/j.landusepol.2020.104491; Meghie Rodrigues, "Politics and the Environment Collide in Brazil: Lula's First Year Back in Office," *Nature*, December 21, 2023, https://www.nature.com/articles/d41586-023-04042-x.

113. Castilho et al., "Invaders," 4; Bart Slob, Gerard Rijk, Matt Piotrowski, "JBS, Marfrig, and Minerva: Material Financial Risk from Deforestation in Beef Supply Chains," Chain Reaction Research, December 2020, https://chainreactionresearch.com/wp-content/uploads/2020/12/JBS-Marfrig-and-Minerva-Material-financial-risk-from-deforestation-in-beef-supply-chains-2.pdf.

114. Sharma and Schlesinger, "The Rise of Big Meat."

115. Ibid.

116. Ibid.

117. Pereira et al., "Brazilian Policy and Agribusiness Damage the Amazon Rainforest"; "The Financiers of Destruction," De Olho Nos Ruralistas, July 2022, https://deolhonosruralistas.com.br/wp-content/uploads/2022/08/Financiers-of-Destruction-2022-EN.pdf, 4–8.

118. Caio Pompeia, *Formação política do agronegócio*, 2021.

119. De Olho Nos Ruralistas, "The Financiers of Destruction," 6–7.

120. Ibid., 4, 6, 7–14.

121. Ibid., 4.

122. Castilho et al., "Invaders," 8.

123. Proposed as PL 3729/2004 and processed as PL 2159/2021, the Bill is under consideration of Brazil's Senate at the time of writing. For discussion of the Bill, see, e.g., "Pacote da Destruição—o que dizem os principais PLs" (EN: Destruction Package—what the main Bills say), Observatorio Do Clima, https://www.oc.eco.br/pacote-da-destruicao-o-que-dizem-os-principais-pls/ (accessed March 27, 2025); Renata Ruaro, Gustavo Henrique Zaia Alves, Livia Tonella, Lucas Ferrante, and Philip Fearnside, "Loosening of Environmental Licensing Threatens Brazilian Biodiversity and Sustainability," *DIE ERDE* 153, 1(2022): 60–64.

124. De Olho Nos Ruralistas, "The Financiers of Destruction," 17.

125. Myanna Lahsen, "Buffers Against Inconvenient Knowledge: Brazilian Newspaper Representations of the Climate-Meat Link," *P2P e Inovação* 4.1 (2017): 59–84.

126. Informed by Joe Fassler, "Hot Air: Five Climate Myths Pushed by the US Beef Industry," *The Guardian*, May 3, 2023, https://www.theguardian.com/environment/2023/may/03/five-beef-industry-myths-busted; Kathryn Clare, Nason Maani, and James Milner, "Meat, Money and Messaging: How the Environmental and Health Harms of Red and Processed Meat Consumption Are Framed by the Meat Industry," *Food Policy* 109, (2022): 102234, https://doi.org/10.1016/j.foodpol.2022.102234; "Big Livestock's Big Greenwash," https://biglivestockgreenwash.com (accessed March 2025).

127. Piers Forster et al., "The Earth's Energy Budget, Climate Feedbacks, and Climate Sensitivity," in *Climate Change 2021: The Physical Science Basis. Contribution of Working Group I to the Sixth Assessment Report of the Intergovernmental Panel on Climate Change*, ed. Valérie Masson-Delmotte et al. (Cambridge University Press, 2023).

128. Reisinger et al., "How Necessary and Feasible Are Reductions of Methane Emissions."

129. E.g., Michaela Herrmann, "The Rise of the 'Climate Friendly' Cow," *DeSmog*, April 26, 2023, https://www.desmog.com/2023/04/26/rise-of-the-climate-friendly-cow/.

130. Melba Newsom, "Turning Hog Waste Into Biogas: Green Solution or Greenwashing?," *Yale Environment 360*, September 9, 2021, https://e360.yale.edu/features/turning-hog-waste-into-biogas-green-solution-or-greenwashing.

131. Sophia Murphy and Ben Lillston, "True or False? Evaluating Solutions for Agriculture and Climate Change," IATP, July 27, 2022, https://www.iatp.org/true-or-false-climate-solutions.

132. See, e.g., "IPCC Special Report on Climate Change and Land," 2019.

133. Jan Dutkiewicz, "The Comforting Lie of Climate-Friendly Meat," *The New Republic*, December 2023, https://newrepublic.com/article/177575/never-trust-green-meat.

134. Shindell et al., *Global Methane Assessment*.

135. Matthew Hayek and Rachael D. Garrett, "Nationwide Shift to Grass-Fed Beef Requires Larger Cattle Population," *Environmental Research Letters* 13, no. 8 (2018): 084005, https://doi.org/10.1088/1748-9326/aad401.

136. Caroline Christen, "Investigation: How the Meat Industry Is Climate-Washing Its Polluting Business Model," *DeSmog*, July 7, 2021, https://www.desmog.com/2021/07/18/investigation-meat-industry-greenwash-climatewash/.

137. Viveca Morris and Jennifer Jacquet, "The Animal Agriculture Industry, US Universities, and the Obstruction of Climate Understanding and Policy," *Climatic Change* 177, no. 41 (2024), https://doi.org/10.1007/s10584-024-03690-w.

138. Clark et al., "Global Food System Emissions."

139. Kyle F. Davis, Jessica A. Gephart, Kyle A. Emery, Allison M. Leach, James N. Galloway, and Paolo D'Odorico, "Meeting Future Food Demand with Current Agricultural Resources," *Global Environmental Change* 39 (2016): 125–132; Springmann et al., "Options for Keeping the Food System."

140. Morris and Jacquet, "The Animal Agriculture Industry"; Jennifer Jacquet, *The Playbook: How to Deny Science, Sell Lies, and Make a Killing in the Corporate World* (Doubleday, 2022).

141. Lazarus et al., "The Climate Responsibilities of Industrial Meat and Dairy Producers."

142. Alexandra Gaucher-Holm, Christine Mulligan, Mary R. L'Abbé, Monique Potvin Kent, and Lana Vanderlee, "Lobbying and Nutrition Policy in Canada: A Quantitative Descriptive Study on Stakeholder Interactions with Government Officials in the Context of Health Canada's Healthy Eating Strategy," *Globalization and Health* 18, no. 54 (2022): 1–12.

143. Clare Carlile, Rachel Sherrington, and Hazel Healy, "Big Ag Delegates More Than Double at COP27," *DeSmog*, November 18, 2022, https://www.desmog.com/2022/11/18/big-agribusiness-delegates-double-cop27/; Rachel Sherrington, Clare Carlile, and Hazel Healy, "Big Meat and Dairy Delegates Triple at COP28," *DeSmog*, December 8, 2023, https://www.desmog.com/2023/12/08/big-meat-dairy-delegates-triple-cop28/; Lisa Held, "At an Annual Sustainability Gathering, Big Ag Describes its Efforts to Control the Narrative," *Civil Eats*, December 9, 2021, https://civileats.com/2021/12/09/at-an-annual-sustainability-gathering-big-ag-describes-its-efforts-to-control-the-narrative/.

144. "COP28 UAE Declaration on Sustainable Agriculture, Resilient Food Systems, and Climate Action," December 2023, https://www.cop28.com/en/food-and-agriculture.

145. Rachel Sherrington, "U.S. Meat Lobby Celebrates 'Positive Outcome' of COP28," *DeSmog*, April 8, 2024, https://www.desmog.com/2024/04/08/us-meat-lobby-celebrates-positive-outcome-cop28/.

146. "Achieving SDG 2 Without Breaching the 1.5°C Threshold: A Global Roadmap," UN Food and Agriculture Organization, 2023, https://www.fao.org/interactive/sdg2-roadmap/en/.

147. Sherrington, "U.S. Meat Lobby Celebrates."

148. Cleo Verkuijl, Jan Dutkiewicz, Laura Scherer, Paul Behrens, Michael Lazarus, Maria José Hötzel, Rebecca Nordquist, and Matthew Hayek, "FAO's 1.5 °C Roadmap for Food Systems Falls Short," *Nature Food* 5 (2024): 264–266.

149. Arthur Nelsen, "UN Livestock Emissions Report Seriously Distorted Our Work, Say Experts," *The Guardian*, April 19, 2024, https://www.theguardian.com/environment/2024/apr/19/un-livestock-emissions-report-seriously-distorted-our-work-say-experts. The original letter can be found at https://www.

universiteitleiden.nl/binaries/content/assets/science/cml/essays/retraction-request-pathways-to-lower-emissions.pdf.

150. Ibid.; see also Arthur Nelsen, "Ex-Officials at UN Farming Body Say Work on Methane Emissions Was Censored," *The Guardian*, October 20, 2023, https://www.theguardian.com/environment/2023/oct/20/ex-officials-at-un-farming-fao-say-work-on-methane-emissions-was-censored.

151. Michael Fakhri, "The Food System Summit's Disconnection From People's Real Needs," *Journal of Agricultural and Environmental Ethics* 35, no. 16 (2022): 1–9.

152. Zach Boren, "Meat Industry Pushes UN Food Summit to Back Factory Farming," *Unearthed*, September 21, 2021, https://unearthed.greenpeace.org/2021/09/21/un-food-systems-summit-meat-climate/.

153. Jennifer Clapp, Indra Noyes, and Zachary Grant, "The Food Systems Summit's Failure to Address Corporate Power," *Development* 64 (2021): 192–198.

154. A leaked draft of the IPCC AR6 WGIII report included the following statements: "A shift to diets with a higher share of plant-based protein in regions with excess consumption of calories and animal-source food can lead to substantial reductions in GHG emissions" and "Plant-based diets can reduce GHG emissions by up to 50% compared to the average emission intensive Western diet." See Lawrence Carter and Crispin Dowler, "Leaked Documents Reveal the Fossil Fuel and Meat Producing Countries Lobbying Against Climate Action," *Unearthed*, October 2021, https://unearthed.greenpeace.org/2021/10/21/leaked-climate-lobbying-ipcc-glasgow/; Michael Thomas, "How Meat and Fossil Fuel Producers Watered Down the Latest IPCC Report," *Distilled*, March 23, 2023, https://www.distilled.earth/p/how-meat-and-fossil-fuel-producers.

155. Morris and Jacquet, "The Animal Agriculture Industry."

156. Ibid.

157. Zach Boren, "Revealed: How the Meat Industry Funds the 'greenhouse Gas Guru,'" *Unearthed*, October 2022, https://unearthed.greenpeace.org/2022/10/31/frank-mitloehner-uc-davis-climate-funding/.

158. Ibid.

159. Ibid.

160. *DeSmog*, "Danish Crown," https://www.desmog.com/danish-crown/ (accessed March 27, 2025).

161. Naomi Oreskes, "Science and Public Policy: What's Proof Got to Do With It?," *Environmental Science and Policy* 7 (2004): 369–383.

162. "The Dublin Declaration," https://www.dublin-declaration.org/ (accessed May 31, 2024); "The Dublin Declaration of Scientists on the Societal Role of Livestock," *Animal Frontiers* 13, no 2 (2023): 10.

163. Damian Carrington, "Revealed: The Industry Figures Behind 'Declaration of Scientists' Backing Meat Eating," *The Guardian*, October 27, 2023, https://www.theguardian.com/environment/2023/oct/27/revealed-industry-figures-declaration-scientists-backing-meat-eating.

164. Helena Bottemiller Evich, "Agriculture Department Buries Studies Showing Dangers of Climate Change," *Politico*, June 23, 2019, https://www.politico.com/story/2019/06/23/agriculture-department-climate-change-1376413; Oliver Millman, "US Federal Department Is Censoring Use of Term 'Climate Change,' Emails Reveal," *The Guardian*, August 7, 2017, https://www.theguardian.com/environment/2017/aug/07/usda-climate-change-language-censorship-emails.

165. Don F. Hadwiger, *The Politics of Agricultural Research* (University of Nebraska Press, 1982).

166. "Public Research, Private Gain: Corporate Influence Over University Agricultural Research," Food and Water Watch, April 2012, https://foodandwaterwatch.org/wp-content/uploads/2021/03/Public-Research-Private-Gain-Report-April-2012.pdf; Gordon Rausser, Leo K. Simon, Reid Stevens, "Public vs. Private Good Research at Land-Grant Universities," *Journal of Agricultural & Food Industrial Organization* 6, no. 2 (2008), https://doi.org/10.2202/1542-0485.1236; G. Edward Schuh, "Revitalizing Land Grant Universities: It's Time to Regain Relevance," *Choices* 1, no. 2 (1986): 6–10.

167. Christopher Hurt, "Industrialization in the Pork Industry," *Choices* 9, no. 4 (1994): 1–5.

168. Trevor J. Smith, "Corn, Cows, and Climate Change: How Federal Agriculture Subsidies Enable Factory Farming and Exacerbate US Greenhouse Gas Emissions," *Washington Journal of Environmental Law & Policy* 9, no. 26 (2019), https://digitalcommons.law.uw.edu/wjelp/vol9/iss1/3.

169. Megan Stubbs, "Agricultural Conservation in the 2018 Farm Bill Congressional Research Service Report R45698," Congressional Research Service, 2019.

170. Anne Schechinger, "New EWG Analysis: Of $7.4B Spent on Two of USDA's Biggest Conservation Programs in Recent Years, Very Little Went to 'Climate-Smart' Agriculture," Environmental Working Group, September 28, 2022, https://www.ewg.org/research/new-ewg-analysis-74b-spent-two-usdas-biggest-conservation-programs-recent-years-very.

171. Silvia Secchi, "What Decades of Policies Aimed at Agricultural Water Pollution Can Teach Us About Agricultural Climate Change Mitigation: A US Perspective," *Frontiers in Sustainable Food Systems* 7 (2023): 1205510, https://doi.org/10.3389/fsufs.2023.1205510.

172. USDA, "Partnerships for Climate-Smart Commodities Project Summaries," https://www.usda.gov/climate-solutions/climate-smart-commodities/projects (accessed March 27, 2025).

173. *Secchi v. United states department of agriculture et al*, 2024, https://dockets.justia.com/docket/district-of-columbia/dcdce/1:2024cv01769/269726.

174. USDA, "Partnerships for Climate-Smart Commodities Project Summaries," https://meatthefacts.eu/.

175. Rena I. Steinzor and Ling-Yee Huang, "Agricultural Secrecy: Going Dark Down on the Farm: How Legalized Secrecy Gives Agribusiness a Federally Funded Free Ride," Center for Progressive Reform, September 15, 2012.

176. D. Lee Miller and Gregory Buren, "CAFOs: What We Don't Know Is Hurting Us," Natural Resources Defense Council, September 2019, https://www.nrdc.org/sites/default/files/cafos-dont-know-hurting-us-report.pdf.

177. See, e.g., Eddy Wax and Emma Anderson, "The Transatlantic Relationship Descends Into a Food Fight," *Politico*, September 29, 2021; Cary Spivak, "Ex-Agriculture Secretary Tom Vilsack Is the Top Paid Executive at Dairy Management," *Milwaukee Journal Sentinel*, December 2, 2019.

178. George Cusworth, Jeremy Brice, Jamie Lorimer, and Tara Garnett, "When You Wish Upon a (GWP) Star: Environmental Governance and the Reflexive Performativity of Global Warming Metrics," *Social Studies of Science* 53 (2022): 3–28.

179. Matthew N. Hayek, Jack Samuel, and Shelby C. McClelland, "Methane Metrics: The Political Stakes," *Nature* 620, no. 7972 (2023): 37–37.

180. Ibid.; Cusworth et al., "When You Wish Upon a (GWP) Star"; Caspar L. Donnison and Donal Murphy-Bokern, "Are Climate Neutrality Claims in the Livestock Sector

Too Good to Be True?," *Environmental Research Letters* 19 (2024): 011001, https://doi.org/10.1088/1748-9326/ad0f75.

181. "Meat and Dairy Facts," Bord BIA, Dairy Industry Ireland, ICMSA, IFA, National Diary Council, and Meat Industry Ireland, https://meatanddairyfacts.ie (accessed March 27, 2025); "BeefUp Sustainability," NCBA, https://beefupsustainability.com (accessed March 27, 2025); "Meat the Facts," European Livestock Voice, https://meatthefacts.eu (accessed March 27, 2025); Fassler, 2023.

182. Barbara E. Willard, "The American Story of Meat: Discursive Influences on Cultural Eating Practice," *Journal of Popular Culture* 36, no. 1 (2003): 105–118; *DeSmog*, "National Cattlemen's Beef Association," https://www.desmog.com/national-cattlemens-beef-association/ (accessed March 27, 2025).

183. USDA employee newsletter, https://www.zimmcomm.biz/usda/HQGreening UpdatesJuly2012.pdf (accessed March 27, 2025).

184. Gretchen Goetz, "USDA Supports Meatless-less Mondays," *Food Safety News*, July 30, 2012, https://www.foodsafetynews.com/2012/07/meatless-less-monday/.

185. Eyder Peralta, "After Uproar, USDA Walks Back 'Meatless Monday' Support," NPR, July 26, 2012, https://www.npr.org/sections/thetwo-way/2012/07/26/157432998/after-uproar-usda-walks-back-meatless-monday-support.

186. E.g., Stanescu, "Cowgate."

187. E.g., Silje Kristiansen, James Painter, and Meghan Shea, "Animal Agriculture and Climate Change in the US and UK Elite Media: Volume, Responsibilities, Causes and Solutions," *Environmental Communication* 15, no. 2 (2021): 153–172; Jose A. Moreno and Núria Almiron, "Representación en la prensa española del papel de la agricultura animal en la crisis climática," *Estudios sobre el Mensaje Periodístico* 27, no. 1 (2021): 349–364; Katherine Sievert, Mark Lawrence, Christine Parker, Cherie A Russell, and Phillip Baker, "Who Has a Beef with Reducing Red and Processed Meat Consumption? A Media Framing Analysis," *Public Health Nutrition* 25, no. 3 (2022): 578–590.

188. Coni Arévalo, Jenny Splitter, and Jo Anderson, "Animal Agriculture Is The Missing Piece In Climate Change Media Coverage," Faunalytics, 2023, https://faunalytics.org/animal-ag-in-climate-media/.

189. Anthony Leiserowitz, Matthew Ballew, Seth Rosenthal, and Jillian Semaan, "Climate Change and the American Diet," Yale Center for Climate Change Communications, February 13, 2020, https://climatecommunication.yale.edu/publications/climate-change-and-the-american-diet/; De Boer and Aiking, "EU Citizen Support for Climate-Friendly Agriculture."

190. Leiserowitz et al., "Climate Change and the American Diet."

191. ProteinPACT, promotional video on website homepage, https://theproteinpact.org/ (accessed September 29, 2023).

192. BBC, "Dutch Tractor Protest Sparks 'Worst Rush Hour,'" *BBC News*, October 1, 2019, https://www.bbc.com/news/world-europe-49891449; Ashoka Mukpo, "How Manure Blew Up in the Netherlands," *Mongabay*, September 6, 2023, https://news.mongabay.com/2023/09/how-manure-blew-up-the-netherlands/.

193. Erik Stokstad, "Nitrogen Crisis Threatens Dutch Environment—and Economy," *Science* 366, no. 6470 (2019): 1180–1181; Mukpo, 2023.

194. Ashoka Mukpto, "In the Netherlands, Pitchforks Fly for an Empire of Cows," *Mongabay*, September 7, 2023, https://news.mongabay.com/2023/09/in-the-netherlands-pitchforks-fly-for-an-empire-of-cows/.

195. Karl Mathiesen, "The Chemist v. the Dutch Farmers," *Politico*, March 9, 2023, https://www.politico.eu/article/johan-vollenbroek-netherlands-nitrogen-

pollution-climate-change-farming/; "Openbare intimidatie', kritiek op uit-spraken Farmers Defence Force-leider," *NOS Nieuws*, June 18, 2022, https://nos.nl/artikel/2433184-openbare-intimidatie-kritiek-op-uitspraken-farmers-defence-force-leider.

196. Peter Dejong, "Dutch Farmers Block Entrances to Supermarket Warehouses," *AP News*, July 2022, https://apnews.com/article/netherlands-blockades-climate-and-environment-1a5ac6c802fcd641f53ff277a7e72194.

197. Jan Douwe van der Ploeg, "Farmers' Upheaval, Climate Crisis and Populism," *Journal of Peasant Studies* 47, no. 3 (2020): 589–605.

198. Ibid.

199. Lisa Peters, "Woede of Marketing? Hoe Een Veevoergigant Boerenprotesten Een Zet Gaf," *NUweekend*, December 27, 2019, https://www.nu.nl/weekend/6020408/woede-of-marketing-hoe-een-veevoergigant-boerenprotesten-een-zet-gaf.html.

200. V. Harmsen, "Agrarische lobby presenteert zich als onafhankelijk onderzoeks-bureau," *Follow The Money*, January 29, 2020. https://www.ftm.nl/artikelen/agrifacts.

201. Dana Kamphorst, Josine Donders, Tineke de Boer, and Nienke Nuesink, "Maatschappelijk Debat Naar Aanleiding van Het PAS Arrest En de Mogelijke Invloed Op Het Natuurbeleid," *WOT Natuur & Milieu*, October 2021, https://edepot.wur.nl/547930.

202. Van der Ploeg, "Farmers' Upheaval."

203. NOS, "Schouten Tegen Protesterende Boeren: Geen Halvering van Veestapel," *NOS Nieuws*, October 1, 2019, https://nos.nl/artikel/2304179-schouten-tegen-protesterende-boeren-geen-halvering-van-veestapel.

204. Van der Ploeg, "Farmers' Upheaval."

205. Margaret Donnelly, "Meat and Dairy Facts Changes PR Advisors," *Irish Independent*, February 4, 2020, www.independent.ie/farming/agri-business/meat-and-dairy-facts-changes-pr-advisors/38922163.html.

206. Núria Almiron, Miquel Rodrigo-Alsina, and Jose A. Moreno, "Manufacturing Ignorance: Think Tanks, Climate Change and the Animal-Based Diet," *Environmental Politics* 31, 4 (2022): 576–597.

207. Jack Hubbard, "New Think Tank Takes on Sustainability, Animal Extrem-ists," *Meatingplace Magazine*, May 9, 2023, https://www.meatingplace.com/Industry/Blogs/Details/109641; Center for Environment and Welfare, https://environmentandwelfare.com/ (accessed September 3, 2023).

208. Eric Lipton, "Hard-Nosed Advice From Veteran Lobbyist: 'Win Ugly or Lose Pretty,'" *New York Times*, October 31, 2014, https://www.nytimes.com/2014/10/31/us/politics/pr-executives-western-energy-alliance-speech-taped.html.

209. Jonathan Latham, "Way Beyond Greenwashing: Have Corporations Captured 'Big Conservation'?," *TruthOut*, March 6, 2012, https://truthout.org/articles/way-beyond-greenwashing-have-corporations-captured-big-conservation/.

210. "Civil Society Calls for JBS to Be Stripped of A Minus Climate Rating," *Mighty Earth*, March 29, 2023, https://www.mightyearth.org/jbs-climate-rating; Jonathan Watts, "Brazilian Meat Firm's A- Sustainability Rating Has Cam-paigners Up in Arms," *The Guardian*, Marcg 30, 2023, https://www.theguardian.com/environment/2023/mar/30/brazilian-meatpackers-a—sustainability-rating-raises-grade-inflation-concerns; Georgina Gustin, "The Department of Agriculture Rubber-Stamped Tyson's 'Climate Friendly' Beef, but No One Has Seen the Data Behind the Company's Claim," *Inside Climate News*, May 8, 2024,

https://insideclimatenews.org/news/08052024/usda-tyson-climate-friendly-beef-claim/.

211. E.g., Jennifer Jacquet, "The Meat Industry Is Doing Exactly What Big Oil Does to Fight Climate Action," *Washington Post*, May 14, 2021, https://www.washingtonpost.com/outlook/the-meat-industry-is-doing-exactly-what-big-oil-does-to-fight-climate-action/2021/05/14/831e14be-b3fe-11eb-ab43-bebddc5a0f65_story.html; Lazarus et al., "The Climate Responsibilities of Industrial Meat and Dairy Producers"; Thomas, 2023.

212. DeSmog, "Agribusiness Database," https://www.desmog.com/agribusiness-database-/ (accessed March 27, 2025).

213. Daina Bray and Thomas M. Poston, "The Methane Majors: Climate Change and Animal Agriculture in U.S. Courts," *Columbia Journal of Environmental Law* 49 (2024): 145–248.

214. Katy Askew, "Climate Labels, Greenwashing and the Row Over the 'Climate-Controlled Pig,'" *FoodNavigator*, October 19, 2021, https://www.foodnavigator.com/Article/2021/10/19/Climate-labels-greenwashing-and-the-row-over-the-climate-controlled-pig.

215. "Doctors Group Petitions FTC to Stop National Cattlemen's Beef Association From Placing Ads That Downplay Beef's Impact on Climate Crisis," *Physicians Committee for Responsible Medicine*, August 2021, https://www.pcrm.org/news/news-releases/doctors-group-petitions-ftc-stop-national-cattlemens-beef-association-placing.

216. Daniel Boffey, "Dutch City Becomes World's First to Ban Meat Adverts in Public," *The Guardian*, September 6, 2022, https://www.theguardian.com/world/2022/sep/06/haarlem-netherlands-bans-meat-adverts-public-spaces-climate-crisis.

217. ClientEarth, "Agricultural giant Cargill Faces Legal Complaint Over Deforestation and Human Rights Failings in Brazil," https://www.clientearth.org/latest/press-office/press/agricultural-giant-cargill-faces-legal-complaint-over-deforestation-and-human-rights-failings-in-brazil/ (accessed September 30, 2023); ClientEarth, "Summary of Complaint to the United States National Contact Point for the OECD against Cargill," May 2023, https://www.clientearth.org/media/puvcgh42/summary-of-clientearth-s-oecd-complaint-against-cargill.pdf.

218. Ibrahim Coulibaly, "The Global Food Crisis: The Right to Decide What We Eat," *The Guardian*, June 2, 2011, https://www.theguardian.com/global-development/poverty-matters/2011/jun/02/food-sovereignty-for-farmers.

219. Natalia Mamonova and Jaume Franquesa, "Populism, Neoliberalism and Agrarian Movements in Europe. Understanding Rural Support for Right-Wing Politics and Looking for Progressive Solutions," *Sociologia Ruralis* 60 (2020): 710–731.

220. JBS Sustainability-Linked Bond Framework, June 2021, 4, https://api.mziq.com/mzfilemanager/v2/d/043a77e1-0127-4502-bc5b-21427b991b22/7b93baf6-49d3-66ec-916f-db73014a99c3.

221. IATP, *DeSmog*, and Feedback Global, "World's Largest Meat Company, JBS, Increases Emissions in Five Years Despite 2040 Net Zero Climate Target, Continues to Greenwash Its Huge Climate Footprint," IATP, April 21, 2022, https://www.iatp.org/media-brief-jbs-increases-emissions-51-percent; IATP and Changing Markets Foundation, 2022.

222. "Mighty Earth Files Complaint with US Securities and Exchange Commission Against JBS 'Green Bonds,'" *Mighty Earth*, January 18, 2023, https://www.mightyearth.org/whistleblower-complaint-to-the-securities-and-exchange-commission-against-jbs/.

223. "Corporate Climate Responsibility Monitor 2022," *New Climate Institute/Carbon Market Watch*, February 2022, 85, https://newclimate.org/sites/default/files/2022-06/CorporateClimateResponsibilityMonitor2022.pdf.

224. JBS 4Q16 and 2026 Financial Statements, March 13, 2017, https://sec.report/otc/financial-report/167764; IATP and Changing Markets Foundation, 2022.

225. Lara Williams, "Meatpacker JBS Listing in NYC Would Be an ESG Nightmare," *Washington Post*, September 5, 2023, https://www.washingtonpost.com/business/energy/2023/09/05/brazilian-meatpacker-jbs-listing-on-nyse-would-be-an-esg-nightmare/9a2531f8-4ba3-11ee-bfca-04e0ac43f9e4_story.html.

226. New Climate Institute/Carbon Market Watch, "Corporate Climate Responsibility Monitor 2022."

227. "National Advertising Review Board Recommends JBS Discontinue 'Net Zero' Emissions by 2040 Claims," National Advertising Review Board, June 20, 2023, https://bbbprograms.org/media-center/dd/narb-jbs-net-zero-emissions.

228. "Attorney General James Sues World's Largest Beef Producer for Misrepresenting Environmental Impact of Their Products," New York State Attorney General, Accessed April 16, 2024, https://ag.ny.gov/press-release/2024/attorney-general-james-sues-worlds-largest-beef-producer-misrepresenting.

229. *The People of the State of New York v. JBS USA Food Company and JBS USA Food Company Holdings*, Complaint, Supreme Court of the State of New York, February 28, 2024, https://ag.ny.gov/sites/default/files/court-filings/jbs-complaint.pdf.

230. See, e.g., Phil McKenna, Georgina Gustin, and Peter Aldhous, "A Texas Dairy Ranks Among the State's Biggest Methane Emitters. But Don't Ask the EPA or the State About It," *InsideClimate News*, August 18, 2023, https://insideclimatenews.org/news/18082023/texas-dairy-among-states-biggest-methane-emitters/; Ben Chugg, Brandon Anderson, Seiji Eicher, Sandy Lee, and Daniel E. Ho, "Enhancing Environmental Enforcement with Near Real-Time Monitoring: Likelihood-Based Detection of Structural Expansion of Intensive Livestock Farms," *International Journal of Applied Earth Observation and Geoinformation* 103 (2021): 102463, https://doi.org/10.1016/j.jag.2021.102463.

CHAPTER 5

Climate Policy Obstruction on the Right and the Far Right

LEAD AUTHORS: DIETER PLEHWE AND JUSTIN FARRELL

CONTRIBUTING AUTHORS: LUCAS ARALDI, ROBERT J. BRULLE, JESSE CALLAHAN BRYANT, WILLIAM CALLISON, KERT DAVIES, RUTH E. McKIE, SOTIRIS MITRALEXIS, AND ALEXANDRU RACU

INTRODUCTION

The political complexity of climate obstruction has increased in recent years, with conservative and far-right movements playing interconnected yet distinct roles. While each movement actively resists climate policy at national and international levels, their respective philosophies and tactics differ significantly. Understanding these similarities and differences is critical to forming a more complete understanding of climate obstruction globally.

This chapter untangles and reveals the overlaps in these intertwined narratives, highlighting the increasing power of far-right movements in climate obstruction and presenting research on ways these movements seek to dismantle international governance and cooperation to hinder meaningful global action on climate change. As these movements grow and as neoliberal conservatism continues to be a deregulatory force in global politics, it is important to carefully catalogue their similarities, differences, and impacts. The chapter begins by examining climate obstruction efforts within contemporary conservativism, focused on its anti-regulatory roots in the United States before turning to a global analysis of its organizational, financial, and cultural linkages. Here we describe how mainstream conservatism has successfully obstructed climate policy by advocating for deregulation of free markets, defending private property rights, and challenging the veracity of scientific consensus.

Dieter Plehwe et al., *Climate Policy Obstruction on the Right and the Far Right*. In: *Climate Obstruction*.
Edited by: J. Timmons Roberts, Carlos R. S. Milani, Jennifer Jacquet, and Christian Downie,
Oxford University Press. © Oxford University Press (2025). DOI: 10.1093/oso/9780197787144.003.0005

Next, we consider the role of the far right in global climate obstruction today, a movement rooted in antipathy toward multiculturalism and cooperation among nation states and in support for ethnopluralism. While often sharing neoliberal conservatism's goal of climate obstruction, the far right tends to take a more radical approach. As we show, the far right's major problem with climate change is not necessarily the science per se, but rather the alleged attempts by elites and global powers to use climate science and climate policy as a Trojan horse to expand global governance and undermine ethnic heritage, cultural identity, and national sovereignty. Finally, we present research on its specific tactics before offering avenues for future research on this evolving global movement.

CONSERVATISM

A network of organizations and advocacy groups has promoted approaches to environmental management that emphasize local control, private property, and individual rights. Over time, these efforts have expanded globally through coordinated initiatives and partnerships focused on shaping public policy and climate discourse.

"Wise Use" Roots

A long history of right-wing anti-environmental ideology, activism, and political action in the United States continues to play a role in the contemporary opposition to effective global climate policy. Nonprofit organizations and their front groups have fought against federal and state environmental regulations protecting everything from wetlands and waterways to endangered species. These organizations have been supported by corporate funders who have an interest in weakening or eliminating environmental regulations[1] and tend to adopt a property-rights-based counternarrative against federal and state environmental regulation.

For example, a landmark event in this history is the Sagebrush Rebellion of the 1970s–1980s, a battle over federal-versus-state land management in Western states that was emblematic of a policy debate over public-versus-private control of nature. Named for the contested land (the sagebrush steppe), the "rebellion" was part of a growing conservative movement that advocated for commercial use if not outright privatization of land under government control and mobilized against federal environmental laws and environmentalism generally.[2] The "wise use" concept emerged from this movement, emphasizing local stewardship and broad individual rights to use nature for human benefit. The Wise Use Movement was founded in 1988

> out of fear that George Bush Sr. was going to live up to his campaign pledge to be "the environmental president." This cabal of anti-environmental activists, organized by federally subsidized industries dependent on public lands, issued a natal document, the Wise Use Agenda. It called for, among other things: drilling the Arctic National Wildlife Refuge, logging Alaska's Tongass National Forest, opening wilderness to energy development, gutting the Endangered Species Act, and privatizing national parks.[3]

Mirroring the strategies of environmental movements, wise use organizations succeeded in mobilizing a wide range of conservative constituencies. For example, the movement in 1994 succeeded in blocking US ratification of the international Convention on Biological Diversity.[4]

This anti-environmental, property-based "guardianship" agenda spread beyond the United States to countries including Canada (with the formation of organizations such as Canadian Women in Timber, Share Canada, etc.) and Australia (Forest Protection Society, etc.).[5] The global expansion also included financial support and promotion of the Wise Use Movement by the Reverend Sun Myung Moon's Unification Church.[6] In addition to advancing wise-use arguments, the conservative conservation movement during this period began to associate environmental activism with terrorism, and contemporary framing of "last generation" climate activism as "eco-terrorism" echoes these earlier versions of that narrative.[7]

Organizational Linkages

The climate policy landscape includes a broad range of think tanks and non-profit organizations with diverse secular and religious backgrounds. Many of these groups are connected through shared funding sources and networks, forming a transnational civil society that engages in coordinated efforts to shape climate governance through a variety of strategies and approaches.

Think Tanks

Conceptual overlap exists between the rhetoric of "free markets" and concepts of "wise use," stewardship, and subsidiarity. US-based think tanks are significant curators and amplifiers of opposition to scientific evidence related to climate change and related policy action such as legislation and regulations.[8] Outside of the United States, these organizations work within a universe of think tanks operating across countries—sometimes connected under large umbrella organizations such as the global Atlas Network, the European Stockholm Network, and the Epicenter Network—to produce and distribute policy reports and otherwise attempt to influence politics and politicians on the issue of climate change.

Most of these organizations and figures oppose ambitious climate policy, in particular the close international cooperation and coordination required to create and implement such policies. Through information dissemination including not just policy reports but also research papers, conferences, and other branded activities, think tanks present quasi-academic positions on the best approaches to tackling the climate crisis. Many of their positions are grounded in the same narrative: positioning climate change not as a science-based issue but as an inherently political topic. Underlying this way of thinking is an economic philosophy of individual liberty and economic and political freedom expressed in a triad of values: property rights, cost-benefit analysis, and local mandate.[9]

Think tanks focused specifically on climate denial—such as the US-based Heartland Institute, the UK-based Global Warming Policy Foundation, and South Africa's Free Market Foundation[10]—challenge the veracity of climate science and attack environmental activists and nongovernmental organizations (NGOs). Other think tanks involved in climate-obstructing activities emphasize market-based mechanisms as a solution to most contemporary social and political problems. According to this worldview, private initiative and the vital role of business will stimulate both a free-market economy and a free-market environmentalism of ideas[11] that will help solve the world's biggest challenges. Some right-wing think tanks in the climate space, such as the Heritage Foundation, promote traditionally conservative values. Others, such as the Mises Institutes, advocate a more libertarian position.[12] Yet all have a fundamental commitment to influence policy and public opinion in favor of economic and political freedoms with limited to no government intervention.

The Atlas Network

The Atlas Network is an important US-based umbrella organization for the neoliberal free-market movement (or "Freedom Movement"[13]), which includes numerous climate obstructionist groups around the world. Founded in 1981 by Antony Fisher, the organization (then named the Atlas Economic Research Foundation) coordinated think tanks Fisher had helped to set up and supported efforts in numerous countries to found new think tanks modelled after the British Institute of Economic Affairs (IEA), which had been the first think tank founded by Fisher, in 1955.[14] Fisher established these and other think tanks on the advice of economist Friedrich August von Hayek. Both were members of the Mont Pèlerin Society of neoliberal intellectuals, which Hayek founded in 1947 and that continues to serve as a pipeline for the board and staff members of Atlas Network partners.[15]

Most think tanks involved in this free-market advocacy network share a basic anti-collectivist and anti-regulatory ideology, which in many cases includes opposition to environmental and climate policymaking. For example, research has revealed the close links between fossil fuel capital, Rupert Murdoch's media empire, and Australian neoliberal think tanks that are part of the network.[16] Recent investigative journalism has documented the role of Atlas Network partners in various countries in a strategic mobilization to criminalize climate justice movements.[17] To this end, Atlas think tanks from diverse regions that have a long history of opposition to the environmental movement have deployed similar strategies and media-savvy campaigns. These tactics have included ad-hominem attacks against climate leaders such as Greta Thunberg, organizations such as Greenpeace, and protest tactics such as sit-down blockades that liken activists to terrorists or Nazis. Atlas partners have also used such "strategy mobility," similar to "policy mobility," to prevent or oppose climate mitigation policies across borders.[18]

Right-wing conservatism remains central to contemporary climate obstructionist mobilizations beyond the United States, and key organizations in the history of the anti-environmental movement[19] (such as the American Legislative Exchange Council) are also partners of the Atlas Network. Atlas is also home to think tanks strongly emphasizing property- and property rights–based forms of conservation such as the Property and Environment Research Center. In 2020, Atlas partner the Austrian Economics Center, in collaboration with two recently founded conservation groups in the United States and the United Kingdom,[20] radicalized the right-wing conservation approach with their book *Green Market Revolution: How Market Environmentalism Can Protect Nature and Save the Planet*. Remarkably, the authors suggest that post-Paris climate policy development is moving in the right direction thanks to the increasing use of market-based instruments, although the book suggests that the privatization of nature must be a precondition of global warming mitigation.[21] Another evolving campaign in Atlas Network circles[22] is the "alternative solution" free-market advocacy of foundation president Rod Richardson, successor of one of the early funders of Atlas think tanks around the world.[23] Richardson created the "clean capitalism without borders" concept, which advocates giving tax breaks for "green" investments secured in international treaties.[24] The Ludwig von Mises Institute, based in the US state of Alabama, is another far-right/libertarian force in climate obstruction that has evolved into a global network of think tanks.[25] Several Mises Institutes in different countries have been active in the Atlas network; according to the results of two global surveys of both networks, about one-quarter of the individuals affiliated with Mises Institutes are also affiliated with Atlas think tanks.[26]

In turn, leading intellectuals involved in Mises Institutes around the world, including economist Hans Hermann Hoppe, founded the Property and Freedom Society. The Society advocates a new version of far-right neoliberalism that opposes almost any state interventionism and supports radical privatization and decentralization. The group's rejection of global policy coordination under the umbrella of the United Nations is similar to that of the anti-systemic far right. Property and Freedom split away from the Mont Pèlerin Society in 2004; the latter supports both culturally progressive and conservative varieties of neoliberalism and, in the context of climate change, is home to supporters of global economic constitutionalism and both supporters and opponents of international policy coordination.[27]

Discourses of Delay

Beyond their free-market rhetoric, a wide range of think tanks has developed other discourses of climate delay,[28] including promoting nontransformative solutions and advocating limited responsibility for addressing climate change. These strategies include the growing emphasis on achieving net-zero emissions, a focus on carbon sequestration rather than de-carbonization, and an emphasis on "discounting" in economic models (which asserts that mitigation can wait because it will be less expensive in the future). These delay-focused forms of discursive obstruction have not only been integrated across the political and corporate policy spheres, but also accepted in the public mind.[29]

These narratives have been echoed by individual members of these think tanks, who spread these ideologically supported positions on climate change via public-facing media appearances.[30] Murdoch media regularly feature content from both free-market conservative and hard-right climate denialist think tanks, companies, and funders in North America, particularly those based in the United States[31] (e.g., the Heartland Institute). They also spotlight speakers and content from think tanks based in Canada (Frontier Centre for Public Policy), Australia (Institute of Public Affairs, Centre for Independent Studies), Europe (Institute of Economic Affairs, the Austrian Economics Center), and South Africa (Free Market Foundation, South African Institute of Race Relations), for example. Consequently, partisan think tanks have ultimately been able to guide "public opinion, political debate, economic discourse and even academic discussions on climate change mitigation" in several countries and policy arenas.[32]

Financial Linkages

Scholarly research on the relationship between conservative foundations and climate obstruction began with a seminal quantitative 2014 analysis of the US Climate Change Countermovement (CCCM).[33] The CCCM is a network of organizations (including think tanks), individuals, and activities dedicated to preventing ambitious climate policy. Since then, research on philanthropic financial support within the political economy of climate obstruction has revealed three themes.

The first theme is opacity: reliable data on foundation support are difficult to obtain. When such data are gathered, it can be difficult to discern the intent and influence of this type of financial support. For example, 74% of donations made through donor-advised funds come from unidentified sources.[34] To learn more about funding patterns, it can be helpful to trace funding flows and board membership. The Dunn Foundation, for example, has funded several think tanks involved in climate denial, including the Heartland Institute and the Cato Institute. Between 2002 and 2017, the latter received more than $7 million from the Dunn Foundation;[35] a foundation member also serves on Cato's supervisory board.

While some such think tanks in the United States reportedly benefit from untraceable contributions from conservative family donations and the libertarian/conservative Donors' Trust,[36] others lack even limited transparency when it comes to their funding sources. Recent research has documented the funding patterns established by brothers Charles and David Koch of corporate giant Koch Industries, who discreetly fund the climate change countermovement by directing money through their family foundations (the Charles Koch Foundation and the David H. Koch Charitable Foundation). The beneficiaries are a network of conservative think tanks and advocacy groups that actively promote climate change skepticism and oppose environmental regulations, often funneling the funds through "dark money" organizations such as Donors Trust to obscure the source of these donations. (This network is sometimes referred to as the "Kochtopus" due to its extensive reach and influence.)[37]

The second theme is concentration. Despite the fact that many contributions to obstructionist think tanks and civil society groups are made anonymously, research has found that a small group of foundations tends to provide the bulk of financial support for organizations involved in climate obstruction efforts:[38] "The top 1% of grant makers account for 67% of grants, and the top 10% of grant makers account for 95% of grants."[39] In Europe, the number of foundations involved in promoting and/or funding climate change denial is smaller than in the United States. To compare the two regions, one can examine the foundations supporting think tanks that are partners of the global Atlas Network. Many Atlas think tanks are involved in climate denial and other forms of obstruction, as discussed earlier. While there is no information on foundation funding in Europe comparable to that available in the United States, a recent survey found forty-eight foundation representatives on the boards of Atlas think tanks in Europe, compared with almost three hundred individuals representing more than two hundred foundations who served on the boards of Atlas think tanks in the United States.[40] A few European foundations in the survey sample are notable for funding Climate Countermovement organizations in Latin America, including the Fundación para el Análisis y los Estudios Sociales in Spain, and the Friedrich Naumann Foundation, for example, funds them in Germany. Both European foundations fund Latin American think tanks. The European arm of the Atlas Network, the International Policy Network, was involved in channeling US funding to European denial organizations.[41]

The third theme is heterogeneity. Scholars have noted the variety of the climate obstruction actors funded. Foundation support flows to many different organizations, with varied levels of focus on climate change; thus, scholars face a challenge in isolating the impact of these flows on climate obstruction when it is one among many issues of concern within the neoliberal and conservative movement.[42]

Cultural Linkages: The Role of Religion

Many major religions embrace ambitious environmental action and endorse climate mitigation, out of concern over, for example, protecting God's creation. There are also some religious linkages to climate obstruction, especially among rank-and-file adherents whose traditions are more closely aligned with right-wing politics.[43] In the United States and Australia, for example, religiosity has been found to be a key factor in the popularization of climate science skepticism.[44,45] Research on spirituality on the Netherlands has mirrored American patterns.[46] Other types of religious organization have been involved in climate obstruction, notably the US-based Cornwall Alliance for the Stewardship of Creation. Its founder, E. Calvin Beisner of the Catholic Acton Institute, described the group as a "network of evangelical theologians, scientists, and economists."[47] Its activities have included organizing a seminar in Rome with the Acton and Heartland institutes to mobilize against the language of "environmental crisis" in Pope Francis's encyclical *Laudato Si.*[48]

Like the Cornwall Alliance's, many of the best-studied Catholic denialist activities have been traced to the United States. Their main themes are reflected in the prominent blog and forum Catholic Answers: climate change is natural, not anthropogenic; anthropogenic climate change is "prudential judgment" (i.e., belonging to the scientific sphere, which is not infallible); climate concern reflects pantheism and neopaganism; support for subsidiarity (the principle of opposing regulatory state centralization); and environmental stewardship (versus ecological thinking).[49] Because information on religion and climate obstruction is more abundant for the United States than for other nations, the chapter will next consider the phenomenon in other regions and countries that underline the need for more attention to the intersection of religion and conservative climate obstruction across the world.

Russia and Romania

In Russia and Romania, denial of anthropogenic climate change can be encountered among Orthodox believers, parties, and political movements such as the Alliance for the Union of Romanians, which claim to ground their ideology in the Orthodox tradition.[50] Though not officially endorsed by the Orthodox leadership, such views are expressed on Orthodox, nationalist, and conservative websites and through the social media accounts of various Orthodox public figures. Both Romanian and Russian Orthodox ultra-conservative platforms including R3Media, Activenews, Russkaya Narodnaya Liniya, and Tsargrad have spread conspiracy theories about climate change. Playing on the anti-semitic "world domination" tropes in the fictional document "The Protocols of the Elders of Zion," Russian Orthodox media have referred to global climate policy strings being pulled by Western interests ("Protocols of the Overseas Wise Men," "Protocols of the Ozone Wise Men," and so on).[51] Thus, while Orthodox dogma and its followers seem to agree that pollution in general is bad, its believers diverge in that they regard climate change as a lie fabricated by "globalist elites" to impose a worldwide dictatorship.

The Russian Orthodox Church has also addressed the ecological crisis in *The Basis of the Social Concept of the Orthodox Church*, its official document of social doctrine published in 2000. The doctrine blames greed and the modern "spiritual crisis" for "an unprecedented and unjustified growth of public consumption" that has brought the Earth to "the verge of a global ecological disaster."[52] But since then, Russian bishops have seldom spoken about the ecological crisis compared with the Ecumenical Patriarch Bartholomew I of Romania. Other than Pope Francis there may be no religious leader more devoted to environmentalism and the fight against climate change than the Ecumenical Patriarch, whose consistent ecological activism earned him the nickname "the Green Patriarch."[53] In stark contrast with Bartholomew's preaching, the Russian bishops make no specific mention of climate change in their discourse,[54] neither acknowledging nor denying it.

In Romania, religious bloggers and social-media influencers rely for their content almost exclusively on conservative and libertarian English-language sources and on conspiracy theories collected from English language websites such as Natural News, The Gateway Pundit, and Zero Hedge. These articles are then translated into Romanian and published on right-wing Orthodox websites such as Cuvântul Ortodox and those mentioned earlier.[55] They also quote renegade scientists such as climate change–denying physicists John Clauser and Victor Manuel Velasco.[56] Such views occasionally find an echo even within the Church hierarchy. For example, in a pastoral letter from 2022, one Romanian bishop stated that "planned pandemics" and "exaggerated climate obsessions" are both instruments of "the great reset."[57]

Latin America

In Latin America, research on religion and climate change is still in its early stages. There currently exist no comprehensive studies on the transnational dimensions of climate denialism among right-wing religious groups in Latin America. A few studies focus on local networks of religious climate denialism in Brazil and these groups' relationships with the government of Jair Bolsonaro, which utilized climate denialism to foster a populist ideology.[58] Much of this work tends to focus on the linkages between recent conservative political movements and the rise of Pentecostal and neo-Pentecostal evangelical movements.[59] In Brazil, for example, the Bolsonarist right-wing movement is supported politically and financially by a conservative network including many Pentecostal evangelicals (a sector that has grown dramatically since the 1990s and today represents around 37% of the population). But the "strategic role of climate denialism" in "blocking processes of environmental governmentalization in Brazil"[60] has also involved the publication and distribution of translated books by small Catholic publishers with ties to agribusiness.[61] Individual government members, such as former Foreign Minister Ernesto Araújo, have used religion to deny climate change.[62] Both Pentecostal and Catholic groups have reinforced the broader climate countermovement in Brazil; for example, the government refused to host COP 26, the annual international UN climate conference, and withdrew key decision-making power from the country's environmental and climate governance institutions.[63]

The ideology behind such right-wing religious obstructionism in Latin America lies at the intersection of markets and nature, postulating that "God is in control," meaning environmental issues should not be a concern for Christians because the Earth's fate is predetermined. Furthermore, the free market is seen as a channel allowing people to "wisely use" nature, which in its undeveloped state is considered savage. The denial discourses of Brazil's religious obstructionists thus differ only slightly from those in the United States and Europe. However, the former is characterized by a stronger focus on discrediting IPCC research and promoting the idea that environmentalism is a communist conspiracy that aims to destroy Western Christian civilization.[64]

THE FAR RIGHT

Revolutionary far-right movements obstruct climate policy not to preserve the status quo, like mainstream conservatism, but to destabilize global cooperation entirely, often as part of broader ethno-nationalist or accelerationist goals.

Attacking and Obstructing Systems

Unlike mainstream and politically empowered conservative movements, for whom climate obstruction is a strategy to maintain the neoliberal, free-market status quo, revolutionary far-right movements often seek to obstruct national and international climate policy as a strategy for disrupting *all* international policy and cooperation.

There is no single definition of the "far right" because, unlike conservatism, its ideological content is heavily dependent on historical and geographical context. One common characteristic of the far right, however, is that it tends to focus on the mystification of race/ethnicity and to promote violence as a movement tactic.[65] Thus, in contrast to conservatism's essential qualities, far-right actors and movements tend to lack a coherent ideology, favoring a system-attacking stance. Ironically, this position can create strange bedfellows: depending on the setting, far-right positions can resemble conservatism (e.g., anti-equality) or progressivism (e.g., anti-corporate). A common feature of right-wing populism is hostility to compromise and balance. The politics of resentment and a perceived antagonism between "good people" and "bad elites," between those who are thought to legitimately belong to the community and others who do not (migrants, the unemployed, minorities, etc.), increasingly overrides the traditional concerns of centrist conservatives. In other words, traditional conservatism—especially its twenty-first-century neoliberal manifestations—is more conceptually straightforward, often focused on anti-regulatory policy obstruction. Far-right culture can be more complex because it is contingent upon the system it is attacking. Thus, any empirical analysis of its actors and movements must be inherently relational and attuned to the specific field on which climate policy is viewed as threat worthy of obstructing.[66] The revolutionary far right today takes its cues from the European New Right that arose in the 1960s. Today's far right is fundamentally "system attacking," or critical of *both* mainstream liberal and conservative ideologies, which they see as operating within a neoliberal consensus.[67] In contrast to the forces of mainstream climate denial, far-right movements use climate obstruction as part of a broader isolationist program driven by conspiratorial thought and ethno-nationalist goals. For example, in the United States since the 1990s, research shows that the mainstream Republican Party has continually sowed climate skepticism in order to maintain a free-market status quo.[68] In contrast, the far-right or "alt-right" movement sees "the anti-environmentalism of the present conservative movement as nothing but a modernist, capitalist . . . [and] Jewish conception that has found fertile ground in the hands of billionaire plutocrats wiping out nature all over the Earth for the sake of higher profit margins."[69]

In the United States, the mainstream right consistently supports international cooperation in military, corporate, and humanitarian affairs, with the exception of climate change. The European spectrum of conservative and right-wing parties tends to be more diverse, but center-right parties typically embrace international climate cooperation whereas Europe's revolutionary far right sees increased international cooperation negatively, advocating instead for a disintegration of global solidarity.

Multiculturalism and Ethnopluralism

The revolutionary far right envisions a world grounded in "ethnopluralism," in which distinct racial or ethnic groups live in separate regions, preserving their unique identities and rejecting global community connectivity.[70] The slogan of Arktos Media, "Making anti-globalism global since 2009," succinctly captures how the revolutionary far right today expands on the idea of ethnically homogeneous nations, as seen in contemporary movements such as Indian Hindutva and Japanese Shinto nationalism. Arktos is known for distributing the work of figures in the global far-right movement and has published books by the prominent American alt-right figure Richard Spencer, the Russian ultra-nationalist Aleksandr Dugin, Alain de Benoist of the French New Right, and Brazilian right-wing philosopher Olavo de Carvalho.[71] The relationship between these revolutionary far-right actors and climate obstruction should thus be thought of as a rejection of the terms of the debate: the global cannot exist.

The authors of the revolutionary far right are united in their critique of what they see as an increasingly "multicultural" world, one where ethnic and national distinctions have become less clear-cut. Eschewing a class-based critique—which for many in the farthest reaches is seen as Marxist ideology invented to sever people from their natural ethnic consciousness—the revolutionary far right turns to cultural explanations for what they see as a trend toward an indistinguishable and homogenous global community.[72]

For many global far-right movements, the supposed threat of a multiculturalist global community is often argued against using some variation on the "Great Replacement Theory," the belief that there is a deliberate effort by obscure global elites, often imagined as Jewish and referred to as a "cabal," to replace white populations in Western countries with non-white and/or immigrant populations. Today, a wide range of multisectoral actors—from Fox News in the United States to the AfD party in Germany—have propagated this belief.[73] In recent years the revolutionary far right has linked the imagined set of "Great Replacement" actors and relationships with climate change, considering it to be one of the many artificial panics and false flags allegedly precipitated by elites. In other words, the charge is no longer that science isn't right, but that climate change itself is part of a conspiracy. Additionally, the basic logic of replacement has found resonance beyond Western contexts. Similar narratives have surfaced within the right-wing Rashtriya Swayamsevak Sangh (RSS), based in India, for example.[74]

Extremist Goals

Beyond advancing replacement theories, the broad global swath of far-right actors exhibits varying responses to anthropogenic climate change. One such approach is "accelerationism," a radical theory of change advocating the deliberate intensification of forces that will destabilize existing systems and precipitate societal collapse, thereby hastening social transformation in line with accelerationists' own ideology and vision of society.[75] Accelerationism is a strategy adopted by radicals on both the left and right. Far-right and neofascist accelerationists view climate collapse—ideally, cultural collapse—as a means of transformation facilitating a return to traditional, often authoritarian, social structures. For example, a section of the manifesto released by a mass shooter in an attack on two mosques in Christchurch, New Zealand in 2019 is titled "Destabilization and Accelerationism: Tactics for Victory." It attempts to justify the murder of fifty-one people as part of a strategy to trigger broader conflicts and speed change. Similarly, a recent study from West Point's Combating Terrorism Center notes that multinational neo-Nazi groups like The Base and Atomwaffen employ accelerationist strategies using extreme violence with the aim of destabilizing and toppling the US federal government.[76]

Extremist groups have always exploited vulnerabilities to shape grievance narratives, but recent research suggests that climate crises (and local governmental responses) provide new avenues for extremist exploitation. For example, a recent study on climate-induced violent extremism in North Africa found that

> violent extremist (VE) groups such as Boko Haram and the Islamic State exploit crises and conflicts resulting from environmental stress to recruit more followers, expand their influence and even gain territorial control. In such cases, climate change may be described as a 'risk multiplier' that exacerbates a number of conflict drivers.[77]

These extreme perspectives thus diverge ideologically from climate obstruction as it is typically understood on the mainstream right, which often employs it to maintain the status quo, especially with regard to the regulatory environment. Instead, the revolutionary far right has begun to integrate climate change into a broader political ideology aimed at promoting far-right ecologies.

For the revolutionary far right, the problem lies not with climate science but with its practical consequence: the need for more coordination between multinational corporate capital and international policymakers. In their logic, the existing resource extraction and distribution networks of global capitalism will lead to a dystopia that will destroy all local ethnic connections to land ("blood and soil") and usher in a New World Order in which sovereign nations are replaced with a totalitarian global government. In addition to examining nonstate extremist actors, such as those involved in the Christchurch attack and a similar 2019 race-based massacre in El Paso, Texas, research has also examined far-right policy platforms related to climate change.[78] Perhaps the most famous example is the rise of "blood and soil" nationalist identities, such as in Nazi Germany, which used the Reich Ministry of Food and Agriculture to anchor an extremist ideology in agrarian connections to nature, place, and the environment. Today, the link between the environment, climate, and nationalism still figures heavily in the rhetoric of French politician Marine Le Pen's Le Front National (FN), especially its emphasis on preserving environmental heritage, or "patriotic ecology," which it frames as a holistic approach to halting the collapse of the French way of life. In a 2022 debate, Le Pen emphasized the urgency of environmental concerns, criticizing globalization for its detrimental impact on the planet and animal welfare and advocating "localism" and "economic patriotism" to enhance the production and consumption of French products.

Strategies and Tactics

An important distinction between mainstream right-wing social forces and revolutionary far-right movements lies in their political strategy and tactics. Whereas mainstream right-wing groups engage directly in policy processes to obstruct climate action, revolutionary far-right movements are often relegated to "metapolitical" strategies, leveraging online platforms, "news" stations, and political spectacles such as rallies to foster skepticism toward international cooperation and climate policies and thereby influence politics indirectly. Internet-based metapolitical obstruction is not merely rhetoric: such far-right ideology has inspired violent attacks on particular groups and activists.

For example, the aforementioned El Paso massacre was a metapolitical spectacle that illustrates how the system-attacking nature of the revolutionary far right has become entangled with environmental ideologies in ways that ideologies of the mainstream right have not. In a manifesto published online before the attack, the shooter wrote:

> The inconvenient truth is that our leaders, both Democrat and Republican, have been failing us for decades . . . the decimation of the environment is creating a massive burden for future generations. Corporations are heading the destruction of our environment by shamelessly over-harvesting resources . . . The government is unwilling to tackle these issues beyond empty promises since they are owned by corporations . . . that also like immigration because more people mean [cheap labor and] a bigger market for their products.[79]

Although revolutionary far-right movements differ substantially across the globe, two lessons can be gleaned from a survey of their shared characteristics: (1) science skepticism and climate policy obstruction are not necessarily linked; and (2) environmental concern is present, albeit in very different forms, across the conservative political spectrum.

CONCLUSION

The chapter's survey of the right is anchored by the role of neoliberalism, highlighting climate obstruction within public-facing institutions often backed by hidden funding from fossil fuel and agribusiness interests. We traced the history of neoliberal narratives linking the early conservation rhetoric of wise use as an alternative to public regulation to present-day rhetoric of free-market environmentalism. We also considered the far right, conceptualized as a revolutionary movement, sometimes at odds with both neoliberalism and conservatism, that attempts to obstruct climate policy as part of a larger effort to disrupt international cooperation itself. The chapter emphasizes the importance of civil society in right-wing obstruction, focusing on the network of actors collaborating through think tanks, foundations, and other organizations. While right-wing narratives differ, civil society networks have offered spaces where the various factions of neoliberalism and other far-right forces intersect. Turning to European and global climate governance and the notion of shared responsibility, we then pointed to the revitalization of reactionary, nationalistic ideologies shared by the diverse sectors of the right.

Religion and spirituality also play a significant role in climate obstruction when they are mobilized against scientific evidence and scientifically grounded policymaking. Here we provided the case studies of Eastern Europe and Latin America to demonstrate how religion can reinforce climate obstruction ideology, despite that many leaders of world religions have made addressing climate change a theological imperative. Finally, we demonstrated how, despite their differing ideologies and goals, the actors and organizations on the right are to some degree unified under "big tent" groups including the Atlas network and the International Democratic Union, amplifying their messages via media outlets that reach successfully across space and time.

Topics for Future Research

Scholars are increasingly focusing attention on the far right and, within that arena, a literature on the relationship between the far right and the environment is growing rapidly.[80] However, much less work is being done on the role of the far right in climate obstruction. This is a critical gap; this chapter lays the groundwork for a basic research program on the subject, which must be undertaken separately from—but in conversation with—the tendency to focus on more traditional, neoliberal conservative climate obstruction efforts.

However, scholars should continue to study the ways in which various movements converge and interact. Far from a sign of weakness or fragmentation, the right's diversity and network overlaps have engendered a growing variety of options and strategies for climate obstruction that coincide with dedicated efforts to communicate across the broadening spectrum. While there are important ideological differences between the more recent far-right obstructionist groups and older right-wing conservative and neoliberal groups, big-tent think tank networks such as Atlas offer an arena in which this synthesizing research might be conducted.

Cultural influences on obstruction also deserve more scholarly attention. While the chapter has focused on the far right's opposition to multiculturalism and the preference for "natural" ethnopluralist enclaves, the role of religion is still largely understudied, especially its more diffuse aspects beyond the institutions and organizations considered here. Researchers should attempt to quantify, or tease apart qualitatively, the influence of causal mechanisms in the relationship between religion and climate obstruction, with a particular focus on the relative importance of political ideology or religious belief in driving outcomes. The role of global religious communication channels such as the Eternal Word TV Network[81] deserve closer attention in conjunction with other commercial right-wing media outlets.

Researchers would do well to expand beyond the focus on examining specific religions or denominations to explore religion's noninstitutionalized forms.[82] For example, beliefs and orientations toward time and temporality (cyclical versus linear; finite versus eternal) and the concept of nature's inherent "sacredness" can have dramatic impacts on the perceived urgency of climate action as well as the level of perceived need to obstruct ongoing climate efforts.

Finally, there is very little research on the organizational and financial linkages between the far right and climate obstruction efforts. The decentralized nature of this movement makes such study difficult, but in addition to examining the organizational and financial networks of national far-right parties, scholars should begin to piece together regional and subnational linkages between movement actors to gain a clearer picture of potential coordination among them.

NOTES

1. David Helvarg, *The War Against the Greens: The "Wise-Use" Movement, the New Right and Anti-environmental Violence* (Sierra Club Books, 1997); Carl Deal, *The Greenpeace Guide to Anti-environmental Organizations* (Odonian Press, 1933).
2. Justin Farrell, *The Battle for Yellowstone: Morality and the Sacred Roots of Environmental Conflict* (Princeton University Press, 2015).
3. Excerpt from Helvarg, *The War Against the Greens*, Grist, https://grist.org/politics/helvarg (accessed February 27, 2025).
4. Andrew Rowell, *Green Backlash: Global Subversion of the Environment Movement* (Routledge, 1996), 30.
5. Rowell, *Green Backlash*, 237f.
6. Rowell, *Green Backlash*, 30.
7. Amy Westervelt and Geoff Dembicki, "Meet the Shadowy Network Vilifying Climate Protestors," September 12, 2023, *DeSmog*, https://www.desmog.com/2023/09/12/atlas-network-vilifying-climate-protestors/.
8. R. E. Dunlap and A. M. McCright, "Organized Climate Change Denial," in *The Oxford Handbook of Climate Change and Society*, ed. John S. Dryzek, Richard B. Norgaard, and David Schlosberg (Oxford University Press, 2011), 144–160; A. M. McCright and R. E. Dunlap, "Challenging Global Warming as a Social Problem: An Analysis of the Conservative Movement's Counter-Claims," *Social Problems* 47, no. 4 (2000): 499–522.
9. Rowell, *Green Backlash*.
10. R. E. McKie, "Climate Change Counter-Movement Neutralization Techniques: A Typology to Examine the Climate Change Counter-Movement," *Sociological Inquiry* 89, no. 2 (2019): 288–316.
11. Dieter Plehwe, "Think Tank Networks and the Knowledge–Interest Nexus: The Case of Climate Change," *Critical Policy Studies* 8, no. 1 (2014): 101–115; Oreskes Naomi and Erik M. Conway, *The Big Myth: How American Business Taught Us to Loathe Government and Love the Free Market* (Bloomsbury, 2023).
12. Dieter Plehwe, Max Goldenbaum, Archana Ramanujam, Ruth McKie, Jose Moreno, Kristoffer Ekberg, Galen Hall, Lucas Araldi, Jeremy Walker, Robert Brulle, Moritz Neujeffski, Nick Graham, Milan Hrubes, and Quinn Slobodian, "The Mises Network and Climate Policy," Policy Briefing, The Climate Social Science Network, July 2021.
13. Brad Lips, "The Freedom Movement," Atlas Network, 2020, https://www.atlasnetwork.org/books/the-freedom-movement.
14. Gerald Frost, *Antony Fisher. Champion of Liberty* (Profile Books, 2008).

15. Marie-Laure Salles-Djelic, "Building an Architecture for Political Influence: Atlas and the Transnational Institutionalization of the Neoliberal Think Tank," in *Power, Policy and Profit. Corporate Engagement in Politics and Governance*, ed. Christina Garsten and Adrienne Sörbom (Edgar Elgar Publishing, 2017), 25–44; Marie-Laure Djelic and Reza Mousavi, "How the Neoliberal Think Tank Went Global: The Atlas Network, 1981 to the Present," in *Nine Lives of Neoliberalism*, ed. Dieter Plehwe, Quinn Slobodian, and Philip Mirowski (Verso, 2020), 257–282.

16. Jeremy Walker, "Freedom to Burn: Mining Propaganda, Fossil Capital, and the Australian Neoliberals," in *Market Civilizations: Neoliberals East and South*, ed. Quinn Slobodian and Dieter Plehwe (Zone Books, 2022), 189–220.

17. Westervelt and Dembicki, "Meet the Shadowy Network Vilifying Climate Protestors."

18. D. Plehwe, "Opposition 'Strategy Mobility': A Dimension Still Missing in the Critical Policy Mobility Literature," *Critical Policy Studies* 17, no. 4 (2023): 637–647.

19. Rowell, *Green Backlash*.

20. The American Conservation Coalition was founded in 2017 (see https://www.acc.eco/), and the British Conservation Alliance was founded in 2019 (see https://en.everybodywiki.com/British_Conservation_Alliance), both accessed February 27, 2025.

21. Christopher Barnard and Kai Weiss (eds.), *Green Market Revolution. How Market Environmentalism Can Protect Nature And Save The World* (Austrian Economics Center and British Conservation Alliance, 2020).

22. Rod Richardson, Global Freedom event alongside Atlas Network Liberty Forum, https://www.youtube.com/watch?v=PKiU_Eu8GnM&t=754s (accessed February 27, 2025).

23. Alejandro Antonio Chafuen, "R. Randolph Richardson (1926–2015): Supporter of Think Tanks and Strategists," *Forbes*, June 25, 2015), https://www.forbes.com/sites/alejandrochafuen/2015/06/25/r-randolph-richardson-1926-2015-supporter-of-think-tanks-and-strategists/.

24. Rod Richardson, "Clean Capitalism Without Borders," LinkedIn, February 6, 2018, https://www.linkedin.com/pulse/clean-capitalism-without-borders-rod-richardson/.

25. Plehwe et al., "The Mises Network and Climate Policy."

26. See "Mises Institutes" in R. E. McKie, *The Climate Change Counter-Movement: How the Fossil Fuel Industry Sought to Delay Climate Action* (Palgrave Macmillan, 2023), 71f.

27. Plehwe et al., "The Mises Network and Climate Policy."

28. W. F. Lamb, G. Mattioli, S. Levi, J. T. Roberts, S. Capstick, F. Creutzig, J. C. Minx, F. Müller-Hansen, T. Culhane, and J. K. Steinberger, "Discourses of Climate Delay," *Global Sustainability* 3, (2020): 1–5

29. McKie, *The Climate Change Counter-Movement*.

30. M. T. Boykoff, "Media and Scientific Communication: A Case of Climate Change," *Geological Society London Special Publications* 305, no. 1 (2008): 11–18; M. Boykoff and J. Farrell, "Climate Change Counter-Movement Organizations and Media Attention in the United States," in *Climate Change Denial and Public Relations: Strategic Communication and Interest Groups in Climate Inaction*, ed. xxx (Routledge, 2019), 121–139; A. M. Petersen, E. M. Vincent, and A. L. Westerling, "Discrepancy in Scientific Authority and Media Visibility of Climate Change Scientists and Contrarians," *Nature Communications* 10 (2019): 3502, https://doi.org/10.1038/s41467-019-09959-4.

31. Edson C. Tandoc, Bruno Takahashi, and Ryan J. Thomas, "Bias vs. Bias," *Journalism Practice* 12, no. 7 (2018): 834–849.

32. Dieter Plehwe, "Think Tank Networks and the Knowledge–Interest Nexus: The Case of Climate Change," *Critical Policy Studies* 8, no. 1 (2014): 101–115.

33. R. J. Brulle, "Institutionalizing Delay: Foundation Funding and the Creation of US Climate Change Counter-Movement Organizations," *Climate Change* 122, no. 4 (2014): 681–694.

34. R. J. Brulle, G. Hall, L. Loy, and K. Schell-Smith, "Obstructing Action: Foundation Funding and US Climate Change Counter-Movement Organizations," *Climatic Change*, 166 (2019): 1–7.

35. "Dunn's Foundation for the Advancement of Right Thinking," *DeSmog*, https://www.desmog.com/dunn-s-foundation-advancement-right-thinking/ (accessed February 27, 2025); Ruth McKie, Atlas Think Tank Main Employers, dataset (De Montfort University, 2023), https://doi.org/10.21253/DMU.22217050.v1.

36. Brulle, "Institutionalizing Delay"; Brulle et al., "Obstructing Action."

37. Jane Mayer, *Dark Money: The Hidden History of the Billionaires Behind the Rise of the Radical Right* (Doubleday, 2016).

38. A. Hertel-Fernandez, T. Skocpol, and J. Sclar, "When Political Mega-Donors Join Forces: How the Koch Network and the Democracy Alliance Influence Organized US Politics on the Right and Left," *Studies in American Political Development* 32, no. 2 (2018): 127–165; J. Farrell, "The Growth of Climate Change Misinformation in US Philanthropy: Evidence from Natural Language Processing," *Environmental Research Letters* 14, no. 3 (2019): 034013, https://doi.org/10.1088/1748-9326/aaf939.
J. Farrell, "Corporate Funding and Ideological Polarization About Climate Change," *Proceedings of the National Academy of Sciences* 113, no. 1 (2016): 92–97.

39. Brulle et al., "Obstructing Action."

40. McKie, Atlas Think Tank Main Employers.

41. "Concealing Their Sources: Who Funds Europe's Climate Change Deniers?," *Corporate Europe Observatory*, December 2010, https://corporateeurope.org/sites/default/files/sites/default/files/files/article/funding_climate_deniers.pdf, 4.

42. While we know from Internal Revenue Service documents (990 forms) that Templeton Foundation provides the largest amount of funds to Atlas Network (https://www.desmog.com/atlas-economic-research-foundation/#h-atlas-network-as-nbsp-recipient) and from the Templeton Foundation grant database (https://www.templeton.org/grants) that it also funds many Atlas Network partner organizations around the world, recipients include think tanks that produce climate denial content and think tanks that do not publish on climate policy related issues, for example.

43. A. Aspelund, M. Lindeman, and M. Verkasalo, "Political Conservatism and Left–Right Orientation in 28 Eastern and Western European Countries," *Political Psychology* 34 (2013): 409–417.

44. Simona-Nicoleta Vulpe, "Belief and Skepticism. Religious Justifications for Vaccine Reluctance and Climate Skepticism," *Sociologie Românească* 19, no. 1 (2021): 69–88.

45. Mark Morrison, Roderick Duncan, and Kevin Parton, "Religion Does Matter for Climate Change Attitudes and Behavior," *PLOS One* 10, no. 8 (2015): e0134868. doi:10.1371/journal.pone.0134868; Miriam Pepper and Leonard Rosemary, "Climate Change, Politics and Religion: Australian Churchgoers' Beliefs about Climate Change," *Religions* 7, no. 5 (2016): 47, doi:10.3390/rel7050047.

46. Bastiaan T. Rutjens and Romy van der Lee, "Spiritual Skepticism? Heterogeneous Science Skepticism in the Netherlands," *Public Understanding of Science* 29, no. 3 (2020): 335–352, doi:0.1177/0963662520908534.

47. Anton Alumkal, *Paranoid Science: The Christian Right's War on Reality* (NYU Press, 2017), 158.

48. Lynn Vincentnathan, S. Georg Vincentnathan, and Nicholas Smith, "Catholics and Climate Change Skepticism," *Worldviews: Global Religions, Culture, and Ecology* 20, no. 2 (2016): 125–149, 135.

49. Vincentnathan et al., Catholics and Climate Change Skepticism.

50. "AUR cere renegocierea Pactului Verde (Green Deal) de guvernul român cu Comisia Europeană," November 18, 2021, https://partidulaur.ro/aur-cere-renegocierea-pactului-verde-green-deal-de-guvernul-roman-cu-comisia-europeana/.

51. Boris Knorre, "The Church and Environmentalists: Friends or Foes?," Russia.Post, February 8, 2023, https://russiapost.info/society/environmentalists.

52. The Russian Orthodox Church, "The Basis of the Social Concept," XIII, 1, 5, https://old.mospat.ru/en/documents/social-concepts/xiii/ (accessed February 27, 2025).

53. John Chryssavgis (ed.), *Patriarch Bartholomew, On Earth as it is in Heaven: Ecological Vision and Initiatives of Patriarch Bartholomew* (Fordham University Press, 2012).

54. The Vatican, "Meeting on 'Faith and Science: Towards COP 26,' Promoted by the Embassies of Great Britain and Italy to the Holy See, Together with the Holy See," Holy See Press Office, 2021, https://press.vatican.va/content/salastampa/en/bollettino/pubblico/2021/10/04/211004a.html.

55. See, for example, Brandon Smith, "Șocul financiar și controlul climei, pe agenda globaliștilor," R3Media, August 4, 2023, https://r3media.ro/socul-financiar-si-controlul-climei-pe-agenda-globalistilor/; Nick Griffin, "Despre încălzirea globală, Cuvântul Ortodox," February 22, 2011, http://www.cuvantul-ortodox.ro/recomandari/de-la-minciuna-incalzirii-globale-la-criza-alimentara-si-infometarea-populatiei-codul-muncii-legalizarea-sclaviei-munca-la-negru-inchisoare/; Adrian Pătrușcă, "Opt dovezi ale escrocheriei încălzismului," July 25, 2023, https://www.activenews.ro/opinii/Opt-DOVEZI-ale-Escrocheriei-Incalzismului-183360.

56. "Un geofizician eminent respinge teoria încălzirii globale și spune că lumea este în pragul unei 'Mini Ere Glaciare,'" Cuvântul Ortodox, February 22, 2011, http://www.cuvantul-ortodox.ro/recomandari/de-la-minciuna-incalzirii-globale-la-criza-alimentara-si-infometarea-populatiei-codul-muncii-legalizarea-sclaviei-munca-la-negru-inchisoare/; Virgil Șerban, "Un om de știință înfruntă isteria: Dr. John Clauser, câștigător al premiului Nobel în 2022 pentru fizică, spune răspicat că 'NU există o criză climatică și schimbările climatice NU provoacă fenomene extreme,'" R3Media, July 31, 2023, https://r3media.ro/un-om-de-stiinta-infrunta-isteria-dr-john-clauser-castigator-al-premiului-nobel-in-2022-pentru-fizica-spune-raspicat-ca-nu-exista-o-criza-climatica-si-schimbarile-climatice-nu-provoaca-fe/.

57. Radu Ursan, "Pastorala incendiară de Crăciun a Episcopului Slatinei: a vorbit împotriva „pandemiilor planificate," a crizelor confecționate și de „obsesii climatice exagerate." PS Sebastian l-a pus pe Soroș pe coperta pastoralei, alături de Hitler și Stalin, R3Media, December 26, 2022, https://r3media.ro/pastorala-incendiara-de-craciun-a-episcopului-slatinei-a-vorbit-impotriva-pandemiilor-planificate-a-crizelor-confectionate-si-de-obsesii-climatice-exagerate-ps-s/.

58. A. Banzatto and R. Tapia, "'Conectando boas intenções com economia sólida': O Acton Institute," OPEU, July 13, 2023, https://www.opeu.org.br/2023/07/13/conectando-boas-intencoes-com-economia-solida-o-acton-institute/; Jean Carlos Hochsprung Miguel, "A 'meada' do negacionismo climático e o impedimento da governamentalização ambiental no Brasil," *Revista Sociedade e Estado* 37, no. 1 (January/April 2022): 219–315.

59. H. Silva, "Os novos atores 'evangélicos' e a conquista do espaço público na América Latina," *Reflexão* 43, no. 2 (2022): 244–263.

60. Miguel, "A 'meada' do negacionismo climático e o impedimento da governamentalização ambiental no Brasil," 293.

61. A. C. de Oliveira, "A 'meada' do negacionismo climático e o impedimento da governamentalização ambiental no Brasil," *Sociedade e Estado* 37, no. 1 (2022): 263–284.

62. Behr, Klaus Ramalho von. A conspiração do clima: populismo e negacionismo climático no início do governo Bolsonaro (2018–2020). Master's thesis, (2022) Universidade de Brasília.

63. Instituto Talanoa. Reconstrução: 401 atos do Poder Executivo Federal (2019–2022) a serem revogados ou revisados para reconstituição da agenda climática e ambiental brasileira, (2022).

64. Miguel, "A 'meada' do negacionismo climático e o impedimento da governamentalização ambiental no Brasil," 305.

65. K. M. Blee, and K. A. Creasap, "Conservative and Right-Wing Movements," *Annual Review of Sociolog* 36, no. 1 (2010): 269–286.

66. Bourdieu's field theory is particularly useful for understanding climate policy obstruction because it theorizes interactions between dominant and subordinate groups within a field of power relations. It assists researchers with providing a clearer understanding of where and how actors mobilize capital (economic, social, cultural, symbolic) to preserve or expand power within the field. Fields also have boundaries, thus delimiting the area of study, and allowing for researchers to identify key actors, institutions, and the rules that govern some fields, but not others. We apply this frame to our current analysis. P. Bourdieu, *Distinction: A Social Critique of the Judgment of Taste* (Harvard University Press, 1987).

67. Jesse Callahan Bryant and Justin Farrell, "Conservatism, the Far-Right, and the Environment," *Annual Review of Sociology* 50 (2024): 273–296.

68. A. M. McCright and R. E. Dunlap, "The Conservative Movements Impact on U.S. Climate Change Policy," *Social Problems* 50 (2003): 348–373.

69. Blair Taylor, "Alt-Right Ecology: Ecofascism and Far-Right Environmentalism in the United States," in *The Far-Right and the Environment*, ed. Bernhard Forchtner (Routledge, 2019), 275–292.

70. Daniel Rueda, "Alain de Benoist, Ethnopluralism and the Cultural Turn in Racism," *Patterns of Prejudice* 55, no. 3 (2021): 213–235.

71. Henrik Arnstad, "Ikea Fascism: Metapedia and the Internationalization of Swedish Generic Fascism," *Fascism* 4, no. 2 (2015): 194–208; Benjamin R. Teitelbaum, War for Eternity: Inside Bannon's Far-Right Circle of Global Power Brokers (Dey Street Books, 2020).

72. M. Varga and A. Buzogány, "The Two Faces of the 'Global Right': Revolutionary Conservatives and National-Conservatives," *Critical Sociology* 48, no. 6 (2022): 1089–1107.

73. Mattias Ekman, "The Great Replacement: Strategic Mainstreaming of Far-Right Conspiracy Claims," Convergence 28, no. 4 (2022): 1127–1143.

74. Shweta Desai, "Inside Hindutva's Great Replacement Conspiracy," GNET, August 10, 2022, https://gnet-research.org/2022/08/10/inside-hindutvas-great-replacement-conspiracy/.

75. M. Loadenthal, "Feral Fascists and Deep Green Guerrillas: Infrastructural Attack and Accelerationist Terror," *Critical Studies on Terrorism* 15, no. 1 (2022): 169–208.

76. A. Newhouse, "The Threat Is the Network: The Multi-Node Structure of Neo-Fascist Accelerationism," *CTC Sentinel* 14, no. 5 (2021): 17–25.

77. M. Bourekba, *Climate Change and Violent Extremism in North Africa* (Barcelona Centre for International Affairs, 2021).

78. Callahan and Farrell, "Conservatism, the Far-Right, and the Environment."

79. K. Campion, "Defining Ecofascism: Historical Foundations and Contemporary Interpretations in the Extreme Right," *Terrorism and Political Violence* 35, no. 4 (2021): 926–944

80. Callahan and Farrell, "Conservatism, the Far-Right, and the Environment."

81. Vincentnathan et al., "Catholics and Climate Change Skepticism," 137f.

82. Justin Farrell, "Environmental Activism and Moral Schemas: Cultural Components of Differential Participation," *Environment and Behavior* 45, no. 3 (2013): 399–423.

CHAPTER 6

Steering the Climate Discourse

Legacy News, Social Media, Advertising,

and Public Relations

LEAD AUTHORS: MELISSA ARONCZYK
AND MAXWELL BOYKOFF
CONTRIBUTING AUTHORS: TRAVIS G. COAN,
MYANNA LAHSEN, HANNA E. MORRIS,
AND CHRIS RUSSILL

Introduction: Steering Obstruction

In the lead-up to the international treaty on climate change known as the Paris Agreement, the Intergovernmental Panel on Climate Change (IPCC) released its fifth report, describing the causes of climate change in unequivocal terms. Anthropogenic greenhouse gas emissions (GHGs), "together with those of other anthropogenic drivers, have been detected throughout the climate system and are *extremely likely* to have been the dominant cause of the observed warming since the mid-20th century."[1] In 2017, *New York Times* opinion writer Bret Stephens challenged those who claimed the "complete certainty" of climate causes. His piece was not meant "to deny climate change of the possible severity of its consequences," he wrote. "But ordinary citizens also have a right to be skeptical of an overweening scientism." Such framing may be perceived as presenting a "balanced" opinion; it is also a way to retain skepticism as an acceptable response to the climate crisis.

Since at least 2004, the pretext of balance in media coverage of global warming has exacerbated the disparity between scientific and public understandings of climate change.[2] Research published in 2004 examining peer-reviewed climate science found that experts had reached agreement (consensus) that

Melissa Aronczyk et al., *Steering the Climate Discourse*. In: *Climate Obstruction*. Edited by: J. Timmons Roberts, Carlos R. S. Milani, Jennifer Jacquet, and Christian Downie, Oxford University Press. © Oxford University Press (2025). DOI: 10.1093/oso/9780197787144.003.0006 humans contribute to climate change.[3] Another 2004 study found that 53% of news articles published between 1988 and 2002 in major American newspapers told a so-called "balanced" story whereby some scientists found that humans contributed to climate change while others continued to argue that humans' role was in doubt.[4] This study—and many others since then—offer explanations for the outsized gap between scientific knowledge and public understanding: the inappropriate deployment of the journalistic norm of balanced reporting combined with the outsized influence of industry groups who use anti-climate action think tanks and other strategies to shape public stories and those quoted in them.[5] While this disparity has diminished (but not disappeared) in these sources since then,[6] it exposes one way in which casting doubt about anthropogenic climate change is a form of climate obstruction.[7]

To be clear: the ongoing disparity between scientific and public understandings of climate change is not a matter of happenstance. Oil and gas producers and their collaborators actively work with legacy news (television, newspapers, radio) as well as social media platforms and public relations and advertising firms to promote disinformation that stymies climate policy. The deliberate production and promotion of climate disinformation and climate obstruction appears via news coverage, journalistic practices, public relations (PR) and strategic influence campaigns, and social media posts as well as in other modes of communication (e.g., blogs, instant messaging) in the public arena. Financial motivations are part of the problem as a result of the corporate control of mainstream media and the importance of advertising revenue to media companies' bottom line.[8] This chapter synthesizes research that explains how climate disinformation circulates through media and contributes to climate obstruction that benefits carbon-based industries and status-quo actors at the cost of society and the environment (see Chapters 5 and 7). It focuses on the discourses and structures that shape the terrain of these practices and then points to areas where this disinformation and obstruction can effectively be countered.

Contrarians in the Mediated Public Sphere

A growing—at times cacophonous—chorus of climate-contrarian voices has gained prominence through multiple media platforms and modes of communication. Contrarian views are promoted by a network of obstructionist organizations and actors with aligned ideological and cultural inclinations (see Chapter 1).[9] These movements often gain attention to their outlier views through media channels.[10] At least four arguments for climate delay can be observed in media communications: (1) shifting the responsibility for climate action; (2) arguing that mitigating climate change is impossible; (3) advancing nontransformative solutions; and (4) emphasizing the downsides of any climate solution.[11] Captured in media portrayals, advertising and PR campaigns, these discourse types demonstrate the growing tendency among disinformation actors to move from overt denial of climate change and its anthropogenic causes to more subtle tactics of delegitimation of climate solutions. Choices about how to frame communications through media, advertising, and PR—shaping perceptions of responsibility and solutions, for example—can channel public attention and action in important ways. In many cases, this agenda setting can foster preferences for system preservation over structural transformation.[12] Mass media enable the formation of political structures and norms that condition public responses to climate change. Ultimately, such obstruction efforts seek to advance special-interest goals and objectives to maintain the status quo or to distract from, delay, and/or deny science—effectively obstructing climate engagement and action.

How Disinformation Spreads in the Mediated Public Sphere

The deliberate circulation of climate disinformation is partly enabled by the unprecedented economic and social power of the global media system. Power structures were further consolidated during the COVID-19 pandemic when many independent and local newsrooms were shuttered as digital media expanded and consolidated.[13] Boosted by new digital technologies and legacy/social media, disinformation campaigns can now operate at greater scale and expanded scope, influencing human experience with increasing control and precision.[14] Technology companies and their platforms allow powerful actors to shape cognition and actions in ways that lack accountability. The use of artificial intelligence (AI) and communications technologies in political campaigns has had profoundly negative consequences for global climate mitigation, even when climate policy obstruction was not an explicit or direct aim.[15]

New media technologies can boost longstanding patterns of control and socioeconomic inequality that are harmful for the environment. Social disparities, a lack of democratic practices, and environmental destruction tend to reinforce one another.[16] To address the problem of disinformation and how it shapes climate and environmental outcomes, it is vital to broaden the scope of what we commonly think of as climate and environmental policy. Collective goals in the name of the public interest, such as democracy, environmental protection, and socioeconomic equity, are threatened by the concentration of ownership and corporate control of communications and digital technologies and how information then becomes subject to financial and political incentives. This concentration is enabled by current policies in the United States and many other countries that are determined by the wealthiest and most powerful players.[17] The concentration of media control in relatively few news outlets and internet companies is a global phenomenon—though concentration can look different depending on ownership types and styles of concentration.[18] These conditions have the potential to produce sociodigital worlds that not only intentionally spread disinformation but also foster values and worldviews that maintain inequality as well as public disempowerment and compliance, thwarting the emancipatory potential of mass communications and digital technologies.[19]

With this backdrop in mind, this chapter explores the political-economic conditions and networks of actors engaged in producing online modes of climate information and disinformation in order to better delineate the scale and scope of the problem and identify possible pathways forward. The authors trace how journalists and other media makers are hampered in their efforts to elucidate climate obstruction and examine how climate obstruction finds visibility through media and PR endeavors, effectively amplifying as opposed to investigating and holding to account contrarian and related outlier perspectives in the public arena. The chapter also notes how journalists and other media actors are responding to these challenges.

OVERVIEW OF ACTORS, SECTORS, AND ORGANIZATIONS

Across national contexts, commitments to economic growth and carbon-intensive industry, along with deeply entrenched technological optimism, have influenced discussions of climate change in the public sphere. A recent study uncovers relationships among conspiratorial beliefs, conservatism, and climate skepticism across nations, concluding, "The greater the vested interests in resisting change, the more incentive there is to engage—and believe—in ideologically driven campaigns of deliberate disinformation about the reality of anthropogenic climate change."[20] A related study of climate contrarianism (here interchangeable with "climate skepticism" or "climate denialism") across twenty-five countries found that "political cultures have emerged that encourage citizens to appraise climate science through the lens of their conservative ideologies."[21] In the United States in particular, the relationship between conservative (or political) ideology and climate contrarianism is "unusually strong and consistent."[22] The authors also observed that nations with the strongest relationships between right-of-center ideology and climate contrarianism "tend to be those whose economies are relatively highly reliant on fossil fuel industries."[23] This finding is consistent with research in Canada, Brazil, and Australia.[24]

Climate obstruction is conducted by traditional/legacy media as well as online.[25] Online networks mirror offline climate obstruction networks and patterns of behavior, connecting disinformation efforts by networks of fossil fuel-supported think tanks and foundations, lobby and PR groups, multinational corporations, "influencers" from scientific and political fields, and media organizations. While most of these networks originated in the United States, in recent years they have expanded to several other countries and have included transnational coordination across China and countries in Latin America, Europe, and Southeast Asia.[26]

Contrarians and other climate obstructionists who gain attention in the United States, China, Germany, Brazil, and in other country contexts have been profiled demographically.[27] Paradoxically, heterogeneous members of climate obstruction networks have also been associated with movements to limit economic growth by evoking climate change as a justification for other issues such as preventing immigration.[28] Research in the Norwegian context found associations between climate skepticism and right-wing nationalism, noting, "Climate change denial is but one facet of a more general complex of resistance to various societal issues such as economic growth, environmental conservation, globalization, governance and relationships to other social groups" (see Chapter 5).[29]

The relationship between climate obstruction organizations and climate contrarianism has been studied extensively in the US context, in part because of the influence of carbon-based industry at the interface of science, policy, and society (see Chapter 1).[30] Emerging analyses of climate coverage in lower- and middle-income countries suggest that climate-skeptic views are granted much less coverage and credibility in these locales. Among the larger and most studied nations, China, India, and Brazil make few references to climate skepticism, for example.[31] In some cases, the focus may be less on fossil fuel emissions and more on carbon pollution from agriculture and other land-use practices. Research on climate coverage in Brazil found that beef production, which has been linked to more than half of Brazil's emissions, was a nearly taboo topic in Brazilian newspapers. Researchers have also looked across Brazil, China, France, India, the United Kingdom, and United States and found that news coverage of skepticism is largely limited to the latter two countries.[32] Elsewhere, disinformation in the Global South has been found to be an ongoing problem (see Chapter 8).[33]

In the Global North, climate obstruction in both public and political spheres is networked across environmentally damaging industries and polluting sectors. With this structure comes asymmetrical power and influence in national politics and policymaking. Advertising campaigns as well as PR efforts have long played a role as the glue joining carbon-based industries, media organizations, and economic sectors, using coalitions, campaigns, and other coordinated processes to question the efficacy of science, news, and political institutions engaging climate-related issues.[34] Fossil fuel PR is embedded in a wide ecosystem of influence that includes trade associations, industry and science advisory councils, think tanks and research institutes, NGOs and foundations, chambers of commerce, and organizational boards.[35] PR firms play several key roles in this network. Among them, there is (1) cross-industrial strategic management, where PR agents engage in intelligence-gathering in the environmental community, at government agencies and in other organizations that influence public policy on environment, climate, and energy issues; and (2) industry-friendly research, public-opinion polling, and data collection to promote clients' viewpoints in the media, in addition to producing scientific, legal, or technical expert material for media circulation. One way various agencies, think tanks, boards and associations are tightly bound with PR firms is through a revolving door strategy (e.g., former government agency administrators are hired by PR firms, former PR firm employees are hired as trade association administrators, and so on).

Media communications aiming to advance climate denial and opposition often amplify existing patterns of polarization.[36] A study of polarizing tweets on X (Twitter) during United Nations climate (COP) summits identified five styles of climate contrarianism, from outright denial ("@COP26 You have been lying to the public and mocking them for decades with your climate scam") to (untrue) rejection of climate solutions ("China is not going to COP26. So what's the point?").[37] In 2022, the Third Working Group of the United Nations Intergovernmental Panel on Climate Change Sixth Assessment Report pointed to "public discourses of media and organized countermovements" as fuel for polarization, with damaging potential for climate action.[38] The technical summary noted that "accurate transference of the climate sciences has been undermined significantly by climate change countermovements, in both legacy and new/social media environments through misinformation."[39] It also noted that "on occasion, the propagation of scientifically misleading information by organised countermovements has fuelled polarisation, with negative implications for climate policy."[40] Further linking obstruction activities and polarisation, the report mentions "political polarisation leading to erosion of environmental governance" (see Chapter 10).[41]

Digital platforms have also been identified as sources of climate disinformation. Meta (Facebook), for instance, has overridden the determinations of its independent fact-checkers for climate science, allowed fossil fuel companies to purchase misleading ads, and limited the transparency and usefulness of its data sharing tool, Crowdtangle, relied upon by journalists and academics to analyze engagement with content on the platform.[42] There are claims that Meta's own Climate Science Center is underequipped to serve its purpose. One report estimated that Google alone received $23.7 million between 2020 and 2022 from the five largest oil companies in the world (ExxonMobil, British Petroleum [BP], Chevron, Shell, and Aramco) to promote their advertising.[43] The nonprofit coalition Climate Action against Disinformation (CAAD) has found that the major tech platforms (TikTok, Meta, YouTube, and X) have become "complicit" actors in the spread of climate denial, since "disinformation is now a guaranteed byproduct, if not a central part, of social media companies' business models."[44] Among several findings in the study,

X ranked last among platforms for its absence of policies on climate disinformation, lack of public transparency mechanisms, and failures of effective policy enforcement.[45]

POLITICAL AND ECONOMIC AIMS, ALLIANCES, AND STRATEGIES

Climate disinformation disseminated through media content, platforms, and formats such as legacy news coverage, social media posts and discussions, advertising campaigns, and corporate promotional strategies often mask the original source and coordinated efforts of these actors.[46]

Macro-Scale Influences

At the *macro scale*, questions of representation in, differential access to, and regulation and ownership configurations of media systems are crucial factors in climate policy obstructionism in and by both alternative and mainstream news media. These are multifaceted processes through which dominant assumptions, beliefs, values, and election outcomes are shaped, often facilitated by public policies that allow private concentration of ownership of media and associated technologies with limited public oversight and participation.[47] Efforts to shape public opinion via media systems, old and new, influence how publics perceive polluting companies and activities. Corporate-funded media tend to privilege certain discourses and make highly unequal access to (and pollution of) environmental resources seem natural and inevitable, even where they are irrational and detrimental to the public interest.[48] One particularly problematic phenomenon is the increased funding of social media influencers by fossil fuel companies to spread a positive image of their sector with the help of PR firms.[49] Research by the international research team *DeSmog* found that more than one hundred influencers have been hired by public relations firms such as Edelman and ad agencies including EssenceMediaCom (owned by WPP, the largest marketing communications services group in the world) to promote oil and gas clients Shell and BP internationally since 2017, from the United States to Malaysia and the South Pole.[50]

Meso-Scale Factors

At the *meso scale*, research has considered how multiple organizations have influenced environmental and climate politics over time. As other chapters in this volume attest, the political power of intermediaries such as trade associations, foundations, think tanks, and universities and research institutes contribute to the maintenance of the corporate political power of fossil fuel producers. These organizations, operating worldwide, have helped to harness the collective power of business and free-enterprise ideology from at least 1970 through the present day. Such power networks wield influence through both discursive and financial means. Many major national and global media companies depend on money from advertising, including from fossil fuel companies.[51] Media ownership is no longer in the hands of just a few wealthy moguls, as was the case some decades ago. The need for capital has pushed media companies to seek multiple sources of investment, leading to more differentiated ownership (still largely within the confines of private ownership groups, distinct from community- or state-owned media).[52] Still, the general reliance by media companies on an advertiser-based business model inhibit more hard-hitting and investigative reporting required to uncover the root causes and impacts of climate change as well as to illuminate the different methods and means of climate obstructionism.[53] To give but one devastating example, climate reporters Amy Westervelt and Matthew Green investigated how seven major news organizations—Bloomberg, *The Economist*, the *Financial Times*, the *New York Times*, *Politico*, Reuters, and the *Washington Post*—featured content and hosted events favorable to fossil fuel companies. In some cases, the media companies even used their internal brand studios to create the content in-house for the fossil fuel corporations.[54]

Micro-Scale Aspects

At the *micro scale*, discourse and behaviors at the individual level shape these spaces of interaction.[55] Most forms of media have an original author as well as an editor who oversees content. These individuals may be influenced by incentives within the media organization or many other factors. Harassment of scientists, journalists, researchers, and climate communicators is also a significant problem.[56] In addition to the suppression or even criminalization of NGO and activist actions, academic researchers and others investigating climate obstruction are experiencing online and offline threats to the continuation of their work. *The Guardian*, a British newspaper, reported in 2023 on "striking similarities in the way governments from Canada and the US to Guatemala and Chile, from India and Tanzania to the UK, Europe and Australia, are cracking down on activists trying to protect the planet."[57] Often, harassment attempts to leverage commonly shared democratic values such as freedom of speech, open debate, and due process to legitimate harmful statements or delay regulatory actions. Politicians and influencers associated with the circulation of disinformation attempt to frame as censorship initiatives such as fact-checking, labeling of content, and enforcement of platform policies on content moderation. Contrarian claims feed journalistic pressure to supply attention-getting, dramatic personal conflicts. Such conflicts draw attention toward decontextualized individual claims-making and away from critical institutional and societal challenges to carbon consumption that call collective behaviors, actions, and decisions to account.

OBSTRUCTIVE DISCOURSES, NARRATIVES, AND THEIR COMMUNICATION

Studies of journalism and climate obstruction tend to focus on how climate science is covered and how this coverage may confuse the impacts and causes of the climate crisis. Journalism scholars illuminate how climate contrarians have long been featured alongside climate scientists in mainstream national news media, such as the BBC in the United Kingdom and CNN in the United States, in the name of balance and fairness.[58] News media, particularly business presses, continue to legitimize climate denial by allowing fossil fuel companies access to mainstream platforms.[59]

Media and Journalism Influences

Climate journalism is prone to an events-based model of reporting that decontextualizes the climate crisis. Extreme heat, floods, and storms tend to be reported as close-ended events as opposed to a part of an unfolding and long-term crisis.[60] This representation prevents a fuller understanding of climate change by cutting out historical contexts, causes, and long-term impacts.[61] Moreover, climate change is not always mentioned in stories of intense storms, violent floods, and severe heat. In some contexts, this may serve to limit recognition of the immediate as well as long-term risks of the fast and slow violences of climate change.[62]

Reporting on the climate crisis tends to apply apocalyptic and/or fatalist framings that prevent deeper considerations of an array of possible responses to climate change proposed by different stakeholders and community members. Like events-based reporting, apocalyptic framing may obscure the root causes of the crisis and obstruct more comprehensive action.[63] This type of media reporting on climate change has the potential to effectively conjure images of "sacrifice zones" and "sacrificed people," often portrayed as poor and/ or from the Global South.[64] Historically marginalized people are rarely featured in mainstream news coverage of climate change in the Global North and, when they do appear, they are often represented as victims or cast in a negative light, as was the case with anti-fracking activists in the United Kingdom.[65]

Partisanship plays a role in climate denial by linking denial with conservative identities. For instance, climate denial and opposition are often propagated by the same social media accounts that spread anti-vaccine sentiment, denial of genocides, or other conspiracy theories and foment disapproval of so-called social and climate justice warriors. Climate denial was supported by the MAGA movement and the closely aligned QAnon movement.[66]

Shaping Ambition

One variety of climate obstruction promotes consensus and compromise. Indeed, some of the most effective undermining of environmental science has come from businesses claiming to work alongside scientists or public policymakers to promote the public interest. PR specialists in particular are trained in how to reach consensus and compromise.[67] Their goal is to influence people by aligning their clients' messages to public values and beliefs in order to create legitimacy for the organization and trust in the message. PR consultants will create public–private partnerships or sponsorships between their clients and environmental organizations or develop benchmarking systems, accounting systems, or certification programs for companies, as indicators of their commitment to environmental protection. One such system, the GHG Protocol, described as a tool to help countries and cities track their climate goals, is conceived, built, and maintained by an alliance of high-carbon industrial multinational organizations, nongovernmental industry-friendly organizations such as the World Business Council on Sustainable Development, and "green" PR experts. Industry also uses the GHG Protocol to communicate that it is self-regulating and sustainable, which helps to ensure that governments are less likely to regulate them.

Examples like the GHG Protocol are important because they force us to move away from conventional definitions that try to classify specific statements or frames as misinformation or disinformation. Portraying themselves as climate-friendly solutions, such efforts weaken the impact of scientific claims that call for a rapid reduction in the causes of climate change; and they reduce trust in the policymaking process, which is made to appear overly stringent, too complicated or economically infeasible (see Chapter 2). The result is weak regulation, if any.

EFFORTS TO EXPOSE/PUSH BACK ON CLIMATE OBSTRUCTION IN THE MEDIA

Researchers have developed typologies and other cataloguing systems for disinformation that not only promote greater awareness of disinformation tactics but also generate disinformation literacy among various publics.[68] A well-known strategy to address climate obstruction in the media is to monitor, document, and report on it, often through exposure of misleading narratives, malign actors, or bad behavior. Climate disinformation has been tracked at significant levels by the network analysis firm Graphika and by the environmental nonprofit organization Friends of the Earth.[69] Scholars across many fields and several media observatories have integrated climate into their operations, including the European Digital Media Observatory. This work informs strategies to "detect and correct" or to "name and shame" and have aided fact-checking by journalists, litigation against corporate actors, labeling of misleading content on platforms, and the identification of coordinated and inauthentic behavior online, and has at times informed governmental deliberations and hearings.[70]

Increasingly, computational models are being used to classify and categorize climate contrarian topics and themes. Early work in this area relied on topic-modeling approaches to examine contrarian discourse in large corpora of texts and network analyses of key elements of the denial countermovement or media coverage of important events.[71] More recently, techniques have been adopted to measure specific frames, claims, and narratives in large datasets of online media.[72] Research that tracked the dynamics of key themes in a sample of key North American conservative think tanks and blogs found that while claims that "outright deny the existence and severity of anthropogenic climate change have remained stable or have declined in relative terms in recent years," claims that offer pseudo-scientific explanations or other alternative explanations for scientific findings, are prominent.[73] Others have demonstrated the link between conservative foundation funding and contrarian discourse.[74] One of the most important benefits of this research is its longitudinal approach. Researchers have developed an extensive taxonomy of specific contrarian claims, as well as an approach for classifying these claims, then applied their computational model to provide a detailed history of climate disinformation over a roughly two-decade period.[75] This research points to the need for specific public education efforts targeted less at overcoming strict denialism than at recognizing false solutions and other industry-sponsored compromise positions.

Moving beyond monitoring climate disinformation, there are important efforts to respond to, or debunk, disinformation on the ground. Media and information literacy—generally understood as the ability to access, analyze, and critically evaluate media messages and their sources[76]—can prevent the susceptibility of individuals to disinformation and fake news.[77] In practice, this means raising awareness of the importance of fact-checking climate news stories, critically evaluating the sources of stories and one's "media diet," and providing resources to check facts and identify climate disinformation.[78] Producers and consumers of news now have valuable resources at their disposal to help check climate facts, from general websites such as Full Fact or PolitiFact to specialist fact-checking organizations such as Climate Feedback. Individuals can also draw on websites such as Skeptical Science to learn about specific climate myths and best practices in the academic literature for debunking and "prebunking" ("inoculating against") climate disinformation. "Prebunking" means preemptively correcting disinformation by pointing out and refuting disinformation before it is disseminated (see Chapter 7). Organizations are working to put insights from media literacy and debunking into practice. On the other hand, emergent forms of AI ratchet up the production of climate disinformation and related efforts to weaken environmental protection. Researchers are anticipating the use of "deep fakes," powerful AI tools that can create fake video scenes or impersonate trusted authorities such as particular politicians and scientists.[79] Balancing the potential of these technologies to serve public interests as well as controlling the dangers they pose depends on immediate measures to wisely and effectively govern their features and use.[80]

Accountability

Accountability is another important tool to address climate denial and delay in both legacy media (newspapers, television, radio) and social media. Legacy media organizations include influential outlets such as the *New York Times*, Fox News, and BBC. Social media platforms such as Meta, X, TikTok, and YouTube help the spread of disinformation online (including AI-generated disinformation) and holding these platforms accountable is a key mechanism for reducing online exposure.[81] A broad coalition of corporations, media companies, and civil society groups are actively calling on the United Nations Framework Convention on Climate Change and CEOs of major social media platforms to adopt a universal definition of climate disinformation and recognize the threat that disinformation on social media plays in derailing climate negotiations (see Chapters 10 and 13).[82] While some progress has been made (e.g., TikTok adopted a program to develop educational content to combat climate misinformation in 2023[83]), many platforms continue to fall short on adopting policies to effectively counter disinformation.[84] Maintaining pressure on social media platforms to strengthen and enforce their policies can help to remove climate disinformation.[85] Importantly, further critical attention must be paid to how people, corporations, and other institutions use and govern these technologies in ways that may also exacerbate the negative effects of human-induced climate change.[86]

Public Awareness

Another strategy for addressing obstruction involves public awareness efforts. This takes many forms, including shaming brands and platforms running digital ads that monetize obstructionist content; calling out blatant examples of greenwashing by fossil fuel companies on social media through "greentrolling"; and identifying specific organizations and important actors within those organizations (see Chapter 7).[87]

Climate shaming has perhaps contributed to the surge of greenwashing as obstructionist actors promote their alignment and support of climate action policies without corresponding shifts in behavior.[88] Researchers have found pervasive evidence of greenwashing by technology companies such as Apple, Google, and Amazon, which propose decarbonized supply chains, "restorative" carbon removal funds, or the development of "carbon-free energy" (see Chapter 2).[89] Market and securities regulators have undertaken studies, organized public comments, and moved to update their guidance, rules, tools, and enforcement actions to deal with the situation, including requirements for mandatory and standardized forms of public disclosure around emissions and net-zero claims. These are often hotly contested. For instance, the Securities and Exchange Commission (SEC) proposal to require Scope 3 carbon emissions disclosures from industry has generated extensive pushback and amplified anti-ESG communication from businesses as a climate obstruction strategy.[90]

Investigations

While corporate shaming is often intended to bring actors into alignment with norms for acceptable climate action, it contributes to wider investigative efforts to improve accountability in the public sphere. Such efforts also serve to expose and delegitimate obstruction.[91] *ExxonMobil* claims that they communicated openly about climate change, but the research suggests otherwise.[92] Other researchers identified a German-based think tank, *EIKE*, and its spokesperson as part of obstructionist efforts by think tanks.[93]

As one broad review of climate litigation efforts surmises, the success of legal challenges requires a shift in widely held societal perceptions regarding the causes of climate harms and role of law in addressing them, or what they call a "superstructure narrative" (see Chapters 7 and 12).[94] As a result, research into climate misrepresentations and obstruction is finding its way into complaints and publicity strategies, and some research explicitly links industry communication to strategies for undermining litigation.[95] These efforts have included hearings and complaints that seek to hold PR and strategic consultancies accountable. In addition, systemic media misrepresentations around climate risk are cited in the legal complaint brought by municipalities in Puerto Rico against several carbon majors.[96] The US-based Climate Investigations Center maintains a list of ongoing climate lawsuits.[97]

Rapid Attribution

Rapid attribution of misleading content and corrective measures to stem disinformation through coordinated research efforts are important to addressing obstruction.[98] This process may lead to diminished trust in transgressive sources as well as lower consumption and belief in their news products. There are daily advisories, weekly and monthly digests, and explainers and reports of climate disinformation, including descriptions of the most frequent narratives, actors, and organizations available online. CAAD has developed backgrounders and synthesized research into guidance on preparing for obstructionist strategies including typologies of common claims, strategies and actors; an archive of profiles of recurrent actors is maintained by DeSmog blog.[99] CAAD also developed an advanced rapid response capability to monitor the information environment for the COP summit for disinformation threats in recent years, issuing daily advisories throughout the meeting to assist journalists in covering stories. Amy Westervelt's Drilled podcast provides an archive of information about climate disinformation.

One of the more effective techniques for addressing climate obstruction in advertising and PR has been to spotlight companies that advertise on websites promoting denial or obstruction and pressure them to stop. Action campaigns by organizations such as Clean Creatives, Check My Ads, and Sleeping Giants have been especially effective in challenging companies to remove ads from disreputable sites and news organizations, including some organizations that circulate climate disinformation, and reports that call out platforms like Meta and Google have generally resulted in action by platforms.

INFORMATION GAPS AND RESEARCH PRIORITIES

Efforts to expose and push back on climate obstruction in the media are underway in many countries around the world.[100] Early efforts by investigative journalists to uncover the ways in which obstructionists use the media to undermine climate science and solutions have evolved into coordinated action to fight the spread of climate disinformation in both legacy and social media. A diverse coalition of journalists, civil society groups, activists, and academic researchers has developed resources for monitoring and exposing deliberate climate disinformation, from online news outlets that specialize in uncovering climate obstruction to open-source databases of the individuals, organizations, corporations, and public relations firms that are fueling climate obstruction in the media and lobbying elected officials (see Chapter 13).[101]

While the obstructionist playbook is well known in terms of its goals, key actors, usual frames, common rhetorical strategies, and funding and distribution networks, there is still low knowledge in several domains. For instance, we lack an understanding of choices made regarding communication platforms used to obstruct climate action in the public sphere. Additional considerations include:

- the scope of PR and strategic consultancy involvement in these networks;[102]
- the cross-platform circulation of obstructionist content including the role of influencers;
- user engagement with such content; and
- the advertising technology systems that monetize this content.

Further research will also need to closely examine the use of generative AI and how to curtail its use for disinformation, including and beyond the problem of climate change.

Additionally, there is a need for research into public and audience engagement with obstructionist content. At present, emphasis is placed on the identification and labeling of deceptive content and its cognitive-behavioral implications. This approach has shaped efforts to prebunk, debunk, and otherwise help individuals avoid some of the logical fallacies or persuasive appeals of disinformation. These tools are valuable but applied most easily to instances of science denial in media education and classroom settings. Other aspects of the media ecosystem are not as easily parsed into true or false statements, including strategies of policy delay, which are often presented as opinions, and image-based or highly emotive communication that limits the effectiveness of logical counterclaims. Importantly, there remains a significant gap in research looking at how climate journalists and activists working in the Global South create different media narratives that may contest these exclusionary ones circulated in the Global North (see Chapter 8).[103] There is also concern that artificial intelligence (AI) tools will facilitate the translation and spread of deliberate disinformation into other languages given the ease, rapidity, and low cost of generating content with these tools.

As our understanding improves about how climate obstruction networks engage in destructive strategies of disinformation in the areas of social media, PR, advertising, and legacy news, more work is needed. Ongoing research must push beyond the detection of highly coordinated or inauthentic activity (which is valuable) to a broader understanding of how climate obstruction can be amplified in more informal and everyday ways through accidental re-circulation of anti-climate rhetoric, infiltration of community groups, or social media "activists" sowing discord, complicating the usual way that obstructionist communication is identified, labeled, and delegitimated online.[104]

Data-driven approaches to monitor obstruction, particularly in social media, often draw on key findings from academic research.[105] For example, leaked documents have suggested that obstruction tactics and networks are extensive and have played decisive roles in electoral success of environmental policy-hostile far-right candidates in countries around the world, including the United States, United Kingdom, and Brazil (see Chapter 5).[106] In short, climate researchers as well as practitioners now have a sophisticated suite of tools to monitor and expose climate disinformation in social and legacy media, but this does not preclude the necessity for more systemic, political-economic change.

CONCLUSION

There are many ongoing efforts to improve media representations of climate change, including investigative studies into the communication strategies of carbon-polluting industries and their enablers, solutions-based journalism that highlight routes toward a "just" green energy transition, and pro-climate advertising and PR campaigns. By extension, some scholars are emphasizing flashpoint events or moments of strategic visibility that can hold outsized importance in processes of social change. Arguably, this has been the approach of obstructionists in targeting key people and moments, especially IPCC reporting or moments of crisis.

Reflecting a broader imbalance in studies of global environmental politics, the great majority of analyses of news coverage of climate change have focused on the largest historical polluting countries in the Global North, in particular the United States and other high-income Anglophone countries.[107] The news media's role in shaping the climate policy agenda in developing countries is thus less studied and understood.[108]

Going forward, climate change media discourses—across legacy and social platforms—must be analyzed in a global context, including the Global South, with attention to particularities of national and regional politics and interests.

As a note of caution, singling out disinformation as well as climate obstruction through media, advertising and PR can run the risk of overlooking shortcomings of institutional structures as well as broader challenges of information, literacy, education, and communication that also deserve careful consideration. As we hurtle forward in a dynamic and rapidly changing world, analyses of media portrayals must continue to adopt a broad scope that includes the political economy, cultural and social systems, and dynamics of power, fostering critical analyses of ownership and control of current media systems and of the increasingly integral and powerful AI technologies.[109]

NOTES

1. IPCC, *Climate Change 2014: Synthesis Report. Contribution of Working Groups I, II and III to the Fifth Assessment Report of the Intergovernmental Panel on Climate Change*, ed. R. K. Pachauri and L. A. Meyer, 2014.

2. J. Marisa Dispensa and R. J. Brulle, "Media's Social Construction of Environmental Issues: Focus on Global Warming—A Comparative Study," *International Journal of Sociology and Social Policy* 23, no. 10 (2003): 74–105.

3. Naomi Oreskes, "The Scientific Consensus on Climate Change," *Science* 306, no. 5702 (2004): 1686–1686.

4. Maxwell T. Boykoff and Jules M. Boykoff, "Balance as Bias: Global Warming and the US Prestige Press," *Global Environmental Change* 14, no. 2 (2004): 125–136; Maxwell T. Boykoff and Jules M. Boykoff, "Climate Change and Journalistic Norms: A Case-Study of US Mass-Media Coverage," *Geoforum* 38, no. 6 (2007): 1190–1204.

5. Maxwell T. Boykoff, *Who Speaks for the Climate? Making Sense of Media Reporting on Climate Change* (Cambridge University Press, 2011).

6. Lucy McAllister, Meaghan Daly, Patrick Chandler, Marisa McNatt, Andrew Benham, and Maxwell Boykoff, "Balance as Bias, Resolute on the Retreat? Updates & Analyses of Newspaper Coverage in the United States, United Kingdom, New Zealand, Australia and Canada Over the Past 15 Years," *Environmental Research Letters* 16, no. 9 (2021): 094008.

7. For example, Michael K. Goodman, Maxwell T. Boykoff, and Kyle T. Evered, *Contentious Geographies: Environmental Knowledge, Meaning, Scale* (Routledge, 2016).; Geoffrey Supran and Naomi Oreskes, "Assessing ExxonMobil's Climate Change Communications (1977–2014)," *Environmental Research Letters* 12, no. 8 (2017): 084019.

8. M. T. Boykoff and T. Yulsman, "Political Economy, Media, and Climate Change: Sinews of Modern Life," *Wiley Interdisciplinary Reviews: Climate Change* 4, no. 5 (2013): 359–371.

9. Robert Brulle, "Institutionalizing Delay: Foundation Funding and the Creation of US Climate Change Counter-movement Organizations," *Climatic Change* 122 (2014): 681–694. We use the terms "climate change countermovement organizations," "contrarian countermovement organizations," "climate change countermovement groups," "contrarian countermovement groups," "climate countermovements," and the like interchangeably.

10. Aaron McCright and Riley Dunlap, "Challenging Global Warming as a Social Problem: An Analysis of the Conservative Movement's Counter-claims," *Social Problems* 47, no. 4 (2000): 499–522; Maxwell Boykoff, "Climate Change Countermovements & Adaptive Strategies: Insights from the Heartland Institute Annual Conferences a Decade Apart," *Climatic Change* 177, no. 5 (2024), https://doi:10.1007/s10584-023-03655-5

11. William Lamb, Giulio Mattioli, Sebastian Levi, Timmons Roberts, Stuart Capstick, Felix Creutzig, Jan Minx, Finn Müller-Hansen, Trevor Culhane, and Julia Steinberger, "Discourses of Climate Delay," *Global Sustainability* 3 (2020): E17.

12. Sandra Harding, "After the Neutrality Ideal: Science, Politics, and 'Strong Objectivity,'" *Social Research* (1992): 567–587; Myanna Lahsen, "We Cannot Afford Not to Perform Constructionist Studies of Mainstream Climate Science," in *Climate, Science and Society*, ed. Z. Baker, T. Law, M. Vardy, & S. Zehr (Routledge, 2024), 29–38.

13. N. Usher, *News for the Rich, White, and Blue: How Place and Power Distort American Journalism* (Columbia University Press, 2021); Jessa Lingel, *The Gentrification of the Internet: How to Reclaim our Digital Freedom* (University of California Press, 2021).

14. Bharat Dhiman, "Key Issues and New Challenges in New Media Technology in 2023: A Critical Review," *Journal of Media & Management* 5, no. 1 (2023): 1–4.

15. See, e.g., Benedetta Brevini, *Is AI Good for the Planet?* (Wiley, 2021).

16. Deborah Rogers, Anantha Duraiappah, Daniela Christina Antons, Pablo Munoz, Xuemei Bai, Michail Fragkias, and Heinz Gutscher, "A Vision for Human Well-Being: Transition to Social Sustainability," *Current Opinion in Environmental Sustainability* 4, no. 1 (2012): 61–73.

17. Robert McChesney, *Digital Disconnect: How Capitalism Is Turning the Internet against Democracy* (The New Press, 2013).

18. Eli Noam and the International Media Concentration Collaboration, *Who Owns the World's Media? Media Concentration and Ownership around the World* (Oxford University Press, 2016) provide a complex picture, showing asymmetries and fragmention of media ownership and concentration as well as different types of ownership and different styles of concentration.

19. Yochai Benkler, Robert Faris, and Hal Roberts, *Network Propaganda: Manipulation, Disinformation, and Radicalization in American Politics* (Oxford University Press, 2018); Anabela Carvalho, "Media(ted) Discourses and Climate Change: A Focus on Political Subjectivity and (Dis)engagement," *WIRES Climate Change* 1, no. 2 (2010): 172–179. Myanna Lahsen, "Buffers Against Inconvenient Knowledge: Brazilian Newspaper Representations of the Climate-Meat Link," *Desenvolvimento e Meio Ambiente* 40 (2017): 17–35; Anabela Carvalho and Tarla Rai Peterson, "Reinventing the Political," in *Climate Change Politics: Communication and Public Engagement*, ed. Anabela Carvalho and Tarla Rai Peterson (Cambria Press, 2012), 1–28.

20. Matthew Hornsey, Emily Harris, and Kelly Fielding, "Relationships Among Conspiratorial Beliefs, Conservatism and Climate Scepticism Across Nations," *Nature Climate Change* 8, no. 7 (2018): 614–620.

21. Ibid. See also Robert Brulle, Jason Carmichael, and J. Craig Jenkins. "Shifting Public Opinion on Climate Change: An Empirical Assessment of Factors Influencing Concern Over Climate Change in the US, 2002–2010," *Climatic Change* 114, no. 2 (2012): 169–188.

22. Hornsey et al., "Relationships Among Conspiratorial Beliefs."

23. Ibid.

24. For Canada, see Erick Lachapelle, Christopher Borick, and Barry Rabe, "Public Attitudes Toward Climate Science and Climate Policy in Federal Systems: Canada and the United States Compared," *Review of Policy Research* 29, no. 3 (2012): 334–357. For Brazil, see Kirsti Jylhä, Clara Cantal, Nazar Akrami, and Taciano Milfont, "Denial of Anthropogenic Climate Change: Social Dominance Orientation helps Explain the Conservative Male Effect in Brazil and Sweden," *Personality and Individual Differences* 98 (2016): 184–187. For Australia, see Kelly Fielding, Brian Head, Warren Laffan, Mark Western, and Ove Hoegh-Guldberg, "Australian Politicians' Beliefs about Climate Change: Political Partisanship and Political Ideology," *Environmental Politics* 21, no. 5 (2012): 712–733.

25. For traditional/legacy media, see Michael Brüggemann and Sven Engesser, "Between Consensus and Denial: Climate Journalists as Interpretive Community," *Science Communication* 36, no. 4 (2014): 399–427. For online media, see Ruth McKie, "Obstruction, Delay, and Transnationalism: Examining the Online Climate Change Counter-Movement," *Energy Research and Social Science* 80 (2021): 102217.

26. Ruth McKie, "Climate Change Counter Movement Neutralisation Techniques: A Typology to Examine the Climate Change Counter Movement," *Sociological Inquiry* 89, no. 2 (2018): 288–316.

27. For examples, see Aaron McCright and Riley Dunlap, "Cool Dudes: The Denial of Climate Change Among Conservative White Males in the United States," *Global Environmental Change* 21, no. 4 (2011): 1163–1172; Jylhä et al., "Denial of Anthropogenic Climate Change."

28. Melissa Aronczyk, "Branding the Nation in the Era of Climate Crisis: Eco-nationalism and the Promotion of Green National Sovereignty," *Nations and Nationalism* 30, no. 1 (2023): 25–38.

29. Olve Krange, Bjørn Kaltenborn, and Martin Hultman, "Cool Dudes in Norway: Climate Change Denial among Conservative Norwegian Men," *Environmental Sociology* 5, no. 1 (2019): 9.

30. Riley Dunlap and Aaron McCright, "Organized Climate Change Denial," in *The Oxford Handbook of Climate Change and Society*, ed. J. Dryzek, R. Norgaard, and D. Schlosberg (Oxford University Press, 2011), 144–160.

31. Chandra Lal Pandey and Priya A. Kurian, "The Media and the Major Emitters: Media Coverage of International Climate Change Policy," *Global Environmental Politics* 17, no. 4 (2017): 67–87.

32. James Painter and Teresa Ashe. "Cross-National Comparison of the Presence of Climate Scepticism in the Print Media in Six Countries, 2007–10," *Environmental Research Letters* 7, no. 4 (2012): 044005.

33. R. McKie, C. Milani, G. Edwards, J. Walz, K. Hochstetler, O. Faruqe, P. Kashwan, P. McElwee, and T. Roberts, *Is Climate Obstruction Different in the Global South? Observations and a Preliminary Research Agenda*, CSSN Position Paper no. 4, Issue. C. S. S. Network, 2021.

34. For examples, see Robert Brulle, Melissa Aronczyk, and Jason Carmichael, "Corporate Promotion and Climate Change: An Analysis of Key Variables Affecting Advertising Spending by Major Oil Corporations, 1986–2015," *Climatic Change* 159 (2020): 87–101.

35. Melissa Aronczyk and Maria Espinoza, *A Strategic Nature: Public Relations and the Politics of American Environmentalism* (Oxford University Press, 2022).

36. Shane Gunster, Darren Fleet, and Robert Neubauer, "Challenging Petro-Nationalism: Another Canada Is Possible?," *Journal of Canadian Studies* 55, no. 1 (2021): 57.

37. Max Falkenberg Alessandro Galeazzi, Maddalena Torricelli, Niccolò Di Marco, Francesca Larosa, Madalina Sas, Amin Mekacher, Warren Pearce, Fabiana Zollo, Walter Quattrociocchi, and Andrea Baronchelli, "Growing Polarization Around Climate Change on Social Media," *Nature*, November 24, 2022, https://www.nature.com/articles/s41558-022-01527-x

38. IPCC, *Climate Change 2022: Mitigation of Climate Change, Working Group III* (Cambridge University Press, 2022). See also Kristoffer Ekberg, Bernhard Forchtner, Martin Hultman, and Kirsti Jylhä, *Climate Obstruction: How Denial, Delay and Inaction Are Heating the Planet* (Taylor & Francis, 2023).

39. IPCC, 2022: *Climate Change 2022: Mitigation of Climate Change. Contribution of Working Group III to the Sixth Assessment Report of the Intergovernmental Panel on Climate Change* (2022), ed. P. R. Shukla, J. Skea, R. Slade, A. Al Khourdajie, R. van Diemen, D. McCollum, M. Pathak, S. Some, P. Vyas, R. Fradera, M. Belkacemi, A. Hasija, G. Lisboa, S. Luz, and J. Malley (Cambridge University Press, 2022), Technical Summary 6.2, 58.

40. Ibid., Chapter 13.4, 127.

41. Ibid., Technical Summary 9, 129.

42. Scott Waldman, "Climate Denial Spreads on Facebook as Scientists Face Restrictions," *Scientific American*, July 6, 2020.

43. Center for Countering Digital Hate, "New Research Shows Google Makes Millions By Greenwashing Big Oil Ahead of COP27," 2022, https://counterhate.com/blog/google-makes-millions-by-greenwashing-big-oil-ahead-of-cop-27/.

44. Michael Khoo, Phil Newell, and Deb Lavoy, "Latest IPCC Reports Underscore Threat of Climate Disinformation," Tech Policy Press, April 4, 2022, https://www.techpolicy.press/latest-ipcc-reports-underscore-threat-of-climate-disinformation/. See also Nathalie Maréchal, Rebecca MacKinnon, and Jessica Dheere, "Getting to the Source of Infodemics: It's the Business Model," Open Technology Institute, May 27, 2020, https://www.newamerica.org/oti/reports/getting-to-the-source-of-infodemics-its-the-business-model/.

45. Climate Action against Disinformation (CAAD), *Climate of Misinformation: Ranking Big Tech*, 2023.

46. For examples, see Justin Farrell, "Network Structure and Influence of the Climate Change Counter-Movement," *Nature Climate Change* 6, no. 4 (2016): 370–374; Max Boykoff, *Creative (Climate) Communications: Productive Pathways for Science, Policy and Society* (Cambridge University Press, 2019); Travis Coan, Constantine Boussalis, John Cook, and Mirjam Nanko, "Computer-Assisted Classification of Contrarian Claims about Climate Change," *Scientific Reports* 11, no. 1 (2021): 22320; Hanna Morris, "Constructing the Millennial 'Other' in United States: Press Coverage of the Green New Deal," *Environmental Communication* 15, no. 1 (2021): 133–143; and Chris Russill, "Alarmism and Accountability in Climate Communication during Extreme Events, *Social Media+Society*, April–June (2023): 1–3.

47. Myanna Lahsen, "Media Reform as Transformation Tool: A Hegemonic Gap in Environmental Research and Policy," *International Journal of Politics, Culture, and Society*, forthcoming.

48. Geoffrey Supran and Naomi Oreskes, "Rhetoric and Frame Analysis of ExxonMobil's Climate Change Communications," *One Earth* 4, no. 5 (2021): 696–719.

49. Maxine Joselow, "Oil Companies Are Hiring TikTok Influencers to Court Young People," *Washington Post*, August 17, 2023.

50. Dimitris Dimitriadis, Joey Grostern, and Sam Bright, "Revealed: Fossil Fuel Giants are Using British Influencers to Go Viral," *DeSmog*, July 27, 2023, https://www.desmog.com/2023/07/27/fossil-fuel-oil-gas-giants-shell-bp-using-british-influencers-to-go-viral/; "How Shell Used a 'Granfluencer' to Promote Its Brand," *DeSmog*, August 15, 2023, https://www.desmog.com/2023/08/15/shell-granfluencer-promote-fossil-fuels-greenwash/.

51. Benedetta Brevini and Graham Murdock (eds.), *Carbon Capitalism and Communication: Confronting Climate Crisis* (Palgrave Macmillan, 2017).

52. Eli Noam, "The Owners of the World's Media," in Noam, *Who Owns the World's Media? Media Concentration and Ownership around the World* (Oxford University Press, 2016), 1180–1242. See also Robert Hackett, Susan Forde, Shane Gunster, and Kerrie Foxwell-Norton, *Journalism and Climate Crisis: Public Engagement, Media Alternatives* (Taylor and Francis, 2017).

53. See, e.g., Timothy Neff, "Transnational Problems and National Fields of Journalism: Comparing Content Diversity in US and UK News Coverage of the Paris Climate Agreement," *Environmental Communication* 14, no. 6 (2020): 730–743.

54. Amy Westervelt and Matthew Green, "Leading News Outlets Are Doing the Fossil Fuel Industry's Greenwashing," *The Intercept*, December 5, 2023, https://theintercept.com/2023/12/05/fossil-fuel-industry-media-company-advertising/.

55. Maxwell Boykoff, *Who Speaks for the Climate? Making Sense of Media Coverage of Climate Change* (Cambridge University Press, 2011).

56. M. Vidal Valero, "Death Threats, Trolling and Sexist Abuse: Climate Scientists Report Online Attacks," *Nature* 616, no. 7957 (2023): 421–422; Global Witness, "Global Hating: How Online Abuse of Climate Scientists Harms Climate Action," 2023, https://www.globalwitness.org/en/campaigns/digital-threats/global-hating/; L. Givatesh, "Young Female Climate Activists Face Hateful Abuse Online. This Is How They Cope," *NBC News*, November 10, 2019, https://www.nbcnews.com/news/world/young-female-climate-activists-face-hateful-abuse-online-how-they-n1079376.

57. See N. Lakhani, D. Gayle, and M. Taylor, "How Criminalisation Is Being Used to Silence Climate Activists Across the World," *The Guardian*, October 12, 2023, https://www.theguardian.com/environment/2023/oct/12/how-criminalisation-is-being-used-to-silence-climate-activists-across-the-world.

58. For examples, see Lucy McAllister, Meaghan Daly, Patrick Chandler, Marisa McNatt, Andrew Benham, and Max Boykoff, "Balance as Bias, Resolute on the Retreat? Updates and Analyses of Newspaper Coverage in the United States, United Kingdom, New Zealand, Australia and Canada Over the Past 15 Years," *Environmental Research Letters* 16, no. 9 (2021): 1–14; Michael Brüggemann and Sven Engesser, "Beyond False Balance: How Interpretive Journalism Shapes Media Coverage of Climate Change," *Global Environmental Change* 42 (2017): 58–67.

59. Michael Brüggemann and Sven Engesser, "Between Consensus and Denial: Climate Journalists as Interpretive Community," *Science Communication* 36, no. 4 (2014): 399–427.

60. Candis Callison, "Refusing More Empire: Utility, Colonialism, and Indigenous Knowing," *Climatic Change* 167, no. 3–4 (2021).

61. For examples, see Candis Callison and Mary Lynn Young, *Reckoning: Journalism's Limits and Possibilities* (Oxford University Press, 2020); and Henrik Bødker and Hanna Morris (eds.), *Climate Change and Journalism: Negotiating Rifts of Time* (Routledge, 2021).

62. For examples, see Nadine Strauss, James Painter, Joshua Ettinger, Marie-Noëlle Doutreix, Anke Wonneberger, and Peter Walton, "Reporting on the 2019 European Heatwaves and Climate Change: Journalists' Attitudes, Motivations and Role Perceptions," *Journalism Practice* 16, no. 2–3 (2022): 462–485. But see Myanna Lahsen, Gabriela de Azevedo Couto, and Irene Lorenzoni, "When Climate Change Is Not Blamed: The Politics of Disaster Attribution in International Perspective," *Climatic Change* 158, no. 2 (2020): 213–233.

63. Callison, "Refusing More Empire," 58.

64. For examples, see Catalina De Onís, *Energy Islands: Metaphors of Power, Extractivism, and Justice in Puerto Rico* (University of California Press, 2021).

65. Morris, "Constructing," 2021; Ella Muncie, "'Peaceful Protesters' and 'Dangerous Criminals': The Framing and Reframing of Anti-Fracking Activists in the UK," *Social Movement Studies* 19, no. 4 (2020): 464–481.

66. Friends of the Earth, "Climate, Clicks, Capitalism, and Conspiracists," August 28, 2020.

67. K. Demetrious, *Public Relations and Neoliberalism: The Language Practices of Knowledge Formation* (Oxford University Press, 2022).

68. Myanna Lahsen, "We Cannot Afford Not to Perform Constructionist Studies of Mainstream Climate Science," in *Climate, Science and Society*, eds. Z. Baker, T. Law, M. Vardy, and S. Zehr (Routledge, 2024), 29–38. "Steering Signification for Sustainability," *Global Sustainability* e15 (2024): 1–10.

69. Climate Action Against Disinformation "Four Days of Texas-Sized Misinformation: Social Media Companies Threaten Action on Climate Change," Friends of the Earth, 2021, https://caad.info/wp-content/uploads/2022/12/Texas_Disinfo_Report_final_v4.pdf; Cristina López G. and SantiagoLokatos, *Los Eco-Ilógicos*, Graphika Report, October 2022, https://graphika.com/reports/los-eco-il%C3%B3gicos.

70. Robert Brulle and Timmons Roberts, "Climate Misinformation Campaigns and Public Sociology," *Contexts* 16, no. 1 (2017): 78–79.

71. Amelia Sharman, "Mapping the Climate Sceptical Blogosphere," *Global Environmental Change* 26 (2014): 159–170; and Hywel Williams, James McMurray, Tim Kurz, and F. Hugo Lambert, "Network Analysis Reveals Open Forums and Echo Chambers in Social Media Discussions of Climate Change," *Global Environmental Change* 32 (2015): 126–138. On media coverage of important events, see Saffron O'Neill, Hywel Williams, Tim Kurz, Bouke Wiersma, and Max Boykoff, "Dominant Frames Evident in Legacy and Social Media Coverage of the IPCC 5th Assessment Report," *Nature Climate Change* 5, no. 4 (2015): 380–385.

72. Constantine Boussalis and Travis Coan, "Text-Mining the Signals of Climate Change Doubt," *Global Environmental Change* 36 (2016): 89–100; Justin Farrell, "Corporate Funding and Ideological Polarization about Climate Change," *Proceedings of the National Academy of Sciences* 113, no. 1 (2016): 92–97; Dominik Stecula and Eric Merkley, "Framing Climate Change: Economics, Ideology, and Uncertainty in American News Media Content from 1988 to 2014," *Frontiers in Communication* 4, no. 6 (2019), DOI:10.3389/fcomm.2019.00006.

73. Travis Coan, Constantine Boussalis, John Cook, and Mirjam O. Nanko, "Computer-Assisted Classification of Contrarian Claims About Climate Change," *Scientific Reports* 11, no. 1 (2021): 22320.

74. Constantine Boussalis and Travis Coan, "Text-Mining the Signals of Climate Change Doubt," *Global Environmental Change* 36 (2016): 89–100.

75. Coan et al., "Computer-Assisted Classification of Contrarian Claims About Climate Change."

76. W. James Potter, "Review of Literature on Media Literacy," *Sociology Compass* 7, no. 6 (2013): 417–435.

77. Maxwell Boykoff, *Creative (Climate) Communications: Productive Pathways for Science, Policy and Society* (Cambridge University Press, 2019).

78. Nozima Muratova, Alton Grizzle, and Dilfuza Mirzakhmedova, *Media and Information Literacy in Journalism: A Handbook for Journalists and Journalism Educators* (UNESCO, 2019).

79. Victor Galaz, Hannah Metzler, Stefan Daume, Andreas Olsson, Björn Lindström, and Arvid Marklund, "Climate Misinformation in a Climate of Misinformation," research brief, Stockholm Resilience Centre (Stockholm University) and the Beijer Institute of Ecological Economics (Royal Swedish Academy of Sciences), 2023, http://arxiv.org/abs/2306.12807.

80. John Naughton, "A Tsunami of AI Misinformation Will Shape Next Year's Knife-Edge Elections," *The Guardian*, August 12, 2023.

81. Kathie Treen, Hywel Williams, and Saffron O'Neill, "Online Misinformation about Climate Change," *Wiley Interdisciplinary Reviews: Climate Change* 11, no. 5 (2020): 1–20.

82. Isabelle Gerretsen, "Laurence Tubiana Leads Call for Crackdown on Climate Misinformation in Glasgow Pact," *Climate Change News*, September 11, 2021.

83. TikTok, "Advancing our Commitment to Sustainabiity and Climate Literacy at COP28," Press Release, November 27, 2023, https://newsroom.tiktok.com/en-us/advancing-our-commitment-to-sustainability-and-climate-literacy-at-cop-28.

84. On responses to climate movement pressure, see Tiffany Hsu, "Pinterest Bans Climate Misinformation from Posts and Ads," *New York Times*, April 6, 2022. On how platforms continue to fall short on adopting policies to effectively counter disinformation, see the Greenpeace report, *In the Dark: How Social Media Companies' Climate Disinformation Problem is Hidden from the Public* (Greenpeace, 2022).

85. For a model of digital regulation, see Amélie Heldt, "EU Digital Services Act: The White Hope of Intermediary Regulation," in *Digital Platform Regulation: Global Perspectives on Internet Governance*, ed. Terry Flew and Fiona Martin (Palgrave Macmillan Cham, 2022), 69–84.

86. Lahsen, "We Cannot Afford Not to Perform Constructionist Studies of Mainstream Climate Science."

87. On corporate contributions to climate change, see Sharon Yadin, *Fighting Climate Change Through Shaming* (Cambridge University Press, 2023). On shaming brands and platforms running digital ads that monetize obstructionist content, see CAAD, *Climate of Misinformation*, 2023. On calling out blatant examples of greenwashing by fossil fuel companies on social media through "greentrolling," see Mary Heglar, "Confessions of a Green Troll," *Orion Magazine*, October 2022.

88. Noémi Nemes, Stephen Scanlan, Pete Smith, Melissa Aronczyk, Stephanie Hill, Simon Lewis, Wren Montgomery, Francesco Tubiello, and Doreen Stabinsky, "An Integrated Framework to Assess Greenwashing," *Sustainability* 14, no. 8 (2022): 4431; Tereza Blazkova, Esben Rahbek Gjerdrum Pedersen, Kirsti Reitan Andersen, and Francesco Rosati, "Greenwashing Debates on Twitter: Stakeholders and Critical Topics," *Journal of Cleaner Production* 427 (2023): 139260.

89. Soh Young In and Kim Schumacher, "Carbonwashing: A New Type of Carbon Data-related ESG Greenwashing," *Social Science Research Network*, 2021.

90. Lesley Clark, "Red States Decry 'Woke Left' SEC Proposal for ESG Investing," *Energy and Environment News*, August 18, 2022.

91. Sharon Yadin, *Fighting Climate Change Through Shaming* (Cambridge University Press, 2023).

92. Geoffrey Supran and Naomi Oreskes, "Assessing ExxonMobil's Climate Change Communications, 1977–2014," *Environmental Research Letters* 12, no. 8 (2017).

93. José Moreno, Mira Kinn, and Marta Narberhaus, "A Discussion of Think Tanks in Climate Obstruction," *International Journal of Communication* 17 (2023): 2077–2086.

94. Friederike Otto, Petra Minnerop, Emmanuel Raju, Luke Harrington, Rupert Stuart-Smith, Emily Boyd, Rachel James, Richard Jones, and Kristian Lauta, "Causality and the Fate of Climate Litigation: The Role of the Social Superstructure Narrative," *Global Policy* 13 (2022): 736–750. See also Amanda Shanor and Sarah Light, "Greenwashing and the First Amendment," *Columbia Law Review* 122, no. 7 (2022): 2033–2118.

95. Supran and Oreskes, "Rhetoric," 2021.

96. Russill, "Alarmism," 2023.

97. Climate Investigations Center, "Climate Lawsuit Index: Complaints with Links to Footnotes and Referenced Documents," July 8, 2020, https://climateinvestigations.org/climate-lawsuit-index-complaints-with-links-to-footnotes-and-referenced-documents.

98. See, e.g., J. E. Hopke, "Tweeting on a Rapidly Warming Planet: Environmental Communication Social Media Research," in *Handbook of Environmental Communication*, eds. Anabela Carvalho and Tarla Rai Peterson (De Gruyter Mouton Publishing, 2024), 172–189.

99. Climate Disinformation Database, *DeSmog*, https://www.desmog.com/climate-disinformation-database/.

100. See N. Almiron, and J. Xifra, *Climate Change Denial and Public Relations: Strategic Communication and Interest Groups in Climate Inaction* (Routledge, 2019); B. Forchtner, *Visualising Far-Right Environments: Communication and the Politics of Nature* (Manchester University Press, 2023); J. Painter, J. Ettinger, D. Holmes, L. Loy, J. Pinto, L. Richardson, L. Thomas-Walters, K. Vowles, and R. Wetts, "Climate Delay Discourses Present in Global Mainstream Television Coverage of the IPCC's 2021 Report," *Communications Earth & Environment* 4, no. 1 (2023): 1–12; M. Poberezhskaya, *Communicating Climate Change in Russia: State and Propaganda* (Routledge, 2015); M. Poberezhskaya, "Blogging About Climate Change in Russia: Activism, Scepticism and Conspiracies," *Environmental Communication* 12, no. 7 (2018): 942–955; and D. Vicente Torrico, "Analysis of Climate Denialism on YouTube: Refuting Instead of Debating," *Catalan Journal of Communication & Cultural Studies* 15 (2023): 217–235 as examples of research.

101. Examples of specialist news coverage on exposing climate obstruction include outlets such as *Inside Climate News*, *DeSmog*, and *Carbon Brief*.

102. Aronczyk and Espinoza, *A Strategic Nature*.

103. Callison and Young, *Reckoning*, 2020; Waqas Ejaz and Adil Najam, "The Global South and Climate Coverage: From News Taker to News Maker," *Social Media+Society* 9, no. 2 (2023): doi.org/10.1177/20563051231177904; Ulrika Olausson and Peter Berglez, "Media and Climate Change: Four Long-standing Research Challenges Revisited," *Environmental Communication* 8, no. 2 (2014): 249–265.

104. Kate Starbird, Ahmer Arif, and Tom Wilson, "Disinformation as Collaborative Work: Surfacing the Participatory Nature of Strategic Information Operations," *Proceedings of the ACM on Human-Computer Interaction* 3, no. 127 (2019): 1–26.

105. CAAD and Center for Countering Digital Hate, *YouTube's Climate Denial Dollars* (Friends of the Earth, 2023).

106. Emily Kubin and Christian von Sikorski, "The Role of (Social) Media in Political Polarization: A Systematic Review," *Annals of the International Communication Association* 45, no. 3 (2021): 188–206.

107. Painter and Ashe, "Cross-National Comparison"; Isak Stoddard, Kevin Anderson, Stuart Capstick, Wim Carton, Joanna Depledge, Keri Facer, Clair Gough, Frederic Hache, Claire Hoolohan, Martin Hultman, Niclas Hällström, Sivan Kartha, Sonja Klinsky, Magdalena Kuchler, Eva Lövbrand, Naghmeh Nasiritousi, Peter Newell, Glen P. Peters, Youba Sokona, Andy Stirling, Matthew Stilwell, Clive L. Spash, and Mariama Williams, "Three Decades of Climate Mitigation: Why Haven't We Bent the Global Emissions Curve?" *Annual Review of Environment and Resources* 46 (2021): 653–689.

108. Chandra Pandey and Priya Kurian, "The Media and the Major Emitters: Media Coverage of International Climate Change Policy," *Global Environmental Politics* 17, no. 4 (2017): 67–87.

109. See for example Robert Epstein Christina Tyagi, and Hongyu Wang, "What Would Happen If Twitter Sent Consequential Messages to Only a Strategically Important Subset of Users? A Quantification of the Targeted Messaging Effect (TME)," *PLOS One* 18, no. 7 (2023): e0284495.

CHAPTER 7

Understanding the Political and Psychological Roots of Climate Misinformation and Its Impact on Public Opinion

LEAD AUTHORS: DOMINIK A. STECUŁA AND JOHN COOK
CONTRIBUTING AUTHORS: ARUNIMA KRISHNA, ADRIAN DOMINIK WÓJCIK, JEAN CARLOS HOCHSPRUNG MIGUEL, MATTHEW HORNSEY, AND SALIL BENEGAL

INTRODUCTION: UNDERSTANDING THE CAUSES AND CONSEQUENCES OF CLIMATE MISINFORMATION

Journalists, academics, and politicians have devoted considerable attention to misinformation since at least 2016, when false information about the US presidential election and the UK Brexit campaign spread rapidly through social media. These developments highlighted how quickly and effectively false information can be disseminated through our digital information ecosystem and the consequences of that spread. Given the urgency of effectively addressing climate change, preventing and addressing misinformation about climate-related issues has become an urgent task.

Misinformation can be defined as false and misleading information, most often spread without the intention to deceive. Disinformation is false and misleading information spread on purpose.[1] Given that intention is very difficult to assess, we use "misinformation" to refer to any falsehoods about climate change, whether or not they are spread with an intent to harm. Misinformation matters because it can influence attitudes and behavior of both the public and political elites. Public opinion is of particular importance in democratic

Dominik A. Stecuła et al., *Understanding the Political and Psychological Roots of Climate Misinformation and Its Impact on Public Opinion*. In: *Climate Obstruction*. Edited by: J. Timmons Roberts, Carlos R. S. Milani, Jennifer Jacquet, and Christian Downie, Oxford University Press. © Oxford University Press (2025).
DOI: 10.1093/oso/9780197787144.003.0007

societies because it provides a link between what the people want, their electoral behavior, and what politicians do on their behalf. The idea of democratic representation is therefore predicated on knowing and understanding public opinion. Misinformation, as part of broader information disorder, can distort and manipulate public opinion on a given issue, affecting not only beliefs and attitudes, but also voting and policy. Misinformation tends to flourish during elections and other events that increase the salience of politics and political action, especially in countries like the United States where political polarization is high, peoples' identities are shaped by partisan attachments, and many Americans process information through a political lens.

Global warming and climate change have been the subject of well-funded misinformation campaigns to cast doubt on climate science and scientists and subsequently to delay climate action.[2] These campaigns have historically been funded by conservative foundations as well as fossil fuel industry and other interest groups with the goal of denying scientific findings regarding human-caused climate change and downplaying the need for mitigation and otherwise addressing the problem.[3]

These campaigns have been found to be at least somewhat effective in making people question the existence of human-caused climate change, especially in the United States.[4] It is unsurprising, then, that the United States is home to the highest proportion of climate deniers and skeptics in the world.[5] There is thus an urgent need for investigations into effective communication strategies to help combat climate change misinformation campaigns, especially in the ever-changing information environment where social media play an increasingly significant role. In the United States, for example, just over one-quarter of Americans got news from social media in 2013. But by 2023, almost half of the population did. In Peru, Thailand, and Kenya, for example, about one-third of the population of each country gets their news on TikTok.[6] In this social media-centric world, where the role of the traditional gatekeepers has been replaced by an algorithm tailored toward virality, it is especially important to understand the dynamics of climate misinformation and its impact on public opinion.

In this chapter, we focus on how public opinion is shaped by misinformation about climate change. Historically, research on this topic has focused explicitly on climate denial, but recently, the umbrella term "climate obstruction" has emerged, highlighting aspects of these efforts beyond explicit climate denial. Although most of the work we have reviewed does not discuss climate obstruction per se, it is still highly relevant because it identifies the core psychological and ideological drivers of public opinion formation and the ways in which public opinion tends to be formed.

Much of this work to date has focused on the United States; thus, many of the examples we cite are from that nation. We do, however, try to capture as broad a view as possible, noting important examples from the Global South.

Box 7.1 on Bolsonaro's Brazil offers a case study of some of the dynamics we discuss in this chapter.

We first discuss the political, ideological, and psychological underpinnings that drive public opinion formation on important societal issues. We outline why people tend to believe what they believe, particularly when it comes to climate change, and demonstrate how and why climate obstruction strategies can be successful. We discuss not only the historical use of climate denial as obstruction, but also show how messages rooted in explicit climate denial have evolved to focus on attacking climate mitigation efforts. We then discuss climate misinformation specifically, centering our analysis on the five key climate beliefs and five key climate disbeliefs, again emphasizing the shift from obstruction via explicit climate denial rhetoric to attacks on mitigation efforts. We ground this discussion in the FLICC framework (which uses *f*ake experts, *l*ogical fallacies, *i*mpossible expectations, *c*onspiracy theories, and *c*herry picking), describing common techniques used in disseminating climate misinformation. We then discuss the concrete impacts of climate misinformation in terms of forming climate beliefs as well as the broader, systemic consequences that make it difficult to act politically and socially to address climate change. We conclude by discussing how the climate misinformation problem can be addressed through communication and education-based interventions.

DRIVERS OF PUBLIC OPINION ABOUT CLIMATE CHANGE

Despite the prevailing scientific consensus on climate change, certain segments of society disbelieve it, denying the mere existence of climate change or its anthropogenic origin.[7,8] Remarkably, these opinions are occasionally expressed by individuals of esteemed scientific reputation (including Nobel Prize laureates), particularly when their expertise does not encompass the discipline of climatology.[9] Thus, even possessing scientific expertise does not shield individuals from climate change misinformation.

That reality may be partially explained by communication study models such as the elaboration likelihood model (ELM) of attitude change.[10] The model assumes that people are motivated to hold correct attitudes and opinions. However, this tendency is at odds with the limited cognitive resources of an average citizen,[11] as the depth of thought possible depends on individual traits and situational factors. One should be able to think deeply about an issue when one is motivated and can process the information at hand—a situation in which new information will be evaluated based on its merits. This ideal scenario is rarely the case for climate change. Although climate change potentially affects a significant portion of Earth's population, thus motivating individuals to seek relevant information, they often do not have the skills or other information to process it correctly and in an unbiased way. Therefore, ELM would predict that climate change information will be processed shallowly, with recipients focusing not so much on the arguments' accuracy but on peripheral features such as the information's source or the emotions it evokes.

Public opinion can be shaped by a variety of factors, depending on the person and the topic. People respond to different arguments based on personal experiences, issue salience, knowledge of the issue, values, partisanship, and a multitude of other forces. In this chapter, we will pay particularly close attention to the impact of elite cues, ideology, the news media, and other relevant factors that scholarship has found play a role in climate opinion formation.

The forces that shape public opinion can be broadly characterized as bottom-up and top-down. One perspective on dissemination of climate change misinformation is that it operates in a bottom-up fashion, originating in the minds of individuals and then spreading through their social networks. Bottom-up misinformation typically arises from genuinely held beliefs rather than an intent to deceive, which can be the product of psychological biases such as confirmation bias, causing people to cherry-pick information that confirms their existing beliefs.[12] An alternative view is that climate change misinformation spreads from the top down: here, vested interests—such as the fossil fuel industry and their affiliated political elites—engage in messaging on climate change that influences the public. Importantly, these top-down and bottom-up approaches are not mutually exclusive.[13]

Public Opinion on Climate Change

The top-down process by which climate misinformation is spread gained widespread prominence with the publication of the book *Merchants of Doubt* by the historians Naomi Oreskes and Erik Conway,[14] although other scholars have also described such processes.[15] The authors argued that the US climate skepticism movement was originally spearheaded by three high-profile American physicists with strong anti-socialist views. Although they represented a small minority within the scientific community, their voices were magnified by the support of various climate change countermovement organizations (e.g., think tanks and other advocacy organizations). In the early 2000s, an analysis of ninety-one such organizations revealed annual funding of just over $900 million, much of it tied to the fossil fuel industry.[16,17] The pro-business, anti-government rhetoric spread by these organizations resonated mostly among US Republicans and were featured prominently in targeted advertisements and business-oriented publications. That messaging was further amplified by conservative media in the United States. The result was an integrated campaign of misinformation.

Climate change became a more salient issue as scholars produced more research about global warming and its consequences. As it entered both political and cultural discourse, polling on it became more commonplace and scholars began to investigate factors driving climate attitudes. Public opinion on climate change in the United States in the 1980s and early 1990s reflected a largely uninformed populace that was generally unaware of the scientific consensus on climate change or the greenhouse effect and had heard little about the issue in news media or from political elites.[18,19] By the late 1990s and early 2000s, polling data indicated that substantial segments of the American public held misinformed rather than uninformed views about the existence of climate change, its causes, and its risks. Moreover, these views were most common among Republican voters, who by the mid-2000s have been far more likely than Democrats to reject the scientific consensus on climate change and oppose action on mitigative emissions-reducing policies.[20,21]

Ideology

By the mid-2000s, the major predictor of climate change opinion among Americans was partisan identification—Democrats accepted scientific consensus whereas Republicans largely did not—and this political polarization was particularly pronounced among highly educated respondents, who generally tend to be more attuned to their respective party positions on issues.[22,23,24] These were the decades during which climate science was drawn into culture wars that describe and prescribe what it means to be left or right in one's politics.

Americans aligned with the Republican party are considerably more skeptical of climate science today than they were in the 1990s. This left–right divide in attitudes toward climate change has been identified in several other countries (although not as starkly as in the United States).[25] Some scholars have identified ideology, values, and general worldviews as the root causes of this polarization of climate attitudes along political lines.[26,27] Their approaches tend to be rooted in psychological research on motivated reasoning[28,29] that suggests people are motivated to reject messages from scientists and experts if they do not conform to their values and identities and even to seek information from contrarian sources to find supporting evidence for these values and identities.[30,31]

A core aspect of political conservatism, for example, is suspicion of "big government," and resistance to regulation of individuals and markets.[32] In the case of climate change, the policy implications of global warming are not compatible with a laissez-faire philosophy.[33,34] Those with an implicit moral suspicion of government regulation therefore find climate science inconvenient because the solutions to climate change frequently imply government intervention. Thus for some conservatives, it might be cognitively easier to reject climate science than to accept its proposed solutions. Indeed, meta-analyses suggest that free-market ideology is one of the biggest predictors of climate denial, suggesting an ideological underpinning to the political divide over climate change (see Chapter 5).[35]

Over time, ideologically driven beliefs about climate change among members of a group can become a collective identity overlaid with moral and political significance. That climate science has become a contested intergroup issue is concerning because it can lead to self-reinforcing and difficult-to-break cycles of intergroup polarization that can hinder action.[36,37]

However, it is difficult to explain the political divide in willingness to believe misinformation about climate change as being driven exclusively by bottom-up ideological concerns. Many Americans are not ideologically consistent, failing to understand "what goes with what" politically and holding views that strong ideologues may view as contradictory.[38,39] A powerful example of this phenomenon in the context of climate change is that market-friendly ways to reduce carbon emissions, such as via carbon pricing and carbon taxes, are considerably less popular among Republicans than among Democrats.[40] Furthermore, ideological explanations struggle to account for the fact that conservatives and Republicans once held views on climate change very similar to those of liberals and Democrats.[41]

Finally, the left–right divide in beliefs about climate change is not universal. There is nothing inherently conservative about climate denial, and in fact, for many years the Republican party has been an outlier on climate change among conservative parties in Western democracies.[42] A more precise typology suggests that in nations where the economy is heavily dependent on fossil fuels (e.g., the United States, Australia, Brazil), belief in climate change tends to be politically polarized, but in nations in which the economy is less dependent on fossil fuels, the left–right divide is weaker or nonexistent.[43] In Germany, for example, mainstream parties across the political spectrum subscribe to the scientific consensus on climate and have become global leaders in the transition from fossil fuels to renewables (albeit this unity has recently been challenged by the growth of the far-right AfD party).[44] That political polarization is most pronounced in nations that have the most to lose economically from the energy transition away from fossil fuels suggests that polarization around climate change is not only a result of bottom-up ideological considerations, but also is shaped by top-down campaigns from economic vested interests and political opportunists. In other words, economic interests invested in protecting the status quo play an important role in shaping public opinion on climate change and ultimately fostering society-wide polarization on the issue.

In seeking to explain how polarization on climate change has shifted over time and across nations, some have focused on the notion of elite cues: that people's beliefs about climate change are influenced by the messages conveyed by senior or high-profile figures within their political communities. For example, aggregates of US polls conducted in the 2000s showed that climate change concern was influenced by top-down messaging (e.g., Congressional press releases from Republicans and Democrats or messages about climate change from political actors transmitted via the mass media) and that these messages were more influential than scientific communication published in academic journals—or even than people's own observations of climate change effects.[45,46] In Australia, polling data over ten years showed that climate skepticism varied reliably as a function of political support for conservative political parties.[47] Furthermore, experiments have shown that when assessing climate policies, people rely on their source of information rather than on the specific content of the policy—thus confirming the so-called party over policy effect.[48] Both Democrats and Republicans tended to more strongly support specific environmental solutions (such as cap-and-trade or carbon tax policies) when they were told that the policy was proposed by representatives of their own party.[49]

Elite cues, like messages from trusted political leaders, play an important role in shaping public opinion about climate change, following top-down models of attitude formation. Public opinion research has consistently shown that public opinion follows the opinions of political leaders, and climate change is no different.[50,51] As noted, during the 1970s in the United States, Republican and Democratic political elites shared similar views on environmental issues and were able to pass many environmental bills with bipartisan agreement—in several cases with near-consensus or supermajorities.[52,53] But politicians from both parties diverged on this issue over time.

Elite divergence on climate change after the Kyoto Protocol saw an asymmetric shift in which Republican politicians began explicitly to deny the existence of climate change, question the validity or consensus of scientific findings supporting it, or claimed that climate change was natural and not affected by human activity.[54] Concurrently, segments of an uninformed public that had heard little about the issue became misinformed as Republican voters turned to trusted in-group elites as sources of information.[55,56] In cases where Republican elites have changed course or presented pro-climate change stances, Republican voters have seemed to show more alignment with the scientific consensus on climate change or increased their support for climate policies.[57,58] However, such cases have been rare or politically costly for Republicans who take such stances: for example, after advocating a bipartisan carbon tax, former Representative Bob Inglis was defeated in the 2010 primary for his seat in the US House of Representatives . Elite cues have also been found to shape public opinion about climate change outside of the US context, for example in the European Union.[59]

In addition to in-group elite messaging effects, out-group cue-taking (responding to messages from opponents) has contributed to the spread of skepticism and misinformation about climate change among Republican voters. Since the 2000s, while Republican elites have largely denied or undermined facts about climate change, Democratic elites have emphasized climate change as central to their policy agenda. As prominent Democratic elites—most notably former Vice President Al Gore and other Congress members such as Senator Bernie Sanders and Representative Alexandria Ocasio-Cortez—have increasingly supported the validity of climate science, the urgency of climate change, and mitigative policy solutions, there has been a backlash from Republicans (particularly those who are better educated or more attentive to news media), who have sought to distance themselves from the views of the opposing party.[60]

The effects of such elite messaging and mass-media framing are greatest in the United States, where scholars have identified large and persistent gaps between supporters of the Republican and Democratic parties in their views about climate change. In other countries and regions that do not see the same extent of polarization among elites on climate change, partisan gaps among the public on climate misinformation belief and endorsement are smaller. However, left–right ideological divisions are still evident in much of Western Europe, where conservatives/respondents who hold strong right-wing worldviews are much more likely to dispute climate science or endorse other misinformation.[61]

News Media

The news media play an important role across any society in informing the public about issues that matter. Especially on scientific issues that are difficult to understand, the media are an important mechanism through which knowledge passes on to the public, given that most members of the public do not read peer-reviewed academic journals. Of course, the public can and does experience the effects of climate change, but the links between extreme weather events and other climate effects need to be pointed out for people to understand the connection.

How the news media cover the issue of climate change is therefore very important. But news organizations have not always provided high-quality coverage of climate that draws these connections. Early work on this topic has shown that the positions of the "merchants of doubt," the networks of fossil fuel advocacy groups and conservative think tanks, garnered traction in the news media. These groups used their political clout to support contrarian scientists in casting doubt on climate change in both their research and in the media.[62,63,64,65] Because of the importance of being seen as unbiased, many journalists were inadvertently complicit in amplifying these contrarian voices in news stories about climate change by providing "balanced" coverage to satisfy professional journalistic norms that give equal weight to various sides of an issue.[66,67] More recent analysis suggests, however, that coverage of climate change has improved since these studies first appeared.[68] At the same time, explicitly conservative news outlets frequently spread messages from these groups in order to undermine the public's confidence in climate science.[69]

More recent work, however, highlights that when the coverage of climate change became the most salient (after the release of the 2006 documentary *An Inconvenient Truth*), coverage of the merchants of doubt wasn't particularly robust. Instead, the American media became increasingly partisan. The news media in the United States focused primarily on Democratic messages on climate change, which tended to be consistently pro–climate science. At the same time, Republican climate messages have been less voluminous and more ambiguous until recently (with more straightforward climate skepticism and denial surfacing after the emergence of the Tea Party, a populist right-wing movement within the Republican Party, in 2009). This research suggests that the news media amplified partisan cues, and in the polarized context of American politics, Republican voters took cues from Democratic elites to reject climate science (see Chapter 6).[70]

A synthesis of the above perspectives suggests that in the spread of misinformation is a collaboration between vested interests at both the individual and collective levels.[71] On the one hand, science has been manipulated by forces that have coached and convinced some conservatives that there is no consensus around climate science (or the solutions to climate change). On the other hand, these misinformation campaigns would likely not be successful if individuals were not motivated to believe them. Individuals also have vested interests—whether they be economic, ideological, or identitarian—that make them prefer certain messages over other messages.

The pernicious influence of misinformation on climate change attitudes, combined with willingness of some societal actors to spread it, creates obstacles that stall climate action. Of course, majority support for a specific policy is no guarantee that said policy will be implemented by the government (see, for example, the lack of policy action on gun control in the United States despite overwhelming support among the American public for policies such as universal background checks). But when coupled with pressure on politicians to act, public support is an important mechanism through which political change can be achieved in democratic societies.

To design effective communication campaigns, understanding the varied audiences for one's message is crucial. Communication scholars and practitioners have long advocated for the development and use of public-segmentation strategies that help classify the broader population into similarly minded groups.[72,73]

Thus, the creation of the typology of disinformation-susceptible publics may be an important step in helping to address climate change misinformation.[74] The typology is a theoretically grounded classification of the broader population into one of four groups: those who are (1) disinformation-immune, (2) disinformation-vulnerable, (3) disinformation-receptive, and (4) disinformation-amplifiers. Drawing on the situational theory of problem solving, scholars have proposed and tested the typology of disinformation-susceptible publics on the issue of climate change.[75]

The central feature of the typology of disinformation-susceptible publics is that individuals who are highly motivated about an issue, hold extreme attitudes about it, and are knowledge-deficient about it (i.e., have previously accepted false information about the topic) can be classified as lacuna publics or disinformation-amplifying publics.[76] Disinformation-amplifiers are conceptualized not only to accept misinformation messages but also to actively amplify and share them with others. Although disinformation-amplifiers are typically very small in number, they tend to be the loudest voices about an issue around which they are motivated.[77] Any efforts to convince disinformation-amplifiers otherwise will likely not be successful. On the other end of the spectrum are disinformation-immune publics, or those who are neither knowledge deficient nor hold extreme attitudes about a topic, in this case climate change. Disinformation-immune publics may benefit most from prophylactic, or preemptive, "inoculations" to forearm them against future climate change misinformation efforts.[78]

Much of the population, however, tends to fall into the categories of disinformation-vulnerable and disinformation-receptive. Disinformation-receptive publics comprise individuals who are knowledge-deficient about an issue and display either high levels of motivation or extreme issue attitudes around it, making them highly susceptible to misinformation. Disinformation-vulnerable publics are conceptualized to be at moderate risk of accepting misinformation and may be the focal point for communication campaigns designed to reduce the negative effects of climate change misinformation. Thus, the typology of disinformation-susceptible publics may provide climate communicators with a starting point to help design interventions to address climate change misinformation. Indeed, it has been noted that "a well-constructed typology can advance other aims of science including prediction, explanation, and understanding, and can offer practitioners insight into how to craft more effective public communication," and the typology of disinformation-susceptible publics has the potential of fulfilling this promise.

Several recent studies also identify other individual-level belief systems and biases that appear to complement and amplify the effects of other factors previously discussed. Anti-intellectualism—or a generalized mistrust of scientific experts—plays an important role in fostering belief in misinformation about climate change. While mistrust of scientific experts is not new and has been well documented across various American political eras,[79,80] anti-intellectual attitudes have become more prevalent among the general public as these views have converged with more populist sentiments[81] and the emergence of populist leaders/political figures in countries such as the United States or Brazil, during the past decade.[82] These populist attitudes are typically associated with a distrust or rejection of scientific experts, and individuals who hold such worldviews also show a greater tendency to seek or disseminate information from nonexperts on climate change and other scientific issues.[83,84,85] These processes may be further amplified by populist rhetoric, which exists across ideological and partisan lines and is prevalent in multiple country contexts.[86] Of course, sometimes mistrust of scientists' statements can be justified, as in the case of eugenics. However, when a consensus is reached through the application of the scientific method, as is the case with climate change, the weight of evidence is squarely on the side of science.[87]

Attitudes about race and gender also shape public views about climate change, motivating both explicit denial of climate change and opposition to mitigation efforts. Multiple processes appear to play a role in this phenomenon. Within the United States, race may have become more strongly associated with climate change through a "spillover effect" during the Obama presidency for voters who frequently viewed Obama and his policy agenda through a racialized lens[88] and through increased awareness or consideration of environmental justice issues in climate action.[89] Gendered differences have also become more salient in a range of country and regional contexts via both conservative elite and media messages that have disproportionately attacked women climate advocates such as Greta Thunberg[90] and amplified gender disparities in the perceived costs of climate mitigation in wealthier nations like the United States or the United Kingdom.[91]

These biases have been amplified by individuals who perceive climate action as a threat to existing capitalist, racial, and gendered hierarchies, and by far-right populist movements in several countries that have also emphasized anti-egalitarian views while dismissing climate change.[92,93,94] Consequently, a range of studies now documents a growing association between individuals' racialized and gendered biases and belief in varying types of misinformation about climate change. While they have become increasingly associated with far-right political parties and populist elites in the United States, Brazil, and parts of Western Europe, such as Spain,[95,96] these biases are evident across partisan and ideological divides and may help explain the prevalence or persistence of climate misinformation in a range of country contexts.

Another dimension of public responses to climate change is explicated in research that explored three types of climate denial: (1) literal (rejection of the existence of climate change), (2) interpretive (downplaying its severity or anthropogenic causes), and (3) implicatory (accepting the basic facts of climate change but ignoring their implications).[97] While literal and interpretive denial involve rejecting the reality of human-caused global warming and the corresponding impacts,[98] implicatory denial describes more subtle psychological responses such as a disconnection between environmental behavior (or lack thereof) and its implications for climate change.

Climate considerations, in other words, interact with many other cultural and political identities and commitments that all citizens have. Ultimately, effectively communicating climate messages to the public and policymakers requires framing that information skillfully in order to consider audience beliefs and identities that might potentially interfere with accepting accurate scientific information.

CLIMATE MISINFORMATION ARGUMENTS

Surveys of climate attitudes have identified five key beliefs about climate change: (1) global warming is real, (2) it is human-caused, (3) experts agree on these two first points, (4) climate impacts are harmful, and (5) there is hope that we can avoid the worst impacts of climate change.[99] Of particular note is the perceived scientific consensus, which has been identified as a gateway belief influencing the other key climate beliefs.[100] However, these positively framed beliefs are mirrored by negatively framed misinformation arguments. As noted in the previous section, a recently developed taxonomy of contrarian claims about climate change has revealed that climate misinformation can be grouped into five main categories, paraphrased as: (1) global warming isn't real, (2) it isn't human-caused, (3) climate impacts aren't bad, (4) climate solutions won't work, and (5) experts are unreliable. These five themes have been described as the five key climate disbeliefs.

The researchers found that from 2000 to 2020, contrarian arguments showed a long-term transition from science-based to solutions-based misinformation. The newer forms negate not so much the existence of climate change, but rather the necessity of swift and decisive action.[101,102,103] This new trend goes under many names: climate delayism, response skepticism,[104] or secondary obstruction.[105] All of those terms describe a new form of climate change misinformation intended to challenge the efficacy of action to address the problem. It comprises four major arguments: denying that climate change may be solved, emphasizing the downsides of mitigating actions, redirecting responsibility for climate change to others, and advocating nontransformative solutions. This new form of climate change misinformation may be challenging to neutralize, as it is often used to "greenwash" the actions of GHG emitters, such as by conducting pro-environmental campaigns while withdrawing from the European Green Deal.[106,107] (Greenwashing is a strategy of misleading public opinion to present entities in a deceptively positive light in terms of their actions on behalf of the environment.) The fifth category of climate misinformation—attacking scientists—is particularly prominent on social media, with 60% of climate misinformation tweets on Twitter/X containing conspiracy theories or ad-hominem attacks on climate actors.[108]

The five climate disbeliefs are content-based in that they describe the content of contrarian arguments. Another important lens through which to understand climate misinformation is the technique-based or logic-based approach, which seeks to understand the misleading rhetorical techniques and reasoning fallacies used in spreading misinformation. A useful tool for understanding this approach is the FLICC framework, which describes five techniques of science denial: (1) *f*ake experts, (2) *l*ogical fallacies, (3) *i*mpossible expectations, (4) *c*herry picking, and (5) *c*onspiracy theories.[109] While these techniques can be deployed as intentional strategies to deceive, they can also result from genuinely held psychological biases. For example, people tend to ascribe greater expertise to those they agree with, which can result in reliance on spokespeople with no expertise or whose expertise is in a nonrelevant domain.[110] People's susceptibility to immediate cues such as warmer or cooler local temperatures make them vulnerable to anecdotal arguments.[111] The difficulty in distinguishing genuinely believed misinformation from intentionally deceptive disinformation means that care must be taken when attributing deceptive motives to misinformation sources.

NEGATIVE IMPACTS OF CLIMATE MISINFORMATION

Climate misinformation is associated with specific negative impacts on public opinion about climate change. Just a few misleading statistics can lower people's acceptance of the reality of climate change.[112] Misinformation casting doubt on the scientific consensus on human-caused climate change—a gateway belief—has been found not only to reduce the degree of consensus someone perceives but also to negatively affect other climate attitudes such as acceptance of the reality of human-caused global warming and support for climate action.[113]

Misinformation has also been found to cancel out attempts to communicate accurate information. When participants received information about the 97% scientific consensus on human-caused climate change along with misinformation casting doubt on the consensus, there was no significant change in the participants' perception of consensus, showing that the consensus message was neutralized by the misinformation. Similarly, when positive frames urging action on climate change appeared alongside a denialist counterframe, belief in the reality of anthropogenic climate change as well as support for reducing carbon emissions decreased.[114] However, the conditions under which factual information is canceled out by misinformation are contextual— one experiment found a canceling effect when misinformation was encountered after the factual message was received, but the factual message was effective if it came after the misinformation was received.[115]

The impact of climate change misinformation isn't limited to the beliefs and attitudes of individuals. For example, misinformation can have detrimental effects on the scientific community itself, by "seeping" non–knowledge-based considerations into scientific research and debates.[116] Similarly, most of the public who are concerned or alarmed about climate change fail to talk about the issue with friends and family.[117] The predominant driver of self-silencing about climate change is pluralistic ignorance—the misconception that most people don't care about climate change.[118] Unfortunately, this process results in a self-reinforcing "spiral of silence,"[119] a consequential outcome given that building social momentum is an important component for building overall momentum for collective action on climate change.[120]

Box 7.1: CASE STUDY: BOLSONARISM IN BRAZIL

Much of this chapter has been built on scholarship findings from the Global North, and particularly the United States. Future work should expand the focus of this research into different political and cultural contexts, especially in the Global South. In this section, we provide a case study of the spread of climate misinformation as a form of obstruction in Brazil under the Bolsonaro administration.

Bolsonarism refers to the political ideology and movement associated with Jair Bolsonaro, who served as president of Brazil for four years (2019 to 2023). Climate denialism in the country during this period did not merely consist of orchestrated disinformation campaigns by Bolsonarist groups on social media platforms. Instead, it can be better understood as a heterogeneous set of practices and discourses. Those networks extended from the digital environment to formal spaces occupied by government authorities in ministries, representatives in the Legislative branch, and reached the groups that took to the streets to defend the government. These actions fundamentally shaped the politics of what many have called the "post-truth" as an authoritarian project of producing ignorance.

Bolsonaro was supported by a coalition of parties and organizations that unified more traditional conservatives, anti-establishment groups, ultranationalists, anti-environment groups, and pro-military sectors. Bolsonaro took advantage of a feature of populism, namely that it is tied to neither the left or the right end of the political spectrum, but instead is motivated by mistrust of "elites" and the superiority of "the people."[121] Misinformation in the Bolsonaro government was also disseminated through "digital populism"—a way of building charismatic power and mobilizing the masses through the circulation of misleading memes and narratives that included climate change misinformation linking it disapprovingly with globalist left-wing ideologies.[122] Climate change denialism found broad acceptance within that worldview, and Bolsonaro's coalition granted it credibility and ultimately made it a part of the government's official discourse and policy on environmental issues.

One researcher has argued that the climate denialism apparatus did not emerge in Brazil with Bolsonarism, but rather found new conditions for growth and the exercise of power within this political movement.[123] Climate denialism first appeared in Brazil in 2007, presenting itself as a conservative discourse in response to the growing concern about climate change as a result of the Intergovernmental Panel on Climate Change's (IPCC) *Fourth Assessment Report* establishing climate change as a scientific fact, the release of Al Gore's documentary *An Inconvenient Truth*, and the first national study by the National Institute for Space Research, which had led to a nationwide effort to promote the development of climate policies. Despite the extensive dissemination of scientific findings in the media, some Brazilian newspapers provided a platform for a minority with opposing opinions that challenged the scientific evidence and promoted climate denialism.

In May 2007, writer Olavo de Carvalho published an article in the *Diário do Comércio* newspaper titled "Science or Charade?" in which he criticized Al Gore's film. In one of the earliest records of climate misinformation in the national media, Carvalho asserted that it was all a "hoax" and associated climate change with "leftist activism" infiltrating international organizations.[124] In 2009, during the political process of reformulating the Brazilian Forest Code, climate denialists José Carlos de Almeida Azevedo (of the University of Brasilia) and Luis Baldicero Molion (of the Federal University of Alagoas) participated as researchers in public hearings organized to discuss whether climate change was related to deforestation and should be considered in debates about the new Forest Code. On that occasion, de Almeida Azevedo, a physicist, addressed the issue, stating that it is "impossible to predict, let alone alter anything related to the climate"; that instead of warming, we are heading for a "new ice age";[125] and that "the Earth's climate is governed by the Sun," therefore, "carbon has no influence on these astronomical phenomena." He also stated that "climate legislation will affect the country" and that

Continued

"what matters to Brazil is what is done here, not what the English or anyone else will tell us to do." In essence, his argument was that the issue of global warming was actually a geopolitical and economic matter filled with scientifically baseless information.

Another line of climate denialism in Brazil pertained to the formation of Christian-conservative political ideology, widespread during the Bolsonaro regime, that encouraged resistance to scientific information. One author has noted that this created a complex situation given that different Christian traditions held different views on politics and the environment.[126] Brazilian authors of denialist books have argued that the IPCC is a political-economic platform that uses scientific fraud for the implementation of what the authors call a "new global order."[127,128,129] According to these authors, the formation of this new global order is expressed in the form of a "trade war" between industrialized countries and less developed countries.

Another mechanism through which Bolsonarism spurred the growth of climate misinformation was the installation of climate denialists in strategic ministries for climate policy, including the Ministry of Foreign Affairs (Itamaraty) and the Ministry of the Environment. As some of his initial administrative actions, the minister of the environment closed the Secretariat of Climate Change and Forests and, in a joint decision with Itamaraty, withdrew Brazil's offer to host the UN Climate Change Conference (COP25) in December 2019. The former foreign minister, Ernesto Araújo, publicly supported a theory of "left-wing globalism" during his tenure and referred to climate change as a global conspiracy he called "climatism."[130]

Events in Brazil from July 2019 through COP25 later that year highlight how climate change denialism is not just a discourse or misinformation strategy but can guide institutionalized practices that undermine environmental policy and science. After Itamaraty sent a diplomat to participate in a conference with climate denialists organized by the Heartland Institute think tank in Washington, D.C., a telegram circulated within the ministry stating "they are putting our way of life at risk. The debate is not about climate change or carbon dioxide. It's not about climate or science. It's about socialism versus capitalism." Then, at COP25, the Brazilian delegation was small, hesitant, and without a clear mission.[131] Unlike at previous meetings, Brazil aligned itself with countries including the United States, Australia, and Saudi Arabia in an attempt to obstruct negotiations. With this decision, the Brazilian delegation declined to continue its position of leadership among developing countries and to demand ambitious CO_2 emissions reduction targets from developed countries.[132]

As demonstrated, the phenomenon of climate denialism in Brazil under Bolsonaro cannot be reduced to "misinformation" or "ignorance." And its consequences are not limited to opinions and attitudes. Under Bolsonaro, deforestation of the Amazon reached a record high, and funding for environmental enforcement was cut.[133] In this context, understanding the complexity and underlying motivations of climate denialism is crucial for promoting an informed and effective debate on environmental issues across the Global South and addressing the challenges posed by an informational and political landscape mired in misinformation.

COMMUNICATION AND EDUCATION-BASED INTERVENTIONS TO COUNTER MISINFORMATION

Research on climate misinformation and how it shapes public opinion has focused on mapping said misinformation and showing the linkage between it and public views on climate change. But there has also been growing interdisciplinary work on fighting climate misinformation.

One option for resisting climate misinformation is to use communication strategies that reduce the motivation for conservatives to embrace it. In nations where the debate on climate action is highly politicized, this means making a conservative-friendly case for such action. The literature reveals several examples of how this can work. For example, if one states that we can respond to climate change using free-market-friendly techniques, then Republicans are less likely to reject the science.[134] If one frames mitigating climate change as something that can help promote "green" technologies, then skeptics are likely to display more pro-environmental intentions than if one just presented the evidence that climate change is real.[135]

Other studies show that when messages about mitigating climate change are framed as a matter of obeying authority or demonstrating patriotism, then conservatives are more likely to think in pro-climate ways than if more traditional messages are used.[136,137] A field study examined the effectiveness of a one-month campaign that used spokespeople who referred to action on climate change as consistent with their conservative values. The campaign increased Republicans' understanding of the existence and causes of climate change by several percentage points.[138] These examples use what is described as "jiu jitsu persuasion"—presenting scientific messages in ways that are congenial to people's underlying motivations and using those identities to capture their attention and leverage change.[139]

Corrective or debunking efforts may also be more effective when the messenger shares a political identity with the targeted audience. Experiments with American survey samples have shown that Republican Party sources may be more effective in correcting climate denialism and increasing overall environmental concern among other Republicans than other political or scientific sources.[140] However, other studies show corrective effects may also be relatively limited in their ability to shape policy preferences: such respondents may be more likely to agree with scientific facts about climate change after being presented with fact-checking, but do not correspondingly update their preferences for mitigative policies.[141]

Finally, simple solutions that promote individual resistance to climate change misinformation may not be sufficient to mitigate the pernicious effects of misinformation on public opinion. Institutional actors such as the fossil fuel industry or conservative think tanks run anti-climate change-related campaigns.[142] Additionally, the funding for anti-environmental campaigns is extensive. The fossil fuel industry is estimated to allocate ten times more funds to lobbying than pro-environmental activists and renewable energy representatives combined.[143]

Therefore, combating climate change misinformation requires focusing on two tasks: (1) enhancing individual immunity to misinformation, and (2) instituting systemic changes that make it harder for misinformation to spread and thrive.[144] These tasks are especially urgent given that fossil fuel companies often engage in actions promoting individual pro-environmental behaviors even as they resist more systemic forms of change in, for example, the law or financial regulations (see Chapters 2 and 3).[145] Even when we accept the inoculation metaphors, people need to be convinced to get inoculated.

CONCLUSION

In this chapter, we first discussed the various forces that shape public opinion on climate change to provide a broad perspective that could help elucidate why public opinion on climate change might be vulnerable to different forms of misinformation. As such, we have focused on the psychological and political dynamics researchers have identified as major forces affecting beliefs and opinions about climate change, such as elite cues, the importance of the news media, and political ideology. We discussed the history of explicit climate denial as a major focus of obstruction and how these messages have evolved to focus on attacking climate mitigation efforts.

Moving on to discussing climate change misinformation specifically, we rooted our analysis in the five key climate beliefs and how misinformation can be summarized by five key climate disbeliefs, again emphasizing the shift from explicit climate denial rhetoric to attacks on mitigation efforts. We complemented this discussion with the FLICC framework, describing misinformation techniques. We then moved on to discussing concrete effects of climate misinformation in terms of climate beliefs, but also broader, systemic consequences that make it difficult to act politically and socially to address climate change. Finally, we concluded by discussing how the climate misinformation problem can be addressed through communication and education-based interventions.

Much of the research on which we have relied in this chapter comes from the WEIRD countries: *western, educated, industrial, rich, democracies*.[146] That reality likely limits the generalizability of these findings to other cultural and political contexts. In order to address this caveat, we presented the case study of Brazil, which has parallels with the United States (in terms of the rise of right-wing populism), but also show considerable differences.

There is still much to be done in researching issues pertaining to climate misinformation and obstruction, as well as the ways in which these societal problems can be addressed and alleviated. As we noted throughout the chapter, future research needs to focus on non-WEIRD countries, including replicating existing studies and typologies in the cultural and political contexts of the Global South. More explicit examination of greenwashing strategies as a form of misinformation should also be the focus of future work.

Another area for future scholarship is the rapid changes to the information environment including the increasing use of social media as a major source of information. Such research will be particularly important in countries that primarily rely on messaging apps such as WhatsApp, which are not the focus of US researchers because the platform is not as widely used in America as in other countries. Relatedly, artificial intelligence (AI) and large language models (LLMs) will likely continue to play an ever-growing role in generating and exacerbating misinformation. Future work should investigate how these tools affect the problem of climate misinformation, as well as the ways in which they might be used as a tool for understanding and responding to misinformation.

Finally, purveyors of climate misinformation increasingly place a strong emphasis on ad-hominem attacks and rely on conspiracy theories, but these methods are understudied in the climate misinformation literature. Future work should draw on insights from the robust literature on conspiratorial ideation and character attacks.

The suggestions in this chapter, along with its explication of the roots and nature of climate change misinformation as a form of obstruction—especially the deeper psychological and political dynamics that allow this phenomenon to affect public opinion—will, we hope, provide a starting point for societies to resist it.

NOTES

1. S. Altay, M. Berriche, and A. Acerbi, "Misinformation on Misinformation: Conceptual and Methodological Challenges," *Social Media+ Society* 9, no. 1 (2023): 20563051221150412.

2. N. Oreskes and E. M. Conway, *Merchants of Doubt: How a Handful of Scientists Obscured the Truth on Issues from Tobacco Smoke to Global Warming* (Bloomsbury, 2011).

3. R. J. Brulle and J. T. Roberts, "Climate Misinformation Campaigns and Public Sociology," *Contexts* 16, no. 1 (2017): 78–79.

4. A. Krishna, "Understanding the Differences Between Climate Change Deniers and Believers' Knowledge, Media Use, and Trust in Related Information Sources," *Public Relations Review* 47, no. 1 (2021): 1–8.

5. B. Teiturier and S. Duhautois, "Climate Change: Citizens Are Worried But Torn Between a Need to Act and a Rejection of Constraints," Ipsos, 2020, https://www.ipsos.com/en/climate-change-citizens-are-worried-torn-between-need-act-and-rejection-constraints.

6. Digital News Report, Reuters Institute for the Study of Journalism, 2023, https://reutersinstitute.politics.ox.ac.uk/digital-news-report/2023.

7. A. Leiserowitz, C. Roser-Renouf, J. Marlon, and E. Maibach, "Global Warming's Six Americas: A Review and Recommendations for Climate Change Communication," *Current Opinion in Behavioral Sciences* 42 (2021): 97–103.

8. B. Tranter and K. Booth, "Scepticism in a Changing Climate: A Cross-National Study," *Global Environmental Change* 33 (2015): 154–164.

9. M. Impelli, "Nobel Prize Winner Who Doesn't Believe Climate Crisis Has Speech Canceled," *Newsweek*, July 24, 2023, https://www.newsweek.com/nobel-prize-winner-who-doesnt-believe-climate-crisis-has-speech-canceled-1815020.

10. R. E. Petty and J. T. Cacioppo, "The Elaboration Likelihood Model of Persuasion," *Advances in Experimental Social Psychology* 19 (1986): 123–205.

11. K. E. Stanovich, *What Intelligence Tests Miss: The Psychology of Rational Thought* (Yale University Press, 2009).

12. J. Cook, "Deconstructing Climate Science Denial," in *Edward Elgar Research Handbook in Communicating Climate Change*, ed. D. Holmes and L. M. Richardson (Edward Elgar Publishing, 2020), 62–78.

13. R. J. Brulle, J. Carmichael, J., and Jenkins, "Shifting Public Opinion on Climate Change: An Empirical Assessment of Factors Influencing Concern Over Climate Change in the US, 2002–2010," *Climatic Change* 114, no. 2 (2012): 169–188.

14. Oreskes and Conway, *Merchants of Doubt*.

15. S. Rampton and J. Stauber, *Trust Us, We're Experts PA: How Industry Manipulates Science and Gambles with Your Future* (Penguin, 2002).

16. R. J. Brulle, "Institutionalizing Delay: Foundation Funding and the Creation of U.S. Climate Change Counter-Movement Organizations," *Climatic Change* 122, no. 4 (2014): 681–694.

17. R. J. Brulle, G. Hall, L. Loy, and K. Schell-Smith, "Obstructing Action: Foundation Funding and US Climate Change Counter-Movement Organizations," *Climatic Change* 166, no. 1 (2021): 17.

18. K. Bowman, E. O'Neil, and H. Sims, *Polls on the Environment, Energy, Global Warming and Nuclear Power* (American Enterprise Institute, 2016).

19. P. J. Egan and M. Mullin, "Climate Change: US Public Opinion," *Annual Review of Political Science* 20 (2017): 209–227.

20. A. M. McCright and R. E. Dunlap, "The Politicization of Climate Change and Polarization in the American Public's Views of Global Warming, 2001–2010," *Sociological Quarterly* 52 (2011): 155–194.

21. E. Merkley and D. A. Stecula, "Party Cues in the News: Democratic Elites, Republican Backlash, and the Dynamics of Climate Skepticism," *British Journal of Political Science* 51, no. 4 (2021): 1439–1456.
22. R. E. Dunlap and A. M. McCright, "A Widening Gap: Republican and Democratic Views on Climate Change," *Environment: Science and Policy for Sustainable Development* 50 (2008): 26–35.
23. R. E. Dunlap, A. M. McCright, and J. H. Yarosh, "The Political Divide on Climate Change: Partisan Polarization Widens in the U.S.," *Environment: Science and Policy for Sustainable Development* 58 (2016): 4–23.
24. J. M. Jones, "Democratic, Republican Confidence in Science Diverges," Gallup, 2021, https://news.gallup.com/poll/352397/democratic-republican-confidence-science-diverges.aspx.
25. M. J. Hornsey, E. A. Harris, and K. S. Fielding, "Relationships Among Conspiratorial Beliefs, Conservatism and Climate Scepticism Across Nations," *Nature Climate Change* 8, no. 7 (2018): 614–620.
26. M. J. Hornsey, "Why Facts Are Not Enough: Understanding and Managing the Motivated Rejection of Science," *Current Directions in Psychological Science* 29 (2020): 583–591.
27. M. J. Hornsey and K. S. Fielding, "Attitude Roots and Jiu Jitsu Persuasion: Understanding and Overcoming the Motivated Rejection of Science," *American Psychologist* 72, no. 5 (2017): 459–473.
28. P. H. Ditto and D. F. Lopez, "Motivated Skepticism: Use of Differential Decision Criteria for Preferred and Nonpreferred Conclusions," *Journal of Personality and Social Psychology* 63, no. 4 (1992): 568.
29. Z. Kunda, "The Case for Motivated Reasoning," *Psychological Bulletin* 108, no. 3 (1990): 480.
30. D. M. Kahan, H. Jenkins-Smith, and D. Braman, "Cultural Cognition of Scientific Consensus," *Journal of Risk Research* 14, no. 2 (2011): 147–174.
31. S. Lewandowsky and K. Oberauer, "Motivated Rejection of Science," *Current Directions in Psychological Science* 25, no. 4 (2016): 217–222.
32. M. J. Hornsey, "The Role of Worldviews in Shaping How People Appraise Climate Change," *Current Opinion in Behavioral Sciences* 42 (2021): 36–41.
33. T. H. Campbell and A. C. Kay, "Solution Aversion: On the Relation Between Ideology and Motivated Disbelief," *Journal of Personality and Social Psychology* 107 (2014): 809–824.
34. G. Dixon, J. Hmielowski, and Y. Ma, "Improving Climate Change Acceptance Among US Conservatives Through Value-Based Message Targeting," *Science Communication* 39, no. 4 (2017): 520–534.
35. M. J. Hornsey, E. A. Harris, P. G. Bain, and K. S. Fielding, "Meta-Analyses of the Determinants and Outcomes of Belief in Climate Change," *Nature Climate Change* 6 (2016): 622–626.
36. K. S. Fielding and M. J. Hornsey, "A Social Identity Analysis of Climate Change and Environmental Attitudes and Behaviors: Insights and Opportunities [Review]," *Frontiers in Psychology* 7 (2016), https://doi.org/10.3389/fpsyg.2016.00121
37. K. S. Fielding, M. J. Hornsey, H. A. Thai, and L. L. Toh, "Using Ingroup Messengers and Ingroup Values to Promote Climate Change Policy," *Climatic Change* 158 (2020): 181–199.
38. C. Achen and L. Bartels, "Democracy for Realists: Holding up a Mirror to the Electorate," *Juncture* 22, no. 4 (2016): 269–275.

39. D. R. Kinder and N. P. Kalmoe, *Neither Liberal Nor Conservative: Ideological Innocence in the American Public* (University of Chicago Press, 2017).

40. D. Amdur, B. G. Rabe, and C. P. Borick, "Public Views on a Carbon Tax Depend on the Proposed Use of Revenue," *Issues in Energy and Environmental Policy* 13 (2014): 1–9.

41. J. A. Krosnick, A. L. Holbrook, and P. S. Visser, "The Impact of the Fall 1997 Debate About Global Warming on American Public Opinion," *Public Understanding of Science* 9, no. 3 (2000): 239.

42. S. Båtstrand, "More Than Markets: A Comparative Study of Nine Conservative Parties on Climate Change," *Politics & Policy* 43, no. 4 (2015): 538–561.

43. Hornsey et al., "Relationships Among Conspiratorial Beliefs."

44. D. Plehwe, "Reluctant Transformers or Reconsidering Opposition to Climate Change Mitigation? German Think Tanks Between Environmentalism and Neoliberalism," *Globalizations* 20, no. 8 (2022): 1277–1295.

45. Brulle et al., "Shifting Public Opinion on Climate Change."

46. J. T. Carmichael and R. J. Brulle, "Elite Cues, Media Coverage, and Public Concern: An Integrated Path Analysis of Public Opinion on Climate Change, 2001–2013," *Environmental Politics* 26 (2017): 232–252.

47. M. J. Hornsey, C. M. Chapman, and J. E. Humphrey, "Climate Skepticism Decreases When the Planet Gets Hotter and Conservative Support Wanes," *Global Environmental Change* 74 (2022): 102492. https://doi.org/10.1016/j.gloenvcha.2022.102492

48. G. L. Cohen, "Party Over Policy: The Dominating Impact of Group Influence on Political Beliefs," *Journal of Personality and Social Psychology* 85, no. 5 (2003): 808–822.

49. L. van Boven, P. J. Ehret, and D. K. Sherman, "Psychological Barriers to Bipartisan Public Support for Climate Policy," *Perspectives on Psychological Science* 13, no. 4 (2018): 492–507.

50. G. S. Lenz, *Follow the Leader? How Voters Respond to Politicians' Policies and Performance* (University of Chicago Press, 2013).

51. J. Zaller, *The Nature and Origins of Mass Opinion* (Cambridge University Press, 1992).

52. J. M. Turner and A. C. Isenberg, *The Republican Reversal: Conservatives and the Environment from Nixon to Trump* (Harvard University Press, 2018).

53. C. R. Shipan and W. R. Lowry, "Environmental Policy and Party Divergence in Congress," *Political Research Quarterly* 54, no. 2 (2001): 245–263.

54. E. Merkley and D. A. Stecula, "Party Elites or Manufactured Doubt? The Informational Context of Climate Change Polarization," *Science Communication* 40, no. 2 (2018): 258–274.

55. Brulle et al., "Shifting Public Opinion on Climate Change."

56. D. L. Guber, "A Cooling Climate for Change? Party Polarization and the Politics of Global Warming," *American Behavioral Scientist* 57, no. 1 (2013): 93–115.

57. S. D. Benegal and L. A. Scruggs, "Correcting Misinformation About Climate Change: The Impact of Partisanship in an Experimental Setting," *Climatic Change* 148, no. 1–2 (2018): 61–80.

58. M. Tesler, "Elite Domination of Public Doubts About Climate Change (Not Evolution)," *Political Communication* 35, no. 2 (2018): 306–326.

59. J. Sohlberg, "The Effect of Elite Polarization: A Comparative Perspective on How Party Elites Influence Attitudes and Behavior on Climate Change in the European Union," *Sustainability* 9, no. 1(2017), Article 1, https://doi.org/10.3390/su9010039.

60. Merkley and Stecula, "Party Cues in the News."
61. A. M. McCright, M. Charters, K. Dentzman, and T. Dietz, "Examining the Effectiveness of Climate Change Frames in the Face of a Climate Change Denial Counter-Frame," *Topics in Cognitive Science* 8, no. 1 (2016): 76–97.
62. R. E. Dunlap and P. J. Jacques, "Climate Change Denial Books and Conservative Think Tanks: Exploring the Connection," *American Behavioral Scientist* 57, no. 6 (2013): 699–731.
63. J. Farrell, "Corporate Funding and Ideological Polarization About Climate Change," *Proceedings of the National Academy of Sciences* 113, no. 1 (2016): 92–97.
64. J. Farrell, "Network Structure and Influence of the Climate Change Counter-Movement," *Nature Climate Change* 6, no. 4 (2016): 370–374.
65. P. J. Jacques, R. E. Dunlap, and M. Freeman, "The Organisation of Denial: Conservative Think Tanks and Environmental Scepticism," *Environmental Politics* 17, no. 3 (2008): 349–385.
66. M. T. Boykoff and J. M. Boykoff, "Climate Change and Journalistic Norms: A Case-Study of US Mass-Media Coverage," *Geoforum* 38, no. 6 (2007): 1190–1204.
67. R. E. Dunlap and A. M. McCright, "Organized Climate Change Denial," in The Oxford Handbook of Climate Change and Society, ed. J. S. Dryzek, R. B. Norgaard, and D. Schlosberg (Oxford University Press, 2011), 144–160.
68. L. McAllister, M. Daly, P. Chandler, M. McNatt, A. Benham, and M. Boykoff, "Balance as Bias, Resolute on the Retreat? Updates & Analyses of Newspaper Coverage in the United States, United Kingdom, New Zealand, Australia and Canada Over the Past 15 Years," *Environmental Research Letters* 16, no. 9 (2021): 094008.
69. Dunlap and McCright, "Organized Climate Change Denial."
70. E. Merkley and D. A. Stecula, "Party Elites or Manufactured Doubt? The Informational Context of Climate Change Polarization," *Science Communication* 40, no. 2 (2018): 258–274.
71. M. J. Hornsey and S. Lewandowsky, "A Toolkit for Understanding and Addressing Climate Scepticism," *Nature Human Behaviour* 6 (2022): 1454–1464.
72. J. N. Kim, L. Ni, and B. L. Sha, "Breaking Down the Stakeholder Environment: Explicating Approaches to the Segmentation of Publics for Public Relations Research," *Journalism & Mass Communication Quarterly* 85, no. 4 (2008): 751–768.
73. A. Leiserowitz, C. Roser-Renouf, J. Marlon, and E. Maibach, "Global Warming's Six Americas: A Review and Recommendations for Climate Change Communication," *Current Opinion in Behavioral Sciences* 42 (2021): 97–103.
74. A. Krishna, "Lacuna Publics: Advancing a Typology of Disinformation Susceptibility Using the Motivation-Attitude-Knowledge Framework," *Journal of Public Relations Research* 33 (2021): 63–85.
75. J.-N. Kim, J. E. Grunig, and L. Ni, "Reconceptualizing the Communicative Action of Publics: Acquisition, Selection, and Transmission of Information in Problematic Situations," *International Journal of Strategic Communication* 4, no. 2 (2010): 126–154.
76. A. Krishna, "Motivation with Misinformation: Conceptualizing Lacuna Individuals and Publics as Knowledge Deficient, Issue-Negative Activists," *Journal of Public Relations Research* 29, no. 4 (2017): 176–193.
77. B. W. McKeever, R. McKeever, A. E. Holton, and J. Y. Li, "Silent Majority: Childhood Vaccinations and Antecedents to Communicative Action," *Mass Communication and Society* 19, no. 4 (2016): 476–498.
78. J. Compton, "Prophylactic Versus Therapeutic Inoculation Treatments for Resistance to Influence, *Communication Theory* 30, no. 3 (2020): 330–343.

79. G. Gauchat, "Politicization of Science in the Public Sphere: A Study of Public Trust in the United States, 1974 to 2010," *American Sociological Review* 77, no. 2 (2012): 167–187.

80. R. Hofstadter, *Anti-Intellectualism in American Life* (Vintage, 1966).

81. E. Merkley, "Anti-Intellectualism, Populism, and Motivated Resistance to Expert Consensus," *Public Opinion Quarterly* 84, no. 1 (2020): 24–48.

82. T. Skocpol and V. Williamson, *The Tea Party and the Remaking of Republican Conservatism* (Oxford University Press, 2016).

83. E. Merkley and P. J. Loewen, "Anti-Intellectualism and the Mass Public's Response to the COVID-19 Pandemic," *Nature Human Behaviour* 5, no. 6 (2021): 706–715.

84. M. Motta, "The Dynamics and Political Implications of Anti-Intellectualism in the United States," *American Politics Research* 46, no. 3 (2018): 465–498.

85. K. Lunz Trujillo, "Rural Identity as a Contributing Factor to Anti-Intellectualism in the US," *Political Behavior* 44, no. 3 (2022): 1509–1532.

86. Merkley, "Anti-Intellectualism, Populism, and Motivated Resistance."

87. N. Oreskes, "Systematicity Is Necessary But Not Sufficient: On the Problem of Facsimile Science," *Synthese* 196, no. 3 (2019): 881–905.

88. S. D. Benegal, "The Spillover of Race and Racial Attitudes Into Public Opinion About Climate Change," *Environmental Politics* 27, no. 4 (2018): 733–756.

89. J. Chanin, "The Effect of Symbolic Racism on Environmental Concern and Environmental Action," *Environmental Sociology* 4, no. 4 (2018): 457–469.

90. S. Benegal and M. R. Holman, "Understanding the Importance of Sexism in Shaping Climate Denial and Policy Opposition," *Climatic Change* 167, no. 3–4 (2021): 48.

91. S. S. Bush and A. Clayton, "Facing Change: Gender and Climate Change Attitudes Worldwide," *American Political Science Review* 117, no. 2 (2023): 591–608.

92. I. Feygina, J. T. Jost, and R. E. Goldsmith, "System Justification, the Denial of Global Warming, and the Possibility of 'System-Sanctioned Change,'" *Personality and Social Psychology Bulletin* 36, no. 3 (2010): 326–338.

93. K. M. Jylhä, C. Cantal, N. Akrami, and T. L. Milfont, "Denial of Anthropogenic Climate Change: Social Dominance Orientation Helps Explain the Conservative Male Effect in Brazil and Sweden," *Personality and Individual Differences* 98 (2016): 184–187.

94. K. M. Jylhä and K. Hellmer, "Right-Wing Populism and Climate Change Denial: The Roles of Exclusionary and Anti-Egalitarian Preferences, Conservative Ideology, and Antiestablishment Attitudes," *Analyses of Social Issues and Public Policy* 20, no. 1 (2020): 315–335.

95. E. Anduiza and G. Rico, "Sexism and the Far-Right Vote: The Individual Dynamics of Gender Backlash," *American Journal of Political Science* 68, no. 2 (2022): 478–493.

96. B. F. Schaffner, M. MacWilliams, and T. Nteta, "Understanding White Polarization in the 2016 Vote for President: The Sobering Role of Racism and Sexism," *Political Science Quarterly* 133, no. 1 (2018): 9–34.

97. K. M. Norgaard, *Living in Denial: Climate Change, Emotions, and Everyday Life* (MIT Press, 2011).

98. T. G. Coan, C. Boussalis, J. Cook, and M. O. Nanko, "Computer-Assisted Classification of Contrarian Claims About Climate Change, *Scientific Reports* 11, no. 1 (2021): 22320.

99. D. Ding, E. W. Maibach, X. Zhao, C. Roser-Renouf, and A. Leiserowitz, "Support for Climate Policy and Societal Action Are Linked to Perceptions About Scientific Agreement," *Nature Climate Change* 1, no. 9 (2011): 462–466.

100. S. van der Linden, A. A. Leiserowitz, G. D. Feinberg, and E. W. Maibach, "The Scientific Consensus on Climate Change as a Gateway Belief: Experimental Evidence," *PLOS One* 10, no. 2 (2015): e0118489.

101. S. B. Capstick and N. F. Pidgeon, "What Is Climate Change Scepticism? Examination of the Concept Using a Mixed Methods Study of the UK Public," *Global Environmental Change* 24 (2014): 389–401.

102. K. Ekberg, B. Forchtner, and M. Hultman, *Climate Obstruction: How Denial, Delay and Inaction Are Heating the Planet* (Routledge, 2023).

103. W. F. Lamb, G. Mattioli, S. Levi, J. T. Roberts, S. Capstick, F. Creutzig, J. C. Minx, F. Müller-Hansen, T. Culhane, and J. K. Steinberger, "Discourses of Climate Delay," *Global Sustainability* 3 (2020): e17, https://doi.org/10.1017/sus.2020.13

104. J. Painter, J. Ettinger, D. Holmes, L. Loy, J. Pinto, L. Richardson, L. Thomas-Walters, K. Vowles, and R. Wetts, "Climate Delay Discourses Present in Global Mainstream Television Coverage of the IPCC's 2021 Report," *Communications Earth & Environment* 4, no. 1 (2023): 118.

105. Ekberg et al., *Climate Obstruction*.

106. A. Cislak, A. Cichocka, A. D. Wojcik, and T. L. Milfont, "Words Not Deeds: National Narcissism, National Identification, and Support for Greenwashing Versus Genuine Proenvironmental Campaigns," *Journal of Environmental Psychology* 74 (2021): 101576, https://doi.org/10.1016/j.jenvp.2021.101576

107. A. Cislak, A. D. Wójcik, J. Borkowska, and T. L. Milfont, "Secure and Defensive Forms of National Identity and Public Support for Climate Policies," *PLOS Climate* 2, no. 6 (2023): e0000146, https://doi.org/10.1371/journal.pclm.0000146

108. C. Rojas, F. Algra-Maschio, M. Andrejevic, T. Coan, J. Cook, and Y.-F. Li, "Augmented CARDS: A Machine Learning Approach to Identifying Triggers of Climate Change Misinformation on Twitter," *Nature Communications Earth & Environment* 5 (2024): Article 1573.

109. P. Diethelm and M. McKee, "Denialism: What Is It and How Should Scientists Respond?," *European Journal of Public Health* 19, no. 1 (2009): 2–4.

110. D. M. Kahan, H. Jenkins-Smith, and D. Braman, "Cultural Cognition of Scientific Consensus," *Journal of Risk Research* 14, no. 2 (2011): 147–174.

111. P. Bergquist and C. Warshaw, "Does Global Warming Increase Public Concern About Climate Change?," *Journal of Politics* 81, no. 2 (2019): 686–691.

112. M. A. Ranney and D. Clark, "Climate Change Conceptual Change: Scientific Information Can Transform Attitudes," *Topics in Cognitive Science* 8, no. 1 (2016): 49–75.

113. S. van der Linden, A. Leiserowitz, S. Rosenthal, and E. Maibach, "Inoculating the Public Against Misinformation About Climate Change," *Global Challenges* 1, no. 2 (2017): Article 1600008.

114. A. M. McCright, R. E. Dunlap, and S. T. Marquart-Pyatt, "Political Ideology and Views About Climate Change in the European Union," *Environmental Politics* 25, no. 2 (2016): 338–358.

115. E. K. Vraga, S. C. Kim, J. Cook, and L. Bode, "Testing the Effectiveness of Correction Placement and Type on Instagram," *International Journal of Press/Politics* 25, no. 4 (2020), 632–652.

116. S. Lewandowsky, N. Oreskes, J. S. Risbey, B. R. Newell, and M. Smithson, "Seepage: Climate Change Denial and Its Effect on the Scientific Community," *Global Environmental Change* 33 (2015): 1–13.

117. A. Leiserowitz, E. Maibach, C. Roser-Renouf, S. Rosenthal and M. Cutler, *Climate Change in the American Mind: May 2017* (Program on Climate Change Communication at Yale University and George Mason University, 2017).

118. N. Geiger and J. K. Swim, "Climate of Silence: Pluralistic Ignorance as a Barrier to Climate Change Discussion," *Journal of Environmental Psychology* 47 (2016): 79–90.

119. E. Maibach, A. Leiserowitz, S. Rosenthal, C. Roser-Renouf, and M. Cutler, *Is There a Climate "Spiral of Silence" in America: March, 2016* (Program on Climate Change Communicatio at Yale University and George Mason University, 2016).

120. I. Fritsche, "It's Getting Dark, But We Will See: Gaining Collective Momentum in Face of Existential Environmental Threat," *Global Environmental Psychology* 2 (2024): 1–16.

121. J.-W. Müller, *What Is Populism* (Penguin, 2017).

122. C. Oliveira, "Bolsonarism Uses Telegram to Destabilize Elections, Affirms Specialist," *Brasil de Fato*, 2022, https://www.brasildefato.com.br/2022/05/19/bolsonarism-uses-telegram-to-destabilize-elections-affirms-specialist.

123. J. A. Miguel, "A 'meada' do negacionismo climático e o impedimento da governamentalização ambiental no Brasil," *Revista Sociedade & Estado UnB* 37, no. 1 (2022): 293–315.

124. O. Carvalho, "Ciência ou palhaçada?," Diário do Comércio, May, 2007, https://olavodecarvalho.org/ciencia-ou-palhacada/.

125. Câmara dos Deputados, Audiência pública sobre o Projeto de Lei nº 1876/1999 – Código Florestal Brasileiro. Comissão Especial do PL 1876/99, November 12, 2009, https://www2.camara.leg.br/atividade-legislativa/comissoes/comissoes-temporarias/especiais/53a-legislatura-encerradas/pl187699/pl187699nt121 109.pdf.

126. R. W. Santos, *Reconfigurações do Ecossistema Religioso diante da Crise Climática Global* (Cadernos do OIMC, 2024).

127. B. Bragança, *Psicose ambientalista: os bastidores do ecoterrorismo para implantar uma 'religião' ecológica, igualitária e anticristã* (Instituto Plínio Corrêa de Oliveira, 2012).

128. P. Bernardin, *O império ecológico ou a subversão da ecologia pelo globalismo* (Vide Editorial, 2015).

129. R. Jakubazko, C. B. L. Molion, and C. P. J. Oliveira, *CO2 aquecimento e mudanças climáticas: estão nos enganando?* (DBO Editores Associados, 2015).

130. E. Araújo, "Sequestrar e perverter," Metapolítica Brasil, November 14, 2018, https://www.metapoliticabrasil.com/post/sequestrar-e-perverter.

131. J. A. Miguel, Negacionismo climático no Brasil. *Coletiva*, Dossiê 27, Crise climática, 2020, https://www.coletiva.org/dossie-emergencia-climatica-n27-artigo-negacionismo-climatico-no-brasil.

132. N. Passarinho, "Como a política ambiental de Bolsonaro afetou imagem do Brasil em 2019 e quais as consequências disso," BBC News Brasil, December 31, 2019, https://www.bbc.com/portuguese/brasil-50851921.

133. G. Sargent and P. Waldman, "Why Bolsonaro's Stunning Loss Should Give Humans a Glimmer of Hope," *Washington Post*, October 31, 2022, https://www.washingtonpost.com/opinions/2022/10/31/jair-bolsonaro-lula-brazilian-election-amazon-putin-ukraine/.

134. T. H. Campbell and A. C. Kay, "Solution Aversion: On the Relation Between Ideology and Motivated Disbelief," *Journal of Personality and Social Psychology* 107 (2014): 809–824.

135. P. G. Bain, M. J. Hornsey, R. Bongiorno, and C. Jeffries, "Promoting Pro-Environmental Action in Climate Change Deniers," *Nature Climate Change* 2 (2012): 600–603.

136. C. Wolsko, "Expanding the Range of Environmental Values: Political Orientation, Moral Foundations, and the Common Ingroup," *Journal of Environmental Psychology* 51 (2017): 284–294.

137. C. Wolsko, H. Ariceaga, and J. Seiden, "Red, White, and Blue Enough to Be Green: Effects of Moral Framing on Climate Change Attitudes and Conservation Behaviors," *Journal of Experimental Social Psychology* 65 (2016): 7–19.

138. M. H. Goldberg, A. Gustafson, S. A. Rosenthal, and A. Leiserowitz, "Shifting Republican Views on Climate Change Through Targeted Advertising," *Nature Climate Change* 11, no. 7 (2021): 573–577.

139. M. J. Hornsey and K. S. Fielding, "Attitude Roots and Jiu Jitsu Persuasion: Understanding and Overcoming the Motivated Rejection of Science," *American Psychologist* 72, no. 5 (2017): 459–473.

140. T. Bolsen, R. Palm, and J. T. Kingsland, "The Impact of Message Source on the Effectiveness of Communications About Climate Change," *Science Communication* 41, no. 4 (2019): 464–487.

141. E. Porter, T. J. Wood, and B. Bahador, "Can Presidential Misinformation on Climate Change Be Corrected? Evidence from Internet and Phone Experiments," *Research & Politics* 6, no. 3 (2019): 2053168019864784.

142. M. E. Mann, *The New Climate War: The Fight to Take Back Our Planet* (Public Affairs, 2021).

143. R. J. Brulle, "The Climate Lobby: A Sectoral Analysis of Lobbying Spending on Climate Change in the USA, 2000 to 2016," *Climatic Change* 149, no. 3–4 (2018): 289–303.

144. N. Chater and G. Loewenstein, "The I-Frame and the S-Frame: How Focusing on Individual-Level Solutions Has Led Behavioral Public Policy Astray," *Behavioral and Brain Sciences* 46 (2022): 1–60.

145. G. Supran and N. Oreskes, "Rhetoric and Frame Analysis of ExxonMobil's Climate Change Communications," *One Earth* 4, no. 5 (2021): 696–719.

146. J. Henrich, S. J. Heine, and A. Norenzayan, "Most People Are Not WEIRD," *Nature* 466, no. 7302 (2010): 29–29.

CHAPTER 8

Climate Obstruction Across the Global South

LEAD AUTHORS: M. OMAR FARUQUE AND RUTH E. McKIE

CONTRIBUTING AUTHORS: LUCAS G. CHRISTEL, CLAIRE DEBUCQUOIS, GUY EDWARDS, PAUL K. GELLERT, RICARDO A. GUTIERREZ, KATHRYN HOCHSTETLER, YIFEI LI, CARLOS R. S. MILANI, ELISABETH MÖHLE, OLUWASEUN J. OGUNTUASE, AND JONATHAN R. WALZ

INTRODUCTION: DEFINING THE GLOBAL SOUTH

The Global South refers to countries, societies, and political economies situated in the "periphery" and "semi-periphery" of global capitalism in the regions of Asia, Africa, Latin America, and Oceania. A shared feature of these countries is their economic and political marginalization compared with their counterparts in the Global North (i.e., the "core" countries in North America and Europe).[1] Although the most common conceptualization of the Global South is a narrow scholarly and policy definition focusing on the nature and dynamics of development in those countries, the term also encompasses geopolitical power relations.[2] Scholars have argued that the Global South "functions as more than a metaphor for underdevelopment. It references an entire history of colonialism, neo-imperialism, and differential economic and social change through which large inequalities in living standards, life expectancy, and access to resources are maintained."[3]

Culturally and politically, the Global South countries are non-Western (multicultural and multinational) societies with different trajectories in state formation. Most countries in the Global South are former colonies of Western powers, and they are usually either excluded from key international

M. Omar Faruque et al., *Climate Obstruction Across the Global South*. In: *Climate Obstruction*. Edited by: J. Timmons Roberts, Carlos R. S. Milani, Jennifer Jacquet, and Christian Downie, Oxford University Press. © Oxford University Press (2025). DOI: 10.1093/oso/9780197787144.003.0008

decision-making processes or play significantly less powerful roles in post-1945 multilateral institutions such as the World Bank and the International Monetary Fund. The term often expresses commonality among developing countries in terms of their conflicting interests with the industrialized countries of the Global North. This North-South conflict feature also applies to global climate governance and politics.[4] However, contemporary political dynamics in the area of global climate governance demonstrate that the Global South countries are socially, politically, and economically diverse, thus making the Global South a fragmented entity.[5] The Global South should therefore be viewed "as a complex and changing set of relations reflecting shifts in the historic world order, and dynamics specific to the contemporary climate regime."[6]

Climate obstruction and delay in the Global South manifest in qualitatively different ways from that in the Global North. State actors in the Global South tend to set climate governance and policy agendas that reflect their structural and economic positions in relation to global climate governance and politics. Many countries in the Global South defend their less ambitious climate actions by invoking arguments underpinned by unequal historical emissions of GHGs and differentiated responsibilities as set out in the 1992 United Nations Framework Convention on Climate Change (UNFCCC). This position is further entrenched by their strong commitment to the "right to development" to ensure energy security, encouraging other actors and organizations in core economic sectors to support policies and actions that fall far short of addressing the climate crisis. Furthermore, the failure of advanced industrial countries of the Global North to deliver their repeatedly promised financial support has created a policy context in the Global South whereby obstructive actors raise the need to fulfill this commitment before they can undertake much-needed climate action. Civil-society actors also invoke narratives of "climate injustice" to draw attention to such failures and, in so doing, indirectly support climate obstruction.

The remainder of this chapter is organized into four sections and a conclusion. It begins by explaining major considerations that distinguish the Global South from the Global North in ways that affect these countries' approach to dealing with climate change, and how these approaches lead to obstruction. Next, we identify key actors and organizations undertaking climate obstruction activities in core economic sectors of the Global South. The next section examines the political aims of these actors, their alliances with other actors and organizations (including those based in the Global North), and the strategies, tactics, and narratives they deploy to undermine climate-related legislative and policy actions. This analysis is followed by a brief discussion of the role of civil-society actors and social movements in contesting the policies and actions of government agencies and carbon majors in the Global South. It also emphasizes the challenges they face in achieving rapid climate action. Finally, the chapter summarizes our key findings on climate obstruction in the Global South and discusses research gaps and policy lessons.

MAJOR CONSIDERATIONS AFFECTING CLIMATE OBSTRUCTION IN THE GLOBAL SOUTH

This section argues that there are five major considerations affecting climate obstruction in the Global South. This section reviews these five different areas and considers what this means for our understanding of climate obstruction in the Global South.

Variations in National Climate Commitments

The Global South countries are heterogenous in terms of their policies to address climate adaptation and mitigation. There are sharp disagreements among stakeholders regarding what constitutes a just transition toward a sustainable energy future, as illustrated by the variety of commitments outlined in the Paris Agreement. Many countries in the Global South are at a disadvantage in terms of economic growth, development, and social justice. These concerns are often given primary consideration in making climate and energy policies. For example, some African stakeholders consider addressing energy poverty at any cost to be a top priority. Decarbonization, in their view, is not an urgent matter for many African countries. Indeed, balancing the goals of ensuring energy security to meet the desired goals of high economic growth (and improving human development) with the need to take climate action is a dilemma for many countries in the Global South. There is also a divide between the elite and the masses in terms of preferred policy choices to address the climate crisis. For example, most people in Colombia support the government's strict measures to combat climate change by moving toward renewable energy systems. However, the elites within and beyond the current ruling class under the Gustavo Petro administration defend the continued use of oil and gas reserves.[7]

Countries in the Global South also differ in their positions on global climate governance negotiations. Although they remain committed to advancing a common goal, power differentials and national interests have created rifts among them. Similarly, although this chapter presents examples of climate obstruction from various regions in the Global South, they also manifest in distinct ways. This reality points to the fact that the Global South is not just a geographical category; but more importantly, a geopolitical and political economic category whose members can be differentiated by their wealth and resources and by their structural position in the world economy and in global governance.

Unequal Vulnerability to Climate Change

While Global South countries are *least* responsible for the historical rise in greenhouse gas (GHG) emissions, scientific evidence shows that countries in the Global South are the *most* vulnerable to the adverse effects of climate change.[8] Yet many of these countries bear significantly less responsibility for the problem's causes than countries in the Global North.[9] At the same time, wealthier countries in the Global North have disproportionately exploited natural resources and damaged institutional capacity in these countries.[10] The consequences of this exploitation have made many such countries more vulnerable than those in the Global North to the effects of climate change. Nations of the Global South also suffer from inadequate resources and technical capacity to counter these historical and ongoing harmful practices. In response to the structural conditions imposed by the Global North, some countries in the Global South have prioritized centralized models of economic development focusing primarily on government-led allocation and development of resources that often lead to the sidelining of climate and other environmental policies.[11]

Differentiated Responsibilities in the International Climate Arena

Despite their limited historical role in causing the climate crisis, Global South countries are entitled to send delegates representing their interests to the annual UNFCCC negotiations and to participate in decision-making on global climate governance strategies.[12] Moreover, all countries in the Global South except Iran, Libya, and Yemen have signed the Paris Agreement committing themselves to Nationally Determined Contributions (NDCs), which require their governments to make self-determined pledges detailing their intended actions to help mitigate and adapt to climate change.[13]

Acknowledging the discrepancies between countries' past roles in rising GHGs emissions and their current capacities to address them, the UNFCCC introduced the concept of "fair share" and "differentiated responsibility" for GHG reductions (Articles 3 and 4). This nonbinding approach to ensuring equity, as laid out in the Paris Agreement, accounts for countries' historical emissions, current resources, and development needs in determining their response to climate change (Article 4). In this way, the process considers how Global South countries differ from those in the Global North, and accordingly suggests their expected contributions moving forward.

Given that countries across the Global South are more vulnerable to climate change than the Global North and have less capacity to respond to it, they have called on the international community to supply specific types of economic and financial aid and technology transfer to support their adaptation and mitigation efforts.[14] In recent years, "loss and damage" has emerged as a "third pillar" of global climate governance.[15] The Intergovernmental Panel on Climate Change (IPCC) describes loss and damage as "harmful impacts or risks that can result from climate change-related slow onset hazards and extreme weather events,"[16] and refers to "the unavoidability and irreversibility of certain climate change impacts and the role played by constraints and limits to adaptation as drivers of adverse outcomes."[17] However, the implementation of this policy has been plagued by the failure of countries in the Global North, which actively sought to avoid such financial obligations, to limit the scope of what is considered loss and damage, and to stall negotiations by aiming to achieve word-perfect policy proposals.[18]

Furthermore, to meet global climate treaty commitments, a just transition and the development of sustainable economies in the Global South must be a priority. In practice, appropriate climate policies in these countries would require implementing, for example, "green" finance initiatives such as green bonds to fund renewable energy projects to help phase out heavily polluting alternatives while also maintaining economic growth. Moreover, this type of climate action "must seek fairness and equity with regards to the major global justice concerns such as (but not limited to) ethnicity, income, gender within both developed and developing contexts"[19] while also minimizing the impacts of climate change. Unfortunately, while the Global South countries claim to recognize the significance of decarbonizing their economies, most of them had not done so by 2019, as illustrated by the absolute rise in their GHG emissions. However, it is important to note that this rise was not distributed equally among countries in the Global South.[20]

The Role of Structural Issues and Geopolitical Shifts in Climate Politics

The existing political, social, and economic conditions in many countries of the Global South have created a complex policy agenda that facilitates climate delay, such that the implementation of policies aiding the energy transition has been slow and its scope is limited. This agenda is driven by traditional notions of development influenced by the market-fundamentalist and neoclassical economic concepts of progress,[21] which in turn shape climate delay by dismissing the urgency of climate action. This ideology also includes the belief that higher levels of environmental degradation due to industrial development will increase economic growth and ultimately empower Global South countries to mitigate GHG emissions and adapt to the impacts of climate change.[22] While this ideology is popular in many Global South countries, discourses of sustainable economic development and post-development also exist. These discourses have served as both a critique of market fundamentalism[23] and a defense of the neo-developmentalism emerging in some Global South countries, which sees an active role for the state in addressing both development and climate change and prioritizes social advancement and inclusivity.[24] Yet a common factor in these various ideologies is the continued use and sometimes expansion of extractive industries for economic development that fail to mitigate the impacts of climate change.

The above discourses are also linked to contemporary geopolitical shifts, as the rise of new development-cooperation mechanisms has lessened the Global South's dependence on the Global North for trade and investment.[25] For example, the BRICS countries (Brazil, Russia, India, China, and South Africa) have the potential to reshape global economic governance in light of the climate crisis.[26] More specifically, emerging environmentally friendly technology in BRICS countries, intra-BRICS trade and investment,[27] and the "new development bank"[28] are changing the model of modernization and development advanced by Western countries.[29] New challenges loom with the expansion of the BRICS+ group, which now includes Saudi Arabia, the United Arab Emirates, Iran, and others. Most importantly, the historical reliance on and the influence of the Global North are fading in these nations. These conditions have changed the landscape in which GHGs emissions have continued to rise over the past one hundred years: emissions in the Global South countries are now increasing, driven by alternative development models, industrialization, and geopolitical relationships. The Global South countries have also caused intentional or unintentional climate obstruction, most often in the name of maintaining their right to economic growth and development, as we will discuss in detail later.

OBSTRUCTIONIST SECTORS, ACTORS, AND ORGANIZATIONS

Various extractive and heavily polluting industries exist across the Global South, with some generating large economic profits for a small monopoly of corporations and/or state actors. This section identifies these key sectors in which powerful actors and organizations play a critical role in shaping climate obstruction.

Multinational Corporations and State Actors

Multinational corporations and state-based actors play a dominant role in climate obstruction activities in the Global South. Across Latin America, the agribusiness sector is one of the most significant contributors to the rise of GHG emissions (see Chapter 4). For example, between 1990 and 2018 agricultural development in carbon-dense tropical forest areas was greatly expanded, driving up emissions from agriculture, forestry, and other land uses (AFOLU) and becoming one of the largest contributors to overall GHG emissions in the region.[30] This ongoing expansion is happening despite that lower-carbon alternatives are already available and/or hold potential to be developed. These conditions align with the activities of climate obstruction actors in the region. For example, in Brazil, the agribusiness sector and other "land-grabbing" enterprises have been key players in Brazilian climate obstruction, delay, and denial.[31] Specific actors include the recent Bolsonaro administration (2019–2023) which oversaw the dismantling of the structures and mechanisms for environmental protection constructed over the previous several decades to boost agricultural development in the Amazon Rainforest.[32]

Rising GHGs emissions in the Global South are also linked to aggressive crude-oil extraction in the Middle East and in North and Northwest Africa, including in several member countries of the Organization of the Petroleum Exporting Countries (OPEC).[33] OPEC serves to unify petroleum-producing countries by activating, coordinating, and aligning petroleum policies that aim to stabilize and protect international oil markets so that they remain secure and efficient.[34] Given the UNFCCC's substantial role in determining global energy trends, its actions and the implementation of global mitigation targets can potentially restrain OPEC members' crude-oil extraction and, as a result, various climate obstruction activities have emerged in these countries (see Chapters 2 and 10).

Asian countries have historically made one of the lowest contributions to rising GHGs emissions. However, between 1990 and 2019 their net emissions expanded exponentially.[35] Due to population growth and an increase in economic activities, the region is expected to continue to increase its share on pace with further industrialization, the expansion of agriculture, and increased consumption patterns. In certain cases, some of these countries, such as China and India, are established coal users and will remain so as they continue to phase in coal-powered power plants.[36] The reliance on and expansion of extractive sectors in many of these countries also lead to delaying the shift toward renewable energy and robust climate action.

Notably, most of the countries phasing out coal are in the Global North and those countries phasing-in coal are in the Global South, particularly in Asia.[37] For example, Indonesia is now one of the world's top contributors to global GHG emissions. Its extractive and agro-extractive activities, specifically coal mines and oil-palm plantations, perpetuate an "extractive regime."[38] In Vietnam and the Philippines, coal remains the primary source of energy, particularly for the electricity sector. Electrification efforts have increased coal consumption in several Southeast and South Asian countries.[39] The share in Vietnam rose to almost 50% before declining since 2021.[40] In the Philippines, coal use continues to rise; its growth has already exceeded 45%.[41] In the South Asian region, Bangladesh has significantly increased the share of coal as a primary energy source for its electricity generation capacities.[42] India's electricity sector also remains a major consumer of coal. More than 48% of its installed power-generation capacity comes from coal-fired power plants.[43] A complex domestic political economy involving mining corporations, public utility companies, and coal-job-dependent communities keeps India as a major consumer of coal, notwithstanding its promise to move toward renewable energy.[44]

In the African continent, South Africa is one of the major economies most dependent on coal for electricity generation, with the energy sector accounting for some 80% of its GHG emissions. Yet despite calls for a transition to renewable energy as early as 1999,[45] wind and solar power remain a small part of South Africa's electricity supply.

In essence, development and industrialization in the Global South are at odds with important elements of climate action. While the use of fossil fuel offers a pathway forward as part of these countries' development, this policy choice is incompatible with the goals of the Paris Agreement and what science shows is needed to ease the climate crisis. Furthermore, it has created a form of carbon lock-in: manufacturing industries, power plants, and economies based on fossil fuels have led to a pattern of investments that cannot easily be altered without a significant impact on development in Global South countries.[46] Ultimately, the entanglement between the fossil fuel sector and these countries' economies serves to restrict robust climate action and a just green transition.

Industry Lobbyists and Business Associations

There is also evidence that industry lobbyists and business associations have influenced climate obstruction in some Global South countries. For example, while the perspectives of industry lobbyists working to undermine or slow progress at the UNFCCC negotiations are often considered authoritative views on various policies, the demands and concerns of environmental and social activists, Indigenous peoples, and subnational actors attending these negotiations are typically ignored or weakened, thus also weakening their ability to influence the proceedings.[47] The industries' authoritative views often align with those of government leaders seeking to protect polluting and extractive activities and are reiterated in national and global forums. The case of Indonesia provides a good example, wherein corporate influence is exercised through business-sector associations, notably the Chamber of Commerce, and industry-wide associations, such as the Indonesian Coal Mining Association (APBI-ICMA).[48] In Bangladesh, an industry group, the Bangladesh Independent Power Producers' Association (BIPPA) downplays the role of renewable energy resources and advocates for more fossil fuel–based power plants to meet the economy's growing energy demands. State agencies operating in an environment characterized by state–business collusive relationships[49] have favored BIPPA's views and offered them various financial incentives denied to entrepreneurs involved with renewable energy projects, thus creating a policy context that obstructs climate actions enabling a transition toward a sustainable-energy future.[50] In India, also, state–business collaboration contributes to "the transition to more fossil fuel energy" in a policy context characterized by "talk renewables, walk coal."[51] Although less-expensive renewable power has begun to undercut the price of coal-fired power, India "face[s] resistance from a coal lobby which controls vast budgets and employs millions."[52] The relationships and interests shared by industry lobbyists, business associations, and the state guide the narratives that play out in the UNFCCC negotiations, diminishing the positions of environmental and social activists, Indigenous peoples, and other actors who seek to accelerate decarbonization.

Labor Movements and Unions

Labor movements and unions in the workplace also play a role in climate obstruction in some Global South countries. For example, in 2018, the organized labor movement in South Africa reversed its positions of support for the implementation of a Renewable Energy Independent Power Producers Procurement Program (REIPPPP), which used auctions to select builders of renewable-electricity plants under long-term power purchase contracts. The labor movement represents one of the largest and most important social bases of the African National Congress (ANC), South Africa's social-democratic political party, which played an essential role in anti-apartheid struggles of the 1990s. While labor initially supported climate action in 2011, they eventually resisted the implementation of the REIPPPP because it specified that electricity generation should be controlled and managed by the public sector rather than via the proposed outsourcing program to private organizations. This position was partly ideological and partly based on a conviction that only the public sector could maintain high-quality and well-paying energy jobs at scale. With comparatively few jobs available in the renewables sector at that time, organized labor in South Africa reoriented to protecting existing coal mining, transport, and power-generation jobs.[53]

POLITICAL AIMS, ALLIANCES, AND NARRATIVES

Actors and organizations engaged in climate obstruction in the Global South have a variety of political aims. Depending on the nature of these goals, they form alliances with other actors and organizations and deploy various narratives to stall climate mitigation efforts. These aims, alliances, and narratives are concerned primarily with maintaining the relationship between fossil fuel–dependent extractive industries that shape domestic policy decisions and building transnational relationships to increase economic development and maintain energy security. This section presents cases from different regions of the Global South to elucidate the political aims and strategies of climate obstruction actors in the sectors discussed earlier and the narratives they deploy to advance their climate obstruction agenda.

Extractive Industry and the Pull of Development Rhetoric

Southeast Asia has embraced the rhetoric of commitment to climate change mitigation goals. Yet there is doubt whether such rhetoric will—or is intended to—be transformed into concrete strategies and actions. For example, climate delay in Southeast Asia is generally shaped by the urgency of development. This ambition includes efforts to bring electricity to whole populations and must contend with the high cost of electricity, the unreliability of transmission, and the problem of power interruptions. Continuing to exploit traditional fossil fuels to meet these growing needs means that the adoption of alternative sources of energy to match the urgency of decarbonization may be sidelined. Climate delay is also shaped by the huge scale of investments needed to develop renewable and sustainable energy sources in the Global South.[54] In parts of Africa, political leaders and the carbon majors continue to echo discourses of climate delay at local and international forums. They assert that the African continent is best equipped to determine how to meet its climate commitments and has the right to chart its own energy path and decide how to balance economic development with sustainability to deliver a better future to its people.[55] This type of obstruction also includes the renegotiation of contracts to procure greater profits from the carbon majors currently operating within their borders, as in the case of Tanzania.[56]

Given that extractive industries are portrayed as the primary tool for development in the Global South, particularly in the elite narratives of governments and corporate executives, it is not surprising to see these narratives implemented in various forms of climate obstruction. One example is the consolidation of South-South cooperation programs. Such programs connect countries in the Global South through knowledge sharing, trade and investment, aid/financial assistance, and technical collaboration.[57] The practice is also promoted as a way to advance sustainable development and challenge their historical dependence on the Global North.[58] Yet it has also opened opportunities to delay decarbonization and accelerated mitigation efforts.

One such example of South-South cooperation is the Belt and Road Initiative (BRI), a Chinese investment project designed in part to incorporate sustainability strategies into fast-growing economies to help align them with member states' NDCs. The project has succeeded in developing new industries and expanding existing ones in many Global South countries.[59] Because the BRI "is not explicitly a sustainable development project that prioritizes climate change and environmental protection,"[60] however, there has been little movement on renewable energy projects while extractive industries have been expanding. Significantly, the top ten recipients of Chinese finance for coal-fired power plants are in Eurasia, including Indonesia, India, Vietnam, and Pakistan. Moreover, in Bangladesh, which has received both Japanese and Chinese financial and technical support, policymakers and climate change experts consider using coal power to be a viable policy to move toward a low-carbon development path, with the government seeking to ensure its energy security by obtaining climate finance from the international community to build more coal-fired power plants.[61] Therefore, while coal may not be considered a low-carbon-development fuel, it is promoted as a viable "transition fuel." The success of such initiatives is likely due to their business-as-usual approach to coal compared with the strict measures taken by several Western countries to phase out coal. In other words, these South-South corporation programs become attractive opportunities because they both expand economic development and opportunity in Global South countries and create pathways to reduce the dependency of the Global South on the Global North, which represents the continuing legacy of colonial and imperialist power. Furthermore, they also provide the opportunity to counter the perceived or realistic lack of capacity from renewable energy sources and tangible policy support for renewable energy development.[62] As such, these investment programs protect their national economic interests, which is portrayed as the best option to reconcile underdevelopment and historical structural inequalities with current climate mitigation responsibilities.

Energy Poverty and Climate Action

The development narratives associated with energy-transition mechanisms have played an important role in creating and facilitating business opportunities, global partnerships, and green financing in the Global South.

For example, in Argentina, carbon-intensive industries such as agriculture, mining, and shale gas development have been discussed as pathways to solve the country's structural economic issues. Moreover, gas is considered to be a transition fuel, contributing to the global decarbonization process even as it supports development.[63] However, this view contradicts the stance of the International Energy Agency that gas can serve only a limited role in a renewables transition.[64] Rhetoric similar to Argentina's was presented by the former chief executive officer of the Nigerian Petroleum Corporation and former OPEC Secretary General, Mohammad Sanusi Barkindo, during the COP26 meeting in Glasgow in 2021:

> The delicate balance between reducing emissions, energy affordability and security requires comprehensive and sustainable policies, with all voices being heard, and listened to. Focusing on only one of these over the others can lead to unintended consequences, market distortions, heightened volatility and energy shortfalls. We need to ensure energy is available and affordable for all; we need to move towards a more inclusive, fair and equitable world in which every person has access to energy, aligned with SDG [Sustainable Development Goal] 7; and we need to reduce emissions.[65]

In this way, attempts to balance socioeconomic needs and climate action serve as a form of obstruction.

Trade Relations and Transnational Influence

While South-South cooperation programs are expanding, the development-focused narratives discussed earlier support the continuation and growth of Global North-Global South financing, influence, and free-trade relationships that result in climate obstruction. Several countries that continue to export, consume, and phase in coal are in the Global South.[66] Therefore, for some countries in the region, choosing to expand coal use or other heavily polluting industries is a priority to ensure economic growth, sometimes in unsustainable ways, because of its importance for global trading partnerships. For example, in Indonesia, coal and palm oil are important commodities for both domestic consumption and for export. Due to deforestation concerns, European markets no longer support palm oil-based biofuels, creating the EUDR (European Union Deforestation Rule), which forbids the sale of several tropical commodities if grown on deforested land; the Indonesian government has interpreted the rule as a form of trade war and a threat to its national sovereignty.[67] Thus, while some countries maintain trading relationships based on commodities, such as coal and palm oil, that contribute to expanding extractive sectors, resistance from powerful transnational forces such as the EUDR serves to partially offset the expansion of these sectors.

Another way the Global North exerts a transnational influence upon the Global South is within the policy sphere. Scholars have identified several members of a coalition of think tanks active within some Global South countries that engage in climate obstruction.[68] Several of these think tanks were part of the now-defunct Civil Society Coalition on Climate Change (CSCCC);[69] a network of organizations formed by a US-based group to challenge the scientific consensus on climate change by creating counterproposals to international efforts and producing media materials to promote climate delay. Located in several countries in the Global South, organizations associated with these think tanks produced op-eds, articles, and policy papers advocating climate delay, mostly reprinted and translated from materials written by US-based authors. One example is the Peruvian think tank Andes Libre, which had previously cited and utilized the work of US-based authors affiliated with Global North organizations considered to be obstructionist.[70] Another example is the Brazil-based Instituto Liberdade, which reproduced the work of the Coalition's first international report in 2007.[71] Similarly, the Nigerian Initiative for Public Policy utilized the work of the CSCCC and promoted the same narratives of climate delay promoted by Global North think tanks noted for climate obstruction (see also Chapter 5). Historically, some of the discourse in these materials was associated with more traditional forms of denialism.[72] More recently, the positions articulated by these think tanks and their affiliated organizations in the Global South have shifted somewhat to frame climate change as best addressed through a free market and decentralized economy with little or no government intervention.[73]

Green Financing and Infrastructure Development

Green financing provided by the Global North to the Global South has operated as a useful tool for some renewable energy development projects. Yet some projects have faced resistance from within these countries due to their seeming incompatibility with achieving climate goals. In Brazil, Ferrogrão, a planned railway line and infrastructure project connecting a series of ports along the Amazon River, was the first project to be approved by the Climate Bonds Initiative (CBI).[74] This project opened up a green-investment pipeline within the framework of a partnership between Brazil's Ministry of Infrastructure,[75] and the country's National Development Bank, with the latter committed to financing up to 70% of the costs. The project was widely promoted at international investor roadshows, kindling the interest of several foreign banks and financial institutions.[76] While Ferrogrão was envisioned as a more sustainable alternative to the existing transport highway, the railway would cause significantly more emissions through extensive deforestation,[77] contravene international law (specifically the ILO Convention 169, also called the Indigenous and Tribal Peoples Convention),[78] and ignore the rights of Indigenous peoples by failing to consult with them about its development.[79] Ferrogrão illustrates how the domain of green finance is filled with outside actors developing and implementing plans that may not be compatible with domestic government structures, policies, and interests. These plans also ignore or fail to adequately consider other environmental impacts,[80] which may result in inseparable ecological and social impacts. Moreover, the case shows that green finance can be controversial, leading some Global South countries to resist the influence of the Global North as they attempt to shape international climate mitigation efforts that affect them.

Climate Vulnerability and Mitigation Responsibility in Global Talks

Historically, climate negotiation forums such as the UNFCCC have been used by some member countries of the Global South as a collective opportunity to stall measures to mitigate the climate crisis. In this obstruction tactic, delegations present their positions in the negotiating process at the UN climate change conferences to spread discourses of climate delay. For example, India's delegation at COP21 in Paris in 2015 argued that less-developed countries like India should be allowed to continue emitting large amounts of carbon to grow their economies, characterizing its position as "climate justice."[81] However, research on the Indian government's actions at home show no evidence of prioritizing the needs of economically poor and socially marginalized communities, whose lives have become more precarious over a quarter-century of steady economic growth that has increased domestic economic inequalities.[82] Indeed, many countries in the Global South have adopted the UNFCCC's Common But Differentiated Responsibilities (CBDR) principle (Article 4) as a strategy to delay climate action under the rubric that mitigation (and decarbonization) is the primary responsibility of the Global North. Furthermore, because global climate negotiations require binding pledges from both the Global North and South, leaders of the Global South, citing principles of CBDR, fairness and equity, and/or climate justice, may use the pressure of commitment-making as a strategy of climate obstruction (see Chapter 10).

The international climate forums provide a vital opportunity to disseminate the accurate narrative that many climate-vulnerable countries in the Global South bear little or no responsibility for addressing carbon emissions because of their historically lower contribution to global warming than advanced industrial countries. However, national delegations from these countries often support their continued reliance on fossil fuels and lack of interest in increasing renewable energy by citing its intermittent nature or their lack of access to new technologies. Such positions have also led to disagreements within negotiating groups representing Global South countries, such as the Group of 77 and China (G77 + China)—especially between major emitters including China, South Africa, India, and Brazil, and the Least Developed Countries and the Alliance of Small Island States (AOSIS), which regard climate change as an existential threat.[83] That is, countries in the Global South are heterogenous, yet share a narrative of economic underdevelopment and poverty.[84] Therefore, while their climate obstruction narratives and actions differ depending on their national interests, some shared interests remain; these can lead to the formulation of arguments that can intentionally or unintentionally stall progress on international climate negotiations.

Climate Policy Agendas and the Public–Elite Divide

In Global North countries, a documented gap exists between public opinion favoring climate action and state-corporate activities to address it. Consequentially, while a public appetite exists to transform the energy sector and act on climate change, the vested interests of corporate actors and national governments serve to obstruct policies, thus undermining the public interest.[85] Similar conflicts between the public and elites or between government and industry over the best actions for mitigating climate change and accelerating decarbonization exist in some Global South countries. For example, in response to high levels of deforestation, Colombia has seen a resurgence of government efforts to reduce deforestation and improve governance and security issues in affected rural areas.[86] On the other hand, the Colombian energy sector, particularly the coal industry, remains powerful and is advancing faster than the transition to renewable energy sources.[87] Moreover, while Colombia declared at COP27 that it planned to stop developing oil and gas,[88] unresolved conflicts remain with conservative lawmakers and the state-owned Ecopetrol company, who defend the use of fracking and the maintenance of oil and gas reserves to protect against the projected loss of thousands of jobs and reduced foreign investment, claiming this is a pathway to decarbonization.[89] This elite narrative challenges the public consensus in Colombia. In 2023, a survey conducted by the European Bank noted that 91% of Colombians believe that the government should implement effective measures to combat climate change, with an overwhelming majority preferring renewable energy development.[90]

Another caveat to this conflict between the public and elites is the justification of climate inaction on the basis of sovereignty and energy independence.

This narrative, promoted by governments and elites in Global North countries, advocates climate policies designed primarily to protect national interests, sometimes even by expanding the national/domestic-based extractive sector.[91] A similar narrative has been adopted by some countries of the Global South, despite public support for climate action. As in the case of Colombia, this type of conflict ultimately creates a stalemate that delays a coordinated international effort to address the climate crisis.

Violent Conflict and Illicit Markets in Fragile States

The United Nations Environment Program (UNEP) considers how environmental degradation and climate change interact with peace and security issues.[92] Addressing environmental and climate crises requires erecting measures to protect the environment as well as to promote peace in politically complex and fragile contexts. In such contexts, denial or delay of climate action is not necessarily purposeful. Rather, as priorities shift, addressing climate change may seem less urgent. For example, recent research on Yemeni conflict showed that "while the parties involved in the conflict encounter climate-related issues, they put it in the back seat to take the so-called rational path to first and foremost arrive at a political arrangement and dissolve the conflict."[93] In such scenarios, policymakers in countries such as Yemen, South Sudan, Congo, and Syria are unable to pursue climate mitigation policies. While the argument that these are obstruction activities is tenuous, the failure to implement climate-related and energy-transition policies in fragile and conflict-ridden political contexts can delay climate action within certain regions or territories.

In addition to conflict-prone and fragile political contexts, illicit markets can contribute to delaying climate change action in the Global South. For example, legitimizing environmental crimes has been a key feature of climate obstruction in Brazil. Such illicit actions have become socially accepted, with criminal actors occasionally counting on corrupt individuals in key institutional positions to grant them amnesty. This trend began in the 1970s during the military dictatorship under Brazilian armed forces (1964–1985), which encouraged the indiscriminate deforestation and occupation of the Amazon Rainforest to defend the country's sovereignty. Politicians, entrepreneurs, military personnel, civil society groups, and religious leaders defended these land-grabbing practices as necessary for the country's development. The advancement of Brazil's carbon-intensive agricultural frontier in the Amazon and the Cerrado biome, both of which are threatened, is still a well-established practice strongly opposed by environmentalists, organizations, and some politicians.[94]

Global Think Tanks and Climate Policy

Another significant aspect of climate obstruction in the Global South is the influence of powerful think tanks such as the Mont Pelerin Society (MPS) and the Atlas Network. Both are global networks of smaller think tanks known for advancing climate change policy denialism. While the role of these think tanks in delaying climate actions in the Global North is well documented in both scholarly literature and media reports,[95] recent research has revealed their role in strengthening market-fundamentalist neoliberal policies and influencing climate delay in the Global South.[96] Individuals and think tanks in the Global South connected with MPS are known for spreading climate delay discourses and climate science denialism. An example is Instituto Liberdade, a Brazilian think tank that promotes the works of Global North-based climate contrarians in Latin America with a goal of influencing climate change policies to stall climate action.[97]

Another key strategy deployed by Atlas-connected think tanks in the Global South is to demonize climate and environmental protesters, characterizing them as extremists or terrorists and "new colonialists."[98] In Africa, for example, the head of Centre for African Prosperity, an Atlas Network affiliate, has emphasized that ensuring access to affordable electricity for vast numbers of people is a pressing priority for African policymakers. According to this narrative, continued extraction and use of fossil fuels using market-driven mechanisms is the way to address energy poverty in Africa. Climate policies that recommend moving away from fossil fuels are viewed as "a death sentence for Africans."[99] In this narrative, climate and environmental activists who oppose continued extraction and use of fossil fuels are characterized as "new colonialists" who wish to impose their choices on African people suffering from the lack of affordable electricity that fossil fuels could solve. Such narratives ignore a viable solution to the problem of ensuring access to affordable energy in Africa, the least electrified continent: a transition toward deploying the vast renewable energy resources endemic to Africa. Research suggests that the major impediment to resolving African energy poverty is the politics of foreign investment and lending to African economies.[100] Addressing this energy-finance gap in harnessing renewable energy resources is the real challenge, a position that contradicts the pro-fossil development views of the neoliberal/market-fundamentalist think tanks affiliated with the Atlas Network (see Chapter 5).[101]

Media Discourses of Climate Delay

While there is a significant body of research on climate misinformation in the Global North[102] (see Chapters 6 and 7), little is known about its spread in Global South countries. Recent research provides some insights into the types of climate change misinformation campaigns disseminated via social media in China. One key element of the misinformation spread on these platforms is the reference to "expert sources" in posts about the environmental and health impacts of climate change that would likely be considered misinformation.[103] In some Indian newspapers, researchers found that the climate issue was framed primarily around the countries' historical and differentiated responsibilities. In particular, these messages emphasize that the Global North is the most responsible for emissions and should therefore be first to lead on decarbonization and global mitigation efforts.[104] These narratives also reflect, for example, the position of India's delegation at the UN climate meetings noted earlier, which has sought to emphasize the historical role of emissions. By emphasizing the failure of the Global North to support countries such as India and reduce emissions, such narratives minimize India's own contribution to GHG emissions and avoid reflecting on its mitigation efforts.[105]

COUNTERING CLIMATE OBSTRUCTION IN THE GLOBAL SOUTH

Since the 1980s, environmental-justice campaigns have grown substantially in the Global South and have gained significant momentum since the 2000s, with a specific focus on the climate crisis.[106] This section highlights campaigns initiated by grassroots communities, environmentalists, and other civil- society actors that illustrate bottom-up challenges to the forces of climate obstruction, with an emphasis on rectifying environmental and climate injustices.

Major Indonesian NGOs such as WALHI (the Indonesian Forum for the Environment) have contested the country's National Strategic Projects (PSNs) which, though they include technological and renewable-energy developments and carbon-offset schemes—prioritize economic growth, job creation, and post-Covid economic recovery over GHG emission reduction.[107] WALHI has mobilized grassroots campaigns to address issues such as the PSNs' lack of environmental impact assessments and their potential for land loss and deforestation through the development of large-scale farms known as food estates.[108] Indonesia's commitment to build more coal-fired power plants for industrial users (captive power plants) demonstrates the lack of tangible success of such campaigns thus far.[109]

There have also been cases of environmental justice campaigns against South-South cooperation projects in the energy sector, which have also failed to achieve their desired outcomes. For example, a decade-long (2010–2020) mobilization by environmentalists against the building of coal-fired power plants and other industrial infrastructures near the Sundarbans, an ecologically fragile UNESCO World Heritage Site, was unable to persuade the Bangladeshi and Indian governments to cancel, halt, or relocate the project.[110]

The project was a joint venture financed by state-owned power corporations of India and Bangladesh, demonstrating how campaigns may be thwarted by the political environment in which they operate—in this case, a hybrid political regime characterized by authoritarianism and extractive political institutions.[111]

There have been some successes. In 2016, environmentalists and other civil-society actors in Sri Lanka mobilized grassroots communities to thwart the construction of a coal-fired power plant financed by India. In addition to the protesters' concern over the environmental and climate impacts of coal power, widespread mobilization against the power plant was galvanized by the land dispossession and security concerns of local communities that had been displaced to establish a high-security zone in the project area. After a Supreme Court ruling in favor of a Fundamental Rights Application submitted by a group of civil society actors led by the Environmental Foundation Limited (EFL), the Sri Lankan government decided not to go ahead with the project.[112] Eventually, in 2022, the Sri Lankan and Indian governments agreed to form a joint venture to build a solar power project instead.[113]

While public concern about climate change and desire to move toward a low-carbon future may be demonstrated in public opinion polls, there are distinct challenges faced by civil-society actors and environmentalists in some countries of the Global South. Kenyan climate justice movements, for example, reveal an interesting dimension of some campaigns to counter climate obstruction. According to one scholar, "Kenyan civil society actors rejected the Global South's climate justice argument because they believe one of its primary components—financial transfers—will be misused or misappropriated."[114] These movements demonstrate a somewhat pessimistic view due to lingering mistrust of their state and its reputation for official corruption as well as questions about their society's capacity to address climate change though technological advances and knowledge alone.

Similarly, there are currently few cohesive and comprehensive campaigns around climate change in Argentina.[115] Key informants from the environmental and academic sectors have held that a counternarrative to climate obstruction discourses has not yet been sufficiently consolidated. In turn, this limitation has been linked to the fragmentation of the environmental sector in Argentina and a lack of urgency among the public and the policy community around resolving the climate emergency. Researchers attribute this problem partly to discourses of national sovereignty and development that appear to stall mitigation efforts and a transition away from extractive industries, particularly because the country's development narrative plays a dominant role in stimulating the short-term economic growth supported by extractive industries.[116]

Finally, in the Global South there are legitimate risks associated with mobilization around climate and environmental injustice. On the front lines of conservation, local and regional civil-society organizations and community groups—including Indigenous peoples—constitute a significant bulwark against local environmental destruction and biodiversity loss. Yet seeking to challenge the actions of climate obstruction activities facilitated by the state or large corporations presents a high risk of physical harm, especially if the defenders are impoverished, rural, Indigenous peoples, and/or women.[117] Indeed, deaths of environmental defenders have been well documented; they frequently face repression when challenging mining and extractive industries, agribusiness expansion, hydropower development, and logging.[118] These risks remain despite the success of some legal suits against harms to environmental defenders, such as the *Kawas v. Honduras* judgment in the Inter-American Court of Human Rights, which ruled that the state is obligated to protect environmental defenders experiencing human rights violations.[119]

CONCLUSION

Our study on the nature and dynamics of climate obstruction in the Global South yields three key findings. First, the legacy of resource and human exploitation has increased vulnerability to climate change in these regions. At the same time, the historical contributions of these countries to rising GHG emissions are less significant than those of the Global North. Recognizing this disparity and the disproportionate share of burden and responsibility, climate delay typically manifests in policy choices seeking to reconcile energy security and development goals with climate action. We suggest that such tactics and discourses of climate delay may be overcome by ensuring that international climate negotiations reduce the implementation barriers to distribution of adequate climate funding, which many Global South actors define as a crucial tool to achieve climate justice.

Second, states, carbon majors, corporations, and other actors frame fossil fuels and extractive industries as the primary instruments for development in the Global South. They also frame them as central to reducing poverty and countering the systematic harms of historical exploitation of many countries. To counter this narrative, we suggest that the Global North must take proactive action in supporting Global South countries to develop the required institutional, technological, and (again) financial capacities to move toward a just and equitable decarbonization process.

Third, sustained mobilizations organized by Indigenous peoples, civil-society actors, and social movements to resist attempts to delay climate action show mixed outcomes in the Global South. There are significant risks for those who challenge the actors and organizations behind these efforts, particularly where political conflict, corruption, or state and corporate crimes are rampant. State agencies and the carbon majors, facing grassroots mobilizations confronting their policies and activities, must allow for the expression of dissent without fear or threat. The institutional practice of democratic deliberation at the local and national levels should also be strengthened to accommodate community concerns.

Research Gaps and Policy Lessons

Further research is needed for academics and policy practitioners to understand and confront the complexity of climate obstruction in the Global South. Scholars have noted several priorities for this research, discussed in this section.[120]

First, researchers should focus on furthering our understanding of developmentalism and the ways development discourses have played a fundamental role in the climate delay narratives and approaches to climate obstruction observed in the Global South. A key component of understanding climate obstruction in the Global North has been identifying the social, political, and economic forces that drive it. Therefore, it is likely that development and alternative- development models that emphasize extractive industries will similarly be entangled with climate obstruction in the Global South.

Second, armed with better insight on development discourses and practices, researchers should identify how they affect domestic-level actions that obstruct climate policies in the Global South as well as if and how they play out in creating stumbling blocks to stronger mitigation efforts in global climate negotiations.

Third, it is also important to examine how political leaders within Global South countries garner political support for stalling or intentionally delaying climate action.

Fourth, there seems to be a disconnection between public opinion in Global South countries and policy implementation. Thus, another important area for future research will be identifying which factors related to climate obstruction shape policy choices despite strong public opinion in favor of climate action.?

Fifth, researchers should also examine the influence of elite interests on both public opinion and policy choice to uncover the forces derailing public support for a transition toward renewable energy in some Global South countries. Such an endeavor might include a critical analysis of media narratives in both mainstream news and on social media. Researchers should also explore the role of the media in science education and climate justice and if and how narratives of denial and delay penetrate the public sphere in the Global South.

Finally, decarbonization rests fundamentally on the development of renewable-energy resources. However, evidence from the Global South reveals that renewable energy projects are often poorly financed and that these investments could potentially cause further environmental and social harm. Moreover, unlike fossil fuel–based energy projects, renewable energy projects do not require many workers and may offer precarious employment opportunities due to the nature of the industry. These issues might eventually produce an unjust energy transition for both new and old energy-sector workers. Moreover, the development of large-scale renewable-energy projects such as solar parks may involve land grabs and dispossession similar to that seen in predatory extractive resource-development projects.[121] Addressing these critical issues and avoiding their consequences require the urgent attention of policymakers.

NOTES

1. See Immanuel Wallerstein, *World-Systems Analysis: An Introduction* (Duke University Press, 2004).
2. Nour Dados and Raewyn Connell, "The Global South," *Contexts* 11, no. 1 (2012): 12–13.
3. Ibid., 13.
4. See J. Timmons Roberts and Bradley C. Parks, *A Climate of Injustice: Global Inequality, North-South Politics, and Climate Policy* (The MIT Press, 2006).
5. David Ciplet and J. Timmons Roberts, "Splintering South: Ecologically Unequal Exchange Theory in a Fragmented Global Climate," in *Ecologically Unequal Exchange: Environmental Injustice in Comparative and Historical Perspective*, ed. R. Scott Frey, Paul K. Gellert, and Harry F. Dahms (Palgrave Macmillan, 2019), 273–305.
6. Ibid., 274.
7. Luke Taylor, "Colombia Announces Halt on Fossil Fuel Exploration for a Greener Economy," *The Guardian*, January 20, 2023, https://www.theguardian.com/world/2023/jan/20/colombia-stop-new-oil-gas-exploration-davos.
8. Richard Marcantonio, Debra Javeline, Sean Field, and Agustin Fuentes, "Global Distribution and Coincidence of Pollution, Climate Impacts, and Health Risk in the Anthropocene," *PLOS One* 16, no. 7 (2021): e0254060; World Meteorological Organization, *State of the Climate in Africa 2021* (WMO, 2022).
9. Harald Fuhr, "The Rise of the Global South and the Rise in Carbon Emissions," *Third World Quarterly* 42, no. 11 (2021): 2724–2746.
10. , Jason Hickel, Christian Dorninger, Hanspeter Wieland, and Intan Suwandi, "Imperialist Appropriation in the World Economy: Drain from the Global South Through Unequal Exchange 1990–2015," *Global Environmental Change* 73, no. 102467 (2022): 1–13.
11. See Prakash Kashwan, *Democracy in the Woods: Environmental Conservation and Social Justice in India, Tanzania, and Mexico* (Oxford University Press, 2017).
12. Sikina Jinnah, "Makers, Takers, Shakers, Shapers: Emerging Economies and Normative Engagement in Climate Governance," *Global Governance* 23, no. 2 (2017): 285–306.

13. UNDP, "What Are the NDCs and How Do They Drive Climate Action?," May 31, 2023, https://climatepromise.undp.org/news-and-stories/NDCs-nationally-determined-contributions-climate-change-what-you-need-to-know.

14. Roberts and Parks, *A Climate of Injustice*.

15. Lisa Vanhala, Elisa Calliari, and Adelle Thomas, "Understanding the Politics and Governance of Climate Change Loss and Damage," *Global Environmental Politics* 23, no. 3 (2023): 1–11.

16. Ibid., 2.

17. Ibid.

18. Danielle Falzon, Fred Shaia, J. Timmons Roberts, Md Fahad Hossain, Stacy-ann Robinson, Mizan R. Khan, and David Ciplet, "Tactical Opposition: Obstructing Loss and Damage Finance in the United Nations Climate Negotiations," *Global Environmental Politics* 23, no. 3 (2023): 95–119.

19. Darren McCauley and Raphael Heffron, "Just Transition: Integrating Climate, Energy and Environmental Justice," *Energy Policy* 119 (2018): 1–7.

20. Fuhr, "The Rise of the Global South."

21. Arif Dirlik, "Developmentalism: A Critique," *Interventions* 16, no. 1 (2014): 30–48.

22. Frederick H. Buttel, "Ecological Modernization as Social Theory," in *The Ecological Modernisation Reader*, ed. Arthus P. J. Mol. David A. Sonnenfeld, and Gert Spaargaren (Routledge, 2009), 123–137.

23. Ibid.

24. Hilal Gezmiş, "From Neoliberalism to Neo-developmentalism? The Political Economy of Post-crisis Argentina (2002–2015)," *New Political Economy* 23, no. 1 (2018): 66–87.

25. Efe Can Gürcan, *Multipolarization, South-South Cooperation and the Rise of Post-hegemonic Governance* (Routledge, 2019).

26. Bas Hooijmaaijers, "China, the BRICS, and the Limitations of Reshaping Global Economic Governance," *The Pacific Review* 34, no. 1 (2021): 29–55.

27. Raul Gouvea, Dimitri Kapelianis, and Shihong Li, "Fostering Intra-BRICS Trade and Investment: The Increasing Role of China in the Brazilian and South African Economies," *Thunderbird International Business Review* 62, no. 1 (2020): 17–26.

28. Niall Duggan, Juan Carlos Ladines Azalia, and Marek Rewizorski, "The Structural Power of the BRICS (Brazil, Russia, India, China and South Africa) in Multilateral Development Finance: A Case Study of the New Development Bank," *International Political Science Review* 43, no. 4 (2022): 495–511.

29. Zaki Laidi, "The BRICS Against the West?," *CERI Strategy Paper* 11, (2011): 1–12.

30. William F. Lamb, Thomas Wiedmann, Julia Pongratz, Robbie Andrew, Monica Crippa, Jos G. J. Olivier, Dominik Wiedenhofer, Guilio Mattioli, Alaa Al Khourdajie, Jo House, Shonali Pachauri, Maria Figueroa, Yamina Saheb, Raphael Slade, Klaus Hubacek, Laixing Sun, Suzana Kahn Ribeiro, Smail Khennas, Stephane de la Rue du Can, Lazarus Chapungu, Steven J Davis, Igor Bashmakov, Hancheng Dai, Shobhakar Dhakal, Xianchun Tan, Yong Geng, Baihe Gu, and Jan Minx, "A Review of Trends and Drivers of Greenhouse Gas Emissions by Sector from 1990 to 2018," *Environmental Research Letters* 16, no. 7 (2021): 073005.

31. Carlos R. S. Milani and Leonildes Nazar Chaves, "How and Why European and Chinese Pro-Climate Leadership may be Challenged by their Strategic Economic Interests in Brazil," *Asia Europe Journal* 20, no. 4 (2022): 403–422.

32. Climate Social Science Network (CSSN), "Dismantling the Environmental State: Actors, Strategies and Discourses Behind the Bolsonaro Attack on the National Environmental Regulation," CSSN Position Paper 3 (2021).

33. Dragana Ostic, Angelina Kissiwaa Twum, Andrew Osei Agyemang, and Helena Adu Boahen, "Assessing the Impact of Oil and Gas Trading, Foreign Direct Investment Inflows, and Economic Growth on Carbon Emission for OPEC Member Countries," *Environmental Science and Pollution Research* 29, no. 28 (2022): 43089–43101.

34. OPEC, "Our Mission," https://www.opec.org/opec_web/en/about_us/23.htm (accessed March 31, 2025).

35. Sharon Shea, "Obstacles to Decarbonization in Southeast Asia," *AseanFocus* 1 (2023): 8–9.

36. Michael Jakob and Jan C. Steckel, *The Political Economy of Coal: Obstacles to Clean Energy Transitions* (Routledge, 2022), 364.

37. Paul K. Gellert and Paul S. Ciccantell, "Coal's Persistence in the Capitalist World-Economy: Against Teleology in Energy "Transition" Narratives," *Sociology of Development* 6, no. 2 (2020): 194–221; Richard York and Shannon E. Bell, "Energy Transitions or Additions?: Why a Transition from Fossil Fuels Requires More than the Growth of Renewable Energy," *Energy Research & Social Science* 51 (2019): 40–43.

38. Paul K. Gellert, "Neoliberalism and Altered State Developmentalism in the Twenty-First Century Extractive Regime of Indonesia," *Globalizations* 16, no. 6 (2019): 894–918.

39. Richard Clark, Noah Zucker, and Johannes Urpelainen, "The Future of Coal-Fired Power Generation in Southeast Asia," *Renewable and Sustainable Energy Reviews* 121 (2020): 109650.

40. Maguire Gavin, "Vietnam's Coal Emissions Primed for Surge After Imports Jumped," Reuters, July 12, 2023, https://www.reuters.com/markets/commodities/vietnams-coal-emissions-primed-surge-after-imports-jump-maguire-2023-07-12.

41. Maguire Gavin, "Philippines Set to Go from Renewable Laggard to Leader in SE Asia," Reuters, March 14, 2023, https://www.reuters.com/markets/commodities/philippines-set-go-renewable-laggard-leader-se-asia-2023-03-14/.

42. Joseph Allchin, "Low-Lying Bangladesh Targets Jump in Coal Use," *Financial Times*, December 29, 2015, https://www.ft.com/content/241059fa-9424-11e5-bd82-c1fb87bef7af.

43. Niti Aayog, "India Climate and Energy Dashboard," https://iced.niti.gov.in/energy/electricity/generation (accessed March 31, 2025).

44. The Economist, "Why Is India Clinging to Coal?," November 16, 2021, https://www.economist.com/the-economist-explains/2021/11/16/why-is-india-clinging-to-coal.

45. Kathryn Hochstetler, *Political Economies of Energy Transition: Wind and Solar Power in Brazil and South Africa* (Cambridge University Press, 2021), 54.

46. BBC News, "Climate Change: Asia 'Coal Addiction' Must End, UN Chief Warns," November 2, 2019, https://www.bbc.com/news/world-asia-50276983.

47. David Ciplet, J. Timmons Roberts, and Mizan R. Khan, *Power in a Warming World: The New Global Politics of Climate Change and the Remaking of Environmental Inequality* (MIT Press, 2015).

48. Greenpeace, "Coalruption: Shedding Light on Political Corruption in Indonesia's Coal Mining Sector," Greenpeace, 2018, https://www.greenpeace.org/static/planet4-indonesia-stateless/2018/12/727d7a2d-coalruption-english-web.pdf.

49. Mushtaq Khan, Mitchell Watkins, and Iffat Zahan, "De-Risking Private Power in Bangladesh: How Financing Design Can Stop Collusive Contracting," *Energy Policy* 168 (2022): 113146.

50. M. Omar Faruque, "Climate Crisis and the Politics of Low Carbon Energy Future in Emerging Economies," paper presented at the annual conference of the American Sociological Association, Philadelphia, 2023.

51. Patrik Oskarsson, Kenneth Bo Nielsen, Kuntala Lahiri-Dutt, and Brototi Roy, "India's New Coal Geography: Coastal Transformations, Imported Fuel and State-Business Collaboration in the Transition to More Fossil Fuel Energy," *Energy Research & Social Science* 73 (2021): 101903; Brototi Roy and Anke Schaffartzik, "Talk Renewables, Walk Coal: The Paradox of India's Energy Transition," *Ecological Economics* 180 (2021): 106871.

52. The Economist, "India's Next Green Revolution," October 20, 2022, https://www.economist.com/leaders/2022/10/20/indias-next-green-revolution.

53. Kathryn Hochstetler, *Political Economies of Energy Transition: Wind and Solar Power in Brazil and South Africa* (Cambridge University Press, 2021).

54. Ryna Cui, Fabby Tumiwa, Alicia Zhao, Deon Arinaldo, Raden Wiranegara, Diyang Cui, Camryn Dahl, Lauri Myllyvirta, Claire Squire, Pamela Simamora, and Nate Hultman, *Financing Indonesia's Coal Phase-Out: A Just and Accelerated Retirement Pathway to Net-Zero* (Institute for Essential Services Reform, 2022).

55. Chika Izuora, "African Oil Producers Back OPEC Output Cut," 2022, https://leadership.ng/african-oil-producers-back-opec-output-cut/.

56. Andrzej Polus and Wojciech Tycholiz, "David versus Goliath: Tanzania's Efforts to Stand Up to Foreign Gas Corporations," *Africa Spectrum* 54, no. 1 (2019): 61–72.

57. United Nations, "South-South Cooperation," 2019, https://www.un.org/development/desa/en/news/intergovernmental-coordination/south-south-cooperation-2019.html.

58. Branislav Gosovic, "The Resurgence of South–South Cooperation," *Third World Quarterly* 37, no. 4 (2016): 733–743.

59. Lihuan Zhou, Sean Gilbert, Ye Wang, Miquel Muñoz Cabré, and Kevin P. Gallagher, "Moving the Green Belt and Road Initiative: From Words to Actions," Working Paper, World Resource Institute, 2018.

60. Wei Yin, "Integrating Sustainable Development Goals into the Belt and Road Initiative: Would It Be a New Model for Green and Sustainable Investment?," *Sustainability* 11, no. 24 (2019): 6991.

61. Government of Bangladesh, *Bangladesh Nationally Determined Contributions (NDCs)* (Ministry of Environment, Forest and Climate Change, 2021).

62. Kelly Sims Gallagher, Rishikesh Bhandary, Easwaran Narassimhan, and Quy Tam Nguyen, "Banking on Coal? Drivers of Demand for Chinese Overseas Investments in Coal in Bangladesh, India, Indonesia and Vietnam," *Energy Research & Social Science* 71 (2021): 101827.

63. Lucas G. Christel, Ricardo A. Gutiérrez and Elisabeth Möhle, "Climate Action and Obstruction in Argentina from the Actors' View," paper presented at the 27th World Congress of Political Science, Buenos Aires, 2023.

64. International Energy Agency, "Natural Gas," https://www.iea.org/energy-system/fossil-fuels/natural-gas (accessed March 31, 2025).

65. OPEC, "Bulletin 2-3/22," 2022, https://www.opec.org/opec_web/static_files_project/media/downloads/publications/OB02_032022.pdf.

66. Yin, "Integrating Sustainable Development Goals."

67. Rilus A. Kinseng et al., "Unraveling Disputes Between Indonesia and the European Union on Indonesian Palm Oil: From Environmental Issues to National Dignity," *Sustainability: Science, Practice and Policy* 19, no. 1 (2023): 2152626.

68. Ruth E. McKie, "Climate Change Counter Movement Neutralization Techniques: A Typology to Examine the Climate Change Counter Movement," *Sociological Inquiry* 89, no. 2 (2019): 288–316.

69. DeSmogblog, "Civil Society Coalition on Climate Change," https://www.desmogblog.com/civil-society-coalition-climate-change/ (accessed March 31, 2025).

70. Ruth E. McKie, *The Climate Change Counter Movement: How the Fossil Fuel Industry Sought to Delay Climate Action* (Springer International, 2023), 139–167.

71. Ibid.

72. Ibid.

73. Ibid.

74. Climate Bonds Initiative, "Climate Bonds Initiative Is the International Organization Working to Mobilize Global Capital for Climate Action," https://www.climatebonds.net/about (accessed March 31, 2025).

75. United Nations Brazil, "Adoção de títulos verdes para concessões ferroviárias é impulsionada por PNUD e Ministério da Infraestrutura," July 28, 2021, https://brasil.un.org/pt-br/137698-ado%C3%A7%C3%A3o-de-t%C3%ADtulos-verdes-para-concess%C3%B5es-ferrovi%C3%A1rias-%C3%A9-impulsionada-por-pnud-e-minist%C3%A9rio.

76. Marianna Duarte De Aragão, "Brazil Hopes Green Debt Will Help Fund $3.1 Billion Amazon-Crossing Railway," *Bloomberg*, November 28, 2019.

77. Rafael Araújo, Juliano Assunção, Arthur Bragança, *Policy Brief: The Environmental Impacts of the Ferrogrão Railroad: An Ex-Ante Evaluation of Deforestation Risks* (Climate Policy Initiative, 2020).

78. International Labour Organization (ILO), *Indigenous and Tribal Peoples Convention (No. 169)* (ILO, 1989).

79. V.a., *Carta Conjunta Kayapó e Munduruku ao Tribunal de Contas da União (TCU)*, February 23, 2021,, https://site-antigo.socioambiental.org/sites/blog.socioambiental.org/files/nsa/arquivos/carta_para_o_tcu.pdf; Isabel Harari, *Indígenas exigem direito à Consulta prévia na fase do planejamento da Ferrogrão*, March 1, 2021, Instituto Socioambiental.

80. Stockholm Resilience Center, "Planetary Boundaries," https://www.stockholmresilience.org/research/planetary-boundaries/the-nine-planetary-boundaries.html (accessed March 31, 2025).

81. Alister Doyle and Tommy Wilkes, "Fueling Its Growth with Coal, India Champions the Poor in Paris," Reuters, December 6, 2015, https://www.reuters.com/article/uk-climatechange-summit-india-idAFKBN0TO0S720151206.

82. Prakash Kashwan, ed., *Climate Justice in India* (Cambridge University Press, 2022).

83. Sjur Kasa, Anne T. Gullberg, and Gørild Heggelund, "The Group of 77 in the International Climate Negotiations: Recent Developments and Future Directions," *International Environmental Agreements: Politics, Law and Economics* 8 (2008): 113–127.

84. Ibid.

85. J. Timmons Roberts and Robert J. Brulle, "Conclusions: Ten Lessons about Climate Obstruction in Europe," in *Climate Obstruction in Europe*, ed. Robert J. Brulle, J. Timmons Roberts, and Miranda C. Spencer (Oxford University Press, 2024), 347–364.

86. World Bank, "Colombia: Leading the Path to Sustainability in Latin America," September 7, 2021, https://www.worldbank.org/en/news/feature/2022/08/31/colombia-leading-the-path-to-sustainability-in-latin-america.

87. Mining Technology, "Coal Production in Colombia and Major Projects," 2023, https://www.mining-technology.com/data-insights/coal-in-colombia/.

88. Union of Concerned Scientists, "Report from COP27: The Fossil Fuel Industry Continues to Block the Path to Climate Justice," November 14, 2022, https://blog.ucsusa.org/delta-merner/report-from-cop27-the-fossil-fuel-industry-continues-to-block-the-path-to-climate-justice/.

89. Rodrigues, Patricia, "Is Colombia One Step Away from a Fracking Ban?," NACLA, February 8, 2023, https://nacla.org/colombia-one-step-away-fracking-ban.

90. European Investment Bank, "9 Colombians in 10 Demand Stricter Climate Policies: New Survey Reveals," September 4, 2023, https://www.eib.org/en/press/all/2023-306-nearly-9-chileans-in-10-demand-stricter-climate-policies-eib-survey-reveals.

91. McKie, *The Climate Change Counter Movement*, 19–50.

92. United Nations Environment Program, "Environment Security," https://www.unep.org/topics/disasters-and-conflicts/environment-security (accessed March 31, 2025).

93. B. Poornima and Rashmi Ramesh, "Yemen's Survival Quandary: The Compounding Effects of Conflict and Climate Obstruction," *Journal of Peacebuilding & Development* 18, no.3 (2023): 264-279

94. Carlos A. Nobre, Gilvan Sampaio, Laura S. Borma, Juan Carlos Castilla-Rubio, José S. Silva, and Manoel Cardoso, "Land-use and Climate Change Risks in the Amazon and the Need of a Novel Sustainable Development Paradigm," *Proceedings of the National Academy of Sciences* 113, no. 39 (2016): 10759–10768.

95. Kristoffer Ekberg, Bernhard Forchtner, Martin Hultman and Kirsti M. Jylhä, *Climate Obstruction: How Denial, Delay and Inaction are Heating the Planet* (Routledge, 2023); Donald Gutstein, *The Big Stall: How Big Oil and Think tanks are Blocking Action on Climate Change in Canada* (James Lorimer & Company, 2018); Nuria Almiron, Jose A. Moreno, and Justin Farrell, "Climate Change Contrarian Think tanks in Europe: A Network Analysis," *Public Understanding of Science 32, no.3* (2023): 268–283.

96. Quinn Slobodian and Dieter Plehwe (eds.), *Market Civilizations: Neoliberals East and South* (Princeton University Press, 2022).

97. McKie, *The Climate Change Counter Movement*.

98. Lyndal Rowlands, Julianna Merullo, Amy Westervelt, and Geoff Dembicki, "How Think Tanks Laid the Groundwork to Criminalize Protest," September 12, 2023, https://drilled.media/news/trfst-atlas.

99. Magatte Wade, "The COP26 Plan to Keep Africa Poor," *Wall Street Journal*, November 26, 2021, https://www.wsj.com/articles/the-cop26-plan-to-keep-africa-poor-climate-change-clean-energy-11637964581.

100. Rebekah Shirley, "The Clean Energy Hub of the Future," YouTube, March 13, 2023, https://www.youtube.com/watch?v=g9VFlXF47f0.

101. Michael Olabisi, Robert B. Richardson, and Adesoji O. Adelaja, "The Next Global Crisis: Africa's Renewable Energy Financing Gap," *Climate and Development* 15, no. 6 (2023): 501–508.

102. Stephan Lewandowsky and Sander Van Der Linden, "Countering Misinformation and Fake News through Inoculation and Prebunking," *European Review of Social Psychology* 32, no. 2 (2021): 348–384; John Cook, "Understanding and Countering Misinformation About Climate Change," in *Research Anthology on Environmental and Societal Impacts of Climate Change*, ed. Information Resources

Management Association (IGI Global, 2022), 1633–1658; Ahmed Al-Rawi, Derrick O'Keefe, Oumar Kane, and Aimé-Jules Bizimana, "Twitter's Fake News Discourses around Climate Change and Global Warming," *Frontiers in Communication* 6 (2021), https://doi.org/10.3389/fcomm.2021.729818.

103. Jianxun Chu, Yuqi Zhu, and Jiaojiao Ji, "Characterizing the Semantic Features of Climate Change Misinformation on Chinese Social Media," *Public Understanding of Science* 32, no. 7 (2023): 845–859.

104. Ranjini Murali, Aishwarya Kuwar, and Harini Nagendra, "Who's Responsible for Climate Change? Untangling Threads of Media Discussions in India, Nigeria, Australia, and the USA," *Climatic Change* 164 (2021): 1–20.

105. Ibid.

106. Paul Almeida, "Climate Justice and Sustained Transnational Mobilization," *Globalizations* 16, no. 7 (2019): 973–979.

107. Ade Thea, "Tiga Kritik Walhi Terhadap Perpres Percepatan Proyek Strategis Nasional," *Hukum Online*, December 1, 2020, https://www.hukumonline.com/berita/a/tiga-kritik-walhi-terhadap-perpres-percepatan-proyek-strategis-nasional-lt5fc5faf7b65e9/.

108. Takeshi Ito, Noer Fauzi Rachman and Laksmi A. Savitri, "Power to Make Land Dispossession Acceptable: A Policy Discourse Analysis of the Merauke Integrated Food and Energy Estate (MIFEE), Papua, Indonesia," *Journal of Peasant Studies* 41, no. 1 (2014): 29–50.

109. Hans Nicholas Jong, "Captive to Coal: Indonesia to Burn Even More Fossil Fuel for Green Tech," *Mongabay News*, August 10, 2023, https://news.mongabay.com/2023/08/captive-to-coal-indonesia-to-burn-even-more-fossil-fuel-for-green-tech/.

110. Jeremy Hance, "Thousands to March Against Coal Plant Threat to Bangladesh's Sundarbans Forest," *The Guardian*, March 2, 2016, https://www.theguardian.com/environment/2016/mar/02/thousands-to-march-protest-coal-plant-threat-bangladeshs-sundarbans-forest.

111. M. Omar Faruque, "Slow Violence, Environmental Governance, and Social Movement Outcomes in a Hybrid Regime," paper presented at the virtual annual conference of the British Sociological Association, 2022.

112. Gz. MeeNilankco Theiventhran,"Energy as a Geopolitical Battleground in Sri Lanka," *Asian Geographer* 41, no. 1 (2024): 21–45; Environmental Foundation Limited (EFL), "Halting the Coal Power Plant in Sampur," https://efl.lk/portfolio/halting-the-coal-power-plant-in-sampur/ (accessed March 31, 2025).

113. Meera Srinivasan, "NTPC Returns to Sri Lanka's Sampur with Solar Project," *The Hindu*, March 12, 2022, https://www.thehindu.com/news/international/ntpc-returns-to-sri-lankas-sampur-with-solar-project/article65217041.ece.

114. Christopher Todd Beer, "Climate Justice, the Global South, and Policy Preferences of Kenyan Environmental NGOs," *The Global South* 8, no. 2 (2014): 98.

115. Christel et al., "Climate Action and Obstruction in Argentina from the Actors' View."

116. Janaina B. Pinto, Arthur Facini and Carlos R. S. Milani, "Climate Change, Denial and Politics of Obstruction in Brazil: Building Conceptual and Empirical Blocks," paper presented at 27th World Congress of Political Science, Buenos Aires, 2023.

117. Dalena Tran, "Gendered Violence Martyring Filipina Environmental Defenders," *The Extractive Industries and Society* 13 (2023): 101211.

118. Kate Hallam, "Environmental Defenders: Murdered, Missing and at Risk," *Socialist Lawyer* 75 (2017): 40–43.

119. Lauri R. Tanner, "Kawas v. Honduras–Protecting Environmental Defenders," *Journal of Human Rights Practice* 3, no. 3 (2011): 309–326.

120. Guy Edwards, Paul K. Gellert, Omar Faruque, Kathryn Hochstetler, Pamela D. McElwee, Prakash Kaswhan, Ruth E. McKie, Carlos Milani, Timmons Roberts, and Jonathan Walz, "Climate Obstruction in the Global South: Future Research Trajectories," *PLOS Climate* 2, no. 7 (2023): e0000241. It is important to note that many contributors of this chapter were also involved in developing those research priorities analyzed in this source.

121. Ryan Stock and Trevor Birkenholtz, "The Sun and the Scythe: Energy Dispossessions and the Agrarian Question of Labor in Solar Parks," *Journal of Peasant Studies* 48, no. 5 (2021): 984–1007; Ryan Stock, "Triggering Resistance: Contesting the Injustices of Solar Park Development in India," *Energy Research & Social Science* 86 (2022): 102464; Priti Gupta, "India's Solar-Powered Future Clashes with Local Life," BBC, October 13, 2022, https://www.bbc.com/news/business-62848096.

CHAPTER 9

Blocking Climate Action at Subnational Levels

LEAD AUTHORS: REBECCA BROMLEY-TRUJILLO, JOSHUA A. BASSECHES, AND MARCELA LÓPEZ-VALLEJO

CONTRIBUTING AUTHORS: LUCAS G. CHRISTEL, ANDREW B. KIRKPATRICK, SIMONE LUCATELLO, AND MARIA ISABEL SANTOS LIMA

INTRODUCTION: THE ROLE OF SUBNATIONAL GOVERNMENTS IN CLIMATE POLICYMAKING

Since the Paris Climate Agreement was adopted in 2015 it has remained clear that subnational governments have an important role to play in climate policy. State/provincial and local governments are key partners in national and international policy to mitigate and adapt to climate change, particularly given their role in the energy and transportation sectors. For example, in the United States, the states' energy sector roles include permitting of fossil fuel production sites as well as regulation of electricity through public utility and public service commissions.[1] Transportation sector roles include setting vehicle emissions standards (such as California's, which are stricter than those of the federal government), jurisdiction over public transit operations and, in many cases, electric vehicle (EV) charging infrastructure.[2,3,4] In addition, cities are the predominant contributors of greenhouse gas emissions and often have jurisdiction over their urban infrastructure, transportation, and land use, key avenues for greenhouse gas (GHG) emissions reductions.[5] Scholars have also suggested that vertical and horizontal integration of climate change policy across and between levels of government is needed to adequately mitigate climate change.[6] This is why state and local governments are important venues for climate policy decision-making. Consequently, climate policy obstruction

Rebecca Bromley-Trujillo et al., *Blocking Climate Action at Subnational Levels*. In: *Climate Obstruction*. Edited by: J. Timmons Roberts, Carlos R. S. Milani, Jennifer Jacquet, and Christian Downie, Oxford University Press. © Oxford University Press (2025). DOI: 10.1093/oso/9780197787144.003.0009 at the subnational level can hamstring national and international efforts, preventing signatories of the Paris Agreement from meeting their mandated targets.

Conversely, in many nations, subnational governments have filled a substantial policy void when national governments have failed to act. For example, several Mexican states developed climate action plans several years before national climate legislation appeared. Canadian provinces also pursued climate policy instruments ahead of their national government, such as the Western Climate Initiative (WCI), an emissions cap-and-trade program between California and Quebec, and Ontario's feed-in tariff program to promote greater use of renewable energy systems.[7] In addition, between the late 1990s and the early 2020s, US climate change policy was made almost exclusively at the subnational level.[8,9] The US federal government only recently made significant policy investments toward a renewable energy transition, and the participation of state and local governments remains necessary to maximize benefits through effective implementation. Given these considerations, understanding subnational climate policy obstruction becomes fundamental to achieving progress in the case of limited-to-nonexistent national action.

The governing context within countries can influence the ways in which climate policy obstruction manifests. For example, whether a country is federal or unitary, or some combination of the two, can shape subnational obstruction. In federal systems, power is shared among national and subnational governments, such as states and cities, while in unitary systems power is highly centralized in a national government and decisions are dictated from the top. Federal arrangements sometimes grant sovereignty or different degrees of autonomy to lower levels of government. As such, subnational governments that have considerable autonomy can in some cases bolster climate policy while undermining it in others.[10] Subnational governments may collaborate with the national government on climate policy design and implementation, innovate as local climate laboratories, or contest policy vacuums at national or international levels.[11,12] They can also obstruct climate policy through lax implementation of federal law or by blocking proactive subnational climate policy proposals.

In subnational governments with less autonomy, climate obstruction operates through complex intergovernmental relations. For example, in cases like Venezuela's and Mexico's where state-owned companies control energy policy, subnational governments that have no ownership over natural resources and energy find it difficult to develop mitigation-oriented climate policies, except via very limited instruments such as energy efficiency. Climate policy obstruction and delay can also be present when subnational legislation is superseded by national or state law, when there is legal ambiguity, or in a policy vacuum. Subnational governments in unitary systems are less equipped than national governments to develop their own climate policies.

Decisions are centralized and the subnational sphere merely implements them as directed. Subnational obstruction can take the form of delay or inaction in the implementation of climate policies. However, there have been cases when subnational governments have tried to develop their own climate policies against national interests. Confrontation or competition may arise in this context. Such is the case in China, where competition between the national carbon market and provincial markets creates uncertainty and a potential for confrontation.[13,14]

This chapter covers subnational climate obstruction in three global regions: North America, Latin America, and Europe. We conclude by identifying the most common forms of obstruction within these regions, avenues for resisting obstruction, and calls for further research, particularly in the Global South, where fewer studies have been conducted.

SUBNATIONAL CLIMATE OBSTRUCTION IN NORTH AMERICA

In this section, we focus on the United States and Canada for two reasons. First, in addition to sharing the largest international land border in the world, the two countries have political economies more comparable with each other's than with Mexico's, with both being significantly wealthier and sharing English as an official language.[15] Second, they share a history of "green bilateralism," cooperating on numerous environmental policies.[16] As a recent example, the northeastern US states and eastern Canadian provinces recently created a "green grid" planning task force.[17]

The United States and Canada are characterized by strong federalist systems in which subnational governments (states and provinces) have a major role to play in policymaking. In the United States, climate change has been on public and governmental agendas since the 1990s; however, the US federal government has been largely unable to pass comprehensive climate mitigation policy until very recently. As such, for more than two decades, US climate change policymaking has been relegated largely to state and local governments.[18,19,20] While state governments have passed climate legislation, there remain a considerable number with limited-to-no climate policies, and even fewer with robust and effective ones.[21]

Nationally, Canada has the advantage of being far less reliant on fossil fuels for electricity production relative to the United States. Its abundant hydroelectricity resources offer a distinct advantage in this respect. Nevertheless, similar to the United States, its provinces are characterized by very different energy economies and political interests, a situation resulting in climate policy obstruction across many of its subnational units.[22] The progression of subnational climate policy in both countries has been severely limited and delayed by factors including public attitudes about climate change, party control in state and provincial governments, and the influence of fossil fuel actors, electric utilities, and other organized interest groups.

Public Opinion on Climate Change

As discussed in detail in Chapter 7, a conservative countermovement in the United States including think tanks, politicians, media organizations, and fossil fuel interests has profoundly shaped public opinion on climate change through a campaign of misinformation.[23,24] Since the late 1990s, party cues from political elites signaling opposition to climate policies have also driven a wedge between Republicans and Democrats on climate change.[25,26] As such, the US public is, on average, less concerned about climate change than individuals in other countries[27] with Republicans being less supportive of climate policy compared with Democrats.[28]

Partisan differences are also apparent when comparing attitudes within and across states and localities (using county-level estimates based on national and state-level data). As of 2021, 57% of Americans believed global warming was caused primarily by humans, but the estimated county-level variation on this question was considerable, ranging from 77% to 44%. When it comes to support for public policy to address climate change, significant variation also exists.[29] For example, on the question of whether the United States should regulate GHGs, an estimated 79% of those living in Alameda County, California support this policy compared with 53% in Loving County, Texas. Importantly, this variation in public opinion has been shown to affect policy adoption within state legislatures; states where public concern is low are significantly less likely to adopt climate mitigation policies.[30]

Research has found that in the United States, climate denial is linked to trust in political leaders who espouse disbelief in anthropogenic climate change, but in Canada, it is related more to political ideology.[31] This study mentions that 21% of Americans and 12% of Canadians expressed climate denial. When looking at Canadian attitudes on climate change, we generally see higher rates of belief in the existence of human-caused climate change and greater support for climate policies than in the United States.[32] Despite this trend, major divides in public opinion exist at the subnational level in Canada similar to what we see in the United States More specifically, other research reports that 87% of Nova Scotians believe climate change exists compared with 66% of residents of Saskatchewan. These findings largely reflect the heavy presence of fossil fuel interests in Saskatchewan. In addition, the authors found considerable variation in support for climate change policies, such as a carbon tax. For example, 70% of residents of Outremont, Quebec, support a carbon tax compared with only 35% of those in Fort-McMurray, a municipality in Alberta where carbon majors are more prevalent.[33]

These differences in attitudes hold important implications for the likelihood of climate change policy adoption within Canadian provinces. For example, Alberta has been reluctant to pursue a renewable-energy transition due in part to lack of public support. Many Alberta residents are skeptical about anthropogenic climate change and often oppose the siting of renewable energy facilities.[34] Similarly, wind energy plans have frequently been shut down due to public outcry in Ontario.[35]

Party Leadership in Subnational Governments

In the United States and Canada, a clear relationship exists between state/provincial party leadership and the adoption or obstruction of climate change policy. While US Republican governors have occasionally passed climate change legislation (e.g., former Gov. Arnold Schwarzenegger in California), the parties have diverged over time, resulting in high elite polarization today.[36,37] As such, when Republicans control a state's legislative body or when they hold the governor's office, they tend to block climate legislation.[38,39,40] For example, in 2024, Republican Governor Youngkin of Virginia vetoed legislation that would have established a "green" bank in the state to facilitate the use of federal grants for renewable energy projects. This situation parallels the polarization seen in Canadian provincial governments, where Conservative leaders have adopted very limited climate policies or blocked more substantial efforts.[41] For example, in early 2024, Alberta Premier Danielle Smith approved a provincial ban on renewable energy projects that would be sited on high-quality agricultural land.[42]

In addition to blocking climate legislation, conservative party leaders have also engaged in policy retrenchment, whereby previous climate mitigation policies are reversed or weakened when party control shifts after an election.[43,44] For example, in the United States, research has documented significant policy reversals in Ohio driven by Republican leadership in the state legislature.[45] In 2019, the state froze its Renewable Portfolio Standard (RPS) for two years and reduced its renewable energy target from 12.5% to 8.5% in addition to providing subsidies for fossil fuel production. In another example, the states of New Jersey and Virginia left a US regional carbon cap-and-trade program, the Regional Greenhouse Gas Initiative (RGGI), soon after electing Republican governors in 2011 and 2023, respectively. After a substantial hiatus, New Jersey returned to RGGI in 2020 under a Democratic governor, Phil Murphy.[46]

We see similar patterns of policy retrenchment in Canada. Though Ontario is usually governed by the Liberal Party, Conservative leader Douglas Ford was elected premier in 2018; as a result, new climate policies, such as the entrance of Ontario into the California-Quebec emissions trading system, were canceled. In contrast with the US subnational level, in Canada parties less often determine climate policy prospects; economic interests can often be far more influential. For example, during Ford's time in office, the phasing out of coal plants and deployment of nuclear energy (projects originating from Liberal governments) continued as pillars of the province's energy transition. It is significant that those provinces with more advanced climate policies (e.g., Quebec, usually run by the Quebecois Party or the Liberals) are also less dependent on fossil fuels for electrification. Nonetheless, even Prince Edward Island (run on wind power), joined a political movement against a federal carbon tax and renewable-fuel regulations when Conservative Premier Dennis King won the majority in Parliament in 2019.[47]

In addition to direct obstruction through the legislative process, Republican or Conservative Party leadership has actively spread misinformation to delay climate policy action under the guise that climate change is not or may not be human-caused. Party leaders have often obscured the benefits of renewable energy to advance fossil fuel interests. For example, in the wake of a severe energy crisis in Texas during the winter of 2021, state Republicans blamed power outages on the supposed poor performance of renewable energy, such as wind and solar, despite clear evidence that fossil fuel systems had suffered significant failures. As such, instead of grappling with concrete ways to improve the Texas electrical grid after the crisis, the state legislature introduced a series of bills that would hamstring the state's renewable energy sector.[48] Similarly, in 2021, the Alberta government initiated a "Public Inquiry into Anti-Albertan Energy Campaigns."[49] These inquiries targeted pro-climate movements, classifying them as being against Alberta's interests and parroting "nationalist" anti-climate propaganda worldwide.

The Fossil Fuel Industry

The fossil fuel industry (coal, oil, and gas and their associated supply chains) is one of the most entrenched in the US and Canadian political systems.[50,51,52] Although the political influence of coal is on the decline relative to oil and gas (in large part because it is no longer able to compete economically with alternatives), the industry remains strong in certain US states such as Kentucky and West Virginia.[53] The oil and gas industries remain formidable in many states. Indeed, in terms of its financial resources, the American Petroleum Institute—a trade association representing oil and gas interests—is "by far the largest" of those that are active on issues related to climate change.[54] Oil and gas trade associations spend a disproportionate amount of their total revenues on politics.[55]

These industries have tended to have an advantage over green interests in state-level policymaking, as shown in a recent qualitative examination of state-level policy retrenchment. For example, a Texas clean energy law was never properly implemented because "after enactment, the Texas Industrial Energy Consumers (TIEC), an alliance of fossil fuel corporations and other industrial companies that rely on cheap energy, intervened aggressively to block this policy at the Public Utility Commission of Texas."[56]

Fossil fuel interests also obstruct climate policy in subnational Canada in similar ways. Alberta and Saskatchewan have entrenched, export-oriented fossil fuel industries that are both critical to the provincial economy and obstruct climate policy at national and subnational levels. For example, influenced by fossil fuel actors, the Alberta government claimed it was not responsible for meeting its own emissions targets.[57] Fossil fuel interests are so important for these so-called carbon provinces that even with the New Democratic Party in office (2015–2019) a "green agenda" and many subnational climate policies in place, exports of refined and crude oil were moved forward by the approval of the Trans Mountain Pipeline from Alberta to British Columbia.[58] Indeed, there is even evidence that these industries have influenced the public education system curriculum in Saskatchewan.[59]

In the United States, there is ample evidence of subnational climate policy obstruction on the part of fossil fuel interests. Even in California—which has long been considered a leader in state-level climate policy[60]—fossil fuel interests have lobbied successfully to maximize their flexibility within the state's climate policy regime, influencing the decision for cap-and-trade to become the centerpiece of the state's implementation of AB 32, its signature 2006 economy-wide GHG emissions-reduction law, and gaining generous treatment for themselves when it came to allowance allocation and compliance flexibility.[61,62] As a result, the policy's overall effectiveness has been called into question.[63] Meanwhile, an analysis of more than 200,000 lobbying and testimony records on bills in seventeen US states found oil and gas industry groups to be among the most likely to take positions opposite to those of environmental groups, and they tended to have success in doing so, especially in more politically conservative US states.[64,65]

In addition, studies have compared the influence of fossil fuel actors across multiple US states to identify institutional factors (i.e., factors apart from these actors' enormous financial resources) that can affect their influence upon policy. One study compared Texas's and Colorado's policy regimes around fracking (hydraulic fracturing, a type of fossil fuel extraction) to determine which state offered more environmental protections and why.[66] It found that Texas is more industry-friendly (and therefore less climate-friendly) than Colorado due to the regulatory capture of the former's Railroad Commission (which regulates the oil and gas industries) and Texas's greater economic dependence on the industry for revenues to support schools and other public programs. Similarly, another study, comparing Colorado and Louisiana, found that Louisiana's lax regulation of the gas industry is due to the privileged, central position of that industry in informal stakeholder processes that then lead to formal policies.[67]

The advent of fracking in the mid-2000s, which employed new techniques for relatively inexpensive horizontal drilling for oil and gas resources, resulted in a boom in US oil and gas production that greatly enriched the fossil fuel industry. In Appalachian states such as Pennsylvania, these new technologies were applied to develop the Marcellus Shale reserves, which the industry had recently discovered. In an interview, a former Pennsylvania state energy regulator explained the political significance of this coupling:

> One thing to understand about gas in Pennsylvania, it was a very small industry pre-shale [before the discovery of Marcellus Shale in particular]. It existed, but it was a very small industry. And now it's the second biggest in the United States and one of the biggest in the world. So the gas producers, the Shale Gas Association, have immense power in the Republican caucuses of the House and the Senate. Immense, immense influence. So the opposition to renewable policy comes principally from the Gas Association in Pennsylvania [whereas historically it had come principally from the coal industry].[68]

Looking at all fifty states, additional research determined that the fracking boom had a statistically significant effect on the weakening of state-level climate policies.[69] Specifically, the states with more fracking potential were more likely to see a weakening of existing policies. In Canada, most fossil fuel companies are foreign-owned, or majority foreign-owned. The influence of international oil companies and global supply chains was found to be a fundamental source of obstruction of climate policies in Canada's "carbon provinces."[70]

Utilities

Scholars have traditionally treated utilities as synonymous with the fossil fuel industry. They are politically powerful, typically monopolies, and, in the United States, have a particularly strong interest in state-level policies because they are regulated primarily at the state level, and have been since the early 1900s when electricity first became commercialized.[71,72,73] Several characteristics of the utility industry, however, differentiate it from the fossil fuel industry.

First, utilities hold monopolies over designated service territories, so they do not compete with one another. Second, the profits of investor-owned utilities (IOUs), which serve three out of four US electricity customers, are not determined by how much electricity they sell but rather by how much infrastructure they build. The amount of infrastructure is determined by what state-level regulators—public utility commissions (PUCs)—allow them to build and the rate of return they allow them to collect for their shareholders based on those capital projects. Third, despite being regulated by individual states, IOUs are typically subsidiaries of parent companies operating in multiple states that own both regulated and nonregulated subsidiaries.[74] This "multilayer subsidiary" form allows the parent company to wield outsized political influence.[75]

Research has found that even if utility subsidiaries are occasionally supportive of renewable energy and/or climate policy,[76] as they can be under certain conditions in certain states, the multilayer subsidiary form has often led to greater overall emissions by their parent companies.[77] Some studies have found unequivocally that IOUs are associated with climate obstruction/denial/delay. For example, in Arizona, where IOUs are vertically integrated, they undermined the state's net-metering policy and renewable-energy targets, and in Ohio, they were active in retrenching the RPS.[78,79] In South Carolina, where IOUs are also vertically integrated, a solar industry lobbyist stated in an interview that Duke and Dominion, the two major IOUs in that state, "were about getting rid of rooftop solar."[80]

On the other hand, other studies have found that IOUs have more nuanced climate policy preferences, occasionally lending their political muscle in support of climate and renewable energy policies,[81] particularly when they are viewed as opportunities to increase returns for their shareholders.[82] Research has found that in California and Massachusetts IOUs were "neutral-to-positive" about economy-wide GHG-reduction legislation, with a former Massachusetts utility commissioner stating, "the utilities . . . didn't care because they didn't own power plants here anymore."[83]

Several US states, including but not limited to California and Massachusetts, adopted electric utility-sector restructuring policies in the 1990s that took IOU monopolies out of the electricity-generation business, opening that business up to competitive generation companies in hopes of reducing costs for consumers. Although the degree to which such restructuring policies have succeeded in reducing costs for ratepayers is unclear,[84] it is likely that these policies altered the IOUs' incentives so as to make them less likely to obstruct certain types of climate policies.[85,86]

And so, whereas oil and gas companies are unequivocally agents of state-level climate policy obstruction, the situation with electric utilities is more nuanced and will require further, state-by-state empirical research. It would appear that variations in the structure of the electric utility sector, as well as the highly unusual utility business model, are major determinants of IOUs' climate policy preferences.[87] Interestingly, and consistent with the logic that greater competition (largely enabled or restricted by state governments) leads to greater participation of renewable-energy generators in the form of independent power producers (IPPs), one study found that states that restructured their utility sectors in the 1990s were more likely to adopt RPS and cap-and-trade programs.[88]

Canadian utilities operate slightly differently from their US counterparts. There are three forms of electric utilities in Canada. The most common are crown corporations (CC) owned by provincial governments, which oversee generation, transmission, system operation, distribution, and retail. CCs are found in British Columbia, Saskatchewan, Manitoba, Quebec, New Brunswick, and Nunavut. The second form consists of private companies running the electricity sector, as in the case of Nova Scotia and Prince Edward Island. The third form is found in Alberta and Ontario, where there are hybrid formats of open wholesale markets and retail competition.[89] (For more information on obstruction in the utilities sector, see Chapter 3.)

In this diverse context, climate obstruction occurs in at least two ways in provincial Canada. The first is when provinces generate electricity with fossil fuels for domestic consumption, selling to Canadian neighbors or for export to the United States. In 2023, the federal government drafted a policy to achieve a national net-zero electricity grid by 2035. Alberta immediately refused to implement it, joining Saskatchewan, New Brunswick, Nova Scotia, and Manitoba.[90] The second form of obstruction ironically occurs when local communities oppose renewable projects due to their alleged environmental impacts, land use concerns, or dissatisfaction with consultation, especially for Indigenous peoples. This is the case with local opposition to wind farm projects in Ontario[91] and big hydroelectricity in British Columbia.[92]

Other Organized Interests and Think Tanks

Real estate developers, local not-in-my-backyard (NIMBY) organizations, think tanks, and environmental conservation groups (occasionally) play an important role in blocking subnational renewable energy projects and other climate policy measures in the United States and Canada.[93,94,95,96] The real estate industry particularly perceives itself to be threatened by state-level policies promoting electrification in the construction of new homes and buildings. Interest groups associated with this industry have sought state-level preemptions of municipal gas hook-up bans and have spearheaded litigation to roll back such policies.[97] A study of interest group pro-climate and anti-climate coalitions in Massachusetts, for example, found real estate groups to be a lynchpin of the typical anti-climate coalition.[98]

Think tanks, in coordination with interest groups, have blocked wind projects initially approved by state and local governments, often through campaigns of misinformation about environmental harms from wind development. For example, the Caesar Rodney Institute has provided financial resources to Protect Our Coast, an interest group advocating against the siting of an offshore wind facility in Ocean City, New Jersey.[99] Protect Our Coast has spread false claims that these wind farms pose a risk to whales. As of this writing, the group plans to file legal challenges that will delay the project and make it so costly as to prevent its construction.[100] Climate denial think tanks also operate in Canada, such as the Vancouver-based Fraser Institute, which hosts important conservative politicians from Alberta and other provinces and advocates against climate change mitigation policy.[101] The institute is funded through donations from private companies and individuals, including one of its largest donors, ExxonMobil.

Other occasional sources of obstruction of renewable energy projects include local organizations and environmental groups that raise environmental concerns with siting. For example, local community and environmental groups united to prevent the Crescent Peak Wind Energy project in Nevada in 2018. These groups argued that the wind farm would be harmful to local bird and bat populations.

SUBNATIONAL CLIMATE POLICY OBSTRUCTION IN LATIN AMERICA

In Latin America, Brazil and Mexico are the top greenhouse gas emitters, accounting for about 60% of the region's emissions, followed by Argentina, Venezuela, Chile, and Colombia, with 25%–30% combined.[102] Despite similar presidential systems, Latin-American countries differ in their subnational government structures, divided into states, departments, municipalities, cities, provinces, and communities, among other subunits. Only Argentina, Brazil, Mexico, and Venezuela are federations. Although subnational political units in unitary countries normally have much less autonomy than in federal ones, some unitary nations in Latin America have granted them *more* autonomy than in federal ones.[103,104]

Three main features define subnational climate obstruction in Latin America. First, is the way in which Latin American territories have been integrated into the global economy. Given these countries' colonial legacies, economic dynamics took the form of enclaves. These economic zones with special productive dynamics have fostered the development of local economic elites whose sectoral interests and environmental preferences have not necessarily coincided with those of national actors. Second, national trends such as weak democratic institutions, lack of accountability and transparency mechanisms, and inefficient judicial systems prone to corruption and mismanagement are magnified at the subnational level.[105,106,107] If gray areas in politics are the rule rather than the exception in Latin America, they are magnified at the subnational level, where the rule of law is often diluted or even nonexistent.[108,109] Third, control over environmental resources also complicates the role of subnational governments in advancing or obstructing climate issues.

On the one hand, when natural resources are centrally managed by national governments, subnational states have limited climate policymaking capacity. On the other hand, when subnational governments have greater powers over natural resources—as in the case of Argentina—there are stronger incentives for climate obstructionism due to their heavy reliance on revenues from extractive industries, despite distributive conflicts that often arise with Indigenous and local communities. These sociohistorical and structural factors have been exploited by political parties and extractive industries to intentionally deny climate change and/or block or delay climate policies.

It is also possible that public opinion may play a role in obstructing subnational climate policy in Latin America; however, a lack of public opinion data outside the national level precludes a definitive answer. Still, the repression of climate activism through violence in Latin America suggests that subnational actors are comfortable with generating fear when it serves their interests, even if it generates public outrage.[110,111]

Party Leadership in Subnational Governments

Several Latin American countries were governed by right-wing parties and/or dictatorships in the twentieth century. In the 1990s, such countries began their transition to democracy, a process involving trial and error with party governance. Evidence shows that right-wing politics in Latin America hinders climate policy, but unlike the case of partisan polarization in the United States or the European Union (EU), even center or left-wing Latin American subnational governments have historically obstructed climate policies. Structural factors related to the international political economy and the pursuit of economic development often lead governments to adopt a discourse that advocates sustainability while sacrificing the environment for the sake of economic growth.[112] This is particularly relevant in the context of subnational governments, where economic dependence on natural-resource extraction and a lack of productive alternatives have meant that extractive activities are viewed as the only viable driver of development. In Brazil, for example, local political economies are highly dependent on extraction royalties (as in Rio de Janeiro and Pará) or agribusiness's economic benefits (as in Mato Grosso and Rio Grande do Sul).

When national parties delay or belittle the importance and urgency of climate action, subnational governments have followed suit, reproducing this obstruction locally. This was the case in Mexico from 2018 to 2023 under Morena, a left-wing populist party, when national energy policies favored fossil fuel extraction, oil refining, and the use of gas. Subnational governments from the same political party, such as in Ciudad del Mexico during Claudia

Sheinbaum's term, included a misleading, pro-climate discourse in their development plans, promising to adopt a cleaner energy mix despite knowing that climate policies would be inoperable within national energy-policy frameworks.

Subnational governments in Bolsonaro's Brazil provide examples of a more subtle form of climate obstruction. Despite the existence of an official climate denial strategy at the national level, some pro-Bolsonaro Amazonian states announced the development of climate action plans, especially those seeking to receive international cooperation funds, such as the UN Reducing Emissions from Deforestation and Forest Degradation in Developing Countries (REDD+), for protection of the Amazon. Despite their calls to action, subnational governments did not follow through with their public statements; many Brazilian cities and states lacked effective local laws, adaptation plans, or emissions inventories, resulting in considerable delay in policy implementation.[113]

Fossil Fuel and Other Extractive Industries

Climate obstruction driven by oil and gas companies and their networks of think tanks and business associations also weakens subnational climate policies in Latin America. The oil-and-gas sector in Latin America is diverse, spanning private, state-owned domestic, and foreign companies, some of which are associated with the region's colonial background. In this arena, national climate goals are typically overshadowed by national strategic development goals.

In some Latin American countries, subnational governments have little influence over fossil fuel industry activities, as nationally owned companies such as PDVSA (Venezuela) and PEMEX (Mexico) drive energy policy. In contrast, Argentina allows its provinces input into energy policies. Though the country has a nationally owned company (YPF), the provinces of Neuquén (53%), Chubut (25%), Santa Cruz (12%), and Mendoza (10%) are its main oil producers.[114] There, subnational dependence on oil royalties, the industry's positive impact on the labor market in these areas, and the strength of oil unions mostly favor obstruction at the subnational level through financial support to electoral campaigns and lobbying in the executive and judicial branches. For example, in Neuquén, Argentina, oil and gas royalties represent 40% of provincial incomes (as of 2022),[115] and the oil and gas sector covers 17% of the total labor market.[116] Guillermo Pereyra, leader of a powerful oil and gas union between 1984 and 2021, developed a prominent political career during those years, serving as provincial deputy, labor subsecretary of Neuquen, and eventually national senator. In each of these positions, Pereyra was an advocate for the oil and gas industry and blocked what he considered to be hostile climate initiatives that may harm workers.[117]

Oil production in Brazil comes mostly from deep waters near the states of São Paulo and Rio de Janeiro. Apart from nationally owned PETROBRAS, international companies including Shell, BP, Statoil, Exxon, and Total are also major oil producers. São Paulo and Rio de Janeiro receive royalties from the oil sector, producing regional inequality and preventing localities from developing strong climate policies.[118] This situation also creates an interesting dynamic for cities like Maricá, which earned the highest oil royalties in 2023. Although Maricá has been governed by PT, the party of President Lula da Silva, for more than fifteen years and has adopted some climate policies, its dependence on the fossil fuel industry represents a roadblock for more effective and ambitious policies aiming to diversify energy sources and create a less oil-centric economic development model.

Climate obstruction also operates in more subtle ways in subnational Latin America. Some research has concluded that through governmental transfers, mining revenue substitutes for local taxes on mineral concessions (annual payments for the land used for mineral extraction and exploration).[119] In other words, hosting mining within their borders makes subnational governments perceive the benefits of the revenue it generates to outweigh the environmental externalities it causes. Instead of collecting a tax, subnational governments receive a portion of the mining income. Similarly, Bolivia and Peru allocate national transfers to subnational governments to compensate them for mining activities.[120] In the Amazon region, transnational companies are permitted to exploit natural resources through, in some cases, questionable licenses to explore Indigenous lands,[121] in exchange for providing jobs and investments in the area as well as income through royalty payments, thereby guaranteeing themselves a level of influence in local politics. This influence allows them to obstruct any effort to adopt climate actions because they are perceived as necessary to the local economy.[122] In Argentina, such a scenario is twofold: while some provinces have passed laws that limit metal mining, others have made mining policy the backbone of their economic policy. The cases of Catamarca and San Juan, illustrate how coordination between state and corporate interests can obstruct environmental initiatives. State and corporate interests in these provinces collaborate against policies that may restrict mining or that would allow direct democracy mechanisms to be used to make decisions about natural resources.[123]

Subnational governments in Latin America exhibit other forms of climate policy gridlock stemming from hybrid forms of governance and local political economies that are highly dependent on mining royalties. Informal mining activities, which operate in a legal gray area, are common, and contribute to large deforestation and GHG emissions rates. "Artisanal" mining (extracting no more than 25 metric tons of minerals per day) is legal in Peru for economic survival. In the Peruvian departments of Madre de Dios, Sur Medio, and Puno, this practice is common, with communities working in mines already exploited by big companies. If artisanal mining groups do not start a formalization process, however, they could be considered illegal.[124] Under this hybrid-policy scenario, developing climate strategies can be difficult to implement and enforce.

Other Organized Interests

Subnational governments in Latin America sometimes support other business actors that obstruct climate policies, such as the forestry industry. In the Brazilian context, the presence of agribusiness is crucial for understanding how climate obstruction operates in the country, as a significant portion of GHG emissions there stem from land use. A notable segment of the political right is closely aligned with this industry, complicating the adoption of bold climate action. At the local level, these actors wield significant influence, particularly in the southern and central-western region of the country that hosts a substantial portion of grain and beef production, thus influencing climate legislation at the subnational level.

Deforestation in the region occurs mostly in the Amazon and the Petén forests, where soy, palm oil, cotton, corn, and/or lumber are profitable commodities.[125] Their production directly influences subnational political dynamics (funding parties, electing friendly politicians, having family members in the judicial branch) where private actors often overpower the national and subnational governments. When stronger environmental laws are established in government-managed land in the Amazon or Petén, agricultural production tends to migrate to other areas, which are then privatized. Private companies then exert substantial influence over subnational governments. For example, the Brazilian region of Cerrado in ten states of the country's center-west, once an important CO_2 sink, is now the powerhouse of the soy industry in Latin America.[126] The result has been surging emissions.

EUROPEAN CLIMATE POLICY OBSTRUCTION

European climate policy obstruction at the subnational level is generally more limited than it is in the United States for two reasons: first, there is a broader consensus among political leaders and the public about the scale of the climate change problem and the range of measures needed to combat it. Europeans overwhelmingly consider climate change a serious problem, ranking it as the third most pressing global issue. According to surveys conducted by the European Union and Eurobarometer, a vast majority (93%) view climate change as a serious issue, with 77% considering it very serious. In general, public opinion strongly supports climate action, with 88% of respondents, including at least 70% in each of the twenty-seven Member States, endorsing minimizing GHG emissions and achieving a carbon-neutral EU economy by 2050.[127] Second, European climate change policy is heavily influenced by the mitigation efforts of the EU, which include ambitious targets and regulatory measures that encompass various sectors including energy, transportation, and agriculture.[128,129]

Despite the top-down approach to climate policy in the region, European climate governance is polycentric, with subnational governments in some countries playing a substantial role in policy adoption and implementation.[130] As such, obstruction can occur when regional, provincial, or local governments' policies or ideologies do not align with the overarching goals set by the European Union.[131] Despite aggressive supranational targets, such as becoming carbon-neutral by 2050, and the introduction of key initiatives such as the European Green Deal,[132] climate policies can become polarized at different levels of government, with political parties obstructing initiatives to, for example, differentiate themselves or to appease a base that may be skeptical of climate change.[133,134]

Misinformation campaigns and public skepticism about climate science have also led to resistance against necessary climate actions at the subnational level in Europe, although so far this has been to a lesser extent than in North America.[135] For example, in Scotland, misinformation campaigns on the benefits of fracking were targeted at both the regional parliament and the public, though this did not result in a change to the ban on fracking.[136] Furthermore, a lack of coordination between government levels can lead to ineffective or delayed policy actions by party actors.[137] Finally, fossil fuel interests play an important role in some European regions, where substantial parts of the economy are driven by fossil fuel production or processing, resulting in the delay of EU policy implementation and a lack of independent action by subnational governments.[138,139]

Party Leadership in Subnational Governments

Whether subnational units obstruct climate policy depends partly on political party leadership and the institutional context. When the same party controls the subnational and national government, we see less obstruction. Similarly, there is less obstruction in unitary states. For example, Belgium has a dual system of federalism whereby both the federal government and subnational units have considerable autonomy. At times, this arrangement has resulted in lax environmental policy implementation as parties shift blame to other levels of government when climate targets are not met.[140] One could think of this as a form of climate obstruction, as it takes advantage of the institutional setting to delay compliance with climate goals.

In another example, subnational regions in the United Kingdom have differed dramatically in how they implemented GHG-reduction standards imposed by the UK Parliament, as well as in their adoption of independent climate change measures, based partly on party politics. In Northern Ireland, the Democratic Unionist Party, a conservative party primarily representing Protestants, blocked a climate change act in the Northern Ireland Assembly for a decade, in part because it held the Agriculture, Environment, and Rural Affairs Ministry for much of that time in Northern Ireland's complicated power-sharing arrangement.[141] Its Climate Change Act was eventually passed in 2022, containing concrete emissions targets for 2030, 2040, and 2050.[142] In contrast, Scotland had passed a similar policy in 2009 and Wales in 2016.[143]

Party leadership can also blunt public and economic resistance to climate action, preventing obstruction despite pushback from some members of the public and business interests. For example, from 2021 to 2023, in North Rhine-Westphalia, Germany, one of the country's most industrialized regions, there were major protests in favor of transitioning from traditional coal and steel industries to less polluting alternatives. This shift is affecting local economies and employment in cities like Essen and Dortmund, which have historically depended on these sectors. The state government, composed of a Christian Democratic Union–Green coalition broadly supportive of Germany's overall climate change plan, chose to promote investments in renewable energy and infrastructure modernization, aiming to position the region as a leader in green technology. The Alternative for Germany Party, which has opposed Germany's and the European Union's climate policies, campaigned on a heavily pro-coal platform, but won only twelve seats in the regional legislature (out of 195), leaving them unable to engage in any meaningful obstruction.[144]

Italy's Veneto and Trentino-Alto Adige regions also face economic diversification challenges from climate action, as they have strong presences in the manufacturing and agriculture sectors. In 2023, local movements and resistance, especially in agricultural subsectors such as wine production, led to increasing pressure on the regional governments to reject some EU directives on climate change.[145] However, both regions were under the control of the Lega party which, while radical-right in orientation, has as of this writing not engaged in climate obstruction, did not respond to the public and sectoral pressure, and continued to comply with the directives.[146]

These examples illustrate the complex interplay of economic, social, and institutional factors as the nations and subnational units of Europe implement EU policies. While some regions are well-positioned to capitalize on the shift toward a fossil fuel–free economy, others face significant challenges that could exacerbate existing disparities and lead to social unrest. Still, this social unrest has yet to translate into obstruction within most subnational governments, due in part to continued support among many party leaders for climate action. The success of Europe's Green Deal will depend largely on addressing economic disparities and ensuring a just transition for all communities.[147]

Fossil Fuel Lobbies in the European Union

Climate obstruction involves a complex network of actors who work actively to impede climate action. These players include influential policymakers in addition to lobbyists, primarily from the fossil fuel sector and automobile industries.[148] The influence of these actors can manifest in the form of weakened environmental regulations, subsidies for fossil fuels, and limited support for renewable energy initiatives. By aligning their interests with those of powerful industries, policymakers contribute to climate obstruction and hinder the transition to a low-carbon economy.[149]

For example, the resistance to climate policies in countries such as Poland and Germany often reflects deeper socioeconomic and cultural concerns affecting local populations. In Poland, coal mining is not just an industry; it's a significant part of the national identity, especially in regions such as Silesia.[150] Coal mines are a major employer, and the industry supports many ancillary businesses. The phase-out of coal therefore threatens to destabilize local economies that depend heavily on mining jobs. Poland's coal-mining sector also enjoys strong political backing, particularly from the Law and Justice party, which garners substantial support in mining regions. Proposals to reduce reliance on coal have been met with resistance from trade unions and local communities that fear job losses and economic decline.[151] However, as Poland is a unitary state, support for coal has surfaced mostly at the national rather than subnational level.

Germany's federal structure means that individual states (such as Länder) have significant autonomy over their energy policies. While federal climate policy is fairly robust in Germany, states such as Bavaria and Baden-Württemberg have been slow to embrace renewable-energy targets, due partly to political leadership that prioritizes fossil fuel–dependent economic interests. Fossil fuel companies and car manufacturers have used Germany's multilevel form of governance to block or slow climate policy, as in the city-state of Hamburg, which produces nearly half of Germany's GHG emissions and opened a new coal-fired power plant in 2015.[152] In regions such as Bavaria, there is significant local opposition to wind turbines, which residents argue would spoil the area's natural and cultural landscape. This opposition is often supported by local politicians who seek to maintain their electoral base by aligning with public and industry sentiment.[153]

CONCLUSION

This chapter demonstrates the pervasiveness of climate obstruction at the subnational level in North America, Latin America, and Europe. Ultimately, climate federalism is a double-edged sword,[154] whereby subnational power provides opportunities for both proponents and opponents of climate policy to advance their goals. While obstruction occurs in all three regions reviewed here, the level and particular forms obstruction takes vary across these global regions. In North America, public opinion—frequently manipulated by political elites supported by fossil fuel actors—is a major factor in subnational climate policy delay. In addition, electric utilities play an outsized role in subnational politics in the United States. In Latin America, the primary source of obstruction comes from extractive firms and agribusiness, who take advantage of the institutional setting to block or delay climate change policy. In Europe, obstruction appears most prevalent at national and supranational levels; however, when subnational obstruction does occur, it commonly takes the form of delayed implementation by political party leadership, often in response to local or regional economic interests. In all three regions discussed here, the distributive politics that creates economic winners and losers across labor and capital can delay climate action.[155,156] Yet in Europe and North America, this type of politics is visible on the surface, seen in protests and overt political actions characteristic of advanced democracies, whereas in Latin America politics tends to be buried under more fundamental governance challenges such as political corruption, hybrid forms of government, and fragile democratic institutions.

The varying forms of obstruction just described lend themselves to different solutions, depending on the context. For instance, in the United States the role of political elites and party polarization largely necessitates a focus on electing and elevating progressive leaders into key subnational government roles. Moreover, climate policy options should be framed to broaden public support, such as discussing the cobenefits of climate policies alongside public health and reducing income inequality.[157,158] When it comes to the US electric utility industry, efforts can be made to restructure the industry to break up vertically integrated utility monopolies.[159] In addition, some have suggested the possible benefits of nationalizing the grid,[160] despite the opposition such a move could ignite. In the Latin American context, it is important to have multistakeholder collaboration that brings economic, community, and environmental interests to the table. Moreover, including clear enforcement mechanisms in climate laws may prevent private interests from delaying their implementation. In Europe, a recent proposal to address the influence of the fossil fuel lobby may help prevent obstruction. The proposal includes a ban on oil and gas lobbying in the name of public health; a similar ban exists for the tobacco industry.[161]

These findings provide important insight into subnational obstruction within a range of global contexts, yet we still have much to learn, especially in the Global South. Analysis of climate obstruction has focused primarily on developed countries, and there is a paucity of published scholarship on subnational levels of government and climate obstruction in developing countries (see Chapter 8). Education and language gaps and the lack of reliable data—or at least, public access to it—have prevented the development of any literature in many regions. In parts of Latin America as well as in Asia, the Middle East, and Africa, data gathering (where it is possible) can be extremely dangerous, and publishing results even more so. Weak federalism, unstable governance, and political violence often preclude the systematic study of this topic in many parts of the world. As such, our overarching findings on global subnational obstruction should be taken as preliminary. Despite the challenges of studying the Global South, climate policy scholars have laid out a clear agenda for such research, which may begin with interviewing key stakeholders in the region.[162] While data challenges abound, qualitative research is a good starting point for identifying the unique characteristics of obstruction in the Global South.

NOTES

1. Jared Heern, "Who's Controlling Our Energy Future? Industry and Environmental Representation on United States Public Utility Commissions." *Energy Research and Social Science*, (2023): 103091.
2. Robert Walton, "New York Makes $100M Available for Schools to Adopt Zero-Emission Buses," Utility Dive, September 29, 2023, https://www.utilitydive.com/news/new-york-makes-100-million-electric-school-buses/695158/.
3. Dan Zukowski, "California Clean Vehicle Rebate Program Expands Statewide." Utility Dive, August 24, 2023, https://www.utilitydive.com/news/california-expands-clean-electric-vehicle-rebate-program-statewide/691894/.
4. Nathan Solis, "17 States Push EPA To Revoke California's Ability to Set Its Own Vehicle Emission Standards." *Los Angeles Times*, May 13, 2022, https://www.latimes.com/california/story/2022-05-13/state-push-epa-on-california-vehicle-emissions-standards.
5. Harriet Bulkeley, Liliana B. Andonova, Michele M. Betsill, Daniel Compagnon, Thomas Hale, Matthew J. Hoffmann, Peter Newell, Matthew Paterson, Charles Roger, and Stacy D. VanDeveer, *Transnational Climate Change Governance* (Cambridge University Press, 2014).
6. Angel Hsu, Amy J. Weinfurter, and Kaiyang Xu, "Aligning Subnational Climate Actions for the New Post-Paris Climate Regime," *Climatic Change* 142, no. 3–4 (April 2017): 419–432.

7. Leah C. Stokes, "The Politics of Renewable Energy Policies: The Case of Feed-In Tariffs in Ontario, Canada," *Energy Policy* no. 56, (2013): 490–500.

8. Barry G. Rabe, "Contested Federalism and American Climate Policy," *Publius: The Journal of Federalism* 41, no. 3 (May 2011): 494–521.

9. Joshua A. Basseches, Rebecca Bromley-Trujillo, Maxwell T. Boykoff, Trevor Culhane, Galen Hall, Noel Healy, David J. Hess, David Hsu, Rachel M. Krause, Harland Prechel, J. Timmons Roberts, and Jennie C. Stephens, "Climate Policy Conflict in the U.S. States: A Critical Review and Way Forward," *Climatic Change* 170, no. 3–4 (2022): 32.

10. Sarah M. Jordaan, Adrienne Davidson, Jamal A. Nazari, and Irene M. Herremans, "The Dynamics of Advancing Climate Policy in Federal Political Systems," *Environmental Policy and Governance* 29, no. 3 (March 2019): 220–234.

11. María Almeida, Huáscar Eguino, Juan Luis Gómez Reino, and Axel Radics, "Decentralized Governance and Climate Change in Latin America and the Caribbean," International Center for Public Policy Working Paper 20-07, November 2022, https://icepp.gsu.edu/files/2022/11/paper2207.pdf.

12. Barry Rabe, "Contested Federalism and American Climate Policy," *Publius: The Journal of Federalism* 41, no. 3 (May 2011): 494–521.

13. Lina Li and Maia Hall, *Climate Governance in China: Policy Diffusion on Emissions Trading in Shanghai and Hubei* (Routledge, 2024).

14. Ye Qi and Tong Wu, "The Politics of Climate Change in China," *WIREs Climate Change*, no. 4 (June 2013): 301–313.

15. GDP (current USD$), World Bank, 2024, https://data.worldbank.org/indicator/NY.GDP.MKTP.CD?most_recent_value_desc=true.

16. Debora VanNijnatten and Mark McWhinney, "Canada-U.S. Green Bilateralism: Targeting Cooperation for Climate Mitigation," North American Colloquium Policy Paper Series, 2022.

17. Daniel Sosland, "International Clean Energy Grid Coordination in the Northeast and Eastern Canada: Starting the Northeast Grid Planning Forum Dialogue," Acadia Center blog, September 19, 2023, https://acadiacenter.org/international-clean-energy-grid-coordination-in-the-northeast-and-eastern-canada-starting-the-northeast-grid-planning-forum-dialogue/.

18. Barry Rabe, "Contested Federalism and American Climate Policy," *Publius: The Journal of Federalism* 41, no. 3 (May 2011): 494–521.

19. Rebecca Bromley-Trujillo, John S. Butler, John Poe, and Whitney Davis, "The Spreading of Innovation: State Adoptions of Energy and Climate Change Policy," *Review of Policy Research* 33, no. 5 (September 2016): 544–565.

20. Leah Stokes, *Short Circuiting Policy: Interest Groups and the Battle Over Clean Energy and Climate Policy in the American States* (Oxford University Press, 2020).

21. Joshua A. Basseches, Rebecca Bromley-Trujillo, Maxwell T. Boykoff, Trevor Culhane, Galen Hall, Noel Healy, David J. Hess, David Hsu, Rachel M. Krause, Harland Prechel, J. Timmons Roberts, and Jennie C. Stephens, "Climate Policy Conflict in the U.S. States: A Critical Review and Way Forward," *Climatic Change* 170, no. 32, (2022), https://doi.org/10.1007/s10584-022-03319-w

22. Douglas Macdonald, *Carbon Province, Hydro Province: The Challenge of Canadian Energy and Climate Federalism* (University of Toronto Press, 2020).

23. Robert Brulle, "Denialism: Organized Opposition to Climate Change AAction in the United States," in *Handbook of U.S. Environmental Policy*, ed. David Konisky (Edward Elgar Publishing, 2020), 328–341.

24. Riley Dunlap and Aaron McCright, "Challenging Climate Change: The Denial Countermovement," in *Climate Change and Society: Sociological Perspectives*, ed. Riley Dunlap and Robert Brulle (Oxford University Press, 2015), 300–332.

25. Robert J. Brulle, Jason Carmichael, and J. Craig Jenkins, "Shifting Public Opinion on Climate Change: An Empirical Assessment of Factors Influencing Concern over Climate Change in the U.S., 2002–2010," *Climatic Change* 114, no. 2 (February 2012): 169–188.

26. Phillip J. Ehret, Aaron C. Sparks, and David K. Sherman, "Support for Environmental Protection: An Integration of Ideological-Consistency and Information-Deficit Models," *Environmental Politics* 26, no. 2 (November 2016): 253–277.

27. Jacob Poushter, Moira Fagan, and Christine Huang, "Americans Are Less Concerned—But More Divided—on Climate Change Than People Elsewhere," Pew Research Center, September 14, 2021, https://www.pewresearch.org/short-reads/2021/09/14/americans-are-less-concerned-but-more-divided-on-climate-change-than-people-elsewhere/.

28. Patrick Egan and Megan Mullin, "Climate Change: US Public Opinion," *Annual Review of Political Science* 20 (2017): 209–227.

29. Jennifer Marlon, Emily Goddard, Peter Howe, Matto Mildenberger, Martial Jefferson, Jennifer Carman, Seth Rosenthal, Eric Fine, Andrew Gillreath-Brown, and Anthony Leiserowitz, "Yale Climate Opinion Maps 2021," Yale Program on Climate Change Communication, December 13, 2023, https://climatecommunication.yale.edu/visualizations-data/ycom-us/.

30. Rebecca Bromley-Trujillo and John Poe, "The Importance of Salience: Public Opinion and State Policy Action on Climate Change," *Journal of Public Policy* 40, no. 2 (October 2018): 280–304.

31. Shelley Boulianne and Stephanie Belland, "Climate Denial in Canada and the United States," *Canadian Review of Sociology* 59, no. 3 (June 2022): 369–394.

32. Erick Lachapelle, Christopher P. Borick, and Barry G. Rabe, "Public Opinion on Climate Change and Support for Various Policy Instruments in Canada and the U.S.: Findings from a Comparative 2013 Poll," *Issues in Energy and Environmental Policy*, no. 11 (June 2014), https://closup.umich.edu/sites/closup/files/uploads/ieep-nsee-2013-fall-canada-us.pdf,

33. Matto Mildenberger, Peter Howe, Erick Lachapelle, Leah Stokes, Jennifer Marlon, and Timothy Gravelle, "The Distribution of Climate Change Public Opinion in Canada," PLOS One 11, no. 8 (August 2016): e0159774, https://doi:10.1371/journal.pone.0159774.

34. Aleksandra Afanasyeva, Debra Davidson, and John Parkins, "Wind Energy Development and Anti-Environmentalism in Alberta, Canada," in *Handbook of Anti-Environmentalism*, ed. David Tindall, Mark Stoddart, and Riley Dunlap (Edward Elgar Publishing, 2022), 329–343.

35. Anahita A. N. Jami and Philip R. Walsh, "The Role of Public Participation in Identifying Stakeholder Synergies in Wind Power Project Development: The Case Study of Ontario, Canada," *Renewable Energy* 68 (August 2014): 194–202.

36. Aaron McCright, Chenyang Xiao, and Riley Dunlap, "Political Polarization on Support for Government Spending on Environmental Protection in the USA, 1974–2012," *Social Science Research* 48 (November 2014): 251–260.

37. Luke Fowler and Jaclyn J. Kettler, "Are Republicans Bad for the Environment?," *State Politics & Policy Quarterly* 21, no. 2 (June 2021): 195–219.

38. Samuel Trachtman, "What Drives Climate Policy Adoption in the U.S. States?," *Energy Policy* 138 (March 2020): 111214, https://doi:10.1016/j.enpol.2019. 111214.

39. Rebecca Bromley-Trujillo, John S. Butler, John Poe, and Whitney Davis, "The Spreading of Innovation: State Adoptions of Energy and Climate Change Policy," *Review of Policy Research* 33, no. 5 (September 7, 2016): 544–565.

40. Rebecca Bromley-Trujillo and John Poe, "The Importance of Salience: Public Opinion and State Policy Action on Climate Change," *Journal of Public Policy* 40, no. 2 (October 30, 2018): 280–304.

41. Shane Gunter, Darren Fleet, and Robert Neubauer, "Challenging Petro-Nationalism: Another Canada Is Possible?," *Journal of Canadian Studies* 55, no. 1 (2021): 57–87.

42. David Ljunggren, "Alberta to Ban Some Renewable Energy Projects, Greens Say Move Is 'Uncertainty Bomb,'" Reuters, February 28, 2024, https://www.reuters. com/world/americas/canadas-alberta-set-ban-renewables-projects-prime-land-report-2024-02-28/.

43. Stokes, *Short Circuiting Policy*.

44. Rebecca Bromley-Trujillo and Mirya R. Holman, "Climate Change Policymaking in the States: A View at 2020," *Publius: The Journal of Federalism* 50, no. 3 (2020): 446–472.

45. Stokes, *Short Circuiting Policy*.

46. Rebecca Bromley-Trujillo, "US States and Climate Policy," in *Elgar Encyclopedia of Climate Policy*, ed. Daniel J. Fiorino, Todd A. Eisenstadt, and Manjyot Kaur Ahluwalia (Edward Elgar Publishing, 2024), 253–256.

47. Cloe Logan, "Why Is This Climate-Friendly Province Backing a Carbon Tax Fight?," *National Observer*, August 30, 2023, https://www.nationalobserver.com/2023/ 08/30/news/why-climate-friendly-province-backing-carbon-tax-fight.

48. Bryan Mena, "Gov. Greg Abbott and Other Republicans Blamed Green Energy for Texas' Power Woes," *Texas Tribune*, February 17, 2021, https://www.texastribune. org/2021/02/17/abbott-republicans-green-energy/.

49. Steve Allan, "Public Inquiry of Anti-Alberta Energy Campaigns," Government of Alberta, 2023, https://www.alberta.ca/public-inquiry-into-anti-alberta-energy-campaigns.

50. Robert J. Brulle, "Institutionalizing Delay: Foundation Funding and the Creation of U.S. Climate Change Counter-Movement Organizations," *Climatic Change* 122, no. 4 (December 2013): 681–694.

51. Truzaar Dordi, Olaf Weber, Ekaterina Rhodes, and Madeleine McPherson, "A Voice for Change? Capital Markets as a Key Leverage Point in Canada's Fossil Fuel Industry," *Energy Research & Social Science* 103 (September 2023): 103189, https://doi:10.1016/j.erss.2023.103189.

52. Stokes, *Short Circuiting Policy*.

53. Lamar Johnson, "Renewable Generation Surpassed Coal for First Time in 2022," *E&E News*, March 29, 2023, https://subscriber.politicopro.com/article/ eenews/2023/03/29/renewable-generation-surpassed-coal-for-first-time-in-2022-00089250.

54. Robert Brulle and Christian Downie, "Following the Money: Trade Associations, Political Activity and Climate Change," *Climatic Change* 175, no. 3–4 (December 2022), https://doi:10.1007/s10584-022-03466-0.

55. Ibid.

56. Stokes, *Short Circuiting Policy*, 126.

57. Macdonald, *Carbon Province, Hydro Province*.

58. Ibid.

59. Emily Eaton and Nick Day, "Petro-Pedagogy: Fossil Fuel Interests and the Obstruction of Climate Justice in Public Education," *Environmental Education Research* 26, no. 4 (2020): 457–473.

60. David Vogel, *California Greenin': How The Golden State Became an Environmental Leader* (Princeton University Press, 2018).

61. Joshua Basseches, "Who Pays for Environmental Policy? Business Power and the Design of State-Level Climate Policies," *Politics and Society* (September 11, 2023), https://doi.org/10.1177/00323292231195184.

62. Joshua Basseches, Kaitlyn Rubinstein, and Sarah Kulaga, "Coalitions that Clash: California's Climate Leadership and the Perpetuation of Environmental Inequality," *The Politics of Inequality* (2021): 23–44.

63. Danny Cullenward and David Victor, *Making Climate Policy Work* (Wiley, 2020).

64. Galen Hall, Joshua Basseches, Rebecca Bromley-Trujillo, and Trevor Culhane, "CHORUS: A New Dataset of State Interest Group Policy Positions in the United States," *State Politics & Policy Quarterly* 24, no. 3 (2024): 1–26.

65. Galen Hall, Trevor Culhane, and J. Timmons Roberts, "Climate Coalitions and Anti-Coalitions." *Energy Research and Social Science* 113 (July 2024): 103562.

66. Charles Davis, "The Politics of 'Fracking': Regulating Natural Gas Drilling Practices in Colorado and Texas," *Review of Policy Research* 29, no. 2 (March 2012): 177–191.

67. Jeffrey Cook, "Research Article: Who's Regulating Who? Analyzing Fracking Policy in Colorado, Wyoming, and Louisiana," *Environmental Practice* 16, no. 2 (June 2014): 102–112.

68. Joshua Basseches interview, May 5, 2023.

69. Sam Zacher, "The US Political Economy of Climate Change: Impacts of the 'Fracking' Boom on State-Level Climate Policies," *State Politics & Policy Quarterly* 23, no. 2 (January 2023): 140–165.

70. Gordon Laxer, "Posing as Canada: How Big Foreign Oil captures Canadian Energy and Climate Policy," The Council of Canadians, CCPA Saskatchewan Office, CCPA BC Office, 2021, https://canadians.org/resource/bigforeignoil/.

71. Joshua Basseches, "Who Pays for Environmental Policy? Business Power and the Design of State-Level Climate Policies," *Politics & Society* (September 2023): 1–43.

72. Trevor Culhane, Galen Hall, and J. Timmons Roberts, "Who Delays Climate Action? Interest Groups and Coalitions in State Legislative Struggles in the United States," *Energy Research & Social Science* 79 (September 2021): 102114, https://doi:10.1016/j.erss.2021.102114.

73. Stokes, *Short Circuiting Policy*.

74. Basseches, "Who Pays for Environmental Policy?"

75. Harland Prechel, *Normalized Financial Wrongdoing* (Stanford University Press, 2020).

76. Harland Prechel, "Neoliberal Organizational and Political-Legal Arrangements and Greenhouse Gas Emissions in the U.S. Electrical Energy Sector," *Sociological Quarterly* 62, no. 2 (March 2020): 209–233.

77. Basseches, "Who Pays for Environmental Policy?"

78. Stokes, *Short Circuiting Policy*.

79. Emily L Williams, Sydney A. Bartone, Emma K. Swanson, and Leah C. Stokes, "The American Electric Utility Industry's Role in Promoting Climate Denial, Doubt, and Delay," *Environmental Research Letters* 17, no. 9 (September 2022): 094026.

80. Joshua Basseches interview, May 18, 2022.

81. Culhane et al., "Who Delays Climate Action?"

82. Basseches, "Who Pays for Environmental Policy?"

83. Ibid.

84. Severin Borenstein and James Bushnell, "The US Electricity Industry after 20 Years of Restructuring," *Annual Review of Economics* 7, no. 1 (August 2015): 437–463.

85. Basseches, "Who Pays for Environmental Policy?"

86. Galen Hall, Trevor Culhane, and J. Timmons Roberts, "Climate Coalitions and Anti-Coalitions: Lobbying Across State Legislatures in the United States," *Energy Research and Social Science* 113.

87. Basseches, "Who Pays for Environmental Policy?"

88. Nathan R. Lee, "When Competition Plays Clean: How Electricity Market Liberalization Facilitated State-Level Climate Policies in the United States," *Energy Policy* 139 (April 2020): 111308, https://doi:10.1016/j.enpol.2020.111308.

89. "Electric Federalism: Policy for Aligning Canadian Electricity Systems with Net Zero," Canadian Climate Institute, May 2022, https://climateinstitute.ca/wp-content/uploads/2022/05/Electric-Federalism-May-4-2022.pdf.

90. Matthew Black, "'Will Not Be Implemented': Alberta Pushes Back against Ottawa's New Clean Energy Regulations," *Edmonton Journal*, August 10, 2023, https://edmontonjournal.com/news/politics/will-not-be-implemented-alberta-pushes-back-against-ottawas-new-clean-energy-regulations.

91. Tanya Christidis, Geoffrey Lewis, and Philip Bigelow, "Understanding Support and Opposition to Wind Turbine Development in Ontario, Canada and Assessing Possible Steps for Future Development," *Renewable Energy* 112 (November 2017): 93–103.

92. "Electric Federalism: Policy for Aligning Canadian Electricity Systems with Net Zero," Canadian Climate Institute, May 2022, https://climateinstitute.ca/wp-content/uploads/2022/05/Electric-Federalism-May-4-2022.pdf.

93. David Iaconangelo, "Gas Ban 2.0: Building Wars," *E&E News*, April 2022, https://subscriber.politicopro.com/article/eenews/2022/04/18/gas-ban-2-0-building-wars-00024201.

94. Culhane et al., "Who Delays Climate Action?"

95. Lawrence Susskind, Jungwoo Chun, Alexander Gant, Chelsea Hodgkins, Jessica Cohen, and Sarah Lohmar, "Sources of Opposition to Renewable Energy Projects in the United States," *Energy Policy* 165 (2022): 112922, https://doi.org/10.1016/j.enpol.2022.112922.

96. Nicolas Graham, "Think Tanks and Climate Obstruction: Atlas Affiliates in Canada," *Canadian Review of Sociology/Revue* 61, no. 2 (May 2024): 1–21.

97. Iaconangelo, "Gas Ban 2.0."

98. Culhane et al., "Who Delays Climate Action?"

99. Isaac Slevin, William Kattrup, and J. Timmons Roberts, "Against the Wind: A Map of the Anti-Offshore Wind Network in the Eastern United States," Brown University Climate and Development Lab, December 2023.

100. Gianna White and Heidi Yeh, "Wind of Change: Overcoming Misinformation in New Jersey's Clean Energy Transition," *Journal of Science Policy and Governance* 24, no. 1, (2024), https://doi.org/10.38126/JSPG240117.

101. Graham, "Think Tanks and Climate Obstruction."

102. "Carbon Emissions," Global Carbon Project, 2023, https://globalcarbonatlas.org/emissions/carbon-emissions/.

103. Gilda Cardoso de Araujo, "Federalism, Decentralization, Intergovernmental Relations, and Educational Provision in Five Latin American Countries," in *Oxford Research Encyclopedia of School Reform*, ed. William T. Pink (Oxford University Press, 2022), https://doi:10.1093/acrefore/9780190264093.013.996.

104. Ronald Watts, "Typologies of Federalism," in *Routledge Handbook of Regionalism & Federalism*, ed. John Loughlin, John Kincaid, and Wilfried Swenden (Routledge, 2013), 19–33.

105. Enrique Leff, "Descolonización del conocimiento eurocéntrico, emancipación de los saberes indígenas y territorialización de la vida," *Utopía y Praxis Latinoamericana* 27, no. 98 (January 2022): 1–20.

106. Daniel Yuichi Kono, "The Politics of Trade and Climate Change," *Politics* (January 2019), https://doi.org/10.1093/acrefore/9780190228637.013.996.

107. Sandra Guzmán, Mariana Castillo, and Alin Moncada, "Financiamiento de Esfuerzos Contra El Cambio Climático En América Latina," *Política, Globalidad y Ciudadanía* 3, no. 5 (June 2016): 66–77.

108. Marcela López-Vallejo and María del Pilar Fuerte-Celis, "Hybrid Governance in Northeastern Mexico: Crime, Violence, and Legal-Illegal Energy Markets," *Latin American Perspectives* 48, no. 1 (December 2020): 103–125.

109. John Cross and Sergio Peña, "Risk and Regulation in Informal and Illegal Markets," in *Out of the Shadows: Political Action and Informal Economy in Latin America*, ed. Patricia Fernández-Kelly and Jon Shefner (Pennsylvania State University Press, 2010), 49–80.

110. "Almost 2,000 Land and Environmental Defenders Killed Between 2012 and 2022 for Protecting the Planet," *Global Witness*, September 13, 2023, https://www.globalwitness.org/en/press-releases/almost-2000-land-and-environmental-defenders-killed-between-2012-and-2022-protecting-planet/.

111. Manu Mediavilla, "Berta Cáceres, defensora ambientalista hondureña. Se cumplen 6 años de su brutal asesinato,"Amnesty International, March 2, 2022, https://www.es.amnesty.org/en-que-estamos/blog/historia/articulo/6-anos-del-asesinato-berta-caceres/.

112. "Is Climate Obstruction Different in the Global South? Observations and a Preliminary Research Agenda," *CSSN Position Paper* 4 (October 7, 2021), https://cssn.org/is-climate-obstruction-different-in-the-global-south-observations-and-a-preliminary-research-agenda/.

113. "Brazil: Case Study on Multi-Level Climate Governance," Climate Chance, 2021, https://www.climate-chance.org/wp-content/uploads/2021/05/brazil-multilevel-climate-governance-climate-chance-1.pdf.

114. "Producción de Petróleo," Gobierno de Argentina, 2022, https://www.argentina.gob.ar/economia/mineria/eiti-portal-de-transparencia-de-las-industrias-extractivas/produccion-de-petroleo.

115. "Las regalías petroleras fueron el recurso estrella de Neuquén en el 2022," Río Negro, 2023, https://www.rionegro.com.ar/politica/las-regalias-petroleras-fueron-el-recurso-estrella-de-neuquen-en-el-2022-2689642/.

116. "En Neuquén hay más de un empleo público cada dos puestos privados," LM Neuquén, 2022, https://www.lmneuquen.com/neuquen/en-neuquen-hay-mas-un-empleo-publico-cada-dos-puestos-privados-n961127.

117. Página 12, "Murió Guillermo Pereyra, histórico dirigente sindical de los petroleros patagónicos," May, 28, 2024, https://www.pagina12.com.ar/740388-murio-guillermo-pereyra-historico-dirigente-sindical-de-los-.

118. Rafaela Nascimento Santos, Luiz Carlos de Santana Ribeiro, and José Ricardo de Santana, "The Effects of Oil Royalties on Regional Inequality in Brazil," *CEPAL Review* 2022, no. 136 (April 2022): 191–208.

119. Mauricio Oyarzo and Dusan Paredes, "Revisiting the Link between Resource Windfalls and Subnational crowding out for Local Mining Economies in Chile," *Resources Policy* 64, (December 2019): 101523.

120. Maria DoloresAlmeida, Huáscar Eguino, Juan Luis Gómez Reino, and Axel Radics, "Decentralized Governance and Climate Change in Latin America and the Caribbean," International Center for Public Policy Working Paper (November 2022): 1–49, https://icepp.gsu.edu/files/2022/11/paper2207.pdf.

121. A. Boadle, "Mina de potássio na Amazônia tem licença estadual questionada pelo MPF," Folha de São Paulo, April 9, 2024

122. Jordan Watts, Patrick Greenfield, and Bibi van der Zee, "The Multinational Companies That Industrialised the Amazon Rainforest," *The Guardian*, June 2, 2023, https://www.theguardian.com/global-development/2023/jun/02/the-multinational-companies-that-industrialised-the-amazon-rainforest.

123. Lucas Christel, "Resistencias sociales y legislaciones mineras en las provincias argentinas. Los casos de Mendoza, Córdoba, Catamarca y San Juan (2003–2009)," *Politico y gobierno* 27, no. 1, (2020), https://ri.conicet.gov.ar/handle/11336/170637.

124. Juana Kuramoto, "La Minería Artesanal e Informal en el Perú," *Mining, Minerals and Sustainable Development* no. 82 (September 2001), https://www.iied.org/sites/default/files/pdfs/migrate/G00731.pdf.

125. Barbara Kuepper, Sarah Drost, Mathew Pietrowski, and Gerard Rijk, "Latin American Palm Oil Linked to Social Risks, Local Deforestation," Chain Reaction Research, December 2023, https://chainreactionresearch.com/wp-content/uploads/2021/12/Latin-American-Palm-Oil-Linked-to-Social-Issues-Local-Deforestation-1.pdf.

126. Andy Robinson, "Muerte por soja en El Cerrado de Brasil," *La Vanguardia*, December 22, 2019, https://www.lavanguardia.com/vida/20191222/472400582187/brasil-amazonia-el-cerrado-deforestacion-soja.html.

127. "Climate Change 538," European Union, 2023, https://europa.eu/eurobarometer/surveys/detail/2954.

128. "State of the Energy Union," European Commission, 2023, https://energy.ec.europa.eu/topics/energy-strategy/energy-union_en.

129. Robert J. Brulle, Timmons Roberts, and Miranda C. Spencer (eds.), *Climate Obstruction Across Europe* (Oxford University Press, 2024).

130. Martin Jänicke and Rudinger Wurzel, "Leadership and Lesson-Drawing in the European Union's Multilevel Climate Governance System," *Environmental Politics* 28, no. 1 (November 2018): 22–42.

131. Sybrig Smit and Takeshi Kuramochi, "Subnational and Non-State Climate Action in the EU: An Overview of the Current Landscape, Emission Reduction Potential and Implementation," *New Climate Institute Working Paper*, December 2020, https://climate.ec.europa.eu/system/files/2021-02/2021_02_05_report_en.pdf.

132. "State of the Energy Union," European Commission, 2023, https://energy.ec.europa.eu/topics/energy-strategy/energy-union_en.

133. Peter Eckersley, Kristine Kern, Wolfgang Haupt, and Hannah Müller, "The Multi-Level Context for Local Climate Governance in Germany: The Role of the Federal States," *Irs Dialog* 3, no. 3 (2021), http://hdl.handle.net/10419/237056.

134. Peter Eckersley, Kristine Kern, Wolfgang Haupt, and Hannah Müller, "Climate Governance and Federalism in Germany," in *Climate Governance and Federalism: A Forum of Federations Comparative Policy Analysis*, ed. A. Fenna, S. Jodoin, and J. Setzer (Cambridge University Press, 2023), 150–176.

135. Jos Delbeke and Peter Vis, *EU Climate Policy Explained* (Routledge, 2016).

136. William Dinan, Victoria Esteves, and Steven Harkins, "Climate Obstruction in Scotland: the Politics of Oil and Gas," in *Climate Obstruction Across Europe*, ed. R. Brulle, J. Roberts, and M. Spencer (Oxford University Press, 2024), 57–79.

137. Sander Happaerts, "Climate Governance in Federal Belgium: Modest Subnational Policies in a Complex Multi-Level Setting," *Journal of Integrative Environmental Science* 12, no. 4, (2015): 285–301.

138. Eckersley, Kristine Kern, Wolfgang Haupt, and Hannah Müller ., "The Multi-Level Context for Local Climate Governance in Germany."

139. Eckersley, Kristine Kern, Wolfgang Haupt, and Hannah Müller, "Climate Governance and Federalism in Germany."

140. Happaerts, "Climate Governance in Federal Belgium."

141. Thomas Muinzer, "Multilevel Drivers: The International Level and the Devolved Level (Northern Ireland, Scotland and Wales)," in *Climate and Energy Governance for the UK Low Carbon Transition* (Springer Nature, 2019), 83–121.

142. "Climate Change Act," Northern Ireland, 2022. https://www.legislation.gov.uk/nia/2022/31/contents/enacted.

143. Muinzer, "Multilevel Drivers."

144. Wilhelmine Preussen, "German Police Remove Activists Protesting Coal Mine Expansion," Politico.eu, January 10, 2023, https://www.politico.eu/article/police-barricade-activist-protest-coal-mine-garzweiler-germany/.

145. Nuran Erkul and Ata Ufuk Seker, "Farmers Across Europe Protest Costly Climate Change Regulations," *AA News*, January, 1, 2024, https://www.aa.com.tr/en/europe/farmers-across-europe-protest-costly-climate-change-regulations/3124624.

146. Mara Bierbach, "EU's Populists at Odds with Climate Protection," *Deutsche Welle*, February 26, 2019, https://www.dw.com/en/climate-protection-where-do-the-eus-right-wing-populists-stand/a-47699703.

147. Philip McCann and Luc Soete, "Place-Based Innovation for Sustainability," Publications Office of the European Union, 2020, https://doi.org/10.2760/250023,JRC121271.

148. Annette Elisabeth Töller, "'Kein Grund zum Feiern! Die Umwelt- und Energiepolitik der dritten Regierung Merkel,'" in *Zwischen Stillstand, Politikwandel und Krisenmanagement: Eine Bilanz der Regierung Merkel 2013–2017*, ed. Reimut Zohlnhöfer and Thomas Saalfeld (Springer, 2018), 569–590.

149. Olli Herranen, "Understanding and Overcoming Climate Obstruction," *Nature Climate Change* 13, no. 6, (June 2023): 500–501.

150. Kacper Szulecki, Tomas Maltby, and Julia Szulecka, "Climate Obstruction in Poland: A Govermental-Industrial Complex," in *Climate Obstruction Across Europe*, ed. R. Brulle, T. Roberts, and M. Spencer (Oxford University Press, 2024), 186–213.

151. Hanna Brauers and Pao-Yu Oei, "The Political Economy of Coal in Poland: Drivers and Barriers for a Shift Away from Fossil Fuels," *Energy Policy* 144 (2020): 111621.

152. Eckersley, Kristine Kern, Wolfgang Haupt, and Hannah Müller, "Climate Governance and Federalism in Germany."

153. Carolina Kyllmann and Julian Wettengel, "Coal Protest in Germany Deepens Rift Between Green Party and Climate Movement," Clean Energy Wire, January 16, 2023, https://www.cleanenergywire.org/news/coal-protest-germany-deepens-rift-between-green-party-and-climate-movement.

154. Roger Karapin, "Federalism as a Double-Edged Sword: The Slow Energy Transition in the United States," *Journal of Environment & Development* 29, no. 1 (November 2019): 26–50.

155. Michaël Aklin and Matto Mildenberger, "Prisoners of the Wrong Dilemma: Why Distributive Conflict, Not Collective Action, Characterizes the Politics of Climate Change," *Global Environmental Politics* 20, no. 4 (November 2020): 4–27.

156. Matto Mildenberger, *Carbon Captured: How Business and Labor Control Climate Politics* (MIT Press, 2020).

157. Leah C. Stokes and Christopher Warshaw, "Renewable Energy Policy Design and Framing Influence Public Support in the United States," *Nature Energy* 2 (2017): 1–6.

158. Parrish Bergquist, Matto Mildenberger, and Leah C. Stokes, "Combining Climate, Economic, and Social Policy Builds Public Support for Climate Action in the US," *Environmental Research Letters* 15, no. 5 (2020): 054019.

159. Basseches, "Who Pays for Environmental Policy?"

160. Sandeep Vaheesan, "The Best Way to Fight Heat Waves and Outages is to Green the Grid: The Fastest Way to Green the Grid, In Turn, Is For Government To Step In," *The New Republic*, July 17, 2024. https://newrepublic.com/article/183786/heat-wave-beryl-outage-public-electrical-grid.

161. Will Dinan, "Oil and Gas Lobbyists Have Deep Pockets and Access to Politicians, But an EU Ban Could Be in the Pipeline," *The Conversation*, February 13, 2024, https://theconversation.com/oil-and-gas-lobbyists-have-deep-pockets-and-access-to-politicians-but-an-eu-ban-could-be-in-the-pipeline-223388.

162. Ruth E. Mckie, Carlos Milani, Paul K. Gellert, Omar M. Faruque, K. Hochstetler, Guy Edwards, D. McElwee Pamela, Jonathan Walz, Prakash Kashwan, and Timmons Roberts, "Climate Obstruction in the Global South: Future Research Trajectories," *PLOS Climate* 2 (2023): 1–5.

CHAPTER 10

Obstruction in the United Nations Framework Convention on Climate Change and the Intergovernmental Panel on Climate Change

LEAD AUTHORS: KARI DE PRYCK AND EDUARDO VIOLA

CONTRIBUTING AUTHORS: STEFAN C. AYKUT, LARISSA BASSO, DANIELLE FALZON, MATÍAS FRANCHINI, FRIEDERIKE HARTZ, HANNAH HUGHES, VINÍCIUS MENDES, CARLOS R. S. MILANI, BRUNA BOSI MOREIRA, GÉRALDINE PFLIEGER, AND EMANUEL SEMEDO

INTRODUCTION: MAKING SENSE OF OBSTRUCTION AT THE INTERNATIONAL LEVEL

One of the most important sites for crafting but also contesting how climate change is to be collectively addressed is the Conference of the Parties (COP) to the United Nations Framework Convention on Climate Change (UNFCCC), held annually in major cities around the globe. For example, in 2023 at COP28 in Dubai, United Arab Emirates, the Climate Action Network (CAN), a large network of civil-society organizations, used the event as an opportunity to bestow daily its inglorious Fossil of the Day award to thirteen countries and one province for their efforts at "doing the most to achieve the least" progress on climate change. CAN has been conferring these awards since the 1990s to denounce climate obstruction in the UNFCCC, but also domestically. It has also awarded countries the Colossal Fossil of the Year to denounce obstruction for the entire duration of the COP.

Kari De Pryck et al., *Obstruction in the United Nations Framework Convention on Climate Change and the Intergovernmental Panel on Climate Change*. In: *Climate Obstruction*. Edited by: J. Timmons Roberts, Carlos R. S. Milani, Jennifer Jacquet, and Christian Downie, Oxford University Press. © Oxford University Press (2025). DOI: 10.1093/oso/9780197787144.003.0010

CAN tends to target countries they believe can be pressured and whose reputation might be negatively affected by being dubbed a Fossil. At COP28, CAN cited Saudi Arabia for resisting "language supporting the just and equitable phase out of fossil fuels and transition to renewables" and "repeated blocking across negotiation tracks."[1] The European Union (EU) also received a Fossil for "ongoing opposition to including Loss and Damage in the negotiations of the New Collective Quantified Goal."[2] The United States (US) was awarded the Colossal Fossil of the Year title for repeatedly "opposing language on the differentiation of fossil fuels" (see Table 10.1).[3]

In this chapter, we review obstruction strategies used by both state and non–state actors (nongovernmental organizations [NGOs], social movements,

Table 10.1 THE CLIMATE ACTION NETWORK'S FOSSIL AWARDS SINCE SIGNING OF THE PARIS AGREEMENT

COP	Fossils of the Day	Colossal Fossil
2023: COP28	Saudi Arabia, EU, Alberta (Canada), Russia, South Africa, Israel, Vietnam, Australia, Norway, Brazil, New Zealand, Japan, US, South Korea	US
2022: COP27	Egypt, Turkey, New Zealand, UAE, Japan, Israel, Russia, USA	US
2021: COP26	New Zealand, Brazil, Saudi Arabia, UK, Poland, IETA, Serbia, France, Australia, Mexico, Czech Republic, US, Norway	Australia, US, UK
2020: Fifth anniversary of the Paris Agreement	Australia, Brazil	US
2019: COP25	Australia, US, EU, Canada, Russia, Brazil, Japan, Bosnia and Slovenia, Belgium	Brazil
2018: COP24	Australia, Egypt, US, Poland, EU, Austria, Norway, UK, Russia, Japan, Germany, Switzerland, Arab Group, Brazil, Saudi Arabia	Poland
2017: COP23	Arab Group, India, Brazil, Germany, Australia, Norway, Canada, EU, US, ICAO, Japan, Kuwait, Developed Countries, France, Poland	US

Continued

Table 10.1 CONTINUED

COP	Fossils of the Day	Colossal Fossil
2016: COP22	Australia, Austria, New Zealand, European Commission, Indonesia, Venezuela, EU, Turkey	
2015: COP21	EU, Umbrella Group, Venezuela, Saudi Arabia, Malaysia, LMDCs, Norway, US, Denmark, Belgium, New Zealand, Japan	

the transnational corporate and national private sectors, etc.) who participate in international climate negotiations—here, obstruction is defined as the efforts by powerful interests to slow or block policies or actions on climate change at the international level. While deliberate obstruction strategies are not the only explanation for limited climate inaction globally, they are an important one.[4]

We focus on the UNFCCC, which provides a basis for negotiation between states with increasing input from non–state actors. This is because the Paris Agreement (drafted at COP21 in 2015) assumes that non–state actors, and particularly the private sector, will play a key role in its implementation. Several initiatives have been launched to allow and promote their engagement, such as the Lima-Paris Action Agenda, the Non–State Actor Zone for Climate Action (NAZCA), and the Marrakech Partnership for Global Climate Action (GCA).[5] We also consider the Intergovernmental Panel on Climate Change (IPCC), a political and scientific body producing reports about the state of scientific, technical, and socioeconomic knowledge on climate change. While independent from the UNFCCC, the IPCC regularly becomes enmeshed in controversy when unresolved issues are deferred to the organization in the hope that it can reset the international dialogue to a more "rational" discourse.[6,7] In particular, the line-by-line approvals of its reports' Summaries for Policymakers (SPMs) have become sites of struggle between governments over what is policy-relevant scientific knowledge to inform the UNFCCC.[8,9] Statements included in the SPMs are considered to have a "soft policy prescriptive" character.[10]

Strategies of obstruction at the international level are regularly denounced but remain underexplored as a research topic. Scholars[11] have only recently provided the first comprehensive framework to assess obstruction in the UNFCCC, distinguishing between tactics to: limit the scope of an issue (e.g., by rejecting or excluding it from the agenda); reduce transparency (e.g., by impeding the collection of information about it); manipulate language around it (e.g., by supporting ambiguous text); and promote nontransformative solutions (e.g., by privileging further discussions over making decisions). While this framework offers valuable insight for tracking obstruction at the international level and should inform future research, there has been insufficient time since its publication for other scholars to apply it in their work. In this assessment of the literature, we highlight areas where the insights from this publication are echoed by other scholars examining obstructive efforts in the UNFCCC and IPCC.

In this chapter, we distinguish between *procedural obstruction*, which targets the negotiation process with the aim of slowing down or derailing the negotiations themselves, and *substantial obstruction*, which targets the substance of the negotiations by, for example, preventing an issue from being placed on the agenda or manipulating the way it is framed, scoped, and defined. After a brief description of important moments in the history of the UNFCCC that set the stage for climate obstruction, we highlight major structural and domestic factors underpinning international obstruction strategies. We then review strategies used by states and non–state actors to obstruct development on key issues (science, mitigation, adaptation, loss and damage, equity and finance) and consider efforts to expose and resist climate obstruction.

THE UNFCCC, A FERTILE GROUND FOR OBSTRUCTION

The UNFCCC was signed in 1992 at the United Nations Conference on Environment and Development Conference in Rio de Janeiro (Rio). It is open to all states and regional economic integration organizations such as the European Union, known as the Parties. The UNFCCC sets the framework under which international climate cooperation occurs and inscribes its key principles—such as the principle of Common but Differentiated Responsibility and Respective Capabilities (CBDR-RC)—distinguishing between Annex I (developed) and non-Annex I (developing) countries, with the former expected to take the lead in addressing climate change.[12]

The first quantified obligation to reduce emissions was inscribed in the Kyoto Protocol at COP3 in 1997 and approved at COP7 in Marrakech in 2001. During the first commitment period (2005–2012), industrialized economies undertook their obligation to reduce greenhouse gas (GHG) emissions by an average of at least 5% against 1990 levels, which reflected the sum of the individual targets pledged by each of them. Abiding by the principle of CBDR-RC, the UNFCCC exempted developing countries from these commitments. The ambition of the Kyoto Protocol was low.[13] On the one hand, the definition of national targets was not based on "objective" criteria, but "on what countries were willing to put forward at the time."[14] On the other hand, not all Parties reached their individual targets, although on paper the agreement achieved its overall goals. Few countries (mainly from the European Union) fulfilled their obligations; the United States did not ratify the Protocol; and Canada withdrew in 2011 when it became obvious that it would not be able to adhere to its target. While a second commitment period (2013–2020) was adopted in 2012, several countries (Japan, Canada, Russia) did not participate.

The Paris Agreement is an indirect recognition of the limitations of the UNFCCC and the Kyoto Protocol crafted in a context in which global emissions had grown by 2% annually between 2000 and 2014. Universal by nature, it sought to commit both Annex I and non-Annex I countries to climate action through the submission of so-called nationally determined contributions (NDCs).[15] The agreement also inaugurated a period of "hybrid multilateralism" by giving non–state actors a more integrated role, encouraging them to contribute to climate action.[16] Non–state actors may differ in their respective lobbying and negotiating behaviors; thus, unpacking the roles played by, for example, the fossil fuel corporate sector, agribusiness, and ultraconservative think tanks versus climate activist networks and environmental NGOs is of paramount importance. Yet almost ten years after the Paris Agreement, there remains a gap between the aspirations stated by the Parties in their NDCs and the level of action that is necessary to meet the Agreement's temperature goal (well below 2 degrees C), raising questions about whether it can be achieved. Challenges to implementing the agreement also affect the credibility of its Parties, raising doubt about whether they are seriously committed to these goals.

Weak implementation creates what one scholar[17] calls a "fertile ground for obstructionists." The weak implementation of the Kyoto Protocol ingrained mistrust against the industrialized world and offered support for arguments to those who claim that these countries have not taken the lead in climate change mitigation as promised. Challenges in implementing the Paris Agreement in a fair and ambitious manner have further eroded trust. Developed countries' failure to mobilize the $100 billion per year in 2020 and 2021 it had promised at COP15 in 2009 to support climate action in developing countries also increased distrust.

Structural Aspects Facilitating Obstruction

States make major decisions during COPs, transnational mega-events and meeting points for state and non–state actors invested in shaping the global response to climate change.[18] The COPs have been increasingly partitioned, divided into a Blue Zone—a space of negotiation but also side events restricted to accredited participants, a Green Zone—a semi-official space where non–state actors can hold side events that is open to the public, and a Fringe space—usually the streets and buildings of the host city where various activities and mobilizations are allowed.[19]

Several aspects of UNFCCC operations contribute to empowering obstructionist behaviors. For example, by not imposing a fair limit to the size of delegations, the UNFCCC has maintained significant asymmetries in participation and negotiating capacities between actors. Countries with high stakes in climate change mitigation often have larger delegations, and are hence better prepared to exert influence on the negotiations.[20] In contrast, low-income states, despite being highly vulnerable to climate change, are disadvantaged because they are "overwhelmingly outmatched in terms of financial resources, political influence, and negotiating capacity."[21] In practice,[22] this means that many of them have smaller delegations and struggle to attend parallel meetings; are less fluent in English and can less effectively get their messages across; have fewer scientific and legal experts involved and encounter more challenges navigating discussions; and have bigger turnover and thus lack institutional knowledge over time. These disadvantages translate into less capacity to negotiate effectively and to oppose obstructionists, especially when negotiations extend into the night. To compensate for such challenges, many countries rely on coalition building.[23]

These inequalities are even more acute in the IPCC because many countries in the Global South lack the infrastructure and resources to partake in global knowledge- production and build a sufficient knowledge base to contribute meaningfully to its assessments. Despite efforts to increase the participation of developing countries in the organization, "the economic and human resources required to conduct IPCC activities means that considerable asymmetries persist."[24] This asymmetry means that their perspectives are not well-represented in the assessment reports, potentially generating biases and greater challenges to implementing climate policy in the Global South (see Chapter 8).

Finally, another aspect of the international climate regime that works in favor of obstructionists is consensus-based decision-making. While consensus is generally praised for giving any Party the right to effectively veto a proposal, in practice, such power is unequally distributed among participants.[25] Studies have noted that consensus can promote lowest- common-denominator outcomes and encourage uncooperative behavior.[26] Obstructionist countries from both the Global North and the Global South have historically maintained inflexible positions and disproportionately used their veto power to slow progress, delay, or block agreement in the UNFCCC[27] and the IPCC.[28,29]

STRUCTURAL AND DOMESTIC FACTORS UNDERPINNING CLIMATE OBSTRUCTIONISM

While international climate negotiations provide a fertile ground for obstruction, some countries and groups of countries obstruct the process more than others. What makes these countries decide that their interests are best served by intentionally slowing or blocking action on climate change? Although this is a complex issue, this section highlights some of the main structural and domestic factors underpinning climate obstruction in the UNFCCC and the IPCC.

Historical Emissions and Responsibilities

The first structural factor in obstruction is a country's historical emissions: nations that have relied heavily on fossil fuels and built their economic and industrial sectors around them may be more reluctant to take a leadership role in the UNFCCC/IPCC, to set ambitious climate goals, and to provide the means of implementation (finance, technology transfer, and capacity building). While there are different ways to define and assess historical responsibility for CO_2 emissions, by all accounting a small number of industrialized countries and emerging economies dominates (Table 10.2). Consider, for example, cumulative CO_2 emissions since 1850, the beginning of the Industrial Revolution: ten countries alone are responsible for almost two-thirds of those total emissions. When considering cumulative territorial emissions since 1990—the standard baseline in the international climate regime—and consumption-based emissions, the same ten countries are listed, although the order changes slightly. Finally, when looking at cumulative emissions per capita and per population, the ranking is different, with large emerging economies falling off the top ten list.

Material Endowments, Values, and Ideas

Another structural factor in obstruction is the material endowments of different countries.[30] Nations whose economies are dependent on fossil fuel exportation or importation may be more reluctant to engage meaningfully in the UFCCC/IPCC negotiations (Table 10.3).

While material endowments are crucial to understanding climate obstructionism, ideas, values, and beliefs are also important to consider, for example, when a country's leader is a climate denier. The election of conservative climate deniers in the United States (George W. Bush and Donald J. Trump),

Cumulative territorial emissions (1850–2021)	Cumulative territorial emissions (1990–2021)	Cumulative emissions: consumption (1990–2021)	Cumulative per capita emissions (1850–2021)	Cumulative emissions per population (1850–2021)
US	China	US	New Zealand	Canada
China	US	China	Canada	US
Russia	India	Russia	Australia	Estonia
Brazil	Russia	Brazil	US	Australia
Indonesia	Brazil	Indonesia	Argentina	Trinidad and Tobago
Germany	Indonesia	Germany	Qatar	Russia
India	Japan	India	Gabon	Kazakhstan
UK	Germany	UK	Malaysia	UK
Japan	Canada	Japan	Congo	Germany
Canada	UK	Canada	Nicaragua	Belgium

Source: Carbon Brief (2021), https://www.carbonbrief.org/analysis-which-countries-are-historically-responsible-for-climate-change and Climate Watch. (The information on per capita GHG emissions is limited to countries with more than one million inhabitants.)

Canada (Stephen Harper), Australia (Scott Morrison), Argentina (Javier Milei), and Brazil (Jair Bolsonaro) diverted these countries' priorities away from leadership positions in the UNFCCC.

Domestic Politics and Vested Interests

Other factors are more closely related to how different actors, especially those against climate regulation, influence decision-making processes at the domestic level. Conflicts between domestic actors that are pro- and anti-climate action can influence the position of a country at the international level, as well as explain changes over time.[31] The United States, for example, has twice complicated climate progress within the international community: by not ratifying the Kyoto Protocol and by withdrawing from the Paris Agreement, both out of the stated fear that the treaties would undermine its economy. According to some scholars,[32] the decision not to ratify the Kyoto Protocol "was almost entirely driven by internal conflicts within the executive branch and the legislature." Others[33] also note that then-President Trump's withdrawal decision "was mainly driven by the U.S. domestic politics and his personal preferences [as a climate denier]."

Table 10.3 TOP TEN COUNTRIES BY EXTRACTION, CONSUMPTION, EXPORT, AND IMPORT OF FOSSIL FUELS 1990–2006 AND 2007–2023

Domestic Extraction		Domestic Consumption		Exports		Imports	
1990–2006	2007–2023	1990–2006	2007–2023	1990–2006	2007–2023	1990–2006	2007–2023
US	China	US	China	Saudi Arabia	Russia	US	China
China	US	China	US	Russia	Australia	Japan	US
Russia	Russia	Russia	India	Australia	Indonesia	Germany	Japan
Canada	Canada	India	Canada	Europe	Saudi Arabia	South Korea	India
Saudi Arabia	India	Canada	Russia	Canada	Canada	Spain	South Korea
India	Saudi Arabia	Germany	Japan	Iran	US	UK	Germany
Australia	Indonesia	Japan	Germany	Indonesia	Iraq	Netherlands	Taiwan
Germany	Australia	UK	South Korea	Venezuela	UAE	Taiwan	UK
Iran	Iran	Poland	Saudi Arabia	Nigeria	Qatar	India	Spain
UK	South Africa	Norway	Iran	UAE	Nigeria	Ukraine	Netherlands

Source: Global Material Flows Database, https://energydata.info/dataset/world-unep-irp-global-material-flows-database.

It is also important to consider the role of vested interests at the domestic level—people or organizations with a financial or personal advantage in delaying or blocking climate action—in obstruction in the UNFCCC/IPCC. Some research[34] emphasizes that various interest groups, particularly those representing the fossil fuel industry and related sectors, have employed deliberate political tactics and lobbying efforts "to maintain laggard positions in countries such as Australia and the United States and to mute leadership in others, such as Germany." In many industrialized countries, notably the United States, industry groups have funded campaigns of misinformation aimed at discrediting the scientific consensus on human-induced climate change or claiming that a swift transition to a low-carbon economy would significantly disrupt established lifestyles.[35,36,37,38,39,40] These groups have been supported by conservative groups that funded denialist think tanks, supported conservatism in academia, and promoted radical free-market ideologies and a positive image of corporations. Conservative media have also played a role in spreading climate denial. While supporting freedom of expression and respect for opposing views, right-wing media have simultaneously promoted the ideas of a minority of denialist scientists, who have subsequently become overrepresented. Denialist columnists and bloggers have become major disseminators of misinformation about climate change (see Chapters 5 and 6).

This denial machine is also active in Global South countries. In many cases, climate denial is associated with nationalism, land politics, and development strategies based on a zero-sum game between growth and environmental protection and has influenced the position of emerging economies in multilateral talks.[41,42,43,44] For example, at COP26, Brazil obstructed progress in negotiations on Article 6, which allows countries to voluntary cooperate to reach emissions-reduction targets, and presented an updated version of its NDC in which the country committed to a target regarded by many experts as an accounting maneuver and manipulation.[45,46] Domestic vested interests have contributed to increased deforestation in the Amazon, reduction in the demarcation of Indigenous lands, the weakening of environmental monitoring bodies—notably the "Brazilian Institute of Environment and Renewable Natural Resources" (IBAMA) and the "Chico Mendes Institute for Biodiversity Conservation" (ICMBio)—and the increased vulnerability of biodiversity in Brazilian biomes. In so doing, they have influenced and discredited Brazil's international climate commitments.[47,48,49,50]

MAJOR OBSTRUCTIONIST COUNTRIES

Most major emitters have held international negotiating positions that could qualify as obstructive. We consider in this section the United States, China, Brazil, India, Russia, and the European Union.

The United States—the world's major "climate power" given its historical emissions and the assets to reduce them—is according to one research team[51] "the nation most significantly undermining the call to action." It has historically alternated between periods during which it blocked advances in the UNFCCC and those in which it contributed to enhance international cooperation. As mentioned, while the United States signed the Kyoto Protocol, it never ratified it, arguing that the agreement would create unfair competition for American companies in favor of China. When Barack Obama became president in 2009, obstruction receded. In 2015, the United States presented a NDC pledging to reduce its emissions by 26%–28% below 2005 levels, an objective that was neither ambitious enough nor fair relative to the country's historical emissions. In 2020, the year in which Parties were to have presented an updated NDC, then-President Trump decided to withdraw from the Paris Agreement. Under the administration of Joe Biden in 2021, the United States rejoined the Agreement. It announced a revised NDC, pledging to reduce its emissions by 50%–52% below 2005 levels by 2030 and to achieve net-zero emissions by 2050. The revised NDC was substantially more ambitious than the previous one, but still insufficient considering the country's responsibility for the largest share in the world's historical GHG emissions. While the climate legislation known as the Inflation Reduction Act (2022) could lead to a 24%–37% decline in US emissions, additional measures are needed if the country is to fulfill the NDC target. While the Act aims to advance the renewable power industry and "green" jobs, it includes concessions for the fossil fuel industry, including an annual minimum area of specified public lands made available for drilling.[52] The Climate Action Tracker (CAT), an independent scientific project that tracks government climate action, thus classifies the US NDC as insufficient.[53]

China' position is more ambiguous. The country's intensive economic growth in the last half-century has placed China as the world's second-largest economy and in direct competition with the United States for global hegemony. In addition, Chinese economic growth and the consequent rise in energy demand have been driven mostly by the abundant use of coal, making China the world's largest polluter. Thus, China has become a key climate actor at the international level, with substantial incumbent responsibilities it has resisted. It hews to a discourse of being a developing country, a characterization that no longer reflects its economic status, and insists on upholding the principles of CBDR-RC. At the same time, according to one author, China's position on CBDR-RC is not "solely driven by concerns about economic growth, maintaining sovereignty, or simply not wanting to address the problem."[54] Namely, China's interests are also based on strongly held moral views about the importance of developed countries taking the lead on climate change. Domestically, China's position is also ambiguous. Beijing produces the majority of the low-carbon energy technologies—especially electric vehicles, batteries, and solar panels—and aims to achieve low-carbon development. Yet coal still provides about three-quarters of China's energy supply including electricity. At COP21 (2015), China presented an NDC that is inconsistent with the world's remaining carbon budget. It has since pledged to be carbon neutral by 2060. The CAT classifies China's NDC as highly insufficient.

India is another fierce defender of the CBDR-RC principle, stressing the country's right to development and its challenges in eradicating poverty. India continuous being a low-/ middle-income country, both in per capita income and emissions, and from the point of view of climate justice has the right to defend the principle (something not as valid for high-/middle-income countries like Brazil and China). In its updated NDC, submitted to the secretariat of the UNFCCC, the country pledged in 2022 to reduce emissions intensity by 45% below 2005 levels by 2030, to increase cumulative renewable grid capacity by 50%, and to increase forest carbon sinks by 2030. It has also pledged to become climate neutral by 2070. Yet implementation is lagging. India is highly dependent on coal for its energy supply—it is the world's second-largest producer and importer of coal. Thus, fluctuations in the global energy markets strongly affect the country and its energy policy, as when it took advantage of decreasing prices and escalated its imports of Russian oil when Western democracies began an embargo after the Russian invasion of Ukraine in 2022. The CAT classifies India's NDC as highly insufficient.

Russia has always been a major obstructionist in the climate regime. Its emissions have been decreasing relative to the 1990 baseline, not as a result of decarbonization policies but rather decreased economic growth.[55] In its 2015 NDC, revised in 2020, Russia pledged to limit its emissions to 70% below 1990 levels by 2030, relying heavily on the capacity of its forests to sequester carbon. Russia has not pledged to reduce the use of fossil fuels, which the country considers key to its development—it is the world's second- largest producer and exporter of oil and natural gas and the third-largest exporter of coal.[56] Even before its invasion of Ukraine, Russia was increasingly wary of multilateral cooperation, arguing that the climate agenda threatens its national security—a doctrine first announced in 2019 and reiterated in 2021.[57] The CAT classifies Russia's NDC as critically insufficient.

Brazil could become a proactive actor in climate change cooperation but has been a reluctant for most of the climate regime's history. Nevertheless, despite a period (2009–2011) during which it joined forces with other proactive actors to advocate climate cooperation and accepted voluntary emissions-reduction targets, Brazil has mostly defended the CBDR-RC principle and tried to exempt itself from responsibility.[58] In 2015, Brazil pledged to reduce emissions by 43% by 2030, without depending on financial transfers from developed countries.

At the time, it was the only major developing country to have a goal of absolute reduction of emissions.[59] In 2021, the administration of Jair Bolsonaro raised the goal to a 50% reduction; however, due to changes in emissions accountancy, the higher percentage actually yielded a lower reduction, in absolute numbers.[60,61] Then, in 2023, under the administration of Luiz Inácio Lula da Silva, Brazil updated its NDC and corrected the error, increasing the reduction to 56%. Since then, the discourse against climate action that had been promoted by Bolsonaro changed and policies to resume deforestation control in the Amazon were reinstated. That year, deforestation in the Amazon was reduced by 20% compared with 2022, a major achievement considering the deterioration of the law-enforcement apparatus that had been allowed to occur during the Bolsonaro administration.[62] Yet Lula's administration has not pursued decarbonization in other sectors. For example, there is a large political divide over exploring oil reserves in the eastern margins of the Amazon, and Brazil is gradually becoming one of the most important exporters of oil, entering OPEC+ (the expanded version of the Organization of the Petroleum Exporting Countries) in 2023.[63] The CAT classifies Brazil's NDC as insufficient.

The European Union has been a consistently proactive actor in the climate regime. For example, in 2015, the European Union pledged to reduce its emissions by 40% compared with 1990 levels and raised its ambition to 55% in 2020.[64,65] However, individual member countries have not always fulfilled their obligations or left important gaps in implementing decarbonization policies. For example, after Russia reduced its exports of natural gas to Europe following the invasion of Ukraine and sanctions imposed by Western economies, many European countries increased their use of coal for electricity supply in 2022, despite an EU target to decrease dependence on fossil fuels.[66] Similarly, the European Green Deal provides a roadmap to reduce emissions from different EU sectors, and different policies have already been approved at the Union level. These policies include a ban on sales of new vehicles using combustion engines after 2035; the European Climate Law, enshrining the 2050 climate neutrality objective; and the European Climate Pact, an agreement to promote efforts to mobilize the public into taking climate action. Yet since the Union and its members share responsibility on climate issues and implementation depends largely on the latter, it has yet to be seen how quickly the gap between regulation and implementation will diminish. For example, in late 2023 and early 2024, major demonstrations of farmers against decarbonization policies in agriculture pushed the European Commission to reconsider several of its recommendations for the agricultural sector (see Chapter 4). Another important variable is the rise of right-wing parties in national parliaments. The CAT classifies the EU's NDC as insufficient.

OBSTRUCTION STRATEGIES BY STATE ACTORS

Obstruction strategies in the UNFCCC and IPCC can be divided into procedural and substantial obstruction. While both strategies are often intertwined, we distinguish them for the sake of clarity.

Procedural Obstruction

Sustained strategies of procedural obstruction have been documented since the establishment of the climate regime. These practices can include abusing rules of procedure, assuming key positions, and taking issues hostage or limiting their scope. Because they have been used repeatedly over time, these obstructive strategies have contributed to spoiling the atmosphere at meetings and creating distrust among participants: stalling the negotiations, weakening their ambition, and occasionally placing the whole regime in jeopardy.[67,68,69,70] These strategies have been documented to occur in both the UNFCCC and the IPCC and have frequently been used by OPEC countries (especially Saudi Arabia and Qatar) and industrialized countries (especially the United States), although other actors' use of them cannot be ruled out.

When abusing rules of procedure, obstructionist countries often take advantage of the tacit norms of deliberation that make it impossible to deny the floor to a Party. The delegation of Saudi Arabia, for example, has repeatedly taken the floor to raise points of order or objections and to propagate their positions under different agenda items.[71] In the IPCC, countries wanting to undermine the organization's influence on climate negotiations have also targeted the procedures used for compiling its reports.[72,73] In 2009, for example, the "Climategate" scandal in which thousands of emails and computer files from a server at the University of East Anglia were hacked and divulged online, as well as some errors found in the IPCC Fourth Assessment Report (AR4), were amplified to discredit the organization. Ahead of COP15 in Copenhagen, a Saudi delegate reportedly said that Climategate would have a "huge impact" on the negotiations as "it appear[ed] from the details of the scandal that there is no relationship whatsoever between human activities and climate change."[74]

A second obstruction strategy used in the UNFCCC/IPCC context is "procedural power,"[75] which involves assuming positions of power in the negotiations. Scholars[76,77] have noted that OPEC countries have chaired the Group of 77 (the largest coalition representing developing countries in the UNFCCC) several times or assumed coordinating roles on several issues. According to another author, "the Chair of the G-77 . . . was filled by a delegate from an OPEC country for six of the eleven years spanning 1994–2004."[78] OPEC countries, and in particular Saudi Arabia, have also been continuously represented in both the COP and IPCC bureaus.[79] While the latter advises the COP president, the former leads the assessment process. Being represented in these bodies is crucial to be kept informed of the latest developments and gain greater control of the process.[80] A related strategy is to join or remain in a negotiating process or treaty for the mere purpose of defending one's interests. Researchers[81] have noted that OPEC countries joined the Kyoto Protocol only once it was confirmed that it would enter into force, to ensure that the group was able to influence negotiations about its implementation. The United States also kept attending meetings of the Kyoto Protocol, sometimes blocking issues and maintaining inflexible positions, despite that it had not ratified it.[82]

A third obstruction strategy in international climate negotiations is bargaining by holding items hostage or limiting the scope of an issue. Research[83,84] has shown how Saudi Arabia sought to link the question of adaptation (a major concern for developing countries) to that of response measures (an agenda item that discusses the potential adverse social or economic impacts of climate change mitigation measures), conditioning progress on the former to consideration of the latter. This strategy resulted in slowing progress on adaptation overall.

Substantial Obstruction

Obstruction strategies can also be observed on matters of substance in international climate negotiations, related to how different issues have been addressed over time in terms of their scope and ambition, but also in the language used to frame them. Many of the strategies of procedural obstruction outlined in the previous section have been used to support substantial obstruction. This section features examples related to science, mitigation (with a focus on energy), adaptation, loss and damage, equity, and climate finance.

Science and the IPCC

Obstructionist countries and their allies have continuously questioned IPCC conclusions, both within the organization and within the UNFCCC. In the IPCC, emphasizing uncertainty in the authors' assessment is a key strategy to weaken the authority of the organization and delegitimize calls for action based on its reports. In the 1990s and 2000s, OPEC countries and the United States focused on the confidence levels assigned to statements on the detection and attribution of climate change (the remit of its Working Group I on the physical science basis), with the aim of postponing discussions about GHG emissions-reduction targets. Negotiations often focused on "nuances of expressions,"[85] such as when the expression "appreciable human influence" on global climate was replaced by "discernible human influence." There have also been instances where countries downplayed the negative impacts of climate change.[86] In the last cycles, obstructionist countries have targeted findings from working groups II (on impact, vulnerability, and adaptation) and III (on mitigation), which assess policy-relevant knowledge for climate action. In AR6, for example, Saudi Arabia sought to dampen the emphasis on emissions reduction and fossil fuel phaseout. China also fought to undermine the focus on 1.5°C by emphasizing uncertainty about the assessment of irreversible impacts resulting from overshooting the target.[87] Finally, developed countries, especially the United States, sought to undermine statements highlighting the finance gap between developed to developing countries.[88]

In the UNFCCC, agenda items related to the IPCC have been another sustained target of OPEC countries, and at times of the United States and the Russian Federation.[89] These countries, especially Saudi Arabia, have often fought to weaken decisions following the release of IPCC reports for fear that the reports' conclusions could underpin calls for more ambitious action. This was most evident after the publication of the Special Report on 1.5°C (SR15) in 2018. At COP24 in Katowice, the US administration aligned with Saudi Arabia, Kuwait, and Russia in opposing "welcoming" the report out of concern that it would convey support for a 1.5°C target and increased ambition.[90] More recently, China, India, and Saudi Arabia opposed noting that the Sixth Assessment Report (AR6) was the "most comprehensive and robust" assessment of climate change and requested the deletion of a sentence linking AR6 to the "best available science."[91]

Mitigation (with a Focus on Energy)

Obstruction of mitigation is sector specific. Some research has documented the reluctance of some states (in particular India, Argentina, and Brazil) to address emissions from agriculture in the UNFCCC.[92] Scholars[93] have also identified strategies by Saudi Arabia and OPEC countries to slow progress on discussions about ways to reduce emissions from international aviation and maritime transport.

Obstruction strategies have been more fully documented in the energy sector. Obstruction is reflected in the near absence of debate over efforts to combat climate change through regulation of energy production and markets.[94,95] Despite being ubiquitous topics in most other climate debates, energy issues have rarely been included in official multilateral climate talks. The basic treaty and negotiation texts of the climate regime either omit discussion of energy issues altogether or frame them in very specific and limited ways, in stark contrast to the way such issues are discussed by civil-society actors and international organizations. Some governments have devised explicit strategies to keep energy questions out of climate negotiations. Historically, the most visible and arguably most important actor in this respect has been Saudi Arabia, which has consistently argued that the UNFCCC is "not an energy treaty"[96] and has successfully blocked progress on energy-related issues. Research shows that[97] Saudi Arabia, Iran and other OPEC members strongly opposed any reference to "CO_2 and energy taxation," "new or increased oil taxation," and "new greenhouse gas taxes" in the Kyoto negotiations. Two decades later, in the Paris negotiations, Saudi Arabia and other oil-exporting countries systematically lobbied against any formulations of a long-term target that contained the word "carbon," such as "decarbonization," "carbon neutrality," or "low carbon economies," so as to avoid a focus on fossil fuels as the main cause of global warming.[98] In the IPCC, Saudi Arabia has repeatedly opposed singling out CO_2 as the main GHG or mentioning fossil fuels as the main source of global GHG emissions.[99]

Questions of fossil fuel phaseout, removal of fossil fuel subsidies, and/or disinvestment have thus been kept out of negotiation documents. In the few cases where fossil fuel regulation did enter climate talks, discursive reframing has occurred.[100] This reframing can be seen in negotiation tracks on "adverse effects of response measures" (which underpins calls for compensation for the prospective economic losses from mitigation policies) and in discussions on market-based approaches and technological fixes within several negotiation tracks of the Kyoto Protocol and Paris Agreement, such as the inclusion of carbon capture and storage (CCS) in the Clean Development Mechanism.[101]

COP26 has been regarded "as a watershed in the adoption and institutionalization"[102] of anti-fossil fuel norms: its cover decision called for accelerating "efforts toward the *phasedown* of unabated coal power and phase-out of inefficient fossil fuel subsidies" [emphasis added]. Although the text was diluted ("phasedown" replaced "phaseout"), "it marks the first time that a COP decision mentions fossil fuels [in the context of phasing out fossil fuel subsidies] and coal as part of the climate problem and as issues that require action from Parties."[103] Countries opposed to the use of phaseout included India and China, which lamented the singling out of coal. Instead, India (unsuccessfully) called for all fossil fuels to be phased down in an equitable manner. At COP28, there was again strong opposition to language on phaseout from the Like-Minded Developing Countries (LMDCs), Arab Group, and some African countries.[104] As a compromise, the Parties agreed to mention "transitioning away from fossil fuels in energy systems."

Strategic linguistic ambiguities were central to reaching an agreement on this issue and it remains unclear whether these developments will lead to major changes in the regulation of energy production and markets. A major loophole is the use of the term "unabated," which could allow for the continuation of fossil fuel extraction if combined with CCS or carbon dioxide removal (CDR), methods whose feasibility and safety are highly uncertain, especially at the scale needed to lower surface temperatures. In addition to advocating adoption of these technologies in the UNFCCC, fossil fuel producers (Saudi Arabia, Norway, the United States, and Japan) also sought to normalize them in IPCC assessments. There have been seven recorded instances in AR6 of Saudi Arabia intervening to add the word "unabated" to sentences referencing CO_2 and GHG emissions reductions, in one instance calling for retaining language "on avoiding unabated fossil fuel emissions" rather than using the phrase "displacing" fossil fuels.[105]

Adaptation

Obstruction of adaptation (the process of adjustment to actual or expected climate impacts) has largely taken place through the relative neglect of the issue in the climate regime compared with mitigation (see Chapter 11). In contrast to mitigation, which has been central to negotiations, adaptation has been treated as something of an afterthought.[106] Adaptation gained more attention in the early 2000s, with several funds being set up in 2001 to support developing countries in writing National Adaptation Plans of Action (NAPAs) and implementing adaptation projects. However, discussions on adaptation were relegated to a work program (the Nairobi Work Programme) in 2005. After years of developed countries (in particular the United States, European Union, and Australia) resisting the addition of a dedicated space in the negotiations to consider adaptation,[107,108] the acceptance of this proposal by Argentina was considered a meaningful advancement of the issue.[109] However, decades later, we can understand the formation of a work program as aligning with a common strategy for delaying action on an issue, with Parties opposing the incorporation of the issue ultimately agreeing only to years of extended discussions rather than concerted decision-making. Mace[110] found that industrialized countries had also been blocking action in the negotiations through assertions of scientific uncertainty about the attribution of impacts to climate change, calls for further studies, concerns about maladaptation, and requirements for adaptation mainstreaming into development planning.

Despite the recognition—in the Cancun Agreements negotiated at COP16 in 2010—that adaptation should be given the same attention as mitigation, adaptation has remained subordinate to mitigation for several reasons. First, adaptation has been framed mainly as an issue to be governed at the national or subnational level.[111] Second, there has been a proliferation of agenda items on adaptation under the UNFCCC, which weakens the issue through fragmentation, and the parameters around what adaptation should entail remain undefined. The fragmentation of adaptation items under the UNFCCC and the lack of clarity over adaptation action going forward, especially under the Global Goal on Adaptation established by the Paris Agreement, mean that adaptation has systematically been made more difficult to track and assess. Third, adaptation has been consistently underfunded relative to mitigation.[112,113] There is also a gap between the adaptation needs expressed by developing countries and the adaptation finance provided by wealthy countries.[114]

Concerns about inadequacies in adaptation finance speak to broader concerns about obstruction in climate finance under the UNFCCC. Wealthy countries have not consistently contributed the climate financing they have promised, most notably failing to meet the goal to mobilize $100 billion per year by 2020. In the negotiations, these same countries have refused to establish clear definitions of climate finance, in particular denying assertions by developing countries that such finance should come in the form of grants (not loans), particularly for adaptation.[115]

Loss and Damage

Similar strategies of obstruction have been employed regarding the definition of loss and damage, or the consequences of climate change that are already felt and that to date have disproportionately affected smaller/lower income/more vulnerable countries. Wealthy countries, particularly the United States, have been obstructing the issue of loss and damage in the climate negotiations since the 1990s, when the Alliance of Small Island States (AOSIS) first raised the issue.[116] The reason the issue of loss and damage, particularly finance for loss and damage, has received so much resistance is likely because it comes closest to raising the topic of liability and compensation in the negotiations. The United States has directly prevented liability and compensation for climate change impacts from entering the decision texts for decades.[117,118,119] Today, while loss and damage has become a key issue under the UNFCCC (through the Warsaw International Mechanism for Loss and Damage and Article 8 of the Paris Agreement) the denial of any basis for the appearance of liability and compensation in UNFCCC decision texts continues to be a fruitful strategy for the United States and supportive Parties in developed nations. Scholars[120] outlined a set of tactics countries have used to obstruct loss and damage finance over the years, which fall into four categories: limiting the scope of agenda items; reducing transparency; manipulating the meaning of textual language; and pushing nontransformative solutions.

Language manipulation over loss and damage has also been well-documented in the IPCC.[121] While the IPCC has engaged with the concept in previous assessments, it was only in AR6 WGII that the loss-and-damage terminology was first agreed to in a SPM, using the wording "losses and damages." In earlier assessment and special reports, several EU countries, the United States, Canada, and other Parties prevented the notion of loss and damage from being mentioned in the SPMs, on the grounds that the concept remained "a political term that has not been defined"[122] and that could lead to "terminological confusion."[123] During the SR15 process, the IPCC authors introduced a glossary entry on loss and damage[124] that separates the political debate around "Loss and Damage" from scientific evidence and projections on "losses and damages.[125] One author argues that such subtle orthographic changes (variations in spelling) paved the way for getting loss-and-damage terminology—in the form of the less political phrase "losses and damages"— included in the AR6 SPMs. On this basis, several countries retroactively recommended against using the term "loss and damage" in the underlying AR6 WGII report.[126] The final corrected official document displays terminology adjustments (e.g., "loss and damage" has become "losses and damages") in several chapters.[127]

These examples highlight how countries have used subtle differences in loss-and- damage orthographies to influence the representation of the issue in IPCC documents. While some appreciated the eventual inclusion of loss-and-damage language in the SPMs, others raised concerns that the term "losses and damages" departs from language agreed to in the Paris Agreement; may result in attempts to depoliticize and dilute loss-and-damage language; could derail research funding; and may undermine efforts by vulnerable countries to have the loss and damage they experience recognized in international climate negotiations.[128]

Equity and CBDR-RC

Obstruction on equity and CBDR-RC consists of either denying the salience of these principles in defining Parties' fair share in solving the climate problem or defining Parties' responsibilities in ways that do not reflect their different contributions to global emissions and economic growth patterns.

The United States has been the fiercest critic of the principles of equity and CBDR-RC in international negotiations. In the 2000s, the country justified its nonratification of the Kyoto Protocol on the ground that it did not include binding emissions- reduction targets for major emitters from the developing world, especially India and China. According to one researcher,[129] "even during more engaged Democratic presidencies, US participation has remained directed toward the flattening of differentiated commitments between developed and developing countries." In 2011 during post-Kyoto negotiations, for example, the country's lead negotiator, Todd Stern, supposedly claimed "if equity is in, we are out."[130]

The Paris Agreement introduced greater differentiation in the climate regime and sought to go beyond the Annex I/non-Annex I dichotomy. While referring to equity and CBDR-RC, the treaty also introduced the phrase "in the light of different national circumstances." This caveat speaks to the country-based approach to climate governance as well as self-differentiation between countries in mitigation obligations under the Agreement. As explained by one scholar,[131] "the qualification of the principle [CBDR-RC] by a reference to 'national circumstances' introduces a dynamic element to the interpretation of the principle. As national circumstances evolve, so too will the common but differentiated responsibilities of States."

This change has not meant that obstruction on this issue has ceased. Global North countries, and in particular the United States, are now working to limit the salience of CBDR-RC and equity in critical conversations around loss and damage, finance, and adaptation, by arguing that these issues should not be dealt with under the Paris Agreement.[132] As part of the Global Stocktake— the process to assess states' and non–state actors' progress toward meeting the goals of the Agreement—many developed countries have emphasized the need to focus on forward-looking plans in an attempt to avoid any assessment of the fairness of their pre-2020 commitments and action.

In reaction to developed countries' minimization of their historical responsibilities, some developing countries continue to fiercely hold to the principles of equity and CBDR-RC, even as they have become major GHG emitters themselves. Many LMDC members, for example, continue to embrace the Annex I/non-Annex I dichotomy, denying the level of responsibility that should be assigned to emerging economies with higher capabilities.[133] China often promotes the discourse that it is still a developing country, which no longer reflects its economic status. Brazil has also often raised the CBDR-RC principle when it worked to its own advantage.[134]

OBSTRUCTION STRATEGIES BY NON–STATE ACTORS

The obstruction strategies of non–state actors are more difficult to trace because, as non-Party stakeholders, they cannot properly engage in formal climate negotiations. This does not make them uninfluential. One study[135] noted that non–state actors can influence negotiations indirectly via domestic channels or even directly at the international level. Their ability to influence is not straightforward and depends on several factors, including the political opportunity structure (e.g., the availability of channels of influence or the decision-making rules), the power of the state(s) supporting their position, and their involvement in transnational coalitions.

Non–state actors in the UNFCCC include nongovernmental organizations (NGOs), intergovernmental organizations (IGOs), and UN bodies and specialized agencies. The UNFCCC clusters NGOs into nine constituencies: business and industry NGOs (BINGO), environmental NGOs (ENGO), farmers and agricultural NGOs (Farmers), Indigenous peoples' organizations (IPO), local government and municipal authorities (LGMA), research and independent NGOs (RINGO), trade union NGOs (TUNGO), a women and gender constituency (WGC), and children and youth NGOs (YUNGO).

Non–state actors use substantial obstruction more frequently than procedural obstruction. The obstructive strategies of industry groups (especially the carbon majors) have been more thoroughly documented than those of other sectors. At the domestic level, because they control energy production, these industry groups are often privileged in negotiations on energy planning and implementation. Policymakers may also anticipate their reaction and take them into account when drafting policies. For example, European industrialists in the 1990s opposed the introduction of a community-wide carbon tax. In the United States, industry groups lobbied negotiators to ensure that the Convention would not include binding emissions-reduction targets. Once established, they pressured them to oppose discussions of a protocol, arguing that it would damage the economy.[136,137]

Increasingly, industry groups have engaged in obstruction at the international level as part of a country delegation or as representatives of an observer organization. The former status gives these actors privileged access to policymakers.[138] Cases have been reported in which employees of state-owned companies accompanied their country delegation to provide technical support. For example, several negotiators from Saudi Arabia had close ties to Saudi Aramco, its national oil company.[139] An analysis by Corporate Accountability, Corporate Europe Observatory, and Global Witness[140] revealed that twenty-nine countries attending COP27 in 2022 had fossil fuel lobbyists within their national delegations, including the United Arab Emirates (UAE) and the Russian Federation. The UAE also successfully nominated Sultan Al Jaber, the CEO of the Abu Dhabi National Oil Company, to be president of COP28.

When attending climate talks as part of an observer organization, the carbon majors are active mainly through the BINGO constituency, such as trade associations, because observer organizations must be NGOs. As representatives of BINGOs, they can network, engage in advocacy, and build coalitions with other non–state actors.[141,142] Two umbrella groups that fit this model were particularly active in the 1990s and early 2000s: the Global Climate Coalition (GCC) and the Climate Council (CC). These groups comprised coal and oil companies, mainly from the United States, and some chemical and car companies (see Chapters 2 and 3). They built transnational alliances with other states, particularly members of OPEC. Other relevant associations (some of which are still active today) have included the International Climate Change Partnership (ICCP), the World Coal Institute (WCI), the International Petroleum Industry Environmental Conservation Association (IPIECA), and the International Chamber of Commerce (ICC). ICC and IPIECA have included representatives of oil companies. Some of their programs have also been chaired by oil companies.[143] The ICCP, IPIECA, and ICC are still registered as observer organizations.

At first, climate denial was prevalent among these groups. The GCC, for example, launched campaigns to discredit IPCC conclusions on climate change detection and attribution. The GCC, supported by well-known American climate skeptics,[144,145] accused the organization in 1996 of corrupting the peer-review process by altering text after it had been formally accepted by governments.[146,147,148] It was also caught giving text to OPEC countries.[149]

In the UNFCCC, representatives of fossil fuel companies fought to have their interests reflected in key decisions. The GCC was in close contact with OPEC states, assisting them by supplying strategic information and political support.[150] Research shows[151] they "have managed to get their positions adopted by many governments, and in several cases incorporated into international documents including the UNFCCC." The ICCP, for example, was crucial in the drafting of Article 4, paragraph 10 of the Convention in collaboration with Australia. The paragraph notes that special consideration should be given to countries whose economies are highly dependent on the production or consumption of fossil fuels. Post-Rio proposals (incremental measures, emission trading, voluntary efforts, etc.) also continued to be, according to the same authors,[152] "clearly in tune with the preferences of the energy industries."

From the 2000s onward, a more pluralistic set of interest groups became involved in the UNFCCC.[153,154] Once it became clear that the United States would not ratify the Kyoto Protocol, the influence of the GCC faded and it was dissolved in 2001.[155] Some corporations and business groups, including oil companies (such as Shell and BP), also became more proactive once the Kyoto Protocol's flexibility mechanisms to help countries meet their emissions-reduction targets were introduced.[156] Interest in the UNFCCC has since increased within the private sector, with BINGOs becoming a dominant group that seeks to influence the agenda and decision-making.[157,158] The lead-up to COP21 in 2015 was "a new moment in business's positioning towards climate negotiation."[159] Much work was done to restore the unity of the private sector. Scholars have observed that[160] "the new 'business voice' proved to be the exact opposite of the initial obstructive and defensive one, and keen to demonstrate that 'business has changed.'" Ahead of the COP, the private sector publicly supported a universal and ambitious agreement. It also called for greater dialogue between the private and public sectors. The World Economic Forum (WEF), in participation with major multinational corporations from the Global North and as a very influential opinion leader in its own right, also began to prioritize climate change mitigation and the energy transition in its programs and public statements.[161]

At COP21, many more business sectors were represented than previously (including extraction, manufacturing, electricity and gas, transportation, and finance and insurance).[162] Multisectoral business associations, especially the World Business Council for Sustainable Development (WBCSD) and the International Emissions Trading Association (IETA) registered more than one hundred delegates each. In this context, the BINGOs have increasingly presented themselves as the providers of the solutions that will drive climate action. Beneath the assumed unity, however, there lay conflicting views and contradictions.[163] For example, there was disagreement over whether an emissions-trading scheme or a carbon tax was a more appropriate response. BINGOs' vision of the energy transition also remains dominated by assumptions about "the self-healing capacities of the market"[164] and least-cost scenarios.

The (official) position of the carbon majors on climate change has changed considerably but remains ambivalent.[165] Research[166] shows that the largest companies accept the risks posed by climate change and seek to offer cost-effective solutions, but that they also do not see the end of fossil fuel energy. Some companies even stress the important role of oil and gas in meeting future demand growth. The same author[167] also notes that many of the proposed solutions are likely to fail to lower net emissions. Many companies support methane gas, a "lower-carbon fuel," and CCS, a technology whose feasibility at large scale is debated. They also stress the need for CDR to reach net-zero emissions, another method whose safety, scalability, and maturity are uncertain.[168]

Several business-driven initiatives have emerged in the last decade, such as the Low Carbon Technology Partnerships initiative (LCTPi), led by WBCSD, and the Oil and Gas Climate Initiative (OGCI). The carbon majors are increasingly acting through the OGCI, a club uniting twelve of the largest firms. Launched at the UN Secretary General's Climate Summit in 2014, the OGCI's stated mission is to accelerate climate action: it promised to invest $1 billion in low-emissions technologies.[169] Major companies among the OGCI membership have committed to ambitious emissions-reduction strategies and attended UNFCCC meetings in increasing numbers. Corporate Accountability, Corporate Europe Observatory, and Global Witness[170] reported that 503 fossil fuel lobbyists were registered at COP26, 636 at COP27, and 2,456 at COP28, creating concern among civil society organizations about the integrity of the negotiation process. The possibility of greenwashing looms large and it remains to be seen how the carbon majors translate their commitments into action.

While the obstructionist strategies of fossil fuel companies have been more systematically studied, those used by representatives of other sectors that attend UNFCCC negotiations have also influenced the talks' direction. For example, civil-society organizations have raised concerns about the domination of agro-business players in discussions related to agriculture, especially through the Global Alliance for Climate-Smart Agriculture (see Chapter 4).[171] The *DeSmog* news platform[172] also reported on how livestock companies mobilized at COP28 to disseminate a pro-meat message throughout the summit.

EFFORTS TO EXPOSE AND RESIST CLIMATE OBSTRUCTION AT THE INTERNATIONAL LEVEL

Efforts exist to expose and oppose climate obstruction within the UNFCCC and beyond. Perhaps the most consistent group in contesting climate obstruction and supporting stronger climate commitments has been AOSIS. Since the start of the climate regime, this organization of small island states has acted as a "single voice" to craft a "niche diplomacy" around common moral arguments and the use of scientific evidence to defend their interests in international negotiations:[173,174] advocating more ambitious emission-reduction targets and raising funds for mitigation and adaptation as well as losses and damages. Environmental NGOs and activists have also played a key role in resisting climate obstruction, naming and shaming the countries blocking progress, as the CAN example in the introduction illustrates.

There have also been coordinated multiactor campaigns to support ambitious climate action at the international level. For example, scholars have[175] identified four distinct schemes aimed at "climatizing" global energy governance inside and outside the global climate regime: (1) carbon pricing, promoted by international organizations such as the Organization for Economic Co-operation and Development (OECD) and the International Energy Agency (IEA); (2) fossil fuel phaseout, discussed by activists and raised in meetings of the Group of Seven advanced economies (G7); (3) anti-extractivism, advocated by social movements in the Global South; and (4) fossil fuel divestment, supported by social movements and NGOs in the Global North. Several countries have also joined these various efforts. Pacific leaders in 2015 joined a call for an international moratorium on the development and expansion of fossil fuel–extracting industries, the Suva Declaration on Climate Change. Since then, calls for a phasing out fossil fuel have followed, leading in 2019 to the Fossil Fuel Non-Proliferation Treaty Initiative, supported by civil society and spearheaded by the small island nations of Tuvalu and Vanuatu.[176]

CONCLUSION

This chapter assessed the available literature to define the extent and methods of climate obstruction in the global climate regime, focusing mainly on the

UNFCCC and IPCC. Defining and identifying procedural and substantial strategies of obstruction by both state and non–state actors, it identified the key tactics used, including slowing down or blocking international climate negotiations as well as controlling and manipulating the way relevant issues are framed and decisions are taken. Major fossil fuel producers and exporters in particular have distinguished themselves by their sustained efforts at impeding progress toward ambitious and fair climate action. The persistence of these efforts over decades seems to echo an observation from 1988:[177] "It is easier to avert international action than to obtain consensus for it."

Obstructionist positions on the international stage are shaped by material endowments but also domestic politics and vested interests. Countries depending on high production, consumption, and/or export of fossil fuels are typically against decarbonization or reluctant to engage in such a transition, as are countries whose leaders adopt anti-climate ideas. The privileged access that some non–state actors (particularly the carbon majors) gain in these countries—sometimes even joining official delegations—can also contribute to reinforcing these positions. At the international level, obstructionist positions continue to be easily sustained in the context of long-standing distrust and suspicion between the Parties to the UNFCCC treaty.

Scholars have highlighted several ways climate obstruction in this arena could be tackled. They note that the UNFCCC and IPCC could change procedures to allow a seven-eighths supermajority voting rule (as was suggested in the past). Such a rule "would capture overwhelming support across the globe, while sidelining a tiny minority of obstructers."[178] Obstructers could also be cited and sanctioned by, for example, being required to provide climate finance or contribute to the UNFCCC and IPCC budgets.

Some scholars[179] have called for greater transparency on who gets to participate in the UNFCCC. They suggest requiring "participating organisations to have plans that address their climate impacts in line with agreed goals."[180] Others[181] suggest establishing conflict-of-interest rules, drawing on best practices from other international forums including the OECD and the World Health Organization. They offer clear recommendations on how to limit the access and engagement of non–state actors holding interests diverging from those of the UNFCCC.

Observers and researchers have a key role to play in tracking and expanding our knowledge of climate obstruction in the UNFCCC and IPCC beyond the usual suspects (Saudi Arabia, the United States, and the fossil fuel industry). For example, more research is needed on climate obstruction on issues related to agriculture, legal and illegal mineral extraction in protected areas, aviation, and shipping. It is also important to learn more about the impact of the increased participation of private actors. Non–state actors differ in their respective lobbying and negotiating behaviors; thus, unpacking the roles played by, for example, the fossil fuel corporate sector, agribusiness, and ultraconservative think tanks versus climate activist networks and environmental NGOs may be a task for future research.

NOTES

1. Climate Action Network, "The Biggest and the Baddest, Colossal Fossil Goes to the USA," https://climatenetwork.org/resource/the-biggest-and-the-baddest-colossal-fossil-goes-to-the-usa/ (accessed June 28, 2024).
2. Climate Action Network, "Fossil of The Day: Attention EU: Loss and Damage is Part of the NCQG," https://climatenetwork.org/resource/fossil-of-they-day-attention-eu-loss-and-damage-is-part-of-the-ncqg/ (accessed June 28, 2024).
3. Climate Action Network, "The Biggest and the Baddest, Colossal Fossil Goes to the USA," https://climatenetwork.org/resource/the-biggest-and-the-baddest-colossal-fossil-goes-to-the-usa/ (accessed June 28, 2024).
4. Isak Stoddard, "Three Decades of Climate Mitigation: Why Haven't We Bent the Global Emissions Curve?," *Annual Review of Environment and Resources* 46, no. 1 (2021): 653–689.
5. Karin Bäckstrand and Jonathan W. Kuyper, "The Democratic Legitimacy of Orchestration: The UNFCCC, Non-State Actors, and Transnational Climate Governance," *Environmental Politics* 26, no. 4 (2017): 764–788.
6. Erland A. T. Hermansen, Bård Lahn, Göran Sundqvist, and Eirik Øye, "Post-Paris Policy Relevance: Lessons from the IPCC SR15 Process," *Climatic Change* 169, no. 7 (2021): 1–18.
7. Felix Schenuit, "Staging Science: Dramaturgical Politics of the IPCC's Special Report on 1.5 °C," *Environmental Science & Policy* 139 (2023): 166–176.
8. Hannah Hughes, Alice Vadrot, Jen Iris Allan, Tracy Bach, Jennifer S. Bansard, Pamela Chasek, Noella Gray, Arne Langlet, Timo Leiter, Kimberly R. Marion Suiseeya, Beth Martin, Matthew Paterson, Silvia Carolina Ruiz-Rodríguez, Ina Tessnow-von Wysocki, Valeria Tolis, Harriet Thew, Marcela Vecchione Gonçalves, and Yulia Yamineva, "Global Environmental Agreement-Making: Upping the Methodological and Ethical Stakes of Studying Negotiations," *Earth System Governance* 10 (2021): 100121.
9. Kari De Pryck, "Intergovernmental Expert Consensus in the Making: The Case of the Summary for Policy Makers of the IPCC 2014 Synthesis Report," *Global Environmental Politics* 21, no. 1 (2021): 108–129.
10. Makane Moïse Mbengue, "The Intergovernmental Panel on Climate Change: A Singular Model of Expertise in the International Legal Order," in *International Law and Environmental Challenges*, ed. Yann Kerbrat and Sandrine Maljean-Dubois (Pedone/Hart, 2011), 97–114.
11. Danielle Falzon, Fred Shaia, J. Timmons Roberts, Md. Fahad Hossain, Stacy-ann Robinson, Mizan R. Khan, and David Ciplet, "Tactical Opposition: Obstructing Loss and Damage Finance in the United Nations Climate Negotiations," *Global Environmental Politics* 23, no. 3 (2023): 95–119.
12. Farhana Yamin and Joana Depledge, *The International Climate Change Regime: A Guide to Rules, Institutions and Procedures* (Cambridge University Press, 2009).
13. Joana Depledge, "The 'Top-Down' Kyoto Protocol? Exploring Caricature and Misrepresentation in Literature on Global Climate Change Governance," *International Environmental Agreements: Politics, Law and Economics* 22, no. 4 (2022): 673–692.

14. Harro van Asselt and Fergus Green, "COP26 and the Dynamics of Anti-Fossil Fuel Norms," *WIREs Climate Change* 14, no. 3 (2023): 31.

15. van Asselt and Green, "COP26 and the Dynamics of Anti-Fossil Fuel Norms."

16. Bäckstrand and Kuyper, "The Democratic Legitimacy of Orchestration."

17. Joana Depledge, "Striving for No : Saudi Arabia in the Climate Change Regime." *Global Environmental Politics* 8, no. 4 (2008): 15.

18. Hannah Hughes and Alice B. M. Vadrot (eds.), 2023. *Conducting Research on Global Environmental Agreement-Making* (Cambridge University Press, 2023).

19. Stefan C. Aykut, Simone Rödder, and Max Braun, *Collaborative Ethnography of Global Environmental Governance* (Cambridge University Press, 2024).

20. Ayse Kaya and Lynne Steuerle Schofield, "Which Countries Send More Delegates to Climate Change Conferences? Analysis of UNFCCC COPs, 1995–2015," *Foreign Policy Analysis* 16, no. 3 (2020): 478–491.

21. David Ciplet, Timmons J. Roberts, and Mizan R. Khan *Power in a Warming World* (MIT Press, 2015), 98.

22. Danielle Falzon, "The Ideal Delegation : How Institutional Privilege Silences 'Developing' Nations in the UN Climate Negotiations," *Social Problems* 70, no. 1 (2021): 185–202.

23. Tracy Bach and Beth Martin, "Negotiations: Navigating Global Environmental Conferences," in *Conducting Research on Global Environmental Agreement-Making*, ed. Alice B. M. Vadrot and Hannah Hughes (Cambridge University Press, 2023), 93–120

24. Hannah Hughes, "Governments," in *A Critical Assessment of the Intergovernmental Panel on Climate Change*, ed. Kari De Pryck and Mike Hulme (Cambridge University Press, 2022), 84.

25. Eva Krick, "The Myth of Effective Veto Power under the Rule of Consensus. Dynamics and Democratic Legitimacy of Collective Decision-Making by 'Tacit Consent,'" *Négociations* 27, no. 1 (2017): 109–128.

26. Luke Tomlinson, *Procedural Justice in the United Nations Framework Convention on Climate Change* (Springer International, 2015), 21.

27. Joana Depledge, "The Opposite of Learning: Ossification in the Climate Change Regime," *Global Environmental Politics* 6, no. 1 (2006): 1–22; Depledge, "Striving for No: Saudi Arabia in the Climate Change Regime," 9–35.

28. Jeremy Leggett, *The Carbon War* (Routledge, 2001).

29. Michael E. Mann, *The Hockey Stick and the Climate Wars* (Columbia University Press, 2012).

30. Navroz C. Dubash, et al., "National and Sub-National Policies and Institutions," in *IPCC, 2022: Climate Change 2022: Mitigation of Climate Change. Contribution of Working Group III to the Sixth Assessment Report of the Intergovernmental Panel on Climate Change*, ed. Priyadarshini R. Shukla, Jim Skea, Raphael Slade, Alaa Al Khourdajie, Renée van Diemen, David McCollum, Minal Pathak, Shreya Some, Purvi Vyas, Roger Fradera, Malek Belkacemi, Apoorva Hasija, Géninha Lisboa, Sigourney Luz, and Juliette Malley (Cambridge University Press, 2022), 1355–1450.

31. Matto Mildenberger, *Carbon Captured: How Business and Labor Control Climate Politics* (MIT Press, 2020).

32. Michael Aklin and Matto Mildenberger, "Prisoners of the Wrong Dilemma : Why Distributive Conflict, Not Collective Action, Characterizes the Politics of Climate Change," *Global Environmental Politics* 20, no. 4 (2020): 18.

33. Hai-Bin Zhang, Han-Cheng Dai, Hua-Xia Lai, and Wen-Tao Wang, "U.S. Withdrawal from the Paris Agreement: Reasons, Impacts, and China's Response," *Advances in Climate Change Research* 8, no. 4 (2017): 222.

34. Stoddard et al., "Three Decades of Climate Mitigation," 661.

35. Yutong Si, Dipa Desai, Diana Bozhilova, Sheila Puffer, and Jennie C. Stephens, "Fossil Fuel Companies' Climate Communication Strategies: Industry Messaging on Renewables and Natural Gas," *Energy Research & Social Science* 98 (2023): 103028.

36. Lucy Michaels and Katharine Ainger, "The Climate Smokescreen: The Public Relations Consultancies Working to Obstruct Greenhouse Gas Emissions Reductions in Europe: A Critical Approach," in *Climate Change Denial and Public Relations: Strategic Communication and Interest Groups in Climate Inaction*, eds. N. Almiron and J. Xifra (Routledge, 2020), 159–177.

37. William K. Carroll, Shannon Daub, and Shane Gunster, "Regime of Obstruction: Fossil Capitalism and the Many Facets of Climate Denial in Canada," in *Handbook of Anti-Environmentalism*, eds. D. Tindall, M. C. J. Stoddard, and R. E. Dunlap (Edgar Elgar Publishing, 2022), 216–233.

38. Matt Huber, "Fueling Capitalism: Oil, the Regulation Approach, and the Ecology of Capital," *Economic Geography* 89, no. 2 (2013): 171–194.

39. Harriet Bulkley and Peter Newell, *Governing Climate Change* (Routledge, 2010).

40. Olli Herranen, "Understanding and Overcoming Climate Obstruction," *Nature Climate Change* 13 (2023): 500–501.

41. Eduardo Viola and Matias Franchini, *Brazil and Climate Change: Beyond the Amazon* (Routledge, 2018), 6–7;

42. Prakash Kashwan, John Chung-En Liu, and Jahnnabi Das, "Climate Nationalisms: Beyond the Binaries of Good and Bad Nationalism," *WIREs Climate Change* 14, no. 2 (2023): e815.

43. Jennifer C. Franco and Saturnino M. Borras Jr., "The Global Climate of Land Politics," *Globalizations* 18, no. 7 (2021): 1277–1297.

44. Caio S. Milagres, Júlia N. Santos, Júlia A. Silva, and Sérgio M. Neto, "Monitoramento das negociações climáticas internacionais: breve relatório da COP 26," Rio de Janeiro: Cadernos do OIMC, 2022, https://obsinterclima.eco.br/wp-content/uploads/2022/04/Cadernos-OIMC-04-2022.pdf.

45. Raoni Rajão, B. Soares Filho, F. Nunes, J. Borner, L. Machado, D. Assis, A. Oliveira, L. Pinto, V. Ribeiro, L. Rausch, G. Holly, and D. Figueira, "The Rotten Apples of Brazil's Agribusiness," *Science* 369, no. 6501 (2020): 246–248.

46. V. Korber Gonçalves and M. Eibs Cafrune, "Brazilian Anti-Indigenous Politics: Tracking Changes on Indigenous Rights Regulation During Bolsonaro's Government," *Revista Direito e Práxis* 14, no. 1 (2023): 436–457.

47. Matias Franchini, Eduardo Viola, and Ana Mauad, "La política ambiental en la era de Bolsonaro: desprotección florestal, destrucción institucional y desprestigio internacional," in *El Brasil de Bolsonaro en español*, ed. Gisela Doval, Juan Lucca, Estevam Iglesias, and Prometeo Livros Buenos Aires (Prometeo Editorial, 2022), 143–160; AO PONTO: Como o Brasil chega à Conferência do Clima da ONU? Entrevistada: Mercedes Bustamante. Entrevistadores: Carolina Morand e Roberto Maltchik. O Globo, podcast, October 28, 2021, https://oglobo.globo.com/podcast/como-brasil-chega-conferencia-do-clima-da-onu-25254253 (accessed November 1, 2021).

48. Brasil Climate Action Hub/ CEBRI, "Uma concertação pela Amazônia—Balanço da primeira semana da COP26 e projeção da segunda semana," 2021, https://www.cebri.org/br/evento/496/uma-concertacao-pela-amazonia-ao-vivo-da-cop26.

49. Cada um por si, vácuo acima de todos. Na ausência de Bolsonaro, governadores brasileiros se articulam na COP26 para levantar recursos para projetos ambientais. Piauí, 2021. https://piaui.folha.uol.com.br/cada-um-por-si-vacuo-acima-de-todos/

50. Café Da Manhã, "COP26: a força da grana nas questões do clima. Entrevistados: Natalie Unterstell, Sergio Arispe, Marina Marçal e Ana Carolina Amaral," *Folha de São Paulo*, November 12, 2021, podcast, https://open.spotify.com/episode/20TpgJuZKeml9idLHQbpKL.

51. Stoddard et al., "Three Decades of Climate Mitigation," 661.

52. "What the Inflation Reduction Act Has Achieved in Its First Year," *The Economist*, August 17, 2023, https://www.economist.com/the-economist-explains/2023/08/17/what-the-inflation-reduction-act-has-achieved-in-its-first-year.

53. The Climate Action Tracker (CAT) is the most important institution measuring NCDs every year, and classifying them in five groups: extremely insufficient, highly insufficient, insufficient, almost sufficient, and 1.5 degrees Paris Agreement compatible. See Climate Action Tracker, https://climateactiontracker.org/, Countries profile, United States, https://climateactiontracker.org/countries/usa/ (last updated April 2024).

54. Sikina Jinnah, "Makers, Takers, Shakers, Shapers: Emerging Economies and Normative Engagement in Climate Governance," *Global Governance: A Review of Multilateralism and International Organizations* 23, no. 2 (2017): 292.

55. Larissa Basso and Eduardo Viola, "Para onde vão os BRICS na transição de baixo carbono?," *Revista CEBRI* 1, no. 4 (2022), https://cebri.org/revista/br/artigo/62/para-onde-vao-os-brics-na-transicao-de-baixo-carbono.

56. Ibid.

57. "Russia," Decree 216/2019 approving the Energy Security Doctrine, Climate Change Laws, 2019, https://www.climate-laws.org/geographies/russia.

58. Viola and Franchini, *Brazil and Climate Change*.

59. Ibid.

60. Natalie Unterstell and N. Martins, "NDC do Brasil: Avaliação da atualização submetida à UNFCCC em 2022," April 7, 2022, https://www.politicaporinteiro.org/2022/04/07/atualizacao-da-ndc-brasileira-vai-contra-acordo-de-paris-ao-nao-elevar-ambicao-climatica/.

61. Eduardo Viola and Larissa Basso, "The First Five Months of Climate Policy of the Lula Administration in Brazil," paper presented at the 27th Congress of the International Studies Association, Buenos Aires, July 15–19, 2023.

62. Brazilian National Institute of Space Research (INPE), 2024; Ministry of Environment and Climate Change, 2024.

63. Eduardo Viola and Larissa Basso, "The Climate Policy (Domestic and International) of the Third Lula Administration," paper presented at the Seminar of the School of International Relations of the Getulio Vargas Foundation, Sao Paulo, May 14, 2024.

64. Claire Dupont and Sebastian Oberthur, "The European Union," in *Research Handbook on Climate Governance*, ed. Karin Bachsrang and Eva Lovrand (Edward Elgar Publishing, 2015), 224–236.

65. Tom Delreux and Frauke Ohler, "Climate Policy in European Union Politics," in *Oxford Research Encyclopedia of Politics*, (Oxford University Press, 2019), https://doi.org/10.1093/acrefore/9780190228637.013.1097.

66. "Coal Production and Consumption Up in 2022," *Eurostat*, https://ec.europa.eu/eurostat/web/products-eurostat-news/w/ddn-20230622-2/ (accessed April 23, 2025).

67. Peter Newell, *Climate for Change: Non-State Actors and the Global Politics of the Greenhouse* (Cambridge University Press, 2000).

68. Suraje Dessai, *An Analysis of the Role of OPEC* (Organization of Petroleum Exporters Countries) *as a G77* (Group for Negotiations at the UN formed originally 77 countries) *Member at the UNFCCC*, report for WWF, 2004.

69. Depledge, "The Opposite of Learning."

70. Stefan C. Aykut and Amy Dahan, *Gouverner le climat ? Vingt ans de négociations internationales* (Presses de Sciences Po., 2015).

71. Depledge, "Striving for No: Saudi Arabia in the Climate Change Regime," 9–35.

72. Kari De Pryck, "Controversial Practices: Tracing the Proceduralization of the IPCC in Time and Space," *Global Policy* 12, no. S7 (2021): 80–89.

73. Hannah Hughes, *The IPCC and the Politics of Writing Climate Change* (Cambridge University Press, 2024).

74. "UN Body Wants Probe of Climate E-mail Row," BBC, December 4, 2009, http://news.bbc.co.uk/2/hi/science/nature/8394483.stm.

75. Depledge, "Striving for No," 9–35.

76. Depledge, "The Opposite of Learning."

77. Dessai, *An Analysis of the Role of OPEC as a G77 Member.*

78. Jon Barnett, "The Worst of Friends: OPEC and G-77 in the Climate Regime," *Global Environmental Politics* 8, no. 4 (2008): 1–8.

79. See also Kari De Pryck and Adèle Gaveau, "Scientists in Multilateral Diplomacy: The Case of the Members of the IPCC Bureau," *Political Anthropological Research on International Social Sciences (PARISS)* 4, no. 1 (2023): 65–105.

80. Hannah Hughes, "Actors, Activities, and Forms of Authority in the IPCC," *Review of International Studies* 50, no. 2 (2024): 333–353.

81. Dessai, *An Analysis of the Role of OPEC as a G77 Member.*

82. Depledge, "The Opposite of Learning."

83. Dessai, *An Analysis of the Role of OPEC as a G77 Member.*

84. Depledge, "The Opposite of Learning."

85. Bert Bolin, *A History of the Science and Politics of Climate Change: The Role of the Intergovernmental Panel on Climate Change* (Cambridge University Press, 2007), 113.

86. Earth Negotiations Bulletin (ENB), "Summary of the 55th Session of the Intergovernmental Panel on Climate Change and the 12th Session of Working Group II: 14–27 February 2022," *International Institute for Sustainable Development* 12, no. 794 (2022): 1–24.

87. Jessica Leigh O'Reilly, Mark Vardy, Kari De Pryck, and Marcela da S. Feital Benedetti, *Inside the IPCC* (Cambridge University Press, 2024).

88. Earth Negotiations Bulletin (ENB), "Summary of the 56th Session of the Intergovernmental Panel on Climate Change and the 14th Session of Working Group III: 21 March–4 April 2022," *International Institute for Sustainable Development* 12, no. 795 (2022): pp 1–32.

89. Newell, *Climate for Change.*

90. Hannah Hughes and Mathew Paterson, "The IPCC and Key Tensions in Global Climate Politics," *Samfundsøkonomen* 2019, no. 4 (2023): 6–14.

91. Joana Depledge, Kari De Pryck, and Timmons Roberts, "Decades of Systematic Obstructionism: Saudi Arabia's Role in Slowing Progress in UN Climate Negotiations," Climate Social Science Network Issue Paper, 2023.

92. Agni Kalfagianni and Sébastien Duyck, *The Evolving Role of Agriculture in Climate Change Negotiations: Progress and Players*, CCAFS Info Note (CCAFS, 2015).

93. Morten K. Flisnes, *Where You Stand Depends on What You Sell: Saudi Arabia's Obstructionism in the UNFCCC 2012–2018* (CICERO, 2019).

94. Thijs Van De Graaf, *The Politics and Institutions of Global Energy Governance* (Palgrave Macmillan, 2013).

95. Aykut and Dahan, *Gouverner le climat ?*

96. Depledge, "Striving for No," 9–35.

97. Joana Depledge, Tracing the Origins of the Kyoto Protocol: An Article-By-Article Textual History, 2000, FCCC/TP/2000/2: 20, https://unfccc.int/resource/docs/tp/tp0200.htm.

98. Susan Biniaz, "Comma But Differentiated Responsibilities: Punctuation and 30 Other Ways Negotiators Have Resolved Issues in the International Climate Change Regime," *Michigan Journal of Environmental & Administrative Law* 6, no. 1 (2016): 19.

99. Depledge, De Pryck, and Roberts, *Decades of Systematic Obstructionism.*

100. Stefan C. Aykut and Monica Castro, "The End of Fossil Fuels? Understanding the Partial Climatisation of Global Energy Debates," in *Globalising the Climate: COP21 and the Climatisation of Global Debates*, ed. Stefan Aykut, Jean Foyer, and Eduard Morena (Routledge Earthscan, 2017), 186.

101. Irja Vormedal, "The Influence of Business and Industry NGOs in the Negotiation of the Kyoto Mechanisms: The Case of Carbon Capture and Storage in the CDM," *Global Environmental Politics* 8, no. 4 (2008): 36–65.

102. Van Asselt and Green, "COP26 and the Dynamics of Anti-Fossil Fuel Norms," 3.

103. Ibid.

104. Earth Negotiations Bulletin (ENB), "Summary of the 2023 Dubai Climate Change Conference: 30 November–13 December 2023," *International Institute for Sustainable Development* 12, no. 842 (2023): pp 1–30.

105. Hughes, *The IPCC and the Politics of Writing Climate Change.*

106. E. Lisa F. Schipper, "Conceptual History of Adaptation in the UNFCCC Process," *Review of European Community and International Environmental Law* 15, no. 1 (2006): 82–92.

107. Schipper, "Conceptual History of Adaptation in the UNFCCC Process," 82–92.

108. Earth Negotiations Bulletin (ENB), "Summary Report 6–18 December 2004," *International Institute for Sustainable Development* 12, no. 160 (2004): 1–16.

109. Earth Negotiations Bulletin (ENB), "Daily Report for 13 December 2004," *International Institute for Sustainable Development* 12, no. 256 (2004): 1–2.

110. Mary J. Mace, "Adaptation under the UN Framework Convention on Climate Change: the international legal framework," in *Fairness in adaptation to climate change*, eds. N. W. Adger, J. Paavola, S. Huq and M.J. Mace (2006): 53—76.

111. Magnus Benzie and Åsa Persson, "Governing Borderless Climate Risks: Moving Beyond the Territorial Framing of Adaptation," *International Environmental Agreements: Politics, Law and Economics* 19, no. 4–5 (2019): 369–393.

112. Barbara Buchner, Angela Falconer, Morgan Hervé-Mignucci, Chiara Trabacchi, and Marcel Brinkman, "The Landscape of Climate Finance," Climate Policy Initiative, 2011.

113. Mizan Khan, Stacy-ann Robinson, Romain Weikmans, David Ciplet, and J. Timmons Roberts, "Twenty-Five Years of Adaptation Finance through a Climate Justice Lens," *Climatic Change* 161 (2019): 251–269.

114. UN Environment Programme (UNEP), "Adaptation Gap Report 2022: Too Little, Too Slow—Climate Adaptation Failure Puts World at Risk," 2002, https://www.unep.org/adaptation-gap-report-2022.

115. Khan et al., "Twenty-Five Years of Adaptation Finance through a Climate Justice Lens," 251–269.

116. For historical developments on loss and damage see Patrícia Galvão Ferreira, "Arrested Development: The Late and Inequitable Integration of Loss and Damage Finance into the UNFCCC," in *Research Handbook on Climate Change Law and Loss & Damage*, ed. Meinhard Doelle and Sara Seck (Edward Elgar Publishing, 2021), 127–148; Linda Siegele, "Loss and Damage under the Convention," in *Research Handbook on Climate Change Law and Loss & Damage*, ed. Meinhard Doelle and Sara Seck, (Edward Elgar Publishing, 2021), 75–99; Lisa Vanhala and Cecilie Hestbaek, "Framing Climate Change Loss and Damage in UNFCCC Negotiations," *Global Environmental Politics* 16, no. 4 (2016): 111–129; Danielle Falzon, J. Timmons Roberts, Fred Shaia, Md. Fahad Hossain, Stacy-ann Robinson, Mizan Khan, and David Ciplet. "Loss and Damage Finance: Four Periods of Obstruction in the UN Climate Negotiations," Climate Social Science Network Briefing, 2022.

117. Daniel M. Bodansky, "The Emerging Climate Change Regime," *Annual Review of Energy and the Environment* 20, no. 1 (1995): 425–461.

118. Irving M. Mintzer and J. Amber Leonard (eds.) *Negotiating Climate Change: The Inside Story of the Rio Convention* (Cambridge University Press, 1994).

119. Vanhala and Hestbeck, "Framing Climate Change Loss and Damage." Link: https://econpapers.repec.org/article/tprglenvp/v_3a16_3ay_3a2016_3ai_3a4_3ap_3a111-129.htm

120. Falzon et al., "Tactical Opposition: Obstructing Loss and Damage Finance in the United Nations Climate Negotiations," 95–119.

121. Friederike Hartz, "'Loss and Damage' to 'Losses and Damages': Orthographies of Climate Change Loss and Damage in the IPCC," *Global Environmental Politics* 23 (2023): 32–51.

122. Intergovernmental Panel on Climate Change (IPCC), "IPCC WGII AR5 Summary for Policymakers (SPM): First-Order Draft: Comments and Responses," 2013, https://archive.ipcc.ch/pdf/assessment-report/ar5/wg2/drafts/WGIIAR5_FODSPM_annotation.pdf, 134.

123. Intergovernmental Panel on Climate Change (IPCC), "IPCC WGI SR15 Final Government Draft Review Comments And Responses—Summary for Policymakers," 2019, https://www.ipcc.ch/site/assets/uploads/sites/2/2019/09/SR15FGD_Summary_for_Policymakers_Comments_and_Responses.pdf, 100.

124. Intergovernmental Panel on Climate Change (IPCC), *Global Warming of 1.5°C: IPCC Special Report on Impacts of Global Warming of 1.5°C above Pre-industrial Levels in Context of Strengthening Response to Climate Change, Sustainable Development, and Efforts to Eradicate Poverty* (Cambridge University Press, 2022), 553.

125. Hartz, "'Loss and Damage' to 'Losses and Damages,'" 32–51.

126. Intergovernmental Panel on Climate Change (IPCC), "IPCC AR6 WGII Second Order Draft Government and Expert Review Compiled Comments Responses (Entire Report)," 2022, https://www.ipcc.ch/report/ar6/wg2/downloads/report/IPCC_AR6_WGII_SOD_CommentsResponses_EntireReport.pdf.

127. Intergovernmental Panel on Climate Change (IPCC), "IPCC Working Group II Contribution to the Sixth Assessment Report—Corrigenda to the Final Draft_Rev2," 2022, https://www.ipcc.ch/report/ar6/wg2/downloads/report/IPCC_AR6_WGII_Corrigenda.pdf.

128. Hartz, "'Loss and Damage' to 'Losses and Damages,'" 32–51.

129. Stoddard et al., "Three Decades of Climate Mitigation," 661.

130. Jonathan Pickering, Steve Vanderheiden, and Seumas Miller, "'If Equity's In, We're Out': Scope for Fairness in the Next Global Climate Agreement," *Ethics & International Affairs* 26, no. 4 (2012): 423–443.

131. Lavanya Rajamani, "Ambition and Differentiation in the 2015 Paris Agreement: Interpretative Possibilities and Underlying Politics," *International and Comparative Law Quarterly* 65, no. 2 (2016): 508.

132. Naveeda Khan, *In Quest of a Shared Planet: Negotiating Climate from the Global South* (Fordham University Press, 2023).

133. Lau Øfjord Blaxekjær, Bård Lahn, Tobias Dan Nielsen, Lucia Green-Weiskel, and Fang Fang, "The Narrative Position of the Like-Minded Developing Countries in Global Climate Negotiations," in *Coalitions in the Climate Change Negotiations*, ed. Carola Klöck, Paula Castro, Florian Weiler, and Lau Øfjord Blaxekjær (Routledge, 2020), 113–135.

134. Kathryn Hochstetler and Eduardo Viola, "Brazil and the Politics of Climate Change: Beyond the Global Commons," *Environmental Politics* 21, no. 5 (2012): 753–771.

135. Tora Skodvin and Steinar Andresen, "Nonstate Influence in the International Whaling Commission, 1970–1990," *Global Environmental Politics* 3, no. 4 (2003): 61–86.

136. Peter Newell and Mathew Paterson, "A Climate for Business: Global Warming, the State and Capital," *Review of International Political Economy* 5, no. 4 (1998): 679–703.

137. Robert J. Brulle, "Advocating Inaction: A Historical Analysis of the Global Climate Coalition," *Environmental Politics* 32, no. 2 (2023): 185–206.

138. Marcel Hanegraaff, "When Does the Structural Power of Business Fade? Assessing Business Privileged Access at Global Climate Negotiations," *Environmental Politics* 32, no. 3 (2023): 427–451.

139. Flisnes, *Where You Stand Depends on What You Sell*.

140. Corporate Europe, Corporate Accountability and Global Witness, "COP27: 100 More Fossil Fuel Lobbyists Than Last Year," November 2022, https:// corporateeurope.org/en/2022/11/cop27-100-more-fossil-fuel-lobbyists-last-year.

141. Michele M. Betsill and Elisabeth Corell, *NGO Diplomacy* (MIT Press, 2008).

142. Kirsten Lucas, "Behind Success Stories: Goal Attainment in Global Trade and Climate Negotiations," *Interest Groups & Advocacy* 8, no. 1 (2019): 44–67.

143. Simone Pulver, "Organising Business: Industry NGOs in the Climate Debates," in *The Business of Climate Change*, ed. Kathryn Begg, Frans van der Woerd, and David Levy (Routledge, 2005), 47–60.

144. Myanna Lahsen, "Experiences of Modernity in the Greenhouse: A Cultural Analysis of a Physicist 'Trio' Supporting the Backlash Against Global Warming," *Global Environmental Change* 18, no. 1 (2008): 204–219.

145. Brulle, "Advocating Inaction," 185–206.

146. Paul N. Edwards and Stephen H. Schneider, "The 1995 IPCC Report : Broad Consensus or 'Scientific Cleansing?,'" *Ecofable/Ecoscience* 1, no. 1 (1997): 3–9.

147. Myanna Lahsen, "The Detection and Attribution of Conspiracies: The Controversy Over Chapter 8," in *Paranoia Within Reason: A Casebook on Conspiracy as Explanation*, ed. George Marcus (University of Chicago Press, 1999), 111–136.

148. Tora Skodvin, *Structure and Agent in the Scientific Diplomacy of Climate Change. An Empirical Case Study of Science-Policy Interaction in the Intergovernmental Panel on Climate Change* (Kluwer Academic Publishers, 2000).

149. Leggett, *The Carbon War*.

150. Newell, *Climate for Change*.

151. Newell and Paterson, "A Climate for Business," 683.

152. Ibid., 688.

153. Pulver, "Organising Business," 47–60.

154. Sarah Benabou, Nils Moussu and Birgit Müller, "The Business Voice at COP21: The Quandaries of a Global Political Ambition," in *Globalising the Climate: COP21 and the Climatisation of Global Debates*, ed. Stefan Aykut, Jean Foyer, and Eduard Morena (Routledge Earthscan, 2017), 57–74.

155. Brulle, "Advocating Inaction," 185–206.

156. Ingvild Andreassen Sæverud and Jon Birger Skjærseth, "Oil Companies and Climate Change: Inconsistencies Between Strategy Formulation and Implementation?," *Global Environmental Politics* 7, no. 3 (2007): 42–62.

157. Miquel Munoz Cabré, "Issue-linkages to Climate Change Measured through NGO Participation in the UNFCCC," *Global Environmental Politics* 11, no. 3 (2011): 10–22.

158. Nagmeh Nasiritousi, Mattias Hjerpe, and Björn-Ola Linnér, "The Roles of Non-State Actors in Climate Change Governance: Understanding Agency Through Governance Profiles," *International Environmental Agreements: Politics, Law and Economics* 16, no. 1 (2016): 109–126.

159. Benabou et al., "The Business Voice at COP21," 57.

160. Ibid.

161. World Economic Forum, "Financing the Transition to a Net-Zero Future," Insignt Report, October 2021, https://www3.weforum.org/docs/WEF_Financing_the_Transition_to_a_Net_Zero_Future_2021.pdf.

162. Benabou et al., "The Business Voice at COP21," 59.

163. Ibid., 57–74.

164. Ibid., 71.

165. Matthew S. Bach, "Is the Oil and Gas Industry Serious About Climate Action?," *Environment: Science and Policy for Sustainable Development* 59, no. 2 (2017): 4–15.

166. Nagmeh Nasiritousi, "Fossil Fuel Emitters and Climate Change: Unpacking the Governance Activities of Large Oil and Gas Companies," *Environmental Politics* 26, no. 4 (2017): 621–647.

167. Ibid., 621–647.

168. Wim Carton, Inge-Merete Hougaard, Nils Markusson, and Jens Friis Lund, "Is Carbon Removal Delaying Emission Reductions?," *WIREs Climate Change* 14, no. 4 (2023): e826.

169. Bach, "Is the Oil and Gas Industry Serious About Climate Action?," 4–15.

170. Corporate Europe, Corporate Accountability and Global Witness, "COP27: 100 More Fossil Fuel Lobbyists Than Last Year."

171. Marie Hrabanski and Jean François le Coq, "Climatisation of Agricultural Issues in the International Agenda Through Three Competing Epistemic Communities: Climate-Smart Agriculture, Agroecology, and Nature-Based Solutions," *Environmental Science & Policy* 127 (2022): 311–320.

172. Rachel Sherrington, "Big Meat Unveils Battle Plans for COP28," *DeSmog*, November 29, 2023, https://www.desmog.com/2023/11/29/big-meat-unveils-battle-plans-for-cop28/.

173. Carola Betzold, Paula Castro, and Florian Weiler, "AOSIS in the UNFCCC Negotiations: From Unity to Fragmentation?," *Climate Policy* 12, no. 5 (2012): 591–613.

174. Nicole Deitelhoff and Linda Wallbott, "Beyond Soft Balancing: Small States and Coalition-Building in the ICC and Climate Negotiations," *Cambridge Review of International Affairs* 25, no. 3 (2012): 345–366.

175. Aykut and Castro, "The End of Fossil Fuels?," 186.

176. See Fossil Fuel Treaty, https://fossilfueltreaty.org/ (accessed June 29, 2024).

177. Newell and Paterson, "A Climate for Business," 690.

178. Depledge et al., *Decades of Systematic Obstructionism*, 15.

179. Nagmeh Nasiritousi, Alexandra Buylova, Mathias Fridahl, and Gunilla Reischl, "Making the UNFCCC Fit for Purpose: A Research Agenda on Vested Interests and Green Spiralling," *Global Policy* 15, no. 2 (2024): 487–494.

180. Nasiritousi et al., "Making the UNFCCC Fit for Purpose," 492.

181. Brook M. R. Dambacher, Matthew T. Stilwell, and Jeffrey S. McGee, "Clearing the Air: Avoiding Conflicts of Interest Within the United Nations Framework Convention on Climate Change," *Journal of Environmental Law* 32, no. 1 (2020): 53–81.

CHAPTER 11

Obstructing Global and Local Climate Change Adaptation

LEAD AUTHORS: LAURA KUHL AND STACY-ANN ROBINSON
CONTRIBUTING AUTHORS: NATALIE DIETRICH JONES, DANIELLE FALZON, ANDREW MALMUTH, MICHAEL MIKULEWICZ, MEG MILLS-NOVOA, MICHELLE MYCOO, MEG PARSONS, M. FEISAL RAHMAN, E. LISA F. SCHIPPER, KIMBERLEY ANH THOMAS, AND EDWARD WALKER

INTRODUCTION: OBSTRUCTION OF ADAPTATION

To date, most scholarship on climate obstruction has focused on the effort to slash greenhouse gas (GHG) emissions, that is, mitigation, aiming to understand how actors delay or block those policies and actions. Significantly less attention has been paid to adaptation obstruction, or how actors delay or block actions that manage and/or reduce the risks and impacts of climate change, both anticipated and already occurring. Despite this disproportionate focus, obstruction is also an observable dynamic in climate adaptation and warrants study for several reasons. First, mitigation and adaptation are complementary activities, and both are essential for mounting an effective response to climate change. Second, climate change impacts are already being experienced globally, affecting individuals, communities, and ecosystems, and adaptation is crucial for addressing these immediate effects and safeguarding lives, livelihoods, and well-being. Third, ensuring that adaptation receives sufficient attention can help address the inequities in climate impacts, providing support and resources to those most in need and least responsible for GHG emissions. Accordingly, we argue that analyzing obstruction using an adaptation lens is helpful for identifying the power and politics that shape its processes and outcomes.

Laura Kuhl et al., *Obstructing Global and Local Climate Change Adaptation*. In: *Climate Obstruction*.
Edited by: J. Timmons Roberts, Carlos R. S. Milani, Jennifer Jacquet, and Christian Downie,
Oxford University Press. © Oxford University Press (2025). DOI: 10.1093/oso/9780197787144.003.0011

Broadly speaking, the goal of adaptation is to "reduce risk and vulnerability to climate change, strengthen resilience, enhance well-being and the capacity to anticipate and respond successfully to change."[1] It is the process of adjustment to actual or expected climate and its effects. In human systems, adaptation seeks to moderate harm or exploit beneficial opportunities.[2] Specific adaptation actions will vary across sectors and regions and be dependent on local conditions. They can range from crop diversification and the introduction of climate-resilient varieties (in agriculture) to rainwater harvesting and building flood management infrastructures (for water resources); from relocation and managed retreat (in coastal zones) to "green" infrastructure and updated building codes (in urban planning), to early warning systems and emergency preparedness (for disaster risk reduction). Adaptation also entails decisions about which actions will support whose well-being and how those measures and strategies will be implemented. Articulating the goals of specific adaptation actions and, by extension, possible adaptation strategies, is therefore inherently political and often contested.[3,4,5] Dimensions such as scope, spatial scale, and time frame are influenced by what is understood as adaptation, and this definition has implications for who designs, funds, and implements adaptation projects or programs.[6,7]

Given the possible benefits of adaptation, it may be difficult to see why actors may work to delay or block these efforts. They do, however, because a given adaptation action itself may not be perceived as being in certain actors' best interest, or it might disrupt existing power dynamics that benefit them. For example, real estate developers might oppose zoning changes, building restrictions, or regulations aimed at reducing vulnerability to climate impacts, as these could limit their development opportunities or increase costs. Tourism operators might resist adaptation measures that they believe could negatively affect their profit margins, such as restrictions on coastal development, changes to beach maintenance practices, or promotion of alternative tourism models. Farmers may resist changes to traditional or Indigenous farming practices, such as crop diversification or reduced water usage, due to potential initial costs, perceived risks, or disruptions to established routines or ways of life. Other forms of resistance can entail actions by marginalized actors against top-down adaptation interventions and can be productive in helping to reduce the actors' climate risk.

Considering these examples, it may be difficult to distinguish between obstruction and resistance. Both are deliberate processes driven by diverse actors operating within adaptation policy and implementation spaces that can result in barriers to the creation and implementation of climate adaptation policy and action. Obstruction can therefore be seen as one type of resistance to adaptation, whereas resistance is a broader concept encompassing a wide set of actions that oppose adaptation.

Though actors can construct barriers to adaptation and obstruction can result in such barriers, obstruction and barriers are conceptually different. In the broader academic literature, an adaptation barrier is a natural, technological, economic, social, or institutional impediment to specific adaptations. It can therefore be difficult to distinguish between them in reality, and much depends on the interests and desires of the actors in power as to whether barriers will be overcome, or those challenges used as a rationale to obstruct adaptation. While there is very little literature on this distinction, this chapter provides an analysis of power and justice that can potentially inform debates on adaptation barriers versus adaptation obstruction.

This chapter aims to shed light on how various actors—governmental, nongovernmental, and private—seek to delay or block adaptation actions at the global and local scales. In the next section, we unpack the tensions between obstruction and resistance and between obstruction and barriers. Next, we describe three types of adaptation obstruction: (1) obstruction through decision-making (direct conflict), (2) obstruction through non-decision-making (agenda setting), and (3) obstruction through hegemony (ideological control). The section that follows addresses adaptation obstruction at the global scale. There, the focus is on the Intergovernmental Panel on Climate Change (IPCC) and the United Nations Framework Convention on Climate Change (UNFCCC), parts of the key organizing regime for climate science and coordination on adaptation, respectively. The section also focuses on donors, due to their role in designing and funding adaptation action in the Global South (for more on the IPCC and UNFCCC, see Chapter 10). Finally, we consider adaptation obstruction at the local scale, examining the roles of actors such as national and local governments, nongovernmental organizations (NGOs), and individuals including private property owners (see Chapter 9 for more on subnational obstruction).

Throughout, the chapter explores how both global and local power dynamics obstruct adaptation. Here, we draw on examples across different sectors and regions, including infrastructure, real estate, coastal development, rural and urban development, agriculture, migration, and managed retreat, and provide examples from both the Global North (e.g., North America, Australia, and Aotearoa-New Zealand) and South (e.g., Asia and Caribbean and Pacific small island developing states [SIDS]), a grouping of particularly vulnerable countries). Given the Global North's outsized contribution to GHG emissions (both historically and currently) and the high vulnerability of the Global South to climate impacts (due to historical exploitation, exposure and lack of adaptive capacity), positions in international climate politics regularly fall along a Global North/Global South divide, with important implications for climate justice.

OBSTRUCTION IS A TYPE OF RESISTANCE TO ADAPTATION

The IPCC's Sixth Assessment Report makes it clear that many climate adaptation projects around the world are not increasing resilience; rather, adaptation interventions can reinforce, redistribute, and create new vulnerabilities.[8] Some coping strategies can be short-term and potentially undermine long-term, sustainable solutions, creating a high risk of maladaptation, whereby people become more, rather than less, vulnerable as a result of the strategy.[9] Importantly, most strategies are not inherently maladaptive or adaptive; it is usually the context in which they are implemented and how they are designed that influences whether they risk creating more damage than improvement.[10] Resistance to these negative impacts and unintended consequences of adaptation policies and projects is necessary to avoid poor project outcomes and maladaptation, which is different from obstruction.

There are many reasons people may legitimately resist government adaptation actions (or inaction). For example, in Brazil, government land-management policies (e.g., dam construction) and inaction (e.g., authorities failing to stop illegal activities) contribute to large-scale deforestation in the Amazon rainforest and the destruction of freshwater systems to make way for industry and infrastructure—not only dams but also agriculture, mining, and more.[11,12,13,14] Such development strategies undermine the adaptive capacity of local communities and resistance may be necessary to avoid injustice and the creation of new environmental harms.

In cases such as these, obstruction is a subset of resistance differentiated by two key factors. First, an action that hinders or stops adaptation that would lead to greater procedural, recognitional, or distributive injustice can be considered obstruction.[15] Second, obstruction may be a calculated and deliberate activity pursued by an actor (a government, an organization, a business, an individual or a group of individuals) with enough power or influence to prevent, delay, or otherwise determine the adaptation process or outcomes, often to promote or protect their own interests.

Powerful Actors Use Barriers to Obstruct Adaptation

As noted, there are many barriers to adaptation, but when powerful interests use these barriers to delay or block adaptation, it is obstruction. Barriers are natural, technological, economic, social, or institutional impediments to specific adaptations for specific actors in their given context that arise from a condition or a set of conditions.[16] These barriers include static obstacles (structural or otherwise) that prevent adaptation, such as institutional lock-in (i.e., once an institution has a particular structure and mode of operation, it is hard for that structure to be changed),[17,18] scientific uncertainty,[19] lack of funding,[20] or neoliberal governance (i.e., due to the hollowing out of public institutions that are critical for public services and a reliance on economic models, which are not always appropriate for adaptation investments).[21] Barriers make adaptation more challenging. For example, competing government-determined socioeconomic development priorities (especially after the COVID-19 pandemic (e.g.[22]), the absence of land-use planning frameworks,[23,24] corruption,[25] and insufficient human resources have been described as barriers to adaptation in Caribbean SIDS such as Jamaica and St. Lucia as well as in the Pacific.[26] However, behind many of these so-called barriers are powerful interests that benefit from the status quo and thus obstruct adaptation.

Although there is a rich literature on adaptation barriers, there is no explicit literature on adaptation obstruction. In the context of the adaptation literature, sometimes the term "barrier" is used inaccurately to describe adaptation obstruction. Using it to describe a deliberate delay or blocking of adaptation, or obstruction, may lead to an assumption that an adaptation action is outside an actor's control and that adaptation is inherently positive but faces obstacles.[27] For example, in the context of urban planning, competing priorities are often cited as barriers to the implementation of zoning and land-use policies designed to assist cities and urban settlements to adapt to the impacts of climate change.[28] However, if the barriers exist as a result of powerful actors' efforts to stop adaptation because the current zoning or land use patterns benefit them, then the creation of such barriers should be considered obstruction.

Adaptation Obstruction Creates Injustice for Marginalized Groups

Adaptation obstruction by powerful actors creates injustice across scales, sectors, and regions. Adaptation often disrupts existing power relations when the actions taken are not in the interests of powerful actors, who may benefit from existing systems that maintain certain groups in positions of vulnerability.[29,30] These actors may actively seek to delay, block, or even redirect such action. This process occurs in three main ways: through decision-making (direct conflict), through non–decision-making (agenda setting), and through hegemony (ideological control).[31]

Adaptation obstruction through decision-making (direct conflict) involves exploitation, dehumanization, and exclusion. Obstruction through non–decision-making (agenda setting) involves earmarking, being left off the agenda, and nominal participation. Obstruction through hegemony (ideological control) involves de facto control, internalized marginality, and nonthreatening participation. Nevertheless, distinguishing between obstruction and resistance along the axes of power and injustice may be challenging in some scenarios due to the complexities of assessing the nature of intersecting power relationships and whether or not adaptation contributes to greater justice.

ADAPTATION OBSTRUCTION AT THE GLOBAL SCALE

Through the intersection of power and injustice, adaptation obstruction dynamics can be observed at all scales. At the global scale, adaptation has been obstructed through the intentional prioritization of mitigation (by powerful actors in the Global North) over adaptation (the priority of the Global South). In the context of multilateral organizations such as the IPCC and UNFCCC, powerful actors have opposed specific definitions of adaptation and obfuscated options for its measurement, particularly where it concerns measuring on-the-ground progress toward globally agreed goals and attracting adaptation finance.[32,33] It has also been argued that the emphasis on defining and measuring adaptation effectiveness has actually obviated discussions on how the measurement process could obstruct adaptation action.[34] This lack of consensus has set the scene for adaptation obstruction. Because adaptation is inherently normative (i.e., what an actor prioritizes depends on what kind of world they envision and the kinds of changes associated with realizing that vision), the process through which authoritative groups like the IPCC or UNFCCC assign adaptation importance and draw the boundaries of what counts as adaptation delineates the adaptation agenda in ways that exclude other priorities and obstruct other potentially transformative actions, particularly on behalf of the most vulnerable peoples and places.[35]

The Intergovernmental Panel on Climate Change

Early IPCC assessment reports emphasized mitigation to reflect the "growing [and overwhelming] concern to sharply reduce greenhouse gas emissions."[36] The First Assessment Report (1990) "discussed both adaptation (sometimes called adjustment) and mitigation (termed limitation at that time)"; however, mitigation was featured more prominently.[37] For the first time, in the Third Assessment Report (2001), adaptation was clearly separated from mitigation, and the Working Groups were renamed to include impacts, adaptation, and vulnerability as part of Working Group II's scope. With Working Group II's contributions to the Fourth, Fifth and Sixth Assessment Reports of 2007, 2014, and 2022, respectively, there was evidence of adaptation becoming a more central goal of climate action—there was more extensive reporting on the topic alongside an emphasis on assessing possible scenarios and impacts. Despite this progress, however, the tension between mitigation and adaptation has remained unresolved in this forum. Although adaptation has grown in visibility, a pattern of sidelining its role and scope in climate action has been firmly established. Additionally, because the IPCC is an intergovernmental body, Member States have considerable sway over the report contents and also approve them. This situation creates opportunities for obstructionist national entities to block and change adaptation-related findings and calls for action.

Additionally, the climate literature itself is not free of bias. For example, it has been observed that academic research on climate adaptation in SIDS has been dominated by the Global North.[38] A recent article[39] found that the most frequently published scholars in the landscape of climate change vulnerability and islands are based in the United Kingdom, Australia, France, and Denmark. Given the fact that most IPCC authors are Caucasian/white/European and the research reflected in the Assessment Reports is conducted largely by Global North scientists, there exists a systematic bias in these reports toward a Global North perspective, which may exclude some values and priorities of actors in the Global South.

The United Nations Framework Convention on Climate Change

The IPCC and UNFCCC have a complementary and interconnected relationship, with the former providing the scientific foundation for the latter's policymaking processes. Much as in the IPCC, Global North countries have regularly worked to prioritize mitigation in the UNFCCC, which relegates other climate objectives, including adaptation, to the backseat.[40,41,42,43] The 1992 UNFCCC text clearly states that its ultimate objective is to stabilize the concentration of GHG emissions in the atmosphere, and thus the entire climate regime under the Convention works to serve this objective. The text also barely addresses adaptation, and when it does, the language is often vague and does not clarify or provide guidance on what it means to adapt to the impacts of climate change.[44,45,46] The 1997 Kyoto Protocol, which sought to clarify national commitments, also focused primarily on climate mitigation rather than adaptation. Its main objective was to reduce GHG emissions from industrialized nations and economies in transition (Global North countries) to combat climate change, thus reinforcing the prioritization of mitigation over adaptation.

There have been several initiatives to advance adaptation in the UNFCCC—for example, the 2005 Nairobi Work Programme on impacts, vulnerability, and adaptation, which sought to improve understanding and assessment of adaptation needs and practices, and the 2010 Cancun Adaptation Framework, which aimed to enhance action on adaptation—but it was not until the 2015

Paris Agreement that the legitimacy of adaptation was elevated. Adaptation was finally given its own article (Article 7) in a legally binding international agreement. However, while the Agreement established a goal for adaptation as "enhancing adaptive capacity, strengthening resilience and reducing vulnerability to climate change," it is not specific, measurable, or time-bound. The Agreement also contains ill-defined mechanisms for accounting and accountability for adaptation actions, and signatories are only encouraged but not required to submit Adaptation Communications outlining their adaptation actions. Notably, throughout the Global Goal negotiations, it has become evident that some Global North countries do not agree that further guidance on adaptation is needed, and they have regularly stalled the addition of substantive targets and indicators for adaptation.[47]

Other obstruction strategies used by the Global North in the context of the UNFCCC have included pointing to scientific uncertainty about the attribution of impacts to climate change, raising concerns about maladaptation, and requiring adaptation to be mainstreamed into development planning to slow negotiations.[48] Global South countries were encouraged to prepare National Adaptation Programmes of Action, but were given guidance only for adaptation planning and not for its implementation.[49,50] Here, an important component of implementation is the provision of finance, so donors are particularly important actors, as they can contribute to or help prevent obstruction of adaptation.

Donors

Due to their unequal power compared to beneficiary countries, donors obstruct adaptation through their failure to provide adequate adaptation finance that meets the needs of the Global South. The provision of "new and additional financial assistance from the Global North to the Global South to offset the costs of climate change impacts" was codified in the 1992 UNFCCC text more than three decades ago. However, adaptation finance still represents less than 10% of all climate finance flows,[51] and even this small fraction is unevenly distributed among the most climate-vulnerable countries.[52] The disbursement of these funds has also been slow and insufficient, considering the colonial histories of donor countries and the responsibility of these industrialized countries for producing the bulk of GHG emissions, as well as the costly impacts of climate change on the developing countries they are obligated to assist.[53,54] Thus the failure of donor countries in the Global North to fulfill their financial obligations and provide needed adaptation finance to countries in the Global South is a form of adaptation obstruction.

Donor countries and multilateral development banks obstruct the flows of adaptation finance in several ways, including by creating onerous requirements for tracking, monitoring, and evaluating these flows.[55,56] They also stipulate how projects are to be designed and implemented, with projects increasingly aiming to meet objectives couched in technical language and jargon developed at the global scale. For example, donors regularly seek to evaluate an adaptation project's "paradigm-shift potential," a term not widely understood or used at the local level and which may not be suitable for supporting local priorities.[57,58] Donors also use investment criteria, such as determining whether funds will be used "efficiently,"[59,60,61] to obstruct adaptation investments that do not align with their interests and priorities.[62] For example, countries with weak institutional capacity may qualify only for preparatory support intended to help them improve their governance practices and planning.[63] Some countries have subsequently secured adaptation finance for concrete adaptation investments or projects, while others are still waiting for sufficient financing to be delivered.[64]

The current international system of adaptation finance clearly privileges mitigation and the priorities and interests of donors, thus enabling obstruction. But despite the power imbalances between donors and recipient countries, the latter are not powerless and can resist donor efforts that do not align with their national priorities. An example is the conflict between the Bangladesh Climate Change Resilience Fund, initially capitalized with funding from various development partners, and the Bangladesh Climate Change Trust Fund, established by the Government of Bangladesh with domestic funds, circumventing Global North donors altogether. Box 11.1 shows how a lack of trust between partners and differences in funding distribution arrangements can obstruct adaptation on the ground.

Box 11.1: RESISTING POWER IMBALANCES BETWEEN ADAPTATION FINANCE DONORS AND RECIPIENT COUNTRIES: AN EXAMPLE FROM BANGLADESH

The Bangladesh Climate Change Resilience Fund (Resilience Fund) was established in May 2010 with a $188 million grant commitment from development partners from Denmark, the European Union, Sweden, the United Kingdom and, later, Switzerland. The World Bank served as the trustee, reflecting the donor-oriented nature of much multilateral climate finance. Around the same time, the Government of Bangladesh had established a similar fund called the Bangladesh Climate Change Trust Fund (Trust Fund) from its own funds. The Government preferred that funding flow through the Trust Fund but accepted the establishment of the Resilience Fund even though its creation suggested a break in trust between the development partners and the Government. Many analysts saw the move as "a failure by the United Kingdom to recognize Bangladesh—one of the world's most climate-vulnerable countries—as an expert on the issue."[65] There were also demonstrations within the country about the World Bank's role as the Fund's trustee.[66]

The lack of trust that shaped the Fund's institutional design, including the use of the World Bank Trust Fund model, led to obstruction of the implementation of proposed projects on the ground, and tensions between the World Bank and donors over the slow delivery of funds.[67,68,69] The Resilience Fund's Governing Council was mandated to make the final selection of projects, placing the World Bank as a key arbiter of these projects.[70] According to one evaluation, the Fund was not successful in its intent or purpose, and nearly $50 million was returned unspent to the donors following a long-running dispute over funding delivery.[71] The evaluation indicated that a lack of commitment by the Government of Bangladesh to the Resilience Fund might have been "due to competition from the nationally-owned" Trust Fund."[72,73] Because the Government's resistance to the initial arrangement of sending the funds through the World Bank was disregarded, the Fund's failure to deliver on its full potential was more likely due to the lack of national buy-in (ownership). This case illustrates the complex interplay of obstruction resulting from both donors and national governments acting in their own interests.

ADAPTATION OBSTRUCTION AT THE LOCAL SCALE

Powerful actors use a variety and combination of natural, technological, economic, social, and/ or institutional barriers to block adaptation. As adaptation is context- and location- specific, much of this obstruction takes place at the local scale, sometimes reflecting the dynamics seen at the global scale. It is therefore important to understand the mechanisms through which this obstruction occurs on the ground and the implications for climate justice.

Obstruction Through Direct Conflict

At the local scale, one type of adaptation obstruction occurs through direct conflict. Direct conflict includes extracting labor and resources from marginalized groups (exploitation), subordinating nondominant cultures—particularly ethnic minorities—through harmful and potentially violent means (dehumanization), and forcibly preventing marginalized groups from voicing their concerns, interests, priorities, or positions (exclusion).[74,75]

The adaptation efforts of Indigenous peoples in particular have faced such obstruction. Indigenous peoples comprise about 6.2% of the global population,[76,77] are among the most vulnerable to the impacts of climate change,[78,79,80,81,82] and have unique expertise and priorities for adaptation. Colonial and non–Indigenous-dominated governments often use the lack of legal recognition of Indigenous self-determination and/or land rights to obstruct their adaptation efforts.[83,84,85,86,87] For example, because the Mohawk community of Kanesatake (located in southern Quebec, Canada) is not legally recognized as an Indigenous people under Canadian law and thus does not have land rights, they are unable to obtain insurance for their properties and cannot relocate despite extreme flood events and sea-level rise.[88] In turn, this situation limits their adaptation options, and moreover allows this and other Indigenous communities to be dispossessed, their ways of life disrupted, and their health and well-being severely undermined, resulting in the dehumanization of these nondominant cultures.[89,90,91,92,93] This subordination limits their capacity to govern and manage their lands and waters in sustainable and adaptive ways, as well as their capacity to access the resources needed to adapt.[94,95,96,97]

Local obstruction to adaptation is also prevalent in particular sectors, such as agriculture, that hold the potential to be divisive and lead to direct conflict. Government and donor initiatives that seek to alter or transform production systems, agrarian economies, and/or lifeways obstruct the traditional adaptive strategies of farmers.[98] Scholars have found resistance when these interventions entail a transition from migratory to sedentary pastoralism (livelihood based on livestock) and agropastoralism (livelihood based on livestock and crops),[99] the refurbishment and expansion of irrigation systems,[100,101] the introduction of climate-smart agricultural techniques,[102] and the intensification and commodification of agricultural systems.[103,104] In these examples, resistance arises from a contested view of what counts as adaptation and the interventions (e.g., agricultural modernization, intensification, and technification in the context of rural and agricultural landscapes) promoted by powerful actors. In the Canadian Okanagan Valley, local farmers fiercely fought the implementation of water- metering technology by the South East Kelowna Irrigation District, which introduced this measure to address ongoing water scarcity because they thought the measures would disproportionately harm some users by limiting access and increasing costs.[105]

Retreat programs and policies are another area where obstruction through direct control can occur. Retreat, or the practice of relocating people, buildings, and other assets away from areas affected by climate change, may be necessary when residual risk is intolerable, other risk reduction measures are not viable, or when the adverse effects of intervention or inaction are unacceptable or irreversible.[106] Retreat can be mandatory or voluntary and driven by state, private enterprise, or community or individual actors, but when this process is driven by government intervention, it is referred to as managed retreat. Projects involving managed retreat have encountered resistance from communities earmarked by the state for relocation. Managed retreat involves difficult sacrifices, and property owners, residents, and land users will have different needs and experience different losses, making the issue a ripe opportunity for the powerful to obstruct the adaptation efforts of marginalized groups.[107]

In Pacific SIDS such as Fiji, moving people from the coast to new inland settlements is often viewed as a last resort because of high costs and the difficulty of maintaining some sociocultural aspects of life in locations that are separate from customary land, making the process inherently complex and socially disruptive.[108] Where affected communities are not involved in the design and implementation of managed retreat, this scenario can result in exploitation, dehumanization and exclusion. In Global South megacities such as Lagos, Nigeria, and Manila, Philippines, retreat programs have been found to marginalize the poor while facilitating class-based displacements[109,110] and racialized forms of climate gentrification.

When the decision-making for managed retreat caters to the interests of the most powerful and excludes the interests of the most marginalized, the responses from these excluded groups can be understood as resistance. Among some Caribbean SIDS, informal urban dwellers have been known to resist government relocation projects as they are suspicious that lands are being cleared not as a form of adaptation but in favor of private investor interests. This type of top-down management can create disempowerment and invoke community resistance and litigation, even in Global North countries such as Aotearoa-New Zealand.[111]

Obstruction Through Agenda-Setting

A second type of adaptation obstruction at the local scale occurs through agenda- setting. In some Global South countries, the constricted space for civil society groups and the lack of financing for their activities have prevented them from being able to influence adaptation strategies and outcomes,[112] which can be understood as obstruction. On the one hand, some environmental NGOs have been able to network and connect with powerful industry groups, setting a pro-adaptation agenda in the Global South and propelling corresponding adaptation efforts.[113] On the other hand, some of these organizations have been perceived as obstructionists by governments, and have been relegated to the margins of adaptation policy discussions.[114] In this context, obstruction through agenda-setting involves designating money and other resources for particular activities that serve the interests of the dominant group (earmarking), raising and/or tabling issues and decisions that reflect only the concerns of the dominant group(s) or culture(s) (left off agenda), and limiting the opportunities for marginalized groups to assert their concerns, interests, priorities, or positions (nominal participation).[115,116]

Earmarking is widespread in urban contexts. For example, governments avoid retrofitting informal urban housing using affordable-housing building codes because this action may send a signal that illegal land occupation and housing are acceptable.[117] At the same time, residents may also resist development or improvement of their settlements because they are concerned that they will come under municipal scrutiny and risk eviction.[118] This tension often leads to money and other resources being earmarked for particular activities that serve the interests of the dominant group (in this case, the local government) and issues and decisions being raised and/or tabled that reflect only the concerns of the dominant group(s) or culture(s).[119] It also demonstrates how resistance and obstruction can unwittingly work together to delay adaptation.

Although earmarking is often understood as the provision of money and resources for certain things and not others, the previous example illustrates how a willingness to ignore regulations can function in a similar way to serve the interests of the dominant group. Such obstruction is embedded in patron-client networks within which the client's access to resources is dictated by the relationship with a powerful patron.[120,121] In SIDS, the interaction of their smallness and patron-client networks can lead to more authoritarian leadership styles in which decisions are made by a small group of powerful actors and may not reflect the interests of vulnerable populations. This process can result in "a practice of executive or familial domination, cultures of compliance, and corruption."[122] Specifically, politicians may be reluctant to devote resources toward adaptation initiatives and seek instead to marshal public support for (potentially politically unpopular) policies such as rezoning, managed retreat, building codes, or retro fits for flood or hurricane resilience.

The privileging of urban adaptation over rural adaptation also allows the concerns of urban elites to be addressed at the expense of those of the rural people left off the agenda. In some cases, adaptation in rural areas is set aside altogether in favor of interventions targeting urban areas.[123,124] People are expected to move from rural to urban areas, either electively and preemptively as a choice to migrate, or forcibly through displacement.[125] This process accords with an imagination of "modern" urban futures, promoted by urban elites generally, that devalues rural livelihoods as traditional.[126] Setting an adaptation agenda focused on urban areas and directing resources toward this vision obstructs the efforts of farmers and other rural actors to adapt autonomously.

Meanwhile, some country governments in the Global South have prioritized highly centralized models of economic development, which have led to "the sidelining of climate and other environmental policies"[127] and limited opportunities for marginalized groups to assert their concerns, interests, priorities, or positions (nominal participation). In countries where formal planning is the norm, urban planning has often remained oriented toward protecting value by adding construction or the fortification of existing high-value physical assets (for example, infrastructure and built cultural heritage, private residential developments) rather than enabling disaster- risk reduction for all, effectively ignoring the needs of the most vulnerable.[128] In other instances, "domestic political and economic elites have centralized policymaking on both environment and development questions, which allows these domestic elites to cynically exploit domestic policymaking processes for maintaining the status quo"[129] (also see[130]). This reality has negative implications for the realization of the financial and technological commitments that countries in the Global North have made to those in the Global South.[131]

Obstruction Through Ideological Control

A third type of adaptation obstruction occurs through ideological control. Ideological control includes actors catering to the interests of the dominant group(s) or culture(s) because they control the money and other resources (de facto control), marginalized actors accepting the superiority of the dominant group(s) or culture(s) (internalized marginality), and dominant actors including marginalized ones in decision-making only when they do not threaten the existing power structures (nonthreatening participation) (following[132]). Ideological control can be exerted through the support or negation of cultural lifeways. Traditional and Indigenous forms of knowledge are increasingly recognized as providing valuable approaches to addressing climate change.[133,134] However, infrastructure may undermine Indigenous livelihoods and practices through which such knowledge is produced and sustained. For this reason, infrastructure and information are two of the several powerful tools through which obstruction by ideological control occurs.

Infrastructure as a Tool of Ideological Control

One of the ways in which dominant groups assert ideological control is through infrastructure. Infrastructure comprises the physical systems (pipes, telecommunication lines, railways, etc.) that facilitate circulation and exchange across space.[135] While flows through these networks may be material (e.g., water, goods, people) or nonmaterial (e.g., ideas, finance), the facilitation of some flows entails the obstruction of others.[136,137] In some cases, literal obstruction may be the goal. Hydropower dams, for example, impound water to generate electricity and in the process impede rivers and their constitutive flows of sediments, fish, and nutrients. The designation of a given flow as necessary (and therefore, deserving of support) or incidental (and subject to sacrifice) is a deeply political, power-laden act that bears profound socioe-cological effects.[138,139] The International Dam in the US—Canada Boundary Waters region, for example, destroyed the wild-rice economy on which local Ojibwe and Métis communities relied for subsistence and cultural identity.[140] Such dynamics have prompted one Diné geographer[141] to describe infrastruc-ture as a "colonial beachhead—a temporal encroachment on Indigenous lands and livelihoods, over time, to augment material and political differences that eventually overwhelms Indigenous nations and curtails certain possibilities." Such concerns have direct consequences for adaptation because infrastruc-ture powerfully mediates people's exposure and sensitivity to climate change impacts.[142] Whether a cyclone, flood, or heat wave becomes a manageable inconvenience or a deadly threat hinges in part on the design and provi-sion of infrastructure used to manage coastal inundation, floods or heat, for example.[143,144]

When adaptation measures primarily affect more privileged or powerful actors, their opposition is obstruction. In communities where coastal real estate is a primary and profitable economic asset, there has been increasing obstruction when determining whether, how, and to what extent development patterns should be restructured to preserve public coastal land and/or avoid public financing to protect private coastal real estate.[145] In the United States, coastal residents routinely oppose revisions to land- use documents and flood maps that incorporate climate change threats, citing the prohibitive costs of flood insurance when rates are adjusted for future risk.[146] In Barbados and the Bahamas, for example, private property owners and developers have signif-icant say in how and which adaptation measures are pursued; they are also governed by minimal oversight.[147] Coastal landowners and hotel investors in Barbados, in particular, have obstructed the implementation of coastal set-back regulations used by the Barbadian Government to reduce the risk of flooding and coastal erosion based on arguments that they will lose valu-able prime real estate and their product is beach tourism.[148] Large developers sometimes bypass the local planning agency, with decisions that support con-tinued coastal developments being made through their personal and business connections with the political directorate.[149]

The politics of coastal adaptation to sea-level rise is shaped not only by private property owners and developers but broader organizational networks of real estate actors, environmental activists, and policymakers. In the state of California, on the west coast of the United States, the California Coastal Commission—a state agency tasked with preserving the coastal zone as an environmental resource—has, in many cases, pushed residents, cities, and counties to plan for the "retreat," or landward movement, of development.

Historically, retreat has been possible on a property-by-property basis.[150] However, in the case of adaptation planning at the city, regional, or state level, a loosely connected but powerfully resourced confederation of wealthy coastal homeowners, property owners' associations, and realty associations has obstructed the commission's attempts to implement robust climate-adaptation policy.

In 2018, in Pacifica, California, the local association of realtors distributed flyers alerting homeowners to what they viewed as the property-value risks of sea-level rise adaptation policy, attempting to exert ideological control by shaping opinions on the issue. Since then, real estate interests' attempts to influence climate adaptation policy have become more formally organized and distributed across the state, with groups such as Smart Coast California (funded and originally organized within the California Association of Realtors) wielding significant influence in decision-making at the local and state levels. These obstruction trends have also been seen on the East Coast of the United States[151] and in Australia.[152,153] In New York, for example, residents obstructing proposed developments in their local area have become an impediment to the siting of affordable housing solutions.[154] While there have historically been alliances between affluent property owners, business communities, and conservation movements to preserve the coast as an environmental and economic resource,[155,156] these alliances are being tested and realigned as sea-level rise sharpens the contradictions between the exchange value of coastal private property and the multifaceted use value (e.g., as a site of unceded sovereignty and stewardship for coastal tribal nations) of coastal public space.

Information as a Tool of Ideological Control

Another way that dominant groups assert ideological control is through the use of information. Various actors obstruct adaptation efforts by focusing on scientific uncertainty, raising fears about declining property values, and situating property owners as the victims of adaptation policies. Such adaptation responses, termed as "reflexive ignorance"[157] involve the rearticulation of climate skepticism around scientific authority and serve to defend specific economic interests and reinforce existing power systems. In these contexts, the use of information as a tool of obstruction is tightly bound with the use of infrastructure as a tool of obstruction.

Due to the partisan nature of adaptation policy in the United States, for example, information (including the denial of climate science) is used as a tool to obstruct adaptation and reflects the politicization of climate change at the local scale. There, the attribution of extreme weather events and related adaptation action in communities depends partially on the party affiliation of local governments.[158,159] Other actors use climate denial to obstruct adaptation as well. The Farm Bureau, a six-million-member farmer's organization, lobbying group, and insurance provider, actively promoted climate denial for decades and refused to acknowledge climate change publicly despite its impact on its members. In 2020, however, the Farm Bureau changed course and formed the Food and Agriculture Alliance, which advocates for elective, incentive-based climate change action. Despite this recent change, advocates argue that the Farm Bureau has constrained the adaptive capacity of US farmers by lobbying for federally subsidized insurance only for major commodity crops such as corn and soybeans, disincentivizing transitions away from large-scale, carbon-intensive, monocrop production.[160] As this example illustrates, information can be a powerful tool for obstruction, reinforcing and enabling other forms of obstruction as well.

CONCLUSION

This chapter focused on how powerful actors obstruct climate change adaptation and how adaptation obstruction creates injustice for marginalized groups. We identified the tensions between obstruction and resistance, and between obstruction and barriers. We also analyzed and illustrated the ways in which adaptation is obstructed at global and local scales. At the global scale where the IPCC, the UNFCCC, and donors operate, adaptation has been obstructed primarily by Global North countries' intentionally prioritizing mitigation over adaptation and other powerful actors' opposing specific definitions of adaptation, which complicate options for measuring on-the-ground progress toward globally agreed goals and attracting adaptation finance. Drawing on a framework that explicates the intersections of power and injustice, we also found that national and local governments, businesses, and individuals including private property owners are obstructing adaptation at the local scale by engaging in various forms of direct conflict, agenda-setting, and ideological control. This type of obstruction, occurring across a wide range of sectors (e.g., real estate, coastal development, rural and urban development, and agriculture) and utilizing multiple strategies (e.g., migration, infrastructure, and managed retreat) in both Global North and Global South countries, delays adaptation action, increases the vulnerability of marginalized groups, and perpetuates injustices.

Several research gaps remain for future studies. These include, but are not limited to, gaining a better understanding of the distinction between obstruction and resistance; the conditions under which adaptation obstruction is most likely to occur and reoccur; and how these conditions might differ across the

Global North and the Global South; the most successful strategies for supporting resistance, especially in the context of Indigenous communities; and the interplay between mitigation obstruction and adaptation obstruction.

NOTES

1. R. Ara Begum, R. Lempert, E. Ali, T. A. Benjaminsen, T. Bernauer, W. Cramer, X. Cui, K. Mach, G. Nagy, N. C. Stenseth, R. Sukumar, and P. Wester, "Point of Departure and Key Concepts," in *Climate Change 2022: Impacts, Adaptation and Vulnerability. Contribution of Working Group II to the Sixth Assessment Report of the Intergovernmental Panel on Climate Change*, ed. H.-O. Pörtner, D.C. Roberts, M. Tignor, E.S. Poloczanska, K. Mintenbeck, A. Alegría, M. Craig, S. Langsdorf, S. Löschke, V. Möller, A. Okem, and B. Rama (Cambridge University Press, 2022), 121–196, 177.
2. Ibid.
3. S. H. Eriksen, A. J. Nightingale, and H. Eakin, "Reframing Adaptation: The Political Nature of Climate Change Adaptation," *Global Environmental Change* 35 (2015): 523–533.
4. M. Mikulewicz, "Politicizing Vulnerability and Adaptation: On the Need to Democratize Local Responses to Climate Impacts in Developing Countries," *Climate and Development* 10, no. 1 (2018): 18–34.
5. M. Pelling, *Adaptation to Climate Change: From Resilience to Transformation* (Routledge, 2010).
6. M. F. Rahman, D. Falzon, S. A. Robinson, L. Kuhl, R. Westoby, J. Omukuti, E. L. F. Schipper, K. E. McNamara, B. P. Resurrección, D. Mfitumukiza, and M. Nadiruzzaman, "Locally Led Adaptation: Promise, Pitfalls, and Possibilities," *Ambio*, 2023, https://doi.org/10.1007/s13280-023-01884-7.
7. IPCC. H.-O. Pörtner, D. C. Roberts, M. Tignor, E.S. Poloczanska, K. Mintenbeck, A. Alegría, M. Craig, S. Langsdorf, S. Löschke, V. Möller, A. Okem, and B. Rama (eds.), *Climate Change 2022: Impacts, Adaptation and Vulnerability. Contribution of Working Group II to the Sixth Assessment Report of the Intergovernmental Panel on Climate Change* (Cambridge University Press, 2022), p. 3056.
8. S. Eriksen, E. L. F. Schipper, M. Scoville-Simonds, K. Vincent, H. N. Adam, N. Brooks, B. Harding, D. Khatri, L. Lenaerts, D. Liverman, M. Mills-Novoa, M. Mosberg, S. Movik, B. Muok, A. Nightingale, H. Ojha, L. Sygna, M. Taylor, C. Vogel, and J. J. West, "Adaptation Interventions and Their Effect on Vulnerability in Developing Countries: Help, Hindrance or Irrelevance?," *World Development* 141 (2121); 105383.
9. E. L. F. Schipper, "Maladaptation: When Adaptation to Climate Change Goes Very Wrong," *One Earth* 3, no. 4 (2020): 409–414.
10. D. Reckien, A. K. Magnan, C. Singh, M. Lukas-Sithole, B. Orlove, E. L. F. Schipper, and E. Coughlan de Perez, "Navigating the Continuum Between Adaptation and Maladaptation," *Nature Climate Change* 13 (2023): 907–918.
11. M. Carneiro da Cunha, R. Caixeta, J. M. Campbell, C. Fausto, J. A. Kelly, C. Lomnitz, C. D. Londoño Sulkin, C. Pompeia, and A. Vilaça, "Indigenous Peoples Boxed in by Brazil's Political Crisis," *HAU: Journal of Ethnographic Theory* 7, no. 2 (2017): 403–426.

12. F.-M. Le Tourneau, "The Sustainability Challenges of Indigenous Territories in Brazil's Amazonia," *Current Opinion in Environmental Sustainability* 14 (2015): 213–220.

13. H. M. Ribeiro and J. R. Morato, "Social Environmental Injustices Against Indigenous Peoples: The Belo Monte Dam," *Disaster Prevention and Management: An International Journal* 29, no. 6 (2020): 865–876.

14. D. Urzedo and P. Chatterjee, "The Colonial Reproduction of Deforestation in the Brazilian Amazon: Violence Against Indigenous Peoples for Land Development," *Journal of Genocide Research* 23, no. 2 (2021): 302–324.

15. Rahman et al., "Locally Led Adaptation."

16. IPCC, "*Impacts, Adaptation and Vulnerability*"

17. L. Groen, M. Alexander, J. P. King, N. W. Jager, and D. Huitema, "Re-Examining Policy Stability in Climate Adaptation Through a Lock-In Perspective," *Journal of European Public Policy* 30, no. 3 (2023): 488–512.

18. A. Taylor, A. Cartwright, and C. Sutherland, *Institutional Pathways for Local Climate Adaptation: A Comparison of Three South African Municipalities*, Agence Francaise Developpement, 2014, https://www.africancentreforcities.net/wp-content/uploads/2014/06/FocalesN18_GB_WEB.pdf.

19. M. Moure, J. B. Jacobsen, and C. Smith-Hall, "Uncertainty and Climate Change Adaptation: A Systematic Review of Research Approaches and People's Decision-Making," *Current Climate Change Reports* 9, no. 1 (2023): 1–26.

20. United Nations Environment Programme, *Adaptation Gap Report 2022*, 2022, https://www.unep.org/resources/adaptation-gap-report-2022.

21. G. Fieldman, "Neoliberalism, the Production of Vulnerability and the Hobbled State: Systemic Barriers to Climate Adaptation," *Climate and Development* 3, no. 2 (2011): 159–174.

22. S. A. Rolle, "An Analysis of Economic and Political Resilience Strategies Adopted by the Bahamas as a Small Island Development State," in *Pandemics, Disasters, Sustainability, Tourism*, ed. I. Bethell-Bennett, S. A. Rolle, J. Minnis, and F. Okumus (Emerald Publishing, 2022), 183–191.

23. M. Mycoo, S. Robinson, C. Nguyen, C. Nisbet, and R. Tonkel III, "Human Adaptation to Coastal Hazards in Greater Bridgetown, Barbados," *Frontiers in Environmental Science* 9, no. 155 (2021): 1–18.

24. S. Robinson, A. Douma, T. Poore, and K. Singh, "The Role of Colonial Pasts in Shaping Climate Futures: Adaptive Capacity in Georgetown, Guyana," *Habitat International* 139 (2023): 102902. https://doi.org/10.1016/j.habitatint.2023.102902.

25. S. Robinson, "Adapting to Climate Change at the National Level in Caribbean Small Island Developing States," *Island Studies Journal* 13, no. 1 (2018): 79–100.

26. S. Robinson and M. Dornan, "International Financing for Climate Change Adaptation in SIDS," *Regional Environmental Change* 17, no. 4 (2017): 1103–1115.

27. A. J. Nightingale, S. Eriksen, M. Taylor, T. Forsyth, M. Pelling, A. Newsham, E. Boyd, K. Brown, B. Harvey, L. Jones, R. Bezner Kerr, L. Mehta, L. O. Naess, D. Ockwell, I. Scoones, T. Tanner, and S. Whitfield, "Beyond Technical Fixes: Climate Solutions and the Great Derangement," *Climate and Development* 12, no. 4 (2020): 343–352.

28. D. Dodman, B. Hayward, M. Pelling, V. Castan Broto, W. Chow, E. Chu, R. Dawson, L. Khirfan, T. McPhearson, A. Prakash, Y. Zheng, and G. Ziervogel, "Cities, Settlements and Key Infrastructure," in *Climate Change 2022: Impacts, Adaptation and Vulnerability. Contribution of Working Group II to the Sixth Assessment Report*

of the Intergovernmental Panel on Climate Change, ed. H.-O. Pörtner, D.C. Roberts, M. Tignor, E.S. Poloczanska, K. Mintenbeck, A. Alegría, M. Craig, S. Langsdorf, S. Löschke, V. Möller, A. Okem, B. Rama (Cambridge University Press, 2022), 907–1040.

29. Rahman et al., "Locally Led Adaptation."
30. Robinson, "Adapting to Climate Change at the National Level."
31. Rahman et al., "Locally Led Adaptation."
32. B. Orlove, "The Past, the Present and Some Possible Futures of Adaptation," in *Adaptation to Climate Change: Thresholds, Values, Governance*, ed. W. N. Adger, I. Lorenzoni and K. O'Brien (Cambridge University Press, 2009), 131–163.
33. B. Orlove, "The Concept of Adaptation," *Annual Review of Environment and Resources* 47 (2022): 535–581.
34. S. Fisher, "Much Ado About Nothing? Why Adaptation Measurement Matters," *Climate and Development* 16, no. 2 (2024): 161–167.
35. Ibid.
36. Orlove, "The Concept of Adaptation," 541.
37. Ibid.
38. S. Robinson, "Climate Change Adaptation in SIDS: A Systematic Review of the Literature Pre and Post the IPCC Fifth Assessment Report," *Wiley Interdisciplinary Reviews (WIREs): Climate Chang* 11, no. 4 (2020): e653, https://doi.org/10.1002/wcc.653.
39. N. A. Zulhaimi, J. J. Pereira, and N. Muhamad, "Global Research Landscape of Climate Change, Vulnerability, and Islands," *Sustainability* 15, no. 17 (2023): 13064, http://dx.doi.org/10.3390/su151713064.
40. J. Depledge, "Striving for No: Saudi Arabia in the Climate Change Regime," *Global Environmental Politics* 8, no. 4 (2008): 9–35.
41. D. Falzon, F. Shaia, J. T. Roberts, M. F. Hossain, S. A. Robinson, M. R. Khan, and D. Ciplet, "Tactical Opposition: Obstructing Loss and Damage Finance in the United Nations Climate Negotiations," *Global Environmental Politics* 23, no. 3 (2023): 95–119.
42. W. Lamb, M. Antal, K. Bohnenberger, L. I. Brand-Correa, F. Muller-Hansen, M. Jakob, J. C. Minx, K. Raiser, L. Williams, and B. Sovacool, "What Are the Social Outcomes of Climate Policies? A Systematic Map and Review of the Ex-Post Literature," *Environmental Research Letters* 15 (2020): 113006.
43. E. L. F. Schipper, "Conceptual History of Adaptation in the UNFCCC Process," *Review of European Community & International Environmental Law* 15, no. 1 (2006): 82–92.
44. M. R. Khan and J. T. Roberts, "Adaptation and International Climate Policy," *Wiley Interdisciplinary Reviews: Climate Change* 4, no. 3 (2013): 171–189.
45. M. J. Mace, "Adaptation Under the UN Framework Convention on Climate Change: The International Legal Framework," in *Fairness in Adaptation to Climate Change*, ed. W. Neil Adger, J. Paavola, S. Huq and M. J. Mace (MIT Press, 2006), 53–76.
46. Schipper, "Conceptual History of Adaptation in the UNFCCC Process."
47. D. Falzon, "How to Track Progress on the Global Goal on Adaptation? A Stock-taking of Parties' Positions on Measurement One Year Into the GlaSS Work Programme," *Climate and Development* 16, no. 10 (2024): 870–879.
48. Mace, "Adaptation Under the UN Framework Convention on Climate Change."
49. S. Huq and M. R. Khan, Equity in National Adaptation Programs of Action (NAPAs): The Case of Bangladesh. *Fairness in Adaptation to Climate Change*, 2006, 181–200.

50. Mace, "Adaptation Under the UN Framework Convention on Climate Change."

51. Climate Policy Initiative, "Global Landscape of Climate Finance: A Decade of Data," 2022, https://www.climatepolicyinitiative.org/publication/global-landscape-of-climate-finance-a-decade-of-data/.

52. Robinson and Dornan, "International Financing for Climate Change Adaptation in SIDS."

53. UNEP, *Adaptation Gap Report 2022.*

54. J. R. Roberts, R. Weikmans, S. Robinson, D. Ciplet, M. Khan, and D. Falzon, "Rebooting a Failed Promise of Climate Finance," *Nature Climate Change* 11 (2021): 180–182.

55. D. Ciplet, D. Falzon, I. Uri, S. Robinson, R. Weikmans, and J. T. Roberts, "The Unequal Geographies of Climate Finance: Climate Injustice and Dependency in the World System," *Political Geography* 99, no. 102769 (2022): 1–12.

56. E. Kalaidjian and S. Robinson, "Reviewing the Nature and Pitfalls of Multilateral Adaptation Finance for SIDS," *Climate Risk Management* 36, no. 100432 (2022): 1–16.

57. L. Kuhl and J. Shinn, "Transformational Adaptation and Country Wwnership: Competing Priorities in International Adaptation Finance," *Climate Policy* 22, no. 9–10 (2022): 1290–1305.

58. E. Friedman, "Constructing the Adaptation Economy: Climate Resilient Development and the Economization of Vulnerability," *Global Environmental Change* 80 (2023): 102673.

59. S. Robinson, J. T. Roberts, R. Weikmans, and D. Falzon, "Vulnerability Based Allocations in Loss and Damage Finance," *Nature Climate Change* 13 (2023): 1055–1062.

60. J. Bertilsson, "Managing Vulnerability in the Green Climate Fund," *Climate and Development* 15, no. 4 (2023): 304–311.

61. L. Kuhl, J. E. Shinn, J. Arango-Quiroga, I. Ahmed, and M. F. Rahman, "The Liberal Limits to Transformation in the Green Climate Fund." *Climate and Development,* 16, no. 4 (2024): 826–837.

62. Ibid.

63. M. Garschagen and D. Doshi, "Does Funds-Based Adaptation Finance Reach the Most Vulnerable Countries?," *Global Environmental Change* 73 (2022): 102450, https://doi.org/10.1016/j.gloenvcha.2021.102450

64. Ibid.

65. K. McVeigh, "Climate Finance Dispute Prompts Bangladesh to Return £13m of UK Aid," *The Guardian*, November 10, 2016, https://www.theguardian.com/global-development/2016/nov/10/climate-finance-dispute-bangladesh-returns-13-million-uk-aid-world-bank.

66. Ibid.

67. Ibid.

68. M. A. B. Siddique, "Donors Shut Down Climate Fund, Bangladesh to Lose $50m," *Dhaka Tribune*, April 5, 2016, https://www.dhakatribune.com/science-technology-environment/climate-change/719/donors-shut-down-climate-fund-bangladesh-to-lose.

69. M. Haque, M. Rouf, and M. Z. H. Khan, *Challenges in Climate Finance Governance and the Way Out* (Transparency International Bangladesh, 2012).

70. Ibid.

71. Siddique, "Donors Shut Down Climate Fund."

72. Ibid.

73. McVeigh, "Climate Finance Dispute."

74. Rahman et al., "Locally Led Adaptation."

75. K. Thomas, "Accumulation By Adaptation," *Geography Compass* 18, no. 1 (2024): e12731, https://doi.org/10.1111/gec3.12731.

76. A. Scheidel, D. Del Bene, J. Liu, G. Navas, S. Mingorría, F. Demaria, S. Avila, B. Roy, I. Ertör, L. Temper, and J. Martínez-Alier, "Environmental Conflicts and Defenders: A Global Overview," *Global Environmental Change* 63 (2020): 102104.

77. A. Scheidel, Á. Fernández-Llamazares, A. H. Bara, D. Del Bene, D. M. David-Chavez, E. Fanari, I. Garba, K. Hanaček, J. Liu, J. Martínez-Alier, G. Navas, V. Reyes-García, B. Roy, L. Temper, M. A. Thiri, D. Tran, M. Walter, and K. P. Whyte, "Global Impacts of Extractive and Industrial Development Projects on Indigenous Peoples' Lifeways, Lands, and Rights," *Science Advances* 9, no. 23 (2023): eade9557.

78. D. K. Bardsley and N. D. Wiseman, "Climate Change Vulnerability and Social Development for Remote Indigenous Communities of South Australia," *Global Environmental Change* 22, no. 3 (2012): 713–723.

79. E. S. Cameron, "Securing Indigenous Politics: A Critique of the Vulnerability and Adaptation Approach to the Human Dimensions of Climate Change in the Canadian Arctic," *Global Environmental Change* 22, no. 1 (2012): 103–114.

80. J. D. Ford, N. King, E. K. Galappaththi, T. Pearce, G. McDowell, and S. L. Harper, "The Resilience of Indigenous Peoples to Environmental Change," *One Earth* 2, no. 6 (2020): 532–543.

81. G. Jackson, K. E. McNamara, and B. Witt, "'System of Hunger': Understanding Causal Disaster Vulnerability of Indigenous Food Systems," *Journal of Rural Studies* 73 (2020): 163–175.

82. D. E. Johnson, K. Fisher, and M. Parsons, "Diversifying Indigenous Vulnerability and Adaptation: An Intersectional Reading of Māori Women's Experiences of Health, Wellbeing, and Climate Change," *Sustainability* 14, no. 9 (2022): 5452.

83. M. Fayazi, I.-A. Bisson, and E. Nicholas, "Barriers to Climate Change Adaptation in Indigenous Communities: A Case Study on the Mohawk Community of Kanesatake, Canada," *International Journal of Disaster Risk Reduction* 49 (2020): 101750.

84. R. Howitt, O. Havnen, and S. Veland, "Natural and Unnatural Disasters: Responding with Respect for Indigenous Rights and Knowledges," *Geographical Research* 50, no. 1 (2012): 47–59.

85. M. Nursey-Bray, R. Palmer, A. M. Chischilly, P. Rist, and L. Yin, "Introducing Indigenous Peoples and Climate Change," in *Old Ways for New Days: Indigenous Survival and Agency in Climate Changed Times*, ed. M. Nursey-Bray, R. Palmer, A. M. Chischilly, P. Rist, L. Yin (Springer, 2022), 1–10.

86. P. Satyal, M. F. Byskov, and K. Hyams, "Addressing Multi-Dimensional Injustice in Indigenous Adaptation: The Case of Uganda's Batwa Community," *Climate and Development* 13, no. 6 (2021): 529–542.

87. S. Veland, R. Howitt, and D. Dominey-Howes, "Invisible Institutions in Emergencies: Evacuating the Remote Indigenous Community of Warruwi, Northern Territory Australia, from Cyclone Monica," *Environmental Hazards* 9, no. 2 (2010): 197–214.

88. Fayazi et al., "Barriers to Climate Change Adaptation in Indigenous Communities."

89. E. Parraguez-Vergara, J. R. Barton, and G. Raposo-Quintana, "Impacts of Climate Change in the Andean Foothills of Chile: Economic and Cultural Vulnerability of Indigenous Mapuche Livelihoods," *Journal of Developing Societies* 32, no. 4 (2016): 454–483.

90. D. Poets, "Settler Colonialism and/in (Urban) Brazil: Black and Indigenous Resistances to the Logic of Elimination," *Settler Colonial Studies* 11, no. 3 (2021): 271–291.

91. United Nations, "United Nations Declaration of the Rights of Indigenous Peoples," 2007, https://www.un.org/development/desa/indigenouspeoples/wp-content/uploads/sites/19/2018/11/UNDRIP_E_web.pdf.

92. Urzedo and Chatterjee, "The Colonial Reproduction of Deforestation in the Brazilian Amazon."

93. S. Veland, R. Howitt, D. Dominey-Howes, F. Thomalla, and D. Houston, "Procedural Vulnerability: Understanding Environmental Change in a Remote Indigenous Community," *Global Environmental Change* 23, no. 1 (2013): 314–326.

94. S. S. Hughes, S. Velednitsky, and A. A. Green, "Greenwashing in Palestine/Israel: Settler Colonialism and Environmental Injustice in the Age of Climate Catastrophe," *Environment and Planning E: Nature and Space* 6, no. 1 (2023): 495–513.

95. A. Löf, "Examining Limits and Barriers to Climate Change Adaptation in an Indigenous Reindeer Herding Community," *Climate and Development* 5, no. 4 (2013): 328–339.

96. Nursey-Bray et al., "Introducing Indigenous Peoples and Climate Change."

97. M. Parsons, J. Nalau, K. Fisher, and C. Brown, "Disrupting Path Dependency: Making Room for Indigenous Knowledge in River Management," *Global Environmental Change* 56 (2019): 95–113.

98. K. Thomas, "Compelled to Compete: Rendering Climate Change Vulnerability Investable," *Climate and Development* 54, no. 2 (2023): 223–250.

99. Mikkel Funder, Carol Mweemba, and Imasiku Nyambe, "The Politics of Climate Change Adaptation in Development: Authority, Resource Control and State Intervention in Rural Zambia," *Journal of Development Studies* 54, no. 1 (2018): 30–46.

100. Carol Hunsberger, Courtney Work, and Roman Herre, "Linking Climate Change Strategies and Land Conflicts in Cambodia: Evidence from the Greater Aural Region," *World Development* 108 (2018): 309–320.

101. Megan Mills-Novoa, "What Happens After Climate Change Adaptation Projects End: A Community-Based Approach to Ex-Post Assessment of Adaptation Projects," *Global Environmental Change* 80 (2023): 102655.

102. Stephen Woroniecki, Hausner Wendo, Ebba Brink, Mine Islar, Torsten Krause, Ana-Maria Vargas, and Yahia Mahmoud, "Nature Unsettled: How Knowledge and Power Shape 'Nature-Based' Approaches to Societal Challenges," *Global Environmental Change* 65 (November 2020): 102132, https://doi.org/10.1016/j.gloenvcha.2020.102132.

103. K. Paprocki. "The Climate Change of Your Desires: Climate Migration and Imaginaries of Urban and Rural Climate Futures," *Environment and Planning D: Society and Space* 38, no. 2 (2020): 248–266.

104. Rhino Ariefiansyah, and Sophie Webber, "Creative Farmers and Climate Service Politics in Indonesian Rice Production," *Journal of Peasant Studies* 49, no. 5 (2022): 1037–1063.

105. P. Shepherd, J. Tansey, and H. Dowlatabadi. "Context Matters: What Shapes Adaptation to Water Stress in the Okanagan?," *Climatic Change* 78, no. 1 (2006): 31–62.

106. C. Hanna, I. White, and B. C. Glavovic, "Managed Retreats by Whom and How? Identifying and Delineating Governance Modalities," *Climate Risk Management*, 31, (2021): 100278.

107. P. Marchman, A. Siders, K. Leilani Main, V. Herrmann, and D. Butler, "Climate Disruption and Planning: Resistance or Retreat?," *Planning Theory & Practice* 21, no. 1 (2020): 125–154.

108. M. Mycoo, M. Wairiu, D. Campbell, V. Duvat, Y. Golbuu, S. Maharaj, J. Nalau, P. Nunn, J. Pinnegar, O. Warrick, G. Anderson, F. A. Cruz, E. Devenish-Nelson, K. Ebi, J. Loehr, R. Mahon, R. McNaught, M. Parsons, J. Price, S. Robinson, and A. Thomas, *Small Islands: Climate Change 2022: Impacts, Adaptation and Vulnerability. Working Group II's contribution to the Sixth Assessment Report* (Intergovernmental Panel on Climate Change, 2022).

109. I. Ajibade, "Planned Retreat in Global South Megacities: Disentangling Policy, Practice, and Environmental Justice," *Climatic Change* 157 (2019): 299–317.

110. I. Ajibade, M. Sullivan, C. Lower, L. Yarina, and A. Reilly, "Are Managed Retreat Programs Successful and Just? A Global Mapping of Success Typologies, Justice Dimensions, and Trade-Offs," *Global Environmental Change* 76 (2022): 102576.

111. C. Jones, "Commissioners Agree: Bay of Plenty Township Unsafe and Residents Must Go," *Stuff*, 2020, https://www.stuff.co.nz/national/politics/local-democracy-reporting/120753289/commissioners-agree-bay-of-plenty-township-unsafe-and-residents-must-go.

112. M. Gereke and T. Brühl, "Unpacking the Unequal Representation of Northern and Southern NGOs in International Climate Change Politics," *Third World Quarterly* 40, no. 5 (2019): 870–889.

113. G. Edwards, P. K. Gellert, O. Faruque, K. Hochstetler, P. D. McElwee, P. Kashwan, R. E. McKie, C. Milani, T. Roberts, and J. Walz, "Climate Obstruction in the Global South: Future Research Trajectories," *PLOS Climate* 2, no. 7 (2023): e0000241, https://doi.org/10.1371/journal.pclm.0000241

114. Ibid.

115. Rahman et al., "Locally Led Adaptation."

116. Thomas, "Compelled to Compete."

117. M. A. Mycoo, "Autonomous Household Responses and Urban Governance Capacity Building for Climate Change Adaptation: Georgetown, Guyana," *Urban Climate* 9 (2014): 134–154.

118. Ibid.

119. K. A. Thomas, "Accumulation by Adaptation," *Geography Compass* 18, no. 1 (2024): e12731.

120. P. Larmour, *Interpreting Corruption: Culture and Politics in the Pacific Islands* (University of Hawaii Press, 2012).

121. W. Veenendaal and J. Corbett, "Clientelism in Small States: How Smallness Influences Patron–Client Networks in the Caribbean and the Pacific," *Democratization* 27, no. 1 (2020): 61–80.

122. Aideen Foley, Laurie Brinklow, Jack Corbett, Ilan Kelman, Carola Klöck, Stefano Moncada, Michelle Mycoo, Patrick Nunn, Jonathan Pugh, Stacy-ann Robinson, Verena Tandrayen-Ragoobur, and Rory Walshe, "Understanding 'Islandness,'" *Annals of the American Association of Geographers* (2023), https://doi.org/10.1080/24694452.2023.2193249.

123. M. Zakir Hossain and M. Ashiq Ur Rahman, "Adaptation to Climate Change as Resilience for Urban Extreme Poor: Lessons Learned from Targeted Asset Transfers Programmes in Dhaka City of Bangladesh," *Environment, Development and Sustainability* 20 (2018): 407–432.

124. V. Niva, M. Taka, and O. Varis, "Rural-Urban Migration and the Growth of Informal Settlements: A Socio-Ecological System Conceptualization with Insights Through a 'Water Lens,'" *Sustainability* 11, no. 12 (2019): 3487.

125. S. Munawar Ishfaq, *Rural-Urban Migration and Climate Change Adaptation: Policy Implications for Pakistan Policy Brief* (International Development Research Centre, 2018).

126. Paprocki, "The Climate Change of Your Desires."

127. Edwards et al., "Climate Obstruction in the Global South," 3.

128. Kimberley Anh Thomas, "Enduring Infrastructure," in *A Research Agenda for Geographies of Slow Violence*, ed. Shannon O'Lear (Edward Elgar Publishing, 2021), 107–122.

129. Edwards et al., "Climate Obstruction in the Global South," 3.

130. P. Kashwan, "Inequality, Democracy, and the Environment: A Cross-National Analysis," *Ecological Economics* 1, no. 131 (2017): 139–151.

131. Edwards et al., "Climate Obstruction in the Global South."

132. Rahman et al., "Locally Led Adaptation."

133. Nicole Klenk, Anna Fiume, Katie Meehan, and Cerian Gibbes, "Local Knowledge in Climate Adaptation Research: Moving Knowledge Frameworks from Extraction to Co-production," *WIREs Climate Change* 8, no. 5 (2017): https://doi.org/10.1002/wcc.475.

134. Kimberley Thomas, R. Dean Hardy, Heather Lazrus, Michael Mendez, Ben Orlove, Isabel Rivera-Collazo, J. Timmons Roberts, Marcy Rockman, Benjamin P. Warner, and Robert Winthrop, "Explaining Differential Vulnerability to Climate Change: A Social Science Review," *WIREs Climate Change* 10, no. 2 (2019): e565, https://doi.org/10.1002/wcc.565.

135. Brian Larkin, "The Politics and Poetics of Infrastructure," *Annual Review of Anthropology* 42, no. 1 (2013): 327–343.

136. Malini Ranganathan, "Storm Drains as Assemblages: The Political Ecology of Flood Risk in Post-Colonial Bangalore," *Antipode* 47, no. 5 (2015): 1300–1320.

137. Kian Goh, "Flows in Formation: The Global-Urban Networks of Climate Change Adaptation,' *Urban Studies* 57, no. 11 (2020): 2222–2240.

138. Wiebe Bijker, "Dikes and Dams, Thick with Politics," *Isis* 98, no. 1 (2007): 109–123.

139. Thomas, "Enduring Infrastructure."

140. Johann Strube and Kimberley Anh Thomas, "Damming Rainy Lake and the Ongoing Production of Hydrocolonialism in the US-Canada Boundary Waters," *Water Alternatives* 14, no. 1 (2021): 135–157.

141. Andrew Curley, "Infrastructures as Colonial Beachheads: The Central Arizona Project and the Taking of Navajo Resources," *Environment and Planning D: Society and Space* 39, no. 3 (2021): 387–404, 388.

142. Isabelle Anguelovski, Linda Shi, Eric Chu, Daniel Gallagher, Kian Goh, Zachary Lamb, Kara Reeve, and Hannah Teicher, "Equity Impacts of Urban Land Use Planning for Climate Adaptation," *Journal of Planning Education and Research* 36, no. 3 (2016): 333–348.

143. Bradley C. Parks and J. Timmons Roberts, "Globalization, Vulnerability to Climate Change, and Perceived Injustice," *Society & Natural Resources* 19, no. 4 (2006): 337–355.

144. J. Ribot, "Vulnerability does not Just Fall from the Sky: Toward Multi-scale Pro-poor Climate Policy," in *Handbook on Climate Change and Human Security*, ed. Michael R. Redclift and Marco Grasso (Edward Elgar Publishing, 2013), 164–199.

145. Tayanah O'Donnell, "Property Rights and Land Use Planning on the Australian Coast," in *Research Handbook on Climate Change Adaptation Policy*, ed. E. C. H. Keskitalo and B.L. Preston (Edward Elgar Publishing, 2019), 403–416.

146. Rebecca Elliott, *Underwater: Loss, Flood Insurance, and the Moral Economy of Climate Change in the United States* (Columbia University Press, 2021).

147. J. Petzold, B. M. W. Ratter, and A. Holdschlag, "Competing Knowledge Systems and Adaptability to Sea Level Rise in the Bahamas," *Area* 50, no. 1 (2018): 91–100.

148. M. Mycoo, "Sustainable Tourism, Climate Change and Sea Level Rise Adaptation Policies in Barbados," *Natural Resources Forum* 38, no. 1 (2014): 47–57.

149. Ibid.

150. Charles Lester, Gary Griggs, Kiki Patsch, and Ryan Anderson, "Shoreline Retreat in California: Taking a Step Back," *Journal of Coastal Research* 38, no. 6 (2022): 1207–1230.

151. Trevor Culhane, Galen Hall, and J. Timmons Roberts, "Who Delays Climate Action? Interest Groups and Coalitions in State Legislative Struggles in the United States," *Energy Research & Social Science* 79 (2021): 102114, https://doi.org/10.1016/j.erss.2021.102114.

152. V. Bowden, D. Nyberg, and C. Wright, "Planning for the Past: Local Temporality and the Construction of Denial in Climate Change Adaptation." *Global Environmental Change*, 57, (2019): 101939.

153. Bowden et al., "'I Don't Think Anybody Really Knows.'"

154. C. Payton Scally and J. R. Tighe, "Democracy in Action?: NIMBY as Impediment to Equitable Affordable Housing Siting," *Housing Studies* 30, no. 5 (2015): 749–769.

155. Elsa Devienne, "Urban Renewal by the Sea: Reinventing the Beach for the Suburban Age in Postwar Los Angeles," *Journal of Urban History* 45, no. 1 (2019): 99–125.

156. David Vogel, *California Greenin: How the Golden State Became an Environmental Leader* (Princeton University Press, 2018).

157. Vanessa Bowden, Daniel Nyberg, and Christopher Wright, "'I Don't Think Anybody Really Knows': Constructing Reflexive Ignorance in Climate Change Adaptation," *British Journal of Sociology* 72, no. 2 (2021): 397–411.

158. H. Boudet, L. Giordono, C. Zanocco, H. Satein, and H. Whitley, "Event Attribution and Partisanship Shape Local Discussion of Climate Change After Extreme Weather," *Nature Climate Change* 10, no. 1 (2020): 69–76.

159. L. Giordono, A. Gard-Murray, and H. Boudet, "From Peril to Promise? Local Mitigation and Adaptation Policy Decisions After Extreme Weather," *Current Opinion in Environmental Sustainability* 52 (2021): 118–124.

160. G. Gustin, J. H. Cushman, and N. Banerjee, "How the Farm Bureau's Climate Agenda Is Failing Its Farmers," Inside Climate News, 2018, https://insideclimatenews.org/news/24102018/farm-bureau-climate-change-denial-farmers-crop-insurance-subsidies-drought-future-at-risk/.

CHAPTER 12

Legal and State Efforts to Address Climate Obstruction

LEAD AUTHORS: GRACE NOSEK, JOANA SETZER, AND BENJAMIN FRANTA

CONTRIBUTING AUTHORS: ALYSSA JOHL, LISA BENJAMIN, SHARON YADIN, WILLIAM W. BUZBEE, AND ARIA KOVALOVICH

INTRODUCTION: KEY TOOLS FOR ADDRESSING CLIMATE OBSTRUCTION

Climate obstruction has been conducted by a complex and diverse group of actors—including industries, politicians and lobbying organizations, PR firms, news media, and more—using numerous strategies, including lobbying, campaign spending, challenging science and scientists, and harnessing social media to disseminate misinformation. As the scope, scale, and profound consequences of organized climate obstruction have become clearer, various approaches to address and mitigate obstruction have emerged, including legal and government strategies. Here we divide legal and government strategies to address climate obstruction into three broad categories: regulation and legislation, litigation, and government hearings and investigations.

We outline the key stakeholders taking legal and government action to address climate obstruction and explain some of the legal tools they use, including consumer protection actions, tort litigation, and congressional investigations. We situate the timeline in the lead-up to and after the 2015 adoption of the Paris Agreement and analyze the potential tensions between addressing climate obstruction and protecting freedom of speech and expression. We include examples from around the globe, but our analysis centers

Grace Nosek et al., *Legal and State Efforts to Address Climate Obstruction*. In: *Climate Obstruction*.
Edited by: J. Timmons Roberts, Carlos R. S. Milani, Jennifer Jacquet, and Christian Downie,
Oxford University Press. © Oxford University Press (2025). DOI: 10.1093/oso/9780197787144.003.0012

on the United States, as it has been an epicenter for both organized climate obstruction[1] and for climate litigation, with more than 1,500 climate lawsuits filed to date, many of which include some element of climate obstruction (Figure 12.1).[2] Cases continue to be filed around the world and in new jurisdictions. The ballooning of climate lawsuits helps illustrate the growing importance of courts and legal tools to addressing climate change and climate obstruction.

Later, we establish three categories of legal action, offer an overview of activities within each category and highlight key examples, outcomes, and possible impacts. Finally, we include a summary of the gaps in the academic literature on legal and government responses to climate obstruction and future research priorities, calling for more sociolegal research on the legal and government responses to climate obstruction.

While outside the focus of our analysis, it is important to note that as legal and regulatory efforts to prevent climate obstruction have proliferated, efforts to *facilitate* climate obstruction have also increased.[3] All three categories of legal strategies covered here—legislation and regulation, litigation, and government hearings and investigations—can similarly be used as tools for climate obstruction, and climate obstructors have often used the law to their advantage. One recent and consequential example of legal action facilitating climate obstruction can be seen in the way governments and corporate actors have used strategic lawsuits against public participation (SLAPP suits)—such as fossil fuel company EnergyTransfer's $300 million suit against Greenpeace USA and Greenpeace International for their alleged role in protests against the Dakota Access Pipeline—as well as new laws and regulations that increase penalties to draconian levels against climate protesters.[4]

KEY STAKEHOLDERS

State and legal actions to address climate obstruction—such as misleading advertising, failure to disclose corporate environmental impacts, and historical deception around climate change—are multifaceted and implicate every level of government (municipal, regional, national, and international) and a variety of private and public actors. For example, national legislators have proposed bills to increase transparency in corporate political spending;[5] municipalities have filed suit against the carbon majors for misleading speech;[6] institutional investors such as pension funds have filed suits in various jurisdictions against corporate actors for discrepancies between internal climate pledges and external lobbying efforts;[7] and nonprofits have brought claims against corporations before administrative bodies charged with protecting consumers and investors.[8] Although not all climate-related legal actions focus on obstruction (many are based on traditional administrative environmental law issues

Figure 12.1:

Number of climate litigation cases globally and in the US (1986–2023). The number of climate lawsuits grew rapidly after 2000, and, since about 2017, dozens of cases based on alleged deception by corporate defendants have been filed.

Source: J. Setzer and C. Higham, *Global Trends in Climate Change Litigation: 2024 Snapshot*, 2024, Grantham Research Institute on Climate Change and the Environment and Centre for Climate Change Economics and Policy, London School of Economics and Political Science, online: https://www.lse.ac.uk/granthaminstitute/wp-content/uploads/2024/06/Global-trends-in-climate-change-litigation-2024-snapshot.pdf.

such as project permitting and proper development of environmental impact assessments), the law's focus on climate obstruction is increasing, and some broad trends in who the key stakeholders are and the legal tools they employ are discernible.

Government actors, particularly those at the national level, have numerous legal tools at their disposal, with the ability to adopt legislation and regulation, to initiate regulatory enforcement actions, to bring both civil and criminal lawsuits against obstructors, and to convene public hearings and investigations. Other actors including private citizens and nongovernmental organizations (NGOs) seeking to target climate obstruction through legal tools have more limited (although still diverse and substantial) options, including filing lawsuits and requesting regulatory enforcement from administrative bodies. Additionally, those actors can petition governments to take legal action against climate obstruction and can also amplify the impact of government action by, for example, raising public awareness of government inquiries into climate obstruction.

Legal action against climate obstruction is directed at an increasingly diverse group of defendants. The first wave of attempts to hold nongovernmental actors accountable for climate obstruction has focused largely on the carbon majors, defined as the world's largest producers of fossil fuels, including companies including ExxonMobil, Shell, BP, Chevron, Peabody, and Total.[9] These attempts have also often targeted trade groups, such as the American Petroleum Institute (API), recognizing the integral role such groups have played in enabling corporate climate obstruction.[10] In recent years, state and legal actions targeting climate obstruction have also focused on a wider universe of corporate actors. For example, in February 2024, the New York State attorney general filed suit against beef producer JBS USA Food Co., alleging the company had misled consumers about its net-zero commitments (see Chapter 4).[11] More recent legal actions have also been focusing on a wider universe of alleged enablers of corporate climate obstruction, including consultancies and public relations firms. For example, in 2023, an Oregon municipality sued McKinsey & Company along with carbon majors and their trade groups, and the US Congress has investigated the role of public relations firms in climate obstruction.[12]

While the United States is and has been an epicenter for climate obstruction and litigation addressing it, new lawsuits and regulatory actions targeting corporate climate obstruction are rapidly proliferating around the globe.[13] In a recent example, the Plains Clan of the Wonnarua People and the Lock the Gate Alliance lodged a complaint with the Australian Competition and Consumer Commission and Australian Securities and Investments Commission alleging that the mining corporation Glencore was engaging in harmful greenwashing around its coal production plans by publicly making claims about ambitious net-zero plans that did not match its actions.[14]

There has been a lag between industries' and other actors' organized climate obstruction efforts and state and legal responses to those efforts, with a significant uptick in efforts to address obstruction since the discovery of internal fossil fuel industry documents pertaining to climate change in around 2015. For example, the first lawsuits attempting to hold companies—particularly fossil fuel companies—directly responsible for the climate harm to communities and individuals caused by their products were filed in the United States in the mid-2000s, but none was successful. For almost a decade, there was a significant lull in the filing of new cases concerned with direct corporate liability for climate harm.[15] That trend changed with the publication of new research in 2014 that attributed more than two-thirds of global greenhouse gas (GHG) emissions directly to the operations of about 100 companies,[16] along with research showing that although companies had denied the existence of climate change, internal corporate knowledge of the phenomenon extended back to at least the 1970s. The attribution research combined with the evidence of climate obstruction formed an evidentiary base for a new wave of climate lawsuits against corporations. As in other legal campaigns to hold corporations accountable, like those responding to lead, asbestos, tobacco, and opioids, knowledge and concealment of harm by corporate defendants are critical components of the case for liability, especially in US courts.

Holding carbon majors legally liable for climate obstruction is distinct from holding them morally or publicly accountable in that it requires a litigant to be able to identify a legally recognized harm and to provide evidence of each element of that harm. Due to the widespread nature of climate impacts, litigants have often had difficulty establishing some elements of their claims against the carbon majors, such as standing (the right to bring a lawsuit about an issue) and causation (connecting the defendants' conduct factually and legally to the harm at issue).[17] Some difficulties around standing and causation are easing as attribution science becomes more sophisticated. Additionally, evidence that the carbon majors knew about the dangers of climate change and how they responded to this knowledge has suggested that they acted intentionally to deceive the public about those dangers rather than merely participating in good-faith scientific debate, and this distinction has led to a new set of legally cognizable claims. Rather than attempting to hold carbon majors strictly liable for their emission of GHGs, many litigants are basing liability on false, misleading, and fraudulent corporate speech by, for example, bringing claims under state consumer-protection laws, public nuisance and other common-law doctrines, and (in the United States) racketeering statutes.[18] This focus on unlawful corporate deception emphasizes the moral and legal culpability of the fossil fuel defendants and helps avoid defenses based on the political question doctrine, which encourages courts to defer to other branches of government in addressing major societal issues such as the regulation of climate pollution.

The abundance of research and understanding regarding internal fossil fuel industry knowledge and concealment of harm have also driven government hearings. All three high-profile examples of government investigations and hearings covered in this chapter were initiated after the uncovering of internal fossil fuel industry knowledge of climate science.

Every level of government and legal body and every legal strategy will have a different set of strengths and limitations. A cross-cutting theme in the state and legal strategies analyzed is the potential tradeoff between speed and flexibility and binding legal outcomes. Another important theme is the potential tension between government efforts to address climate obstruction tactics and legal protections for freedom of speech and expression. For example, there would be important hurdles and limitations to any legal effort to curtail certain key strategies of climate obstruction such as political spending and lobbying.[19]

REGULATION AND LEGISLATION

Regulation and legislation are critical tools for addressing climate change obstruction. The landscape of climate legislation is vast and rapidly increasing: as of September 2023, there were at least 2,166 climate policies and 1,066 climate laws worldwide.[20] Research has identified the short-term and long-term impacts of climate legislation on GHG emissions.[21] A study examined 133 countries from 1999 to 2016 and found that each new law reduced annual carbon dioxide (CO_2) emissions per unit of gross domestic product by 0.78% nationally in the short term (during the first three years) and by 1.79% in the long term (beyond three years). These results are driven by parliamentary acts and by countries with a strong rule of law. Cumulative CO_2 emissions savings during the study period amounted to 38 GT or 38,000 MT CO_2, the equivalent of one year's worth of global CO_2 output.

There are many different approaches through which regulation and legislation can address climate obstruction. Here, we have organized legal tools into categories for ease of analysis. First, regulators and legislators can overcome obstruction by regulating GHG emissions directly, as we briefly discuss later. These binding provincial, national, and international regulatory schemes are the kind of government actions that have historically faced fierce campaigns to defeat them. Thus successful passage and effective enforcement of such laws can be seen as a powerful legal win against climate obstruction.

Second, regulators and legislators can employ a suite of tools to increase the transparency of and provide critical information on corporate climate actions to the public, consumers, and shareholders. Such laws and regulations are often designed to protect and empower consumers and investors and prevent greenwashing (discussed later) as well as fraud. In a similar vein, regulators can also use their platforms and expertise to shame corporate climate obstructors and their enablers.

Third, because the legislative process itself is often a key target of climate obstruction, regulators and legislators can also use legal mechanisms to protect the legislative process from undue corporate influence, revolving-door conflicts of interest, and bribery and other forms of corruption.

In addition to considering the most appropriate types of regulation and legislation for addressing climate obstruction, it is also important to understand how binding, or legally enforceable, such regulatory schemes may be. A second point, which will be briefly touched on later, is the important and growing role of courts in either upholding or striking down regulation and legislation aimed at climate obstruction.

Regulatory Tools Aimed at Reducing GHG Emissions Directly

Legislation and regulation can target emissions reductions through many different regulatory mechanisms,[22] including standards and targets (e.g., for emissions levels, energy efficiency, or regulating specific sectors), incentives (e.g., tax credits or subsidies for low-carbon technologies) or by setting carbon-pricing mechanisms (e.g., carbon taxes or cap-and-trade systems). Currently, there is only a small body of literature that examines legal mechanisms around achieving net-zero pledges and policies, or balancing the amount of GHG emitted by removing an equal amount of GHGs from the atmosphere.[23] Existing literature assessing the governance mechanisms required to attain net zero focuses on the effectiveness of corporate governance amid divergent models across different jurisdictions that try to regulate net-zero claims and pledges made by corporations and financial firms.[24] While the significance of legal frameworks (encompassing contracts, legislation, and enforceable regulation) is recognized at both corporate and country levels, research on what these regulatory pathways might look like is mostly absent. A stock-take in 2023 showed that economy-wide net-zero targets enshrined in domestic legislation have substantially increased in the 2020s, while most regulations governing companies' net-zero targets were still limited to disclosure rules.[25]

Regulatory Tools Aimed at Information and Transparency

In addition to regulating domestic or corporate greenhouse gas emissions directly, regulators have a suite of tools to increase transparency around corporate emissions for the public, consumers, and investors.

Regulation Targeting Greenwashing

Greenwashing is defined as acts that mislead investors, consumers, or the public regarding environment-related impacts or outcomes or overstate an actor's environmental performance.[26] Climate-washing is a form of greenwashing focused on misleading climate-related acts. This type of greenwashing can create a false sense of confidence in some proposed climate solutions, potentially leading to complacency and delaying more effective solutions.[27] Key greenwashing techniques include misleading communication on climate issues; deceptive environmental, social, and governance (ESG) credentials based on certifications, ratings, or audits on corporate performance across a range of important criteria that may be unreliable; and false or misleading emissions-reduction or carbon-offsetting claims related to the use of potentially ineffective carbon credits to meet net-zero or other voluntary emission-reduction pledges.

Little to no specific legislation currently exists aimed at combating climate-related greenwashing. The behaviors associated with this risk are often covered under existing laws, which typically enforce administrative or civil penalties rather than levying criminal sanctions. Consumer-protection laws and, in some cases, securities laws can address greenwashing, while the emerging human rights and environmental due-diligence laws, such as those emerging from the European Court of Human Rights, Inter-American Court of Human Rights, and International Court of Justice, could further help mitigate these risks.

Efforts to address greenwashing are being intensified by regulators, enforcement agencies, and legislators, however. Some jurisdictions have recently established more rigorous standards for making climate claims. In the European context, for example, in February 2024 the European Council adopted new, EU-wide rules empowering consumers for the green transition, known as the Green Transition Directive, which amends the Unfair Commercial Practices Directive and the Consumer Rights Directive. These amendments are expected to prevent businesses from making unsubstantiated climate-related assertions. For example, claims of carbon neutrality will be deemed unfair if they rely solely on carbon offsetting schemes. Claims about future goals, such as reducing carbon emissions by a certain percentage by a target date, will face stringent requirements. Member States are required to incorporate the Directive into national law by the end of March 2026, and to apply the new measures from September 2026. Until then, many practices inconsistent with the forthcoming standards are already deemed unlawful under current national legislation.

A variety of information-based regulatory tools with a shaming or monitoring component, such as databases, registries, labels, rankings, ratings, lists, regulatory postings, and announcements, and corporate disclosure obligations are currently being developed and implemented in various jurisdictions worldwide,[28] and some of these address climate obstruction. For example, the UK Environment Agency posts information regarding civil penalties imposed under all UK climate change laws and regulations on a governmental website, as well as company-specific information about performance and breaches of voluntary climate agreements established between the Agency and specific firms.[29] Without such public accountability, the voluntary agreements might otherwise create a false impression of compliance with climate laws and regulations and misrepresentation of beyond-compliance corporate activities.

Such regulatory tools serve not only to inform, educate, warn, and nudge various stakeholders, but also they can inform anti-obstruction efforts[30] by exposing and condemning firms' anti-climate practices and substantial carbon footprints, thereby countering corporate greenwashing, climate denial and delay, and other forms of corporate climate obstruction.[31] For example, the Netherlands Authority for Consumers and Markets (ACM) declared that the fashion company H&M has engaged in greenwashing by making "unclear and insufficiently substantiated sustainability claims," including its use of terms like "ecodesign" and "conscious."[32] Regulatory efforts to highlight firms' anti-climate behavior can be regarded as "regulatory climate shaming," aiming to encourage action by corporate stakeholders, such as consumers, investors, and creditors, against named and shamed firms.[33]

Regulatory climate-shaming tools can play a crucial role in combating climate obstruction by utilizing credible government information sharing in a way that counters corporate disinformation and publicly assigns liability to industries and companies that often deny or shift responsibility.[34] Regulatory shaming is also highly suitable for fighting climate-focused greenwashing—a practice centered around enhancing corporate reputation—by capitalizing on corporate reputational vulnerabilities and disseminating reliable information about actual corporate climate performance and greenwashing practices.[35] For example, due to their significant noncompliance with climate-risk disclosure rules, the European Central Bank (ECB) is contemplating the public listing of banks that engage in greenwashing or repeatedly fail to fully disclose their climate risks.[36]

Greenwashing makes it harder for regulators and the public to hold companies accountable for their role in climate change.[37] For instance, fossil fuel companies have flaunted misleading ESG ratings and advertise their misleading "carbon offsets" to consumers, investors, creditors, and other stakeholders in order to maintain their climate-friendly reputation and evade financial penalties from said stakeholders.[38] In 2022, the Carbon Disclosure Project, a private rating agency, gave JBS, the world's biggest meat producer, an A- rating, which may be highly misleading for investors (see Chapter 4).[39] In contrast, state regulatory publications of carbon footprints, environmental infringements, or misleading net-zero statements of companies such as JBS are likely to be perceived by their audience as more credible and reliable than publications by private rating agencies because they originate from authoritative bodies of government.[40] In this vein, state climate obstruction shaming is also likely to prove more effective than climate obstruction shaming conducted by nonstate actors due to the general trustworthiness and credibility often attributed to governmental publications (see Chapter 13). Direct regulatory shaming of firms for climate obstruction is a policy still mostly in the development stage.

In addition, greenwashing practices create the impression that companies successfully self-regulate and that legislation or governmental regulation is not needed.[41] Thus, direct and indirect climate obstruction methods may undermine both public critique of corporations and government regulation.[42] Legal research has found that this situation is an often-overlooked meta-regulation problem that significantly differs from the conventional noncompliance problem wherein firms simply violate climate laws and regulations.[43] Indirect climate obstruction is also a more subtle tactic than regulatory noncompliance or anti-regulatory litigation, making it more difficult for the public and regulators to identify its use, let alone address it.

Regulatory Tools Aimed at Addressing Undue Corporate Influence Over the Legislative Process

Regulation can help prevent undue influence over and obstruction of the legislative process.[44] Obstruction in the passing and enforcement of such regulation may therefore arise when individuals or groups have interests (professional, personal, monetary, or otherwise) that are not aligned with the requirements set in draft or existing regulations, and these people or organizations try to obstruct the passage or implementation of climate-related regulations. A recent study about corruption and integrity risks in climate solutions provides insights into some of the ways through which such obstruction can manifest.[45]

The passing and enforcement of regulation can be obstructed as a result of conflicts of interest or revolving doors, whereby individuals move between positions in public office and private organizations.[46] Conflicts of interest and revolving doors can cause corporate interest groups to have an outsized influence in public debates about the priority given to different approaches to the low-carbon transition. Efforts to obstruct climate legislation also extend to the international level. For example, more than six hundred fossil fuel lobbyists attended the 2022 UN climate change conference COP27 in Egypt, raising concerns about the influence of vested interests on global climate policy.[47] In response to such criticism and a demand for greater transparency in the COP process, the United Nations Framework Convention on Climate Change (UNFCCC) Secretariat revised its registration system to require conference attendees to disclose their affiliations, including any paid or unpaid relationships.

The risk of conflicts of interest may, in theory, be managed through existing codes of conduct and by law in some jurisdictions. Within the European Union, for example, several procedures have been established to prevent conflicts of interest and demonstrate the impartiality of political representatives, civil servants, and external contractors.[48] However, there is little evidence of guidance and legislative safeguards aimed at addressing actual or potential conflicts of interest arising in the implementation of climate action specifically, or of action by key watchdog organizations. Despite efforts to improve transparency and accountability, addressing conflicts of interest in law and policymaking remains a significant challenge. Transparency and clear guidelines in decision-making processes are essential to ensure that climate regulations are not shaped by competing and conflicting interests.

BARRIERS TO ESTABLISHING REGULATORY SCHEMES AIMED AT CLIMATE OBSTRUCTION: A US CASE STUDY

Various countries have established regulatory mechanisms to deter deceptive practices and obstruction or to increase accountability.[49] The United States offers an interesting case study in which there is not only an absence of regulation to deter climate obstruction, but also legal challenges limiting the extent to which regulation or litigation can address climate obstruction, including comparatively more robust protections for corporate speech than in other jurisdictions.[50]

In the United States, most federal climate policy is implemented under the Clean Air Act, the 2022 Inflation Reduction Act (which uses spending and tax incentives to catalyze climate progress), through energy laws, or through securities laws governing corporate disclosure obligations. These laws delegate authority for enforcement to federal agencies, mainly the US Environmental Protection Agency (EPA) and the Securities and Exchange Commission (SEC).

Apart from securities laws, however, none of these laws and regulations deal directly with the problem of climate obstruction. In addition, under US federalism, states can have their own laws, passed by state legislatures, agencies, or handed down by state courts.[51] Therefore, due to a combination of the lack of specific federal climate obstruction enactments and preserved state powers, government climate obstruction responses have had to rely on a mixture of federal and state laws, including statutory, regulatory, and common law.

However, other facets of US law pose challenges to federal and state regulation and litigation over climate obstruction. First, at the federal level, the US Supreme Court has since 2021 embraced and strengthened the "major questions doctrine."[52] Under this doctrine, any agency newly asserting powers, especially with major consequences, needs to root the action in a "clear statement" of authorization.[53] This doctrine is central to industry challenges to new regulations or laws. For example, the 2022 climate-disclosure regulation proposed by the SEC was challenged under this major question doctrine. Second, the First Amendment of the US Constitution creates robust protections for political speech and dissent and protects against governmental constraints of speech or compelled speech. The US Supreme Court has also recently expanded its protections of corporate speech and activities. As a result, oil and gas companies and aligned scholars seek to argue that their past public relations and misinformation campaigns about climate change should be protected as political speech and dissent.[54] However, corporate miscommunication or fraud in connection with consumers and corporate statements to investors have long been viewed as legitimately subject to regulation.[55] Recent work argues that the First Amendment should not preclude regulation of false or misleading corporate claims (i.e., greenwashing) about companies' products and environmental performance.[56]

Given the barriers to government action and few tools aimed specifically at climate obstruction, some of the most potent regulatory tools available are long-standing state consumer-protection and corporate disclosure laws and regulations.[57] Many cases have used these legislative and regulatory hooks to bring climate obstruction claims. Examples include the cases of *People v. Exxon-Mobil* (California) and *Commonwealth v. ExxonMobil* (Massachusetts).[58] Some similar claims have also been made with the US Federal Trade Commission, which has the authority to regulate misleading marketing.[59] After two years of public debate and just weeks after the historic adoption of its Climate Disclosure Rules in March 2024, the SEC voluntarily stayed the implementation of these rules, pending judicial review. Prior to this development, some scholars anticipated that courts might uphold the requirements for financial transparency, while others questioned the enforceability of broader corporate policy disclosure requirements related to social welfare issues.[60]

Litigation has become an important strategy in efforts to address climate obstruction. A recent report found that as of June 2024 at least 2,666 climate-related cases had been filed in at least 65 jurisdictions worldwide since the 1980s, up from 884 cases by 2017 and 1,550 cases by 2020.[61] These cases have been filed before a range of judicial venues, including domestic, international, and regional courts as well as tribunals, quasi-judicial bodies, and other adjudicatory bodies such as Special Procedures at the United Nations and arbitration tribunals. In interpreting these case numbers, it is important to note that only some of these cases are pending, that only some seek to support climate action, and that the cases vary widely in terms of causes of action, remedies sought, and the scope of potential impact. Approximately 50–100 ongoing cases directly address climate obstruction globally.

Although climate litigation is still concentrated in the Global North, recent reports indicate a noticeable increase in cases in the Global South.[62] This trend is evident in Brazil, one of the countries with the largest number of climate litigation cases in the world.[63] The number of climate cases against the Brazilian government increased significantly in response to the socioenvironmental setbacks suffered between 2019 and 2022.[64] A landmark moment occurred in July 2022 when Brazil's highest constitutional court acknowledged the Paris Agreement as a human rights treaty in *PSB et al. v. Brazil* (on Climate Fund) (ADPF 708). This case, which addressed the government-induced inaction of the Agreement's Climate Fund, underscored the integration of climate change and human rights in the legal arena.[65]

In addition to such high-profile cases, routine climate litigation also challenges specific projects and their GHG emissions, including disputes over environmental licensing and civil liability for climate-related environmental damages. Notably, more than half of Brazilian climate litigation involves land-use change and forests, reflecting the country's GHG emissions profile. These cases introduced novel legal arguments and frameworks that could influence efforts to combat greenwashing and protect community rights. Among them, *Amorema and Amoretgrap v. Sustainable Carbon*,[66] accuses companies of exploiting the concept of "social carbon credits" in the Amazon without delivering real benefits to local communities, misusing community names, images, and cultural heritage, and misrepresenting their socioenvironmental commitments. This case highlighted the complex and specific impacts of greenwashing, which include not only misleading the public about a company's environmental practices but also obscuring their severe human-rights violations. The diversity of legal approaches and jurisdictions involved in greenwashing suits underscores the need for a broad understanding of the strategies employed in this type of litigation.

The remainder of this section provides an overview of legal actions addressing obstruction of climate action, including suits alleging long-standing historical deception by major fossil fuel producers and ongoing misleading advertising and greenwashing. Political obstruction aimed directly at influencing governments (e.g., through legislative lobbying) often falls outside the purview of such litigation due to strong protections for government petitioning, even when deception can be proved.

Litigation Addressing Historical Deception by Fossil Fuel Producers

It is no coincidence that the number of climate lawsuits seeking to hold corporations accountable for climate harms has grown in parallel with expanding research in climate attribution science, carbon accounting associating specific amounts of emissions with individual companies, and historical evidence of these companies' deceptions around climate change. Since the mid-2010s, at least fifty-nine climate liability cases have been filed against companies globally, mostly against fossil fuel producers, with at least twenty of them instigated by US cities and states, most of which are still ongoing at the time of this writing.[67] Within this wave of corporate liability cases, there is considerable variation in terms of the relief sought.[68] "Retrospective" polluter-pays cases, such as *Lliuya v. RWE*[69] (filed 2015 in Germany), focus on the causal connection between a company's past GHG emissions and climate change, and seek financial damages based on that historical responsibility. "Prospective" corporate framework cases, such as *Milieudefensie v. Shell*[70] (filed 2019 in the Netherlands), focus on what companies should do now and in the future based on the global consensus around the need to rapidly reduce emissions. Both retrospective and prospective arguments have also begun to be used together, as in the case of *Asmania et al. v. Holcim*,[71] a lawsuit filed in 2022 by a group of Indonesian islanders before a court in Switzerland, where claimants' requested relief includes both a court order requiring the company to rapidly reduce emissions to align with the goals of the Paris Agreement and a request that the company pay a proportionate share of the costs of adapting the island to climate impacts.

Most litigation seeking redress for historical climate deception is ongoing in the United States. Since 2017, eight states (California, Connecticut, Delaware, Massachusetts, Minnesota, New Jersey, Rhode Island, and Vermont), the District of Columbia, numerous county and municipal governments (including those in California, Colorado, Hawaii, Maryland, New Jersey, New York, Oregon, South Carolina, and Puerto Rico), two indigenous tribes (the Shoalwater Bay Indian Tribe and the Makah Indian Tribe), and an industry trade association (the Pacific Coast Federation of Fishermen's Associations, Inc.) have all filed lawsuits to hold major oil and gas companies accountable for deceiving consumers, policymakers, the media, and public about the role of fossil fuels in climate change.[72] These cases are similar in narrative and legal strategy to previous successful legal campaigns to hold corporate actors accountable in the United States, including those against asbestos, tobacco, and opioid producers.

Cases that address historical deception have been brought almost exclusively under state law in state courts and, until 2023, were tied up in procedural battles over whether the cases should be heard in federal courts. Despite efforts by the industry defendants, six federal appeals courts and at least twelve federal district courts ruled that the cases should proceed in state courts,[73] and twice in 2023 the US Supreme Court denied appeals on the issue by the industry defendants.[74] In October 2023, the Hawaii Supreme Court affirmed the trial court's denial of the industry defendants' motion to dismiss a case filed by the capital city of Honolulu, marking one of the first times one of these cases could proceed to legal discovery and paving the way for a jury trial.[75] Another case filed in 2019 by Massachusetts Attorney General Maura Healey is in a similar posture.[76]

These cases generally point to historical evidence showing that carbon producers including ExxonMobil, Chevron, Shell, and BP, as well as trade associations such as API, understood the damaging climatic effects of their products at least as far back as the 1970s and in some cases earlier, but failed to warn the public about the danger, concealed the industry's own knowledge, affirmatively misrepresented the science of climate change, downplayed the harms caused by fossil fuels, and downplayed the industry's climate-damaging activities. As a result, these cases allege, climate change is worse than it otherwise would be and the adaptation costs and damages it has caused are higher than if the defendants had acted honestly and lawfully. Causes of action supporting these cases include public nuisance; product liability; unlawful, unfair, or fraudulent business practices; and, in some cases, RICO (the Racketeer Influenced and Corrupt Organizations Act).[77] Remedies sought include compensatory and punitive damages, establishment of abatement funds, and injunctive and other equitable relief in which the companies would be held to certain actions.

Generally, these cases allege that the defendants are responsible for a wide range of climate-related damages, including extreme heat, droughts, water shortages, wildfires, public health injuries, massive storms, flooding, agricultural damage, sea level rise, coastal erosion, damage to ecosystems, and biodiversity disruption.[78] A few cases, in contrast, are still based on damages from specific events (and liability in these cases is based on historical deception). One such case was Multnomah County, Oregon's, based on the 2021 Pacific Northwest heat dome, which scientists have found was virtually impossible without human-caused climate change.[79] Another was a class action (a type of lawsuit organized by a group of people who raise similar legal and factual issues) filed by sixteen Puerto Rican municipalities for damages from the 2017 hurricane season, which devastated the island.[80]

The total scale of potential liability for the defendants is not yet known but could be in the range of trillions of dollars or more, and the implications for addressing historical and ongoing obstruction of climate action are significant.

Litigation Addressing Ongoing Deception of Consumers

A related but distinct class of legal action is focused on remedying recent and ongoing unlawful deception of consumers. These greenwashing suits are often based on consumer-protection and/or truth-in-advertising laws and in many cases do not require a showing of harm but instead only that the consumer-facing statements in question are false or tend to mislead. The remedies in such cases are often civil penalties and/or judicial orders to cease the misleading activity.

Because many countries have truth-in-advertising laws, greenwashing complaints have been filed in a variety of jurisdictions. Since 2017, at least thirteen states, counties, and municipalities in the United States have filed state consumer-protection actions against fossil fuel producers, alleging recent and ongoing deception aimed at consumers regarding the harms of fossil fuel products, the climate-positive activities of fossil fuel producers, and the viability and efficacy of advertised climate solutions, among other things. At least eight of these governmental plaintiffs have sued under theories of both consumer protection and climate-damages recovery as described earlier. Other greenwashing lawsuits filed in the United States include a class action filed in 2023 in California state court against Delta Airlines for allegedly misleading consumers about "carbon-neutral flights"[81] and a case filed in 2022 by the environmental law nonprofit ClientEarth against gas utility Washington Gas for allegedly misleading consumers about the climate impacts of fossil gas (since dismissed).[82]

Outside the United States, complaints about greenwashing by fossil fuel producers have been filed before various courts and regulatory bodies, including the United Kingdom's Advertising Standards Authority and the Netherlands Advertising Code Committee. For example, in 2022 the Advertising Standards Authority ruled that poster advertisements from the bank HSBC were misleading because they touted the bank's positive climate-related activities while omitting its financing of fossil fuel development,[83] and in 2023 the same body ruled that some of oil major Shell's advertisements were misleading because they portrayed the company as focusing a large proportion of its business on renewables rather than fossil fuels.[84] Additionally, in 2023, the European Consumer Organisation, or BEUC (consisting of twenty-two member groups across eighteen countries), filed a complaint before the European Commission and consumer-protection authorities against seventeen airlines for allegedly misleading carbon-offsetting schemes.[85]

Litigation Addressing Corporate Policy

Another series of cases filed primarily in Europe seeks to obtain court orders compelling fossil fuel producers and other corporations to align their business plans with the Paris Agreement. These cases address climate obstruction insofar as they seek to prevent further delay in decarbonization and sometimes reference the defendants' history of deception and obstruction in their legal arguments. Buoyed by a successful suit against Royal Dutch Shell at The Hague District Court in the Netherlands in 2021 (which was appealed and partly overturned),[86] similar suits have been filed against Total in France (dismissed),[87] bank BNP Paribas in France (ongoing),[88] Volkswagen in Germany (dismissed),[89] and carbon major Eni in Italy (ongoing).[90] The cases against Total and Eni draw upon historical research showing longstanding knowledge by the defendants of the climate harms caused by fossil fuels. These suits are generally based on open legal standards (legal standards that are determined based on the particular situation presented to the court) under various duty-of-care laws, so their prospects for success may evolve as climate damages emerge and societal norms regarding decarbonization shift.

International Court and Human Rights–Based Actions

Various human-rights-based proceedings around the world have the potential to affect climate obstruction. In 2023, for example, the UN Committee on the Rights of the Child issued a comment pertaining to climate change and the Convention on the Rights of the Child, one of the most widely adopted international treaties in the world.[91] The strongly worded comment noted, among other things, that "delaying a rapid phase out of fossil fuels will result in higher cumulative emissions and thereby greater foreseeable harm to children's rights."[92]

Cases and petitions seeking to clarify the responsibility of states with regard to climate change and human rights are currently pending before the Inter-American Court of Human Rights, the European Court of Human Rights, and the International Court of Justice.[93] In 2022, a coalition of youth submitted a report and request to the International Criminal Court to open an investigation of BP senior executives for allegedly concealing BP's knowledge of its role in climate change, misleading the public, and delaying action to prevent and address climate change; the court has not yet acted on it.[94]

Government hearings and investigations connected to climate obstruction represent a relatively new tactic and are far rarer than lawsuits or regulations. These instruments have the potential to bypass limitations of the judicial and legislative processes outlined earlier, including barriers to standing and delays in court cases. They also have drawbacks, including the inability to impose legally binding outcomes. Three of the most high-profile examples of government hearings and investigations on climate obstruction (covered in detail later) are the Philippines Commission on Human Rights' National Inquiry on Climate Change, which produced its final findings in 2022; a series of investigations by members of the US Congress between 2019 and 2021; and a European Parliament public hearing on climate denial in 2019.[95] Scholarly analyses have highlighted the potential of these investigations to spur further climate litigation[96] and their potential replicability across jurisdictions.[97]

Finally, like other legal tools, government investigations and hearings can also be used to facilitate climate obstruction. For example, in 2019 the Canadian province of Alberta initiated a Public Inquiry Into Anti-Alberta Energy Campaigns to investigate the alleged spread of disinformation by groups opposing fossil fuel extraction. The inquiry ultimately found no evidence of illegal activity.[98]

The Philippines Commission on Human Rights' National Inquiry on Climate Change

The Philippines Commission on Human Rights announced its National Inquiry on Climate Change at COP21 in December 2015. It took an innovative approach, structuring the proceedings as a noncontentious fact-finding mission.[99] The investigation publicized evidence of and testimony on climate harms to vulnerable constituents across both the Global South and Global North. It also connected those harms to the activities of the carbon majors.

Oil and gas companies had argued that the Commission's jurisdiction was limited to adjudicating civil and political rights and therefore did not encompass climate change, and must be limited to ruling on the activities of companies located within the Philippines. The Commission rejected those arguments, deciding that its mandate did extend to climate change and furthermore that climate change caused the violation of the human rights of the people of the Philippines. The Commission decided, however, that the inquiry would be transnational and centered on dialogue rather than being adversarial.[100]

The process the Commission adopted also had some limitations, including that the conclusions of the investigation were recommendations only. Historically, the reluctance of human-rights bodies to impose legally binding obligations on companies stems in part from uncertainty as to whether corporations are subjects of international law and the traditional understanding that the concept of human rights concerns the relationship between states and their citizens.

The effects of Typhoon Haiyan (Yolanda) in 2013 caused the deaths of at least six thousand people in the Philippines, and the impacts of that storm and the role of carbon majors were the focus of the inquiry. The investigation was based in the Philippines, but the panel held hearings in New York and London as well as the Philippines. The investigation used the law as an expository tool, connecting the activities of carbon majors with people's real-world experience of the devastating impacts of climate change in the Global South.[101]

The Commission invited forty-seven oil and gas companies to be part of the investigation and published their written responses. Most of the responses from carbon majors did not comment on the issue of their contribution to climate change or its impacts, and simply referred the Commission to their sustainability reports and/or climate change policies, relying heavily on their corporate social responsibility (CSR) initiatives to justify their activities.[102] The carbon majors did not otherwise participate actively in the investigation, although they did send representatives to sit in on the hearings.[103] In contrast to this relative silence, the victims and survivors of climate change actively participated in the proceedings.

The inquiry featured individual testimonies from the survivors of Super Typhoon Haiyan, and the Commission also included voices from across the Global South. Filipino-Americans in New York compared their experiences with Hurricane Sandy to those of the survivors of Super Typhoon Haiyan. The inquiry was framed as a dialogue connecting the climate vulnerable in both the Global South and the Global North, seeking to portray a global experience of human suffering that connects the climate vulnerable across boundaries in an era of climate change. The Commission's work therefore wove a narrative of both suffering and solidarity in the face of increasing climate-induced impacts.[104]

The investigation was also unique in its focus on corporate liability, soliciting testimony from legal experts on the ability of companies to be held liable for incidents of loss and damage connected to climate change.[105] These expert testimonies articulated the potential for liability for companies under tort (civil harm) and corporate law, and provided a roadmap of how companies could ensure due diligence as well as remedial action on climate change.

In 2022, the Commission issued a report on the responsibilities of the carbon majors in the context of climate change.[106] One of its findings was that the companies "engaged in willful obfuscation and obstruction to prevent meaningful climate action."[107] The report documented the "massive climate denial" campaign perpetrated by the carbon majors, including strategies designed to undermine climate science and international climate laws such as the Kyoto

Protocol, and stated that these efforts may violate existing constitutional provisions.[108] Specifically in relation to climate obstruction, the Commission stated that the carbon majors should desist from all activities that undermine climate science.[109] In addition, it said that these companies should cease funding lobbyists, politicians, pseudo-scientists, trade associations, and other organizations that disseminate false information about climate change and climate science.[110] The Commission noted that while no international legal liability for climate harms currently exists, these companies had a clear moral responsibility to minimize climate harm, and the Commission recommended that each state pass legislation and establish liability through its court systems. While many of these recommendations have yet to be adopted, the outcome of the investigation and the process of elevating the voices of the vulnerable provide a model that other human rights commissions can adopt and follow.

US Congressional Investigations of Carbon Majors' Misinformation Campaigns

In the 2010s, members of the US Congress (its legislative branch) took a series of steps to respond to climate obstruction, including sending letters to fossil fuel companies inquiring whether they contributed funding to climate denial research and urging the Department of Justice and SEC to investigate ExxonMobil's undermining of climate science.[111] Over nine years, Senator Sheldon Whitehouse of Rhode Island gave more than 270 speeches on climate inaction.[112] Beginning in 2019, members of Congress initiated a series of public hearings and investigations connected to climate obstruction.[113]

The highest profile of these investigations was the 2021–2022 House Committee on Oversight and Reform's investigation into the fossil fuel industry campaign of misinformation about the climate crisis.[114] This investigation, led by representatives Ro Khanna and Carolyn Maloney, focused on the oil and gas industry's modern-day deception. Over the course of the next year, the Committee obtained, via subpoena, a trove of internal documents from the API, BP, the US Chamber of Commerce, Chevron, ExxonMobil, and Shell related to their climate-related advertising campaigns, corporate strategy, and funding for third-party organizations.[115]

The Committee held three hearings addressing the industry's contemporary climate delay tactics. In the first, the six CEOs from the oil companies and trade associations testified. They all admitted that climate change was a serious threat but refused to invest more resources into clean energy or to stop lobbying against climate policy.[116] At the end of the Congress's term, the Committee released more than 1,500 pages of documents and two memoranda explaining key findings.[117] These documents illustrated the oil majors' modern playbook of delay and obstruction techniques, which included: (1) promoting oil and gas expansion; (2) making relatively tiny investments in clean energy research and deployment; (3) heavily advertising the latter minor investments; and (4) attempting to block any significant climate policy that arises.[118] Parallel investigations undertaken by other committees into the industry's use of public relations firms and voluntary methane leak-detection programs similarly found much talk and little action on addressing its role in climate change.[119]

In conjunction with the political and coalition-building work of thousands of policymakers and activists, these investigations put the industry on the defensive and may have contributed to creating political space for the passage of the Inflation Reduction Act, which invested billions of dollars into climate-related programs across the United States. The investigations also revealed how much work remains. When congressional Republicans won the majority position in the US House in January 2022 and declined to continue the climate investigation, Rep. Khanna admitted that the Committee had run out of time to finish reviewing the relevant documents.[120] While this series of congressional investigations demonstrated the potentially profound impact of public hearings on addressing climate obstruction, it also reveals how such investigations can be curtailed by a shift in political power.

European Parliament Hearing on Climate Change Denial

In March 2019, the European Parliament held a ground-breaking public hearing at which European lawmakers heard expert testimony on climate change denial.[121] The hearing featured speakers on the role and evolution of climate deniers more broadly, as well as on the specific role of ExxonMobil in spreading climate misinformation.[122] ExxonMobil was invited to attend the public hearing and respond to the expert testimony but declined, citing restrictions connected to ongoing climate litigation in the United States.[123] Instead, the company sent a private memo to the Members of the European Parliament (MEPs), later leaked, which attempted to discredit the scientific research upon which part of the expert testimony was based.[124] ExxonMobil's actions drew negative political and media attention, and certain MEPs unsuccessfully attempted to revoke ExxonMobil's lobbying privileges at the European Parliament.[125] Research by Geoffrey Supran, a witness at the hearing, highlighted how ExxonMobil's attempts to influence lawmakers and undermine scientific research that make the company accountable for misleading the public was yet another example of the company manufacturing uncertainty around climate science (i.e., obstruction).[126]

CONCLUSION

State and legal efforts already play a significant role in addressing climate obstruction and are growing rapidly. Regulation, litigation, and government-led investigations are three key areas of activity, each with their own advantages and limitations. The academic literature assessing these activities remains sparse, with most commentary and analysis thus far provided by journalists. Overall, the literature suggests that state and legal efforts to address climate obstruction are expanding and may play an increasingly key role going forward.

Both academic and journalistic research have key roles to play in advancing our understanding of the impact of climate regulation and litigation. Nearly all cases seeking liability for GHG pollution rely on research quantifying historical emissions by defendants.[127] Historical research on corporate malfeasance and scientific research on climate damages also play important roles in advancing litigation.[128] Thus areas where further studies would inform and enable effective climate litigation include historical research on corporate and governmental obstruction of climate action; social scientific research on greenwashing, false solutions, and misleading advertising; scientific research showing causal links between climate change and various types of damages; and research on viable decarbonization pathways to inform Paris-Agreement-aligned injunctions.[129] Similarly, research on obstruction efforts in the regulatory arena would aid in efforts to pass and enforce effective regulations, and much research remains to be done using documents and information obtained through congressional and other government-led investigations, as well as studying the impact of such investigations.

Actors such as NGOs and the media also have a key role to play in supporting and amplifying legal efforts against climate obstruction. As previously discussed, the carbon majors are attempting to frame themselves as corporate leaders who will be indispensable to our climate future. Thus state and non-state legal actors are uniquely suited to challenging this industry framing and exposing historical and continued climate deception as well as explicitly quantifying the harms borne by the public due to such corporate actions.[130] The more frequently and thoroughly these legal actions are covered in the media, the harder it will be for the carbon majors to claim leadership on climate change and leverage their social and political capital to obstruct climate action.

NOTES

1. Riley E. Dunlap and Aaron M. McCright, "Challenging Climate Change," in *Climate Change and Society: Sociological Perspectives*, ed. Riley E. Dunlap and Robert J. Brulle (Oxford University Press, 2015), 300, 318–319.

2. Joan Setzer and Catherine Higham, "Global Trends in Climate Change Litigation: 2023 Snapshot," Grantham Research Institute on Climate Change and the Environment and Centre for Climate Change Economics and Policy, London School of Economics and Political Science, 2023, https://www.lse.ac.uk/granthaminstitute/wp-content/uploads/2023/06/Global_trends_in_climate_change_litigation_2023_snapshot.pdf.

3. Jacqueline Peel and Hari M. Osofsky, *Climate Change Litigation: Regulatory Pathways to Cleaner Energy* (Cambridge University Press, 2015), 19–20.

4. Grace Nosek, "The Fossil Fuel Industry's Push to Target Climate Protesters in the US," *Pace Environmental Law Review* 38 (2020): 53–108; Dunja Mijatović, Commissioner for Human Rights of the Council of Europe, "Crackdowns on Peaceful Environmental Protests Should Stop and Give Way to More Social Dialogue," June 2, 2023, https://www.coe.int/en/web/commissioner/-/crackdowns-on-peaceful-environmental-protests-should-stop-and-give-way-to-more-social-dialogue; Nikhil Dutta, *Protecting Activists from Abusive Litigation: SLAPPS in the Global South and How to Respond*, International Center for Not-for-profit Law, 2020, https://www.icnl.org/wp-content/uploads/SLAPPs-in-the-Global-South-vf.pdf.

5. For the People Act of 2021, H.R. 1, 117th Cong. (2021), https://www.congress.gov/bill/117th-congress/house-bill/1/text#toc-H9491B200900E40AC926EDFF1299F1F7B.

6. *City of Hoboken v. Chevron Corp.*, 45 F. 4th 699 (3rd Cir. 2022), cert. denied, 143 S. Ct. 2483 (May 15, 2023); *City of Hoboken v. Exxon Mobil Corp.*, 558 F. Supp. 3d 191 (D.N.J. 2021), aff'd sub nom. *City of Hoboken v. Chevron Corp.*, 45 F. 4th 699 (3rd Cir. 2022), cert. denied, 143 S. Ct. 2483 (May 15, 2023); *Mayor & City Council of Baltimore v. BP P.L.C.*, 388 F. Supp. 3d 538 (D. Md. 2019), aff'd, 31 F. 4th 178 (4th Cir. 2022), cert. denied, 143 S. Ct. 1795 (April 24, 2023).

7. *Danish AkademikerPension and the Church of England Pensions Board v. Volkswagen.*

8. Setzer and Higham, "Global Trends in Climate Change Litigation: 2023 Snapshot."

9. Richard Heede, "Tracing Anthropogenic Carbon Dioxide and Methane Emissions to Fossil Fuel and Cement producers, 1854–2010," *Climatic Change* 122 (2014): 229–241.

10. *Minnesota v. Am. Petroleum Inst.*, No. 20-1636, slip op. (D. Minn. March 31, 2021), aff'd, 63 F. 4th 703 (8th Cir. 2023), petition for cert. pending, No. 23-168 (August 22, 2023);Robert Brulle and Christian Downie, "Following the Money: Trade Associations, Political Activity and Climate Change," *Climatic Change* 175, no. 3 (2022): 1–19.

11. Setzer and Higham, "Global Trends in Climate Change Litigation: 2023 Snapshot"; *People v. JBS USA Food Co.*

12. *County of Multnomah v. Exxon Mobil Corp.*; House Natural Resources Committee Staff Hearing Report (September 14, 2022), https://www.congress.gov/117/meeting/house/115094/documents/HHRG-117-II15-20220914-SD007.pdf.

13. Setzer and Higham, "Global Trends in Climate Change Litigation: 2023 Snapshot."

14. "ACCC/ASIC Complaint re Glencore," September 2, 2022, Environmental Defenders Office, https://www.edo.org.au/wp-content/uploads/2023/02/Glencore-complaint-FINAL.pdf.

15. Geetanjali Ganguly, Joana Setzer, and Veerle Heyvaert, "If at First You Don't Succeed: Suing Corporations for Climate Change," *Oxford Journal of Legal Studies* 38, no. 4 (2018): 841–868.

16. Heede, "Tracing Anthropogenic Carbon Dioxide and Methane Emissions."

17. Michael Burger, Jessica Wentz, and Radley Horton "The Law and Science of Climate Change Attribution," *Columbia Journal of Environmental Law* 45, no. 1 (2020): 57, 129, 149–208.

18. William C. Tucker, "Deceitful Tongues: Is Climate Change Denial a Crime?," *Ecology Law Quarterly* 39 (2012): 831; Jessica A. Wentz and Benjamin Franta, "Liability for Public Deception: Linking Fossil Fuel Disinformation to Climate Damages," *Environmental Law Reporter* 52 (2022): 10995–10997.

19. See, e.g., Kathie M. d'I. Treen, Hywel T. P. Williams, and Saffron J. O'Neill, "Online Misinformation About Climate Change," *Wiley Interdisciplinary Reviews: Climate Change* 11, no. 5 (2020): e665; Amanda Shanor and Sarah E. Light, "Greenwashing and the First Amendment," *Columbia Law Review* 122, no. 7 (2022): 2033–2118.

20. Climate Change Laws of the World Database, London School of Economics and Political Science, Grantham Research Institute on Climate Change and the Environment, https://climate-laws.org/ (accessed April 1, 2025).

21. S. M. S. U. Eskander and Sam Fankhauser, "Reduction in Greenhouse Gas Emissions from National Climate Legislation," *Nature Climate Change* 10 (2020): 750–756.

22. Eloise Scotford and Stephen Minas, "Probing the Hidden Depths of Climate Law: Analysing National Climate Change Legislation," *Review of European, Comparative & International Environmental Law* 28, no. 1 (2019): 67–81; Benjamin K. Sovacool, "Clean, Low-Carbon But Corrupt? Examining Corruption Risks and Solutions for the Renewable Energy Sector in Mexico, Malaysia, Kenya and South Africa," *Energy Strategy Reviews* 38 (2021): 100723, https://doi.org/10.1016/j.esr.2021.100723.

23. Albert C. Lin, "Making Net Zero Matter," *Washington and Lee Law Review* 79, no. 2 (2022): 679–766; X. Tan, L. Kong, B. Gu, A. Zeng, and M. Niu, "Research on the Carbon Neutrality Governance Under a Polycentric Approach," *Advances in Climate Change Research* 13, no. 2 (2022): 159–168; D. Merner, L. Benjamin, W. Ercole, I. Keuschnigg, J. Kunik, K. Martínez Toral, L. Peterson, J. Setzer, K. Sokol, A. Tandon, and K. Turowski, "Comparative Analysis of Legal Mechanisms to Net-Zero: Lessons from Germany, the United States," *Brazil, and China, Carbon Management* 15, no. 1 (2024), https://doi:10.1080/17583004.2023.2288592.

24. Tom erb, Bob Perciaspepe, Verena Radulovic, and M. Niland, "Corporate Climate Commitments: The Trend Towards Net Zero," in *Handbook of Climate Change Mitigation and Adaptation*, ed. M. Lackner, B. Sajjadi, and W. Y. Chen (Springer, 2022), 2985–3018; J. Sarra, *From Ideas to Action: Governance Paths to Net Zero* (Oxford University Press, 2020).

25. Kaya Axelsson, Richard Black, Peter Chalkley, Thomas Hale, Nick Hay, Frederic Hans, Angel Hsu, Camilla Hyslop, Takeshi Kuramochi, John Lang, Natasha Lutz, Alexis McGivern, Silke Mooldijk, Natalie Short, Steve Smith, and Zhi Yi Yeo, "Net Zero Stocktake 2023: Assessing the Status and Trends of Net Zero Target Setting Across Countries, Subnational Governments and Companies," (Report, Net Zero Tracker (NewClimate Institute, Oxford Net Zero, Energy and Climate Intelligence Unit and DataDriven EnviroLab, 2023)), https://ca1-nzt.edcdn.com/Reports/Net_Zero_Stocktake_2023.pdf?v=1696255114.

26. Lisa Benjamin, Akriti Bhargava, Benjamin Franta, Karla Martínez Toral, Joana Setzer, and Aradhna Tandon, "Climate-Washing Litigation: Legal Liability for Misleading Climate Communications," Policy Briefing, The Climate Social Science Network, 2022, https://cssn.org/wp-content/uploads/2022/01/CSSN-Research- Report-2022-1-Climate-Washing-Litigation-Legal-Liability-for-Misleading-Climat e-Communications.pdf.

27. Ibid.

28. Sharon Yadin, *Fighting Climate Change through Shaming* (Cambridge University Press, 2023), 47–52.

29. UK Government, Climate Change Civil Penalties, https://www.data.gov.uk/dataset/13c0893a-049a-4608-9f9b-7f268a71f15a/climate-change-civil-penalties (accessed August 31, 2023).

30. Sharon Yadin, "Regulatory Shaming and the Problem of Corporate Climate Obstruction," *Harvard Journal on Legislation 60*, no. 2 (Summer 2023): 337, 360–361.

31. Nosek, "The Fossil Fuel Industry's Push to Target Climate Protesters in the U.S."

32. "Dutch Regulator Says H&M Ads Include Unsubstantiated Sustainability Claims," TFL, September 13, 2022, https://www.thefashionlaw.com/dutch-regulator-says-hm-ads-include-unsubstantiated-sustainability-claims/#:~:text=Dutch%20market%20regulator.-,In%20the%20wake%20of%20an%20investigation%20by%20the%20Netherlands%20Authority,conscious%2C%E2%80%9D%20the%20Swedish%20fast%20 fashion.

33. Yadin, *Fighting Climate Change through Shaming.*

34. Christophe Bonneuil, Pierre-Louis Choquet, and Benjamin Franta, "Early Warnings and Emerging Accountability: Total's Responses to Global Warming, 1971–2021," *Global Environmental Change* 71 (2021): 102386.

35. Yadin, "Regulatory Shaming and the Problem of Corporate Climate Obstruction," 343.

36. Martin Arnold, "ECB Accuses Eurozone Banks of 'White Noise' on Climate Risks," *Financial Times*, March 14, 2022, https://www.ft.com/content/aaa06d90-0356-44b4-b637-0e47c9003ba4.

37. Yadin, "Regulatory Shaming and the Problem of Corporate Climate Obstruction," 362.

38. Mei Li, Gregory Trencher, and Jusen Asuka, "The Clean Energy Claims of BP, Chevron, ExxonMobil and Shell: A Mismatch Between Discourse, Actions and Investments," *PLOS One* 17 (February 2022), https://doi:10.1371/journal.pone.0263596.

39. Mighty Earth, "Civil Society Calls for JBS to be Stripped of A Minus Climate Rating," March, 29, 2023, https://mightyearth.org/article/jbs-climate-rating/.

40. Yadin, "Regulatory Shaming and the Problem of Corporate Climate Obstruction," 370.

41. William F. Lamb, Giulio Mattioli, Sebastian Levi, J. Timmons Roberts, Stuart Capstick, Felix Creutzig, Jan C. Minx, Finn Müller-Hansen, Trevor Culhane, and Julia K. Steinberger, "Discourses of Climate Delay," *Global Sustainability* 3 (2020): e17, https://doi:10.1017/sus.2020.13.

42. Ibid.

43. Yadin, "Regulatory Shaming and the Problem of Corporate Climate Obstruction," 360.

44. This subsection is based on Tiffany Chan, Laura Ford, Catherine Higham, Shirley Pouget and Joana Setzer, "Corruption and Integrity Risks in Climate Solutions: An Emerging Global Challenge," Grantham Research Institute on Climate Change and the Environment and Centre for Climate Change Economics and Policy, London School of Economics and Political Science, 2023.

45. Ibid.

46. Nicholas E. Sanchez, "NGOs Reveal 71 'Revolving-Door' Cases at Fossil-Fuel Giants," *EU Observer*, October 25, 2021, https://euobserver.com/rule-of-law/153317.

47. Global Witness, "636 Fossil Fuel Lobbyists Granted Access to COP27," 2022, https://www.globalwitness.org/en/campaigns/fossil-gas/636-fossil-fuel-lobbyists-granted-access-cop27/.

48. European Union European Data Protection Supervisor, "Prevention of Conflics of Interest," https://www.edps.europa.eu/data-protection/data-protection/reference-library/prevention-conflicts-interest_en (accessed April 22, 2025)

49. Catherine Higham, Alina Averchenkova, Joana Setzer, and Arnaud Koehl, "Accountability Mechanisms in Climate Change Framework Laws," 2023, Grantham Research Institute on Climate Change and the Environment and Centre for Climate Change Economics and Policy. London School of Economics and Political Science.

50. Sean P. Flanagan, "Up in Smoke—Commercial Free Speech in the United States and the European Union: Why Comprehensive Tobacco Advertising Bans Work in Europe, but Fail in the United States," *Suffolk University Law Review* 44 (2011): 212–213; Micah L. Berman, "Commercial Speech Law and Tobacco Marketing: A Comparative Discussion of the United States and Canada," *American Journal of Law & Medicine* 39, no. 2–3 (2013): 218–236.

51. See William W. Buzbee, "Federalism Hedging, Entrenchment, and the Climate Challenge," *Wisconsin Law Review* 2017, no. 6 (2017): 1037–1113 (analyzing how federal law's preservation of space for state climate action can lock in progress and test policies); William W. Buzbee, "Asymmetrical Regulation: Risk, Preemption, and the Floor/Ceiling Distinction," 82 *New York University Law Review* 82, no. 6 (2007): 1547 (exploring differences between preemptive federal regulation setting unitary ceilings versus floors that leave room for additional state action).

52. Daniel T. Deacon and Leah Litman, "The New Major Questions Doctrine," *Virginia Law Review* 109, no. 5 (2023): 1011–1014.

53. *West Virginia v. EPA*, 142 S. Ct. 2587 (2022) (for the first time naming the major questions doctrine and using it to invalidate EPA climate regulation).

54. Grace Nosek, "Climate Change Litigation and Narrative: How to Use Litigation to Tell Compelling Climate Stories," *William & Mary Environmental Law and Policy Review* 42 (2018): 766–768.

55. Robert L. Kerr, "From Sullivan to Nike: Will the Noble Purpose of the Landmark Free Speech Case Be Subverted to Immunize False Advertising," *Communication Law and Policy* 9 (2004): 545–549; Katherine G. Horner, "Does the First Amendment Protect Fossil Fuel Companies' Public Speech?," *Environmentaal Law Reporter (ELI)* 53 (2023): 10053–10055

56. Shanor and Light, "Greenwashing and the First Amendment"; Shannon M. Roesler, "Evaluating Corporate Speech About Science," *The Georgetown Law Journal* 106 (2017): 499–513.

57. Gabrielle B. Guliana, "The States Are the Answer to Requests for A Specific ESG Disclosure Mandate, Not the SEC . . . Yet," *The University of Toledo Law Reivew* 54, no. 2 (2023): 263.

58. Ben Clapp and Casey J. Snyder, "Climate Change Litigation Trends 2015–2020," *Natural Resources & Environment* 36, no. 1 (2021): 3–4; Kendall Savage Dunne, "A Rising Tide: Exploring the Viability of Pursuing Climate Change Accountability Through Securities Fraud Litigation," *Penn State Law Review* 127 (2022): 319–328.

59. Albert C. Lin, "Making Net Zero Matter," *Washington & Lee Law Review* 79 (2022): 679–767 (connecting FTC authority and state consumer law statutes and analyzing how they are used to police deceptive climate claims and could be used to meet net zero climate policies).

60. Nick G. Schwake, "What Is Material? Creating A Conducive Environment for Climate-Related Information Disclosure in the Oil and Gas Industry," *Iowa Law Review* 108 (2023): 1959.

61. Joana Setzer and Catherine Higham, "Global Trends in Climate Change Litigation: 2024 Snapshot," Grantham Research Institute on Climate Change and the Environment and Centre for Climate Change Economics and Policy, London School of Economics and Political Science, 2024, https://www.lse.ac.uk/granthaminstitute/wp-content/uploads/2024/06/Global-trends-in-climate-change-litigation-2024-snapshot.pdf; United Nations Environment Programme, "Global Climate Litigation Report: 2023 Status Review," https://www.unep.org/resources/report/global-climate-litigation-report-2023-status-review.

62. Setzer and Higham, "Global Trends in Climate Change Litigation: 2023 Snapshot," 11–18; UNEP, "Global Climate Litigation Report: 2023 Status Review," 6–21; Jacqueline Peel and Jolene Lin, "Transnational Climate Litigation: The Contribution of the Global South," *American Journal of International Law* 113, no. 4 (2019): 679–726; Césaw Rodríguez-Garavito, "Human Rights: The Global South's Route to Climate Litigation," *AJIL Unbound* 114 (2020): 40–44; Joana Setzer and Lisa Benjamin, "Climate Litigation in the Global South: Constraints and Innovations," *Transnational Environmental Law* 9, no. 1 (2020): 77–101.

63. Brazilian Climate Litigation Platform (Research Group Law, Environment and Justice in the Anthropocene / Grupo de Pesquisa Direito, Ambiente e Justiça no Antropoceno). The database brings together Brazilian cases directly related to climate change, with 72 cases as of October 2023. See https://www.juma.nima.puc-rio.br/en/base-dados-litigancia-climatica-no-brasil (accessed April 1, 2025).

64. Danielle Moreira (coord)., "Boletim da Litigância Climática no Brasil—2023," JUMA, 2023, https://www.juma.nima.puc-rio.br/_files/ugd/a8ae8a_297d7c0470044a49bba5c325973675cb.pdf.

65. Maria A. Tigre and Joana Setzer, "Human Rights and Climate Change for Climate Litigation in Brazil and Beyond: An Analysis of the Climate Fund Decision," *Georgetown Journal of International Law* 54, no. 4 (2023): 593–637.

66. "*Amorema and Amoretgrap v. Sustainable Carbon and others (Carbon Credits and Extractive Reserves)*," Sabin Center for Climate Change Law Global Climate Change Litigation database, https://climatecasechart.com/non-us-case/amorema-e-amoretgrap-vs-sustainable-carbon-and-others-carbon-credits-and-extractive-reserves/ (accessed April 22, 2025).

67. Setzer and Higham, "Global Trends in Climate Change Litigation: 2023 Snapshot," 35.

68. Joana Setzer, "The Impacts of High-Profile Litigation against Major Fossil Fuel Companies," in *Litigating the Climate Emergency: How Human Rights, Courts, and Legal Mobilization Can Bolster Climate Action, Globalization and Human Rights*, ed. César Rodríguez-Garavito (Cambridge University Press, 2022), 206–220.

69. "*Luciano Lliuya v. RWE AG*," Sabin Center for Climate Change Law Global Litigation database, https://climatecasechart.com/non-us-case/lliuya-v-rwe-ag/ (accessed April 22, 0205).

70. "*Milieudefensie et al. v. Royal Dutch Shell plc.*," Sabin Center for Climate Change Law Global Litigation database, https://climatecasechart.com/non-us-case/milieudefensie-et-al-v-royal-dutch-shell-plc/ (accessed April 22, 2025).

71. "*Asmania et al. vs Holcim*," Sabin Center for Climate Change Law Global Litigation database, https://climatecasechart.com/non-us-case/four-islanders-of-pari-v-holcim/ (accessed April 22, 2025)

72. Climate Accountability Lawsuits, Center for Climate Integrity, https:// climateintegrity.org/cases (accessed April 1, 2025).

73. *Rhode Island v. Chevron Corp.*, 292 F. Supp. 3d 142 (D.R.I. 2019), aff'd sub nom. *Rhode Island v. Shell Oil Products Co.*, 35 F. 4th 44 (1st Cir. 2022), cert. denied, 143 S. Ct. 1796 (April 24, 2023); *Massachusetts v. Exxon. Mobil Corp.*, No. 19-12430-WGY, slip op. (D. Mass 2020); *Connecticut v. Exxon Mobil Corp.*, No. 3:20-cv-1555, slip op. (D. Conn. June 2, 2021), appeal docketed, No. 21-1446 (2nd Cir. argued September 23, 2022); *Delaware v. BP America Inc.*, 578 F. Supp. 3d 618 (D. Del. 2022), aff'd sub nom. *City of Hoboken v. Chevron Corp.*, 45 F. 4th 699 (3rd Cir. 2022), cert. denied, 143 S. Ct. 2483 (May 15, 2023); *City of Hoboken v. Exxon Mobil Corp.*, 558 F. Supp. 3d 191 (D.N.J. 2021), aff'd sub nom. *City of Hoboken v. Chevron Corp.*, 45 F. 4th 699 (3rd Cir. 2022), cert. denied, 143 S. Ct. 2483 (May 15, 2023); *Platkin v. Exxon Mobil Corp.*, No. 22-cv-06733, slip op. (D.N.J. June 20, 2023); *City of Oakland v. BP P.L.C.*, 325 F. Supp. 3d 1017 (N.D. Cal. 2018), vacated, 969 F.3d 895 (9th Cir. 2020) (remanding to state court), cert. denied, 141 S. Ct. 2776 (June 14, 2021); *City of Oakland v. BP P.L.C.*, Nos. 3:17-cv-06011 & 3:17-cv-06012, slip op. (N.D. Cal. 2022) (granting renewed motion to remand), appeal docketed, No. 22-16810 (November 25, 2022); *Mayor & City Council of Baltimore v. BP P.L.C.*, 388 F. Supp. 3d 538 (D. Md. 2019), aff'd, 31 F. 4th 178 (4th Cir. 2022), cert. denied, 143 S. Ct. 1795 (April 24, 2023); *City of Annapolis v. BP P.L.C.*, Nos. SAG-21-0072, SAG-21-01323 (D. Md. September 29, 2022), appeal docketed sub nom. *Anne Arundel County v. BP P.L.C.*, Nos. 22-2082, 22-2101 (4th Cir. October 25, 2022); *City of Charleston v. Brabham Oil Co.*, No. 2:20-CV-03579, slip op. (D.S.C. July 6, 2023); *Minnesota v. Am. Petroleum Inst.*, No. 20-1636, slip op. (D. Minn. March 31, 2021), aff'd, 63 F. 4th 703 (8th Cir. 2023), petition for cert. pending, No. 23-168 (August 22, 2023); *City & County of Honolulu v. Sunoco LP*, No. 20-CV-00163-DKW-RT (D. Haw. February 12, 2021), aff'd, 39 F. 4th 1101 (9th Cir. 2022), cert. denied, 143 S. Ct. 1795 (April 24, 2023); *County of San Mateo v. Chevron Corp.*, 294 F. Supp. 3d 934 (N.D. Cal. 2018), aff'd, 32 F. 4th 733 (9th Cir. 2022), cert. denied, 143 S. Ct. 1797 (April 24, 2023); *Board of County Commissioners v. Suncor Energy (U.S.A.), Inc.*, 405 F. Supp. 3d 947 (D. Colo. 2019), aff'd 25 F. 4th 1238 (10th Cir. 2022), cert. denied, 143 S. Ct. 1795 (April 24, 2023); *District of Columbia v. Exxon Mobil Corp.*, 640 F. Supp. 3d 95 (D.D.C. 2022), appeal docketed, No. 22-7163 (D.C. Cir. argued May 8, 2023).

74. See *Supreme Court Order List*, 143 S. Ct. 1795-97 (April 24, 2023) (denying cert. in cases originally filed by Rhode Island, Baltimore, Honolulu, Maui, San Mateo, and Boulder); *Supreme Court Order List*, 143 S. Ct. 2483 (May 15, 2023) (denying cert. in cases originally filed by Delaware and Hoboken).

75. *City & County of Honolulu v. Sunoco LP, et al.*, No. 1CCV-20-0000380 (Haw. October 31, 2023).

76. *City & County of Honolulu v. Sunoco LP, et al.*, No. 1CCV-20-0000380 (Cir. Ct. Haw. March 9, 2020); *Commonwealth v. Exxon Mobil Corp.*, No. 1984CV03333 (Mass. Super. Ct. October 24, 2019).

77. Title IX of the Organized Crime Control Act of 1970, 18 U.S.C. § 1961–1968.

78. See, e.g., *Complaint, The People of the State of California v. Exxon Mobil Corporation et al.*, September 15, 2023, 30–31, https://climateintegrity.org/uploads/media/ calif-v-exxon-complaint-sf.pdf.

79. Sjoukje Y. Philip et al., "Rapid Attribution Analysis of the Extraordinary Heat Wave on the Pacific Coast of the US and Canada in June 2021," *Earth System Dynamics* 13, no. 4 (2022), https://doi.org/10.5194/esd-13-1689-2022.

80. Center for Climate Integrity, Municipalities of Puerto Rico, November 2022, https://climateintegrity.org/cases/municipalities-of-puerto-rico.

81. See *Berrin v. Delta Air Lines Inc.*, Sabin Center for Climate Change Law, https://climatecasechart.com/case/berrin-v-delta-air-lines-inc/ (accessed April 1, 2025).

82. See *Client Earth v. Washington Gas Light Co.*, Sabin Center for Climate Change Law, https://climatecasechart.com/case/client-earth-v-washington-gas-light-co/ (accessed April 1, 2025).

83. Advertising Standards Authority, ASA Ruling on HSBC UK Bank plc, October 19, 2022, https://www.asa.org.uk/rulings/hsbc-uk-bank-plc-g21-1127656-hsbc-uk-bank-plc.html.

84. Advertising Standards Authority, ASA Ruling on Shell UK Ltd t/a Shell, June 7, 2023, https://www.asa.org.uk/rulings/shell-uk-ltd-g22-1170842-shell-uk-ltd.html.

85. "KLM Climate Claims Targeted in EU-Wide Consumer Complaint," *Dutch News*, June 22, 2023, https://www.dutchnews.nl/2023/06/klm-climate-claims-targeted-in-eu-wide-consumer-complaint/.

86. See *Milieudefensie et al. v. Royal Dutch Shell plc.*, Sabin Center for Climate Change Law, https://climatecasechart.com/non-us-case/milieudefensie-et-al-v-royal-dutch-shell-plc/ (accessed April 1, 2025).

87. See *Notre Affaire à Tous and Others v. Total*, Sabin Center for Climate Change Law, https://climatecasechart.com/non-us-case/notre-affaire-a-tous-and-others-v-total/ (accessed April 1, 2025).

88. See *Notre Affaire à Tous Les Amis de la Terre*, and *Oxfam France v. BNP Paribas*, Sabin Center for Climate Change Law, https://climatecasechart.com/non-us-case/notre-affaire-a-tous-les-amis-de-la-terre-and-oxfam-france-v-bnp-paribas/ (accessed April 1, 2025).

89. "Court Dismisses Greenpeace-Led Climate Case Against Volkswagen, Plaintiffs to Appeal," Reuters, February 14, 2023, https://www.reuters.com/business/autos-transportation/court-dismisses-greenpeace-led-climate-case-against-volkswagen-plaintiffs-appeal-2023-02-14.

90. See *Greenpeace Italy et. Al. v. ENI S.p.A.*, the Italian Ministry of Economy and Finance and Cassa Depositi e Prestiti S.p.A., Sabin Center for Climate Change Law, https://climatecasechart.com/non-us-case/greenpeace-italy-et-al-v-eni-spa-the-italian-ministry-of-economy-and-finance-and-cassa-depositi-e-prestiti-spa (accessed April 1, 2025).

91. See "United Nations Committee on the Rights of the Child," August 22, 2023, General comment No. 26 on children's rights and the environment with a special focus on climate change, https://docstore.ohchr.org/SelfServices/FilesHandler.ashx?enc=6QkG1d%2FPPRiCAqhKb7yhsqIkirKQZLK2M58RF%2F5F0vHrWghm hzPL092j0u3MJAYhyUPAX9o0tJ4tFwwX4frsfflPka9cgF%2FBur8eYD%2BEeDm uoVnVOpjkzwB9eiDayjZA.

92. Ibid., Section V(A).

93. See "Request for an Advisory Opinion on the Climate Emergency and Human Rights Submitted to the Inter-American Court of Human Rights by the Republic of Colombia and the Republic of Chile," January 9, 2023, https://www.corteidh.or.cr/docs/opiniones/soc_1_2023_en.pdf; European Court of Human Rights, "Climate Change," February 2022, https://www.echr.coe.int/documents/d/echr/fs_climate_change_eng; International Court of Justice, "Request for an Advisory Opinion, Obligations of States in Respect of Climate Change," March 29, 2023,

https://www.icj-cij.org/sites/default/files/case-related/187/187-20230412-app-01-00-en.pdf.

94. *NZ Students for Climate Solutions and UK Youth Climate Coalition v. Board of BP*, Sabin Center for Climate Change Law, https://climatecasechart.com/non-us-case/nz-students-for-climate-solutions-and-uk-youth-climate-coalition-v-board-of-bp/ (accessed April 1, 2025).

95. Geoffrey Supran, Stefan Rahmstorf, and Naomi Oreskes "Assessing ExxonMobil's Global Warming Projections," *Science* 379 (2023), https://doi:10.1126/science.abk0063 at 2.

96. Joana Setzer and Catherine Higham, "Global Trends in Climate Change Litigation: 2022 Snapshot," Grantham Research Institute on Climate Change and the Environment and Centre for Climate Change Economics and Policy, London School of Economics and Political Science, 2022, 13 ("With the publication by the Commission on Human Rights of the Philippines of the final report on the responsibility of the Carbon Majors for climate-related human rights harms, further cases are likely to follow in jurisdictions with strong human rights regimes").

97. Keely Boom, Iman Prihandono, and Nadirsyah Hosen, A Mandate to Investigate the Carbon Majors and the Climate Crisis: The Philippines Commission on Human Rights Investigation, *Australian Journal of Asian Law* 23 (2022): 73.

98. Patrick McCurdy, "The Rise and Fall of the Synthetic: The Mediatization of Canada's Oil Sands," *International Journal of Cultural Studies* 26, no. 4 (2023): 427, 439.

99. Joana Setzer and Lisa Benjamin, "Climate Litigation in the Global South: Constraints and Innovations," *Transnational Environmental Law* 9, no. 1 (2020): 77–101; J. Setzer and L. Benjamin, "Climate Litigation in the Global South: Filling in Gaps," *AJIL Unbound*, 114 (2020): 56–60.

100. Commission of Human Rights of The Philippines, National Inquiry On Climate Change Report 2022, https://chr.gov.ph/wp-content/uploads/2022/12/CHRP_National-Inquiry-on-Climate-Change-Report.pdf, 4–5.

101. Lisa Benjamin, "The Responsibilities of Corporations: New Directions in Environmental Litigation," in *Research Handbook on Transnational Environmental Law*, ed. Veerle Heyvaert and Leslie-Anne Duvic-Paoli (Edward Elgar Publishing, 2020), 229–247.

102. For example, see Megan Darby, "Carbon Majors Respond to Climate and Human Rights Inquiry," *Climate Home News* (October 27, 2016), https://www.climatechangenews.com/2016/10/27/carbon-majors-respond-to-climate-and-human-rights-inquiry/ (accessed April 22, 2025).

103. It should also be noted that this more informal legal process raises contentious issues of due process, evidence and discovery.

104. Lisa Benjamin, "Carbon Major Companies and Liability for Loss and Damage," in *Research Handbook on Climate Change Law and Loss & Damage*, ed. Meinhard Doelle and Sara Seck (Edward Elgar Publishing 2021), 391–409.

105. For example, Professor Cynthia Williams and David Estrin testified at the Human Rights Commission hearings in New York on September 28, 2018.

106. Commission on Human Rights of the Philippines, "National Inquiry on Climate Change Report," 2022, https://www.ciel.org/wp-content/uploads/2023/02/CHRP-NICC-Report-2022.pdf.

107. Ibid., 104.

108. Philippines Human Rights Commission (n 4), 105–109.

109. Ibid., 131.

110. Ibid.

111. "Markey, Boxer, Whitehouse Query Fossil Fuel Companies, Climate Denial Organizations on Science Funding," Senator Ed Markey, February 25, 2015, https://www.markey.senate.gov/news/press-releases/markey-boxer-whitehouse-query-fossil-fuel-companies-climate-denial-organizations-on-science-funding; "Rep. Lieu, DeSaulnier Request Investigation into Allegations of ExxonMobil Deception on Climate Change," Representative Ted Lieu, October 16, 2015, https://lieu.house.gov/media-center/press-releases/reps-lieu-desaulnier-request-investigation-allegations-exxonmobil; "Congressman Lieu Leads Letter to SEC Calling for an Investigation of ExxonMobil on Climate Science," Representative Ted Lieu, November 3, 2015, https://lieu.house.gov/media-center/press-releases/congressman-lieu-leads-letter-sec-calling-investigation-exxonmobil.

112. Edward Fitzpatrick, "Senator Whitehouse Delivers 279th and Final 'Time to Wake Up' Climate Change Speech," *Boston Globe*, January 27, 2021, https://www.bostonglobe.com/2021/01/27/metro/senator-whitehouse-delivers-279th-final-climate-change-speech/.

113. "Examining the Oil Industry's Efforts to Suppress the Truth about Climate Change," Subcommittee on Civil Rights and Civil Liberties, Committee on Oversight and Reform, October 23, 2019, https://www.govinfo.gov/content/pkg/CHRG-116hhrg38304/html/CHRG-116hhrg38304.htm.

114. "Oversight Committee Launches Investigation of Fossil Fuel Industry Disinformation on Climate Crisis," Committee on Oversight and Reform, September 16, 2021, https://oversightdemocrats.house.gov/news/press-releases/oversight-committee-launches-investigation-of-fossil-fuel-industry.

115. Ben Lefebvre and Zack Colman, "House Oversight Committee Accuses Oil Companies of 'Lying' About Climate Actions Politico," December 9, 2022, https://www.politico.com/news/2022/12/09/oversight-memo-oil-companies-climate-impact-00073248.

116. Maxine Joselow and Dino Grandoni, "Big Oil CEOs Testify Before House Oversight Committee," *Washington Post*, October 28, 2021, https://www.washingtonpost.com/climate-environment/2021/10/28/oil-executives-testimony-live-updates/.

117. "Memorandum Re: Investigation of Fossil Fuel Industry Disinformation," Chairwoman Carolyn B. Maloney and Chairman Ro Khanna, September 14, 2022, https://oversightdemocrats.house.gov/sites/democrats.oversight.house.gov/files/2022.09.14%20FINAL%20COR%20Supplemental%20Memo.pdf; "Oversight Committee Releases New Documents Showing Big Oil's Greenwashing Campaign and Failure to Reduce Emissions," Chairwoman Carolyn B. Maloney and Chairman Ro Khanna, December 9, 2022, https://oversightdemocrats.house.gov/news/press-releases/oversight-committee-releases-new-documents-showing-big-oil-s-greenwashing.

118. Lucas Thompson, "Oil Companies 'Could Doom Global Efforts' Around Climate Change, House Committee Finds," *NBC News*, December 9, 2022, https://www.nbcnews.com/science/environment/oil-companies-doom-global-efforts-climate-change-house-committee-finds-rcna59443.

119. "Hearing Report: The Role of Public Relations Firms in Preventing Action on Climate Change," House Natural Resources Committee Democrats, September 14, 2022, https://democrats-naturalresources.house.gov/hearing-report-the-role-of-public-relations-firms-in-preventing-action-on-climate-change; "Seeing CH_4 Clearly: Science-Based Approaches to Methane Monitoring in the Oil and

Gas Sector," House Committee on Science, Space, and Technology, June 2022, https://democrats-science.house.gov/imo/media/doc/science_committee_majority_staff_report_seeing_ch4_clearly.pdf.

120. Amy Westervelt, "Subpoenaed Fossil Fuel Documents Reveal an Industry Stuck in the Past," *The Intercept*, December 14, 2022, https://theintercept.com/2022/12/24/oil-gas-climate-disinformation/.

121. European Parliament Committees, "Joint PETI and ENVI Public Hearing on Climate Change Denial," https://www.europarl.europa.eu/committees/en/product/product-details/20190313CHE06141 (accessed April 1, 2025).

122. European Parliament Committees, "Draft Programme Joint Hearing 'Climate Change Denial'" https://www.europarl.europa.eu/cmsdata/161827/Programme_PETI-ENVI%20Hearing%20on%20Climate%20Change%20Denial.pdf (accessed April 20, 2025).

123. Geoffrey Supran and Naomi Oreskes, "Letter from Geoffrey Supran and Naomi Oreskes to Adina-Ioana V˘alean, MEP (Chair, Committee on the Environment, Public Health and Food Safety) and Cecilia Wikström, MEP (Chair, Committee on Petitions) of European Parliament," April 2, 2019, https://perma.cc/725N-ZJLG); Arthur Neslen, "ExxonMobil Faces EU Parliament Ban After No Show at Climate Hearing," *The Guardian*, March 22, 2019, https://www.theguardian.com/business/2019/mar/22/exxonmobil-faces-eu-parliaments-ban-after-climate-hearing-no-show.

124. Ibid.; Geoffrey Supran and Naomi Oreskes, "Reply to Comment on 'Assessing ExxonMobil's Climate Change Communications (1977–2014)'Supran and Oreskes (2017 Environ. Res. Lett. 12 084019)", *Enrivomental Research Letters* 15, no.11 (2020): 2.

125. The Greens in the European Parliament, "Deactivation of ExxonMobil's lobby badges pursuant to Rule 116a(3) of the European Parliament's Rules of Procedure," March 21, 2019, https://www.greens-efa.eu/files/doc/docs/3c155f5187f87e587f8f32dd8af2ac4c.pdf; Frédéric Simon, "Exxon Lobbyists Allowed to Keep EU Access Badges," *Euractiv*, April 12, 2019, https://www.euractiv.com/section/climate-environment/news/exxon-lobbyists-allowed-to-keep-eu-access-badges/.

126. Supran and Oreskes, "Letter from Geoffrey Supran and Naomi Oreskes to Adina-Ioana V˘alean"; Geoffrey Supran, "ExxonMobil Misled the Public. Now They're Trying to Mislead the European Parliament," *Euractiv*, April 8, 2019, https://perma.cc/U3VY-EPWU.

127. See Climate Accountability Institute, Carbon Majors 2018 Data Set Released December 2020, https://climateaccountability.org/carbonmajors_dataset2020.html (accessed April 1, 2025).

128. Jessica Wentz, Delta Merner, Benjamin Franta, Alessandra Lehmen, and Peter C. Frumhoff, "Research Priorities for Climate Litigation," *Earth's Future* 11, no. 1 (2022), https://agupubs.onlinelibrary.wiley.com/doi/full/10.1029/2022EF002928.

129. Ibid.

130. Grace Nosek, "Climate Change Litigation and Narrative: How to Use Litigation to Tell Compelling Climate Stories," *William & Mary Environmental Law and Policy Review* 42 (2018): 733.

CHAPTER 13

Confronting Climate Obstruction

The Role of Civil Society and Non–State Actors

LEAD AUTHORS: JENNIE C. STEPHENS AND SHARON YADIN
CONTRIBUTING AUTHORS: LAURENCE L. DELINA, LOUISE
M. FITZGERALD, FRANCISCO GARCIA-GIBSON, FERGUS
GREEN, NOEL HEALY, DAVID HESS,
TARIRO KAMUTI, CRISTIANA LOSEKANN,
DAVID MICHAELS, SONJA SOLOMUN, AND
ROBIN TSCHOETSCHEL

INTRODUCTION: EFFORTS TO RESIST CLIMATE OBSTRUCTION

Resistance to climate obstruction is gaining momentum globally as civil society attempts to hold powerful actors accountable for intentionally delaying the transformative action required for a climate-stable future. This volume defines climate obstruction as efforts intended to slow or block policies or actions on climate change that are commensurate with the current scientific consensus of what is necessary to avoid dangerous anthropogenic interference with the climate system. The previous chapter on litigation laid out ways in which the courts and governments are attempting to fight climate obstruction; in contrast, this chapter reviews the many efforts to confront climate obstruction outside legal systems, which encompass civil society and non–state actor strategies and tactics to expose, confront, and resist climate obstruction.

As climate obstruction expands in complexity and sophistication,[1] understanding the range of efforts deployed to confront it is important to strengthen coalitions and build resistance. Norm-changing initiatives to counter climate

Jennie C. Stephens et al., *Confronting Climate Obstruction*. In: *Climate Obstruction*. Edited by: J. Timmons Roberts, Carlos R. S. Milani, Jennifer Jacquet, and Christian Downie, Oxford University Press. © Oxford University Press (2025). DOI: 10.1093/oso/9780197787144.003.0013 obstruction attempt to stigmatize and delegitimize those who engage in the practice and to revoke their social license. Because actions that reveal and confront climate obstruction within individual companies, industries, and governments and among networks of powerful actors can weaken these players' effectiveness, an improved understanding of these strategies and tactics holds powerful potential to disrupt it.

Various civil-society and non–state actors actively resist climate obstruction. Key actors include climate and environmental NGOs, climate activists, local communities, corporations, financial organizations, international organizations, researchers, journalists, and educational institutions. Many are part of formal and informal coalitions collaborating to share insights and provide mutual support.

These actors direct their efforts at multiple target audiences, including the public, the media, policymakers, corporations, and civil-society organizations. For example, educational initiatives and journalistic endeavors expose climate obstruction primarily by informing the public while indirectly targeting policymakers and corporations. In contrast, some NGO and journalistic efforts to expose climate obstruction focus explicitly on government actors such as policymakers, program administrators, regulators, and the courts. Another dimension of resisting climate obstruction is directing efforts at individual corporations and industry networks. This approach involves publicly naming and shaming specific companies and/or performing research and journalistic investigations to expose industry delay strategies. This approach can also include working with media organizations to persuade them to adopt policies that fight climate mis-/disinformation.[2] The various strategies also ripple beyond their primary targeted audiences, exerting influence on other actors.

This review focuses on strategies and tactics explicitly designed and motivated to counter climate obstruction. Although the scale and breadth of efforts to confront and resist climate obstruction have grown steadily since the 1980s, when significant investments were made in denying climate science,[3] this chapter narrows its focus to the period following the 2015 Paris Agreement.

A Framework for Analysis

Efforts to resist climate obstruction have expanded rapidly since 2015, due partly to the rapid expansion of climate obstruction itself.[4] While there is ample research on climate action in civil society across various countries,[5,6] there is a dearth of research examining how civil society is using counter–climate obstruction initiatives. To date, there exists minimal analysis of activities aimed at combating climate disinformation, efforts to resist climate denial, initiatives focused on confronting instances of "climate-washing"—the deliberate misrepresentation of climate actions and commitments—and projects designed to counter other manipulative strategies and tactics aimed at blocking or delaying climate policies and actions.

The chapter aims to address this gap by presenting a framework for categorizing dominant strategies and their respective tactics deployed by civil society to confront climate obstruction. Based on a review of climate obstruction scholarship, reports, news articles, and prominent movements, actions, and practices, we identify three main strategies: (1) social exposure to reveal and expand awareness about climate obstruction, (2) lobbying to advocate for the prevention of climate obstruction, and (3) campaigns and protests to mobilize public concern about climate obstruction. This framework provides a way to characterize diverse, dynamic, nonlinear, and context-dependent strategies.

As this chapter shows, these strategies and tactics are employed by a wide range of actors spanning North and South America, Europe, Africa, and Asia. The examples provided offer a snapshot of recent efforts to fight climate obstruction and the diverse mechanisms developing in this arena. Characterizing these examples within the framework encourages comparative reflection upon the value of various strategies and highlights some prominent synergies among them. The chapter concludes by underscoring the need for future research and expanded civil-society initiatives focused on confronting climate obstruction.

STRATEGIES TO COUNTER CLIMATE OBSTRUCTION

Non–state actors have deployed a wide array of strategies to address climate obstruction. One way to categorize them is by distinguishing between institutional and extra-institutional strategies. Institutional strategies operate within the confines of existing institutions and norms, while extra-institutional strategies transcend these boundaries, often taking on more radical and disruptive approaches.[7] This review includes both prominent institutional strategies for reducing climate obstruction (social exposure and lobbying) and extra-institutional strategies (such as campaigns and protests deploying tactics such as rallies, civil disobedience, and site occupations). The two categories sometimes generate tensions or divisions among actors and within coalitions. However, these differing approaches can also work synergistically, such as when more radical groups open opportunities for actors pursuing more conventional strategies.[8] Extra-institutional strategies and tactics often emerge when institutional strategies appear to be ineffective or inadequate.[9] Strategies tend to evolve and adapt in response to shifts in coalition composition, goals, and intended audience.[10] Also, the use of specific strategies can prompt and facilitate the use of other strategies and tactics. For example, social exposure, such as naming and shaming tactics, can reinforce lobbying efforts and encourage campaigns and protests. Although not exhaustive, this overview covers the most prominent strategies.

Social Exposure

One prominent strategy for confronting climate obstruction is unveiling it and expanding awareness of its perpetrators and their methods. Our framework refers to this reputation-based strategy of revoking social license (community approval of a company or institution) as *social exposure*. Social exposure leverages scholarly research, media coverage and monitoring, and naming and shaming to confront climate obstruction.

Research

Many actors and organizations have been investigating climate obstruction to expand social exposure, including researchers from universities, think tanks, and NGOs (including contributors to this volume). Recent publications have, for example, provided a framework for identifying the discourses of delay[11] and misleading public claims of fossil fuel companies.[12] Some research is designed to expose all kinds of climate obstruction and thereby encourage transparency. Other research focuses on uncovering and understanding particularly egregious forms of climate obstruction (through naming and shaming). For example, researchers have produced evidence revealing the funding sources behind climate denial campaigns,[13] the environmental and social impacts of fossil fuel investments,[14,15] and the compliance of governments, industries, and regulatory agencies with climate-related commitments.[16] These efforts have exposed instances of obstruction, including greenwashing, noncompliance, and other forms of resistance to climate action.[17]

Research has also revealed how the fossil fuel industry promotes misinformation through its communication channels and by strategically investing in higher education and university partnerships.[18,19,20] These insights have inspired and served as a basis for many of the other strategies to confront climate obstruction that this chapter discusses, including lobbying and campaigns for governmental efforts and legal actions (see Chapter 12).

Research has also been critical for illuminating carbon offsetting as a delay tactic, critiquing the validity of net-zero claims, and exploring the problematic power dynamics of allowing corporate interests to participate in climate policy discussions.[21,22,23]

Notably, insight sharing among climate obstruction researchers was facilitated by an international conference on fossil fuel supply and climate change at Oxford University in 2016. Spearheaded by the Stockholm Environment Institute, this conference was subsequently held in 2018, 2020, and 2022, and has provided an avenue for exchanging findings on climate obstruction.

While climate obstruction research is conducted by both universities and nongovernmental organizations (NGOs), NGO research has been critically important for exposing climate obstruction because fossil fuel interests have invested heavily in university-based climate research, which has in turn constrained academic research on climate obstruction. The term "academic capture" describes how industry has influenced university climate initiatives in this way.[24] For example, since 2020, the Climate Social Science Network (CSSN), the international network of researchers coordinating this volume, has facilitated and supported international collaboration on research that resists climate obstruction by revealing and exposing it.[25] Nonetheless, fossil fuel phaseout-related research remains an intellectual "no-fly zone" in many universities due to the significant influence of powerful corporate interests on university research agendas.[26]

Multiple examples of nonacademic research uncovering climate obstruction exist worldwide. The Australia Institute, an influential public policy think tank in Australia, has played a leading role in revealing the part industry lobbyists have played in delaying climate policy. The US Center for Climate Integrity advocates accountability and counters disinformation from fossil fuel companies. The UK-based Institute for Strategic Dialogue (ISD) has provided empirical research on climate mis-/disinformation and made recommendations for policymakers, internet and social media companies, and civil stakeholders to resist obstruction. The Climate Policy Initiative has produced several influential publications on obstruction in Brazil.[27] Such think tanks serve as crucial intermediaries between research and policymaking, offering evidence-based analysis of climate obstruction and providing policy recommendations to confront and restrain climate obstruction.

NGOs also study the suppression of climate scientists and their research to document the dissemination of climate disinformation. The suppression and targeting of climate scientists have been widespread, particularly in government administrations hostile to energy-transition policies.[28] During the first Trump administration in the United States (2017–2021), for example, climate scientists were fired and forced into retirement. Censorship was also widespread.[29] NGOs and research organizations such as the Union of Concerned Scientists,[30] the Sabin Center, and the Climate Science Legal Defense Fund (CSLDF) have closely monitored this suppression.[31] CSLDF has not only tracked but also provided support to climate scientists who have experienced suppression. The censorship of climate scientists also tends to coincide with the suppression of climate science itself. The Environmental Data and Governance Initiative[32] was founded to prevent the loss of crucial datasets during the first Trump administration. Since its inception, the Initiative has become a leading watchdog organization monitoring data availability and dataset changes.

NGOs also engage in research activities to track climate finance and corporate investments; these data are then used to confront climate obstruction. This crucial work pinpoints the origins and destinations of funding that either supports climate action or contributes to climate obstruction.[33] For example, BankTrack.Org monitors private banks and their fossil fuel investments globally, providing information to strengthen campaigns to challenge climate obstruction in the fossil fuel industry. Fossil Free Indexes offer data and analysis on fossil fuel investments within financial indexes, aiding climate-conscious investors in making informed decisions.[34] In the United States, the Rainforest Action Network campaigns against institutions funding environmentally destructive projects and climate denial.[35] Globally, 350.Org, Greenpeace, Oil Change International, and Global Witness track fossil fuel industry investments, counteract industry-driven misinformation, and challenge efforts to present fossil gas as a climate-friendly option. The UK-based Reclaim Finance concentrates on investments supporting climate denial and harmful fossil fuel activities. The Climate Investigations Center scrutinizes funding related to climate denial and its policy impacts, particularly within the United States.

NGO research by 350.org, Oil Change International, Oil Watch International, Amazon Watch, and Friends of the Earth on the opposition to continued build-out of fossil fuel infrastructure has also documented and opposed climate obstruction by, for example, highlighting the fossil fuel industry's promotion of technological solutions to avoid the phase-out of fossil fuels.[36] For example, a coalition of NGOs produced approximately fifty reports on the environmental risks, absence of economic benefits, issues of environmental justice, and influence of politics during the campaigns against the Atlantic Coast Pipeline in the United States.[37]

Research efforts among NGOs also extend beyond specific campaigns, encompassing broader studies exploring the feasibility of an energy transition. For example, Oil Change International produced a report in 2023 on how Japan's fossil fuel finance threatens to derail Asia's energy transition.[38] Additionally, NGOs have published studies examining the safety and environmental risks associated with existing energy technologies such as pipelines. These reports provide valuable public information and serve as a counterpoint to claims of fossil fuel developers. Simultaneously, these studies enhance the credibility of opponents as these campaigns progress into regulatory hearings and litigation.

NGOs also mediate between the scientific research community and the media, civil society, and governmental bodies. In the United States, the Climate Science Rapid Response Team connects climate scientists with journalists and government officials to swiftly deliver scientifically grounded information.[39] In Brazil, the Instituto Clima Info,[40] Climate Observatory,[41] and the Serrapilheira Institute[42] combat climate change mis/disinformation by offering accurate science-based information to the media, the public, governmental bodies, and other NGOs.

Media Coverage and Media Monitoring

Media actors, including journalists, NGOs, individuals, and media organizations, are working to counter climate obstruction through media coverage. Among them are well-established and widely recognized news publications specializing in exposing corporate climate obstruction while promoting public awareness of greenwashing, climate denial, and disinformation. Investigative journalism also involves research (as discussed earlier), but this chapter distinguishes it from media-related efforts to capture the breadth of media engagement in countering climate obstruction. Media actors wield significant influence in the fight against climate obstruction, especially in packaging and disseminating climate change information to broad audiences.[43] Media coverage—publishing news articles and opinion pieces of relevant events and issues—is a counterobstruction strategy and a part, or outcome, of other strategies discussed here. For example, media coverage can be a tool of investigative journalism to reveal climate obstruction. Media coverage can also result from protests and campaigns, shaming tactics, and research publications that attract media attention.

Journalists and media organizations contribute to countering climate obstruction by debunking climate misinformation, promoting climate literacy, and raising awareness about innovative climate solutions and sustainable practices. For example, over time, climate scientists, activists, and other non–state actors have effectively leveraged the media to clarify the link between climate change and adverse weather conditions.[44] These may be effective strategies to overcome the fossil fuel industry's effort to utilize the media: an analysis of climate change coverage by prominent US news outlets, including the *New York Times, Wall Street Journal*, the *Washington Post*, and the Associated Press, revealed a diminishing trend in the portrayal of climate change issues in ways that dissuade people from participating in the fight against the crisis.[45] This analysis suggests the potential for media actors to play a broader role in countering climate obstruction. Importantly, Indigenous journalists and media also counter climate obstruction by exposing systemic biases and reporting on issues not covered by the mainstream media.[46]

Media actors also counter climate mis/disinformation[47] and expose tactics of delay and inactivism.[48] Some news outlets, like *The Guardian*, reveal the role and influence of corporate climate obstruction of this type. Their 2021 climate crimes series investigated the type of tactics used by fossil fuel companies to evade responsibility for their contributions to global heating.[49] One noteworthy investigation by the Canadian Broadcasting Corporation exposed corporate obstruction through revelations of the multinational pipeline and energy company Enbridge's sponsorship of the University of Calgary.[50] The media coverage revealed direct attempts by fossil fuel energy corporations to shape public discourses on climate change.

Organizations also operate ancillary websites or newsletters to identify and debunk climate crisis mis/disinformation. For example, Climate Nexus,[51] a strategic communications group founded in 2011, was a prominent New York-based oil and gas industry watchdog until it was shut down abruptly in June 2024. They produced the *Nexus Hot News*, a newsletter featuring a "Denier Roundup" that reported climate mis/disinformation. Similarly, *DeSmog*,[52] founded in 2006, publishes regular reports and investigations on climate deniers and obstruction and aims to provide accurate fact-based information regarding climate misinformation campaigns. Similarly, the Australian website Skepticalscience.com,[53] launched in 2007, specializes in providing rebuttals to arguments by actors associated with climate science attacks and denialism. Other news outlets, such as Drilled News, a global multimedia project focused on climate accountability, produce materials providing in-depth exposés of climate disinformation purveyors. Politico's *E&E News* has also published pieces exposing climate obstruction, including an October 2020 article revealing that American automotive companies knew fifty years ago that vehicle emissions cause climate change.[54]

Despite these efforts, media actors face significant challenges in combating climate obstructionism. One pressing challenge relates to the journalistic inclination of presenting "balanced" coverage of issues, which involves giving a platform to climate countermovement perspectives in climate-related stories.[55] Additionally, media actors may inadvertently amplify climate mis/disinformation through their reporting when they repeat misleading claims by those engaged in climate obstruction.[56] Media actors must also navigate a public arena where "authenticity," or how easy it is to believe information, is increasingly valued over "facticity," the degree to which the claims made can be verified.[57] This situation is particularly troubling when climate denial and misleading claims gain traction as perceived authentic and truthful narratives.[58]

Monitoring the media to promote climate-responsible reporting is another strategy for resisting obstruction. Because climate mis/disinformation circulates widely on social media, often amplified by algorithmic systems and inadequate policies to control climate mis/disinformation among platform companies,[59] monitoring by NGOs and other non-state actors has emerged as a prominent countermeasure. Campaigns led by organizations such as the Conscious Advertising Network,[60] which focuses on severing the economic ties between advertising and harmful online content, have successfully pressured web platforms including Google, Pinterest, and Facebook to revise their monetization and climate misinformation policies.

Similarly, the Climate Action Against Disinformation (CAAD) coalition, comprising more than fifty leading climate organizations focused mainly on combatting the spread of mis/disinformation on tech platforms,[61] monitors climate mis/disinformation spread on major technology platforms. CAAD also provides recommendations for policymakers, platform companies, and civil stakeholders on combatting media mis/disinformation. The coalition has exposed specific communications tactics and rhetoric used in climate obstruction as well as offered a comprehensive definition of "climate disinformation," calling for its implementation by tech platforms including Meta, Google, and X (Twitter).

University researchers are also coordinating their expertise to monitor and resist climate obstruction in the media. For example, Climate Feedback, a network of scientists, analyzes and critiques media coverage of climate change by fact-checking and labeling the accuracy of online climate-related content.[62]

Naming and Shaming

"Naming and shaming" climate obstruction refers to civil-society actors' exposure of individuals, organizations, and countries[63] for their actions, omissions, or decisions that obstruct climate policies or may impede climate action.[64] This tactic targets firms,[65] corporate executives, and policymakers engaging in climate-washing, anti-climate litigation, anti-climate organizational policies, and climate denial to hold obstructors accountable and deter ongoing or future climate obstruction.[66] Shaming centers on causing reputational harm to those engaged in obstruction[67] and includes a moral message exposing the deceitful nature of climate obstructionists to motivate action to counter them.[68] Shaming climate obstructionists holds these particular entities accountable for impeding climate policies and action, thereby encouraging other actors, (including shareholders, suppliers, creditors, customers, advertisers, litigators, regulators, policymakers, and employees) to sanction them. The significance of shaming also arises from the law's limited capacity to address climate obstruction. Shaming thus functions as a form of non-state regulation and a governance tool discouraging climate obstruction. It can also inspire various institutional and extra-institutional strategies and tactics to further resist climate obstruction, some of which are discussed in this chapter, such as campaigns and protests.[69]

Climate obstruction shaming differs from the broader concept of climate shaming. While the latter refers to shaming people and organizations, such as corporations, for their carbon footprint,[70] the former targets people and organizations explicitly involved in denying climate change or delaying climate action or policy.[71] Climate obstruction shaming is more intricate because it requires clear explanations to the public or relevant audiences about the meaning and implications of climate obstruction tactics, which are often complex.[72]

Another challenge lies in substantiating and communicating the existence and methods of climate obstruction. For example, shaming policymakers for not enacting climate laws or implementing lax climate policies due to successful anti-climate lobbying by the fossil fuel industry requires both proof and skillful argumentation. Lobbying efforts are often covert, so establishing a causal connection between lobbying and policies can be arduous. Making a compelling case for why climate obstruction is perilous and destructive may also pose communication challenges, as not everyone may directly perceive the threat posed by the climate crisis.[73]

Another conceptual distinction between climate obstruction shaming and climate obstruction "faming" should be made. Efforts by civil society to respond to climate obstruction can also be conducted by recognizing firms, officers, policymakers, and states that are performing well in their commitments and actions to expose and fight climate obstruction. Shaming and faming tactics can also be combined by, for example, ranking and exposing the best and worst actors in either climate obstruction or fighting climate obstruction.

For example, the international NGO Greenpeace has dedicated decades to shaming fossil fuel companies for their false statements and misinformation.[74] In 2001, Greenpeace embarked on a series of reports and web projects, including the "Decade of Dirty Tricks" report and the web project "ExxonSecrets," which exposed the climate denial activities of carbon majors including ExxonMobil.[75] Greenpeace continues its efforts to expose and shame climate deniers, including "polluters, lobbyists, and politicians standing in the way of progress."[76] They operate a "PolluterWatch" to hold "polluters, their lobbyists, and the politicians who work with them accountable for poisoning the climate debate and blocking much-needed environmental regulations."[77] Greenpeace also tracks and shames business magnates the Koch brothers and their affiliated organizations for financially supporting nearly one hundred groups that attacked climate change science and policy from 1997 to 2018.[78] Their website features a list shaming these groups, signaling to stakeholders associated with them that they should distance themselves or face reputational repercussions.[79] This form of shaming has led to increased scrutiny and resistance from individuals and organizations in the United States[80] and the establishment of the organization UnKoch My Campus, a campaign to disrupt corporate power on US campuses and communities, as discussed later.

ExxonMobil was also a focus of a coordinated climate obstruction shaming campaign initiated by a coalition of US-based NGOs that harnessed social media activists, most notably using the hashtag #ExxonKnew.[81] The campaign, which began in 2015 based on investigative reporting by *Inside Climate News*, the *Los Angeles Times*, and Columbia University's Graduate School of Journalism,[82] focused on exposing what ExxonMobil knew as early as the late 1960s about global warming and its link to burning fossil fuels and how, despite that knowledge, it continued business as usual and misled the public and policymakers.[83] As part of the campaign, activists also initiated and participated in a mock trial during the 2015 United Nations Conference of the Parties (COP21). Titled "Exxon vs. the People," it highlighted the company's wrongdoing and deceit.[84]

Another notable example involves the Energy and Policy Institute, a watchdog organization exposing attacks against renewable energy and countering mis/disinformation by fossil fuel and utility interests.[85] One of the Institute's primary tactics is to publicly name and shame specific fossil fuel producers, utilities, and automakers that have been aware of climate change for decades while simultaneously supporting climate mis/disinformation campaigns.[86] The NGO InfluenceMap also conducts climate obstruction shaming by evaluating and ranking businesses based on whether their lobbying efforts align with the goals of the Paris Agreement or are focused on counterobstructionist goals (see the following section).[87]

Efforts to encourage transparency, including ranking systems like that used by the US-based organization Law Students for Climate Accountability, are also a form of shaming. The group rates and ranks US law firms on a scale from A (colored green) to F (colored red) based on their engagement in fossil fuel work over five years.[88,89] Similarly, CAAD rates and ranks social-media platforms according to their climate disinformation policies.[90] This rating system incentivizes tech corporations to enhance their standards, catalyzes policymakers (including regulators) into advocating legislation mandating the adoption of these standards, and increases public awareness regarding the role platforms play in facilitating the dissemination of climate misinformation.[91]

Climate obstruction shaming also targets specific countries. One example is the Fossil of the Day Award from the Climate Action Network (CAN), an international, Germany-based network of hundreds of civil-society organizations.[92] First bestowed in 1999, the award is presented at COPs to countries that are "doing the most to achieve the least" and "doing their best to be the worst" in climate negotiations and climate action.[93] Numerous governments have been shamed in their national media for receiving the award, including

Australia, which was dubbed the "Colossal Fossil" at COP 26 for its ongoing embrace of fossil fuels under its then-conservative government.[94]

Lobbying

A second strategy for countering obstruction involves lobbying, which refers to directly advocating policy or regulatory changes to policymakers through two main tactics: *disinformation rebuttal* and *counterbalancing*. Disinformation rebuttal attempts to persuade policymakers that obstructionist claims are false. It includes refuting disinformation about climate science, the costs of climate policy, and the severity of climate risks.[95] Counterbalancing involves making offers and/or threats to policymakers with the intention of directly countering obstructionists' influence over those or other policymakers.[96]

These tactics are used at different stages in the policymaking process. Counterobstructionist lobbying often targets the policy-drafting stage. At this stage, counterobstructionist lobbyists persuade policymakers to resist diluting climate policy proposals, such as attempts to introduce loopholes that significantly reduce a policy's climate mitigation potential. At other times, it targets the legislative voting stage to ensure that policymakers understand climate obstruction efforts before they vote. This lobbying can be influential because sometimes the fate of climate legislation is determined by a few votes.

Counterobstruction lobbying has been used by businesses, NGOs, and related coalitions with varying success. Counterobstruction lobbying by companies can be particularly prevalent when divisions are present within a sector or industry in which some firms, for example, rely on emissions-intensive technologies and others use more environmentally friendly technologies.[97] Within the automobile industry, for example, companies well positioned to profit from electric vehicles (EVs) have engaged in counterobstruction lobbying to encourage regulations that favor EVs in response to climate obstruction lobbying by other automotive companies who want to maintain high demand for their fossil fuel-based internal combustion engines. For example, Tesla cited the Federal Chamber of Automotive Industries in Australia in 2024 for making false claims about the impact that different fuel efficiency standards for passenger cars would have on emissions and car prices.[98]

Another example of counterobstructionist lobbying was employed by a coalition of businesses and NGOs that lobbied in response to the European Commission's proposal to ban (fossil fuel) internal-combustion cars and vans by 2035. In early 2023, the Commission was considering diluting the proposal under pressure from Germany, which highlighted the potential costs to its large auto industry (see Chapter 3).[99] In response, the coalition addressed a joint letter to the Commission's president, Ursula von der Leyen, requesting no changes to the proposal.[100] The letter used disinformation-rebuttal tactics (highlighting the ban's air-quality and other environmental benefits as a counter to the costs highlighted by Germany) and counterbalancing tactics (warning that weakening the ban would hamper the signatory companies' decarbonization plans and undermine business trust in the policymaking process). In the end, Germany's demands were excluded from the legislation, but the Commission did agree to include them in subsequent technical legislation.[101]

Businesses often engage in counterobstructionist lobbying through regional, national, and international business associations. Many of these associations form partnerships with business-oriented NGOs such as CERES (an advocacy organization working to accelerate the sustainability transition by working with investors and companies to change markets and sectors), EDF+Business (an environmental organization that partners with businesses to help them make their business practices more sustainable), the We Mean Business Coalition (a network that catalyzes business and policy action to halve emissions by 2030), and The Climate Group (an international nonprofit founded in 2002 that works with businesses and governments to accelerate climate action). These organizations all coordinated efforts in producing the joint letter in the example above.

NGOs also engage independently from businesses in lobbying efforts to counter climate obstruction. However, it is crucial to acknowledge the significant disparity in funding available to support counterobstruction lobbying by NGOs compared with the resources possessed by businesses engaged in climate obstruction.[102] NGOs often overcome their comparatively limited resources through forming coalitions including transnational networks.[103] An example is the Natural Resources Defense Council (NRDC) which, in coalition with other NGOs, lobbied intensely for the passage of the 2022 climate policy known as the US Inflation Reduction Act (IRA).[104] Against fears sown by the US Chamber of Commerce and other organizations, NRDC and its lobbying partners used disinformation-rebuttal tactics such as providing credible and detailed reports to congressional committee members illustrating how the policy would create investment and jobs in states that faced the closure of fossil fuel infrastructure.[105,106]

One approach to counter climate obstruction lobbying is the development of guidelines and standards to constrain corporate climate lobbying. While these standards sometimes require lobbying efforts themselves, they constitute another form of monitoring to resist climate obstruction (similar to the media tracking discussed previously).[107] One of the most prominent guides and standards in this regard is the Global Standard on Responsible Climate Lobbying, which has been incorporated into the Climate Action 100+ Net Zero Benchmark. These standards mandate that companies lobby for science-based climate policies and refrain from obstructing them. Investors sometimes use these standards to establish expectations for companies. The Climate Action Network (CAN) also promotes standards for including fossil fuel, industrial-scale forestry, and agribusiness industries in United Nations Framework Convention on Climate Change (UNFCCC) negotiations.[108]

Campaigns and Protests

Campaigns and protests that contest climate obstruction are another important category of strategies. These strategies are classified as extra-institutional because they involve actions that transcend the boundaries of current institutions and tend to be more radical and disruptive than the other strategies. Campaigns and protests often build on evidence from research that revealed climate obstruction and then develop a specific demand that people can rally around to delegitimize or disassociate from the climate obstruction. Many campaigns and protests to counter climate obstruction have focused on governments and actors aligned with the fossil fuel sector, who are known to be blocking climate action through spreading misinformation and other tactics. Others are focused explicitly on resisting fossil fuel development and infrastructure, not just because fossil fuels are the most significant contributor to climate change but also because the fossil fuel industry has been strategically investing in climate obstruction for decades.[109] This section categorizes campaigns into three types based on their main objectives (though some may overlap): (1) shifts in industry norms and practices, (2) sunsetting industries, and (3) stopping specific projects or seeking remediation for them.

Shifting Norms

In the pursuit of shifting industry norms and practices related to climate obstruction, civil-society groups have used campaigns and protests to raise awareness and frame the fossil fuel industry's economic and political activities as morally objectionable. This effort contributes to the emergence of "anti-fossil fuel norms."[110] The goal of norm change is to stigmatize and delegitimize the fossil fuel industry and those supporting it, effectively revoking their "social license to operate."[111] Stigmatized sectors tend to have fewer allies and weaker public support than legitimate industries, making it more feasible to enact laws and policies that restrict those industries. Thus, such campaigns can indirectly help counter the industry's obstructionist activities.[112]

Targeted activities include campaigns on fossil fuel investments, subsidies,[113] and new projects and infrastructure.[114] Campaigns related to investments typically pressure organizations to "divest" their capital from fossil fuel companies. For example, Fridays for Future, the global youth climate movement inspired by activist Greta Thunberg's school strike in Sweden,[115] demands an end to government subsidies for fossil fuels.[116] Divestment campaigns at universities and public pension funds demand an end to investments in companies profiting from continued fossil fuel expansion, and some target directly companies that seek to deny climate science or delay efforts to mitigate emissions.[117] The divestment campaign can be understood economically and as a coordinated effort to delegitimize fossil fuels and the fossil fuel industry's climate obstruction strategies.[118,119]

The higher-education sector has played a key role in the fossil fuel divestment movement, in part because of the strong opposition among students and faculty on college campuses to the fossil fuel industry's climate obstructionism.[120] One example of a coordinated counterobstructionism campaign in higher education is the aforementioned UnKoch My Campus, a nonprofit organization founded by student activists in 2013 in the United States in response to the realization that donations from the Koch family, who were associated with climate obstruction and fossil fuel support, were exerting influence over college curricula, research agendas, and the hiring and firing of faculty on multiple campuses.[121] The organization investigates and audits relationships between wealthy donors, corporations, and educational institutions to reveal strategic investments promoting private interests of the fossil fuel industry and provides training on how to resist them.[122] UnKoch My Campus supports university campaigns and protests to resist the fossil fuel industry's efforts toward "academic capture" designed to advance their climate obstruction agenda.[123]

While students have been the primary driving force behind university-based fossil fuel divestment campaigns,[124] faculty and academic staff have also contributed to the movement.[125] Student participation in campaigns to change their universities' investments can create, in some cases, an important educational impact by shifting the universities' core values.[126] Beyond fossil fuel divestment, campaigns in some universities, including Brown[127] in the United States and Cambridge[128] in the United Kingdom, have expanded to call for disassociation from companies that promote disinformation or engage in obstructionist lobbying. The Fossil Free Research campaign and organization also urges universities to reject fossil fuel industry funding for their research.[129]

Sunsetting Industries

Resisting decades of climate obstruction by fossil fuel companies also includes efforts to sunset (phase out or terminate) fossil fuel extraction or use within a specific industry or region. By sunsetting these industries or reducing the fossil fuel investments of leading companies within them, their capacity and motivation to engage in climate obstruction can also be reduced. For example, in the United States, the "Beyond Coal" campaign has leveraged research exposing the coal industry's climate obstruction efforts to accelerate the closure of many coal-burning electricity-generation facilities. This campaign has expanded to include gas as well as coal, as some utilities transition from coal to gas while continuing to engage in climate obstruction.[130] Fossil fuel phaseout campaigns highlighting the climate obstruction of the fossil fuel industry are also being mounted in specific geographic regions. For example, building on decades of efforts to protect the rainforest, dating back to the work of Chico Mendes, Brazil's "No More Oil" campaign advocates for oil-free zones by bolstering the anti-oil agenda in communities negatively affected by the industry and its climate obstructionism.[131] In the Amazon region, these campaigns often involve direct action, including occupying extraction infrastructure sites and staging protests at government offices. The mis/disinformation of the industry are frequently highlighted in these campaigns, which sometimes include alliances among Indigenous communities. These alliances can be fragile, and Indigenous communities may have divisions over resisting extraction or seeking its benefits.[132]

Stopping Specific Projects

In addition to broader campaigns aimed at countering obstruction by sunsetting industries and phasing out fossil fuels, numerous local campaigns have emerged to counter false claims to stop local fossil fuel extraction and infrastructure projects or secure remediation for the damage caused (see Chapter 12 for further discussion on legal actions). Although these campaigns often highlight the climate impacts associated with expanding extractive capacity and infrastructure,[133] they also include opposition to local environmental injustices and other adverse effects. Understanding complex coalition dynamics requires research on the tensions that can emerge between national or global organizations focused on energy transitions and climate obstruction and local organizations more concerned with local environmental, cultural, and economic effects of fossil fuel infrastructure.[134]

These diverse campaigns and protests can be interconnected, with localized mobilizations against a single project expanding into broader regional sunsetting initiatives. A prime example was the local dissent against constructing a coal-fired power plant in Bohol, an island province in the central Philippines.[135] Led by a coalition of citizens, businesspeople, and members of the clergy who recognized the climate obstruction strategies of the coal plant developers, this movement employed various tactics that prompted the Bohol government to prohibit any future coal-based development in the province: mounting social media campaigns, publishing position papers, and organizing public events.[136] Such campaigns can transcend local opposition struggles by forming alliances with national and international organizations. Such shifts in campaign scale often involve a change in goals and frames, shifting the focus toward global issues such as climate change and human rights.[137]

Since 2020, for example, a rapidly growing global network of policymakers, scholars and activists has been working to counter fossil fuel interests' extensive climate obstruction by advocating a global fossil fuel nonproliferation treaty.[138] Fossil-free zones are also gaining traction within communities and organizations, including at some universities.[139] Another emerging campaign is the Fossil Free Careers campaign, led by People & Planet UK. This campaign urges universities to discontinue their recruitment partnerships with oil, gas, and some mining companies, discouraging students from entering extractivist industries.[140]

Another global campaign to counter climate obstruction is Clean Creatives, a movement of advertisers, public relations (PR) professionals, and their clients who are committed to cutting ties with the fossil fuel industry, motivated in part by evidence that "fossil fuel producers have been able to mislead people about climate change with the support of advertising agencies who have helped them spin false or deceptive narratives."[141] Recognizing how the fossil fuel industry has relied on advertising to promote climate obstruction, this campaign invites individuals and organizations to take a pledge not to work for the fossil fuel industry. Consequently, it resists climate obstruction by constraining who will collaborate with and provide services to support climate obstruction.

CONCLUSION

This chapter has identified three distinct yet overlapping and often synergistic strategies for countering climate obstruction. This categorization of efforts to confront climate obstruction provides a framework to assess and compare such efforts. The social exposure category includes research, media coverage, media monitoring, and naming and shaming to reveal and expand awareness about climate obstruction. The lobbying category includes advocacy efforts to prevent climate obstruction. Campaigns and protests comprise a broad range of coordinated collective actions to mobilize public concern about climate obstruction. Each of these strategies and tactics is dynamic, nonlinear, and context-dependent. This framework offers a structure to advance further theoretical and empirical analysis of civil society's role in responding to climate obstructions. The examples reviewed here provide insights into the diversity of efforts to confront climate obstruction in recent years, and the various mechanisms that are developing in this arena.

Additional research is needed to determine the effectiveness, subsequent mobilization, and overall impact of various strategies and tactics to resist climate obstruction. Assessing the effectiveness of efforts to resist climate obstruction is challenging. For example, some strategies may not be effective initially but could provide the building blocks for other strategies to be used over time. Given the adaptability of different tactics used to resist climate obstruction, their calibration to specific targets, and their historical context within various campaigns, additional assessments of the emergence of coalitions and synergies among different types of actors engaged in various strategies would be valuable.

Evaluating the effectiveness of resistance to climate obstruction should also consider the role of contextual conditions in outcomes and explore the risks and likely consequences associated with various strategies and tactics in other places and contexts. In regions characterized by authoritarian states and weak civil-society institutions, some strategies and tactics may inadvertently lead to repression and weaken mobilization efforts. Recognizing the scope and scale of targets, ranging from local communities to subnational, national, and regional governments to international organizations and multinational corporations, should also be an integral part of future analysis of resistance to climate obstruction.

As climate instability worsens, many of the same powerful actors and institutions promoting climate obstruction also support other forms of misinformation to fuel polarization and expand divisiveness. In this context, revealing, resisting, and confronting climate obstruction is increasingly entangled with the complex geopolitics of fossil fuel power.[142]

NOTES

1. Sonya Ahamed, Gillian Galford, Bindu Panikkar, Donna Rizzo, and Jennie C. Stephens, "Carbon Collusion: Cooperation, Competition, and Climate Obstruction in the Global Oil and Gas Extraction Network," *Energy Policy* 190 (2024): 114103, https://doi:10.1016/j.enpol.2024.114103.
2. Mei Li, Gregory Trencher, and Jusen Asuka, "The Clean Energy Claims of BP, Chevron, ExxonMobil, and Shell: A Mismatch Between Discourse, Actions, and Investments," *PLOS One* 17, no. 2 (2022): e0263596.
3. Chapter 2 of this volume.
4. Chapter 1 of this volume.
5. Jennifer Hadden, *Networks in Contention: The Divisive Politics of Climate Change* (Cambridge University Press, 2015).
6. David Schlosberg, John S. Dryzek, and Richard B. Norgaard (eds.), *The Oxford Handbook of Climate Change and Society* (Oxford University Press, 2011).
7. Ming-sho Ho, "The Dialectic of Institutional and Extra-Institutional Tactics: Explaining the Trajectory of Taiwan's Labor Movement," *Development and Society* 44, no. 2 (2015): 247–273.

8. Todd Schifeling and Andrew J. Hoffman, "Bill McKibben's Influence on US Climate Change Discourse: Shifting Field-Level Debates Through Radical Flank Effects," *Organization & Environment* 32, no. 3 (2019): 213–233.

9. Dana R. Fisher and Sohana Nasrin, "Climate Activism and Its Effects," *Wiley Interdisciplinary Reviews: Climate Change* 12, no. 1 (2021): e683, https://doi:10.1002/wcc.683.

10. Benjamin K. Sovacool, David J. Hess, and Roberto Cantoni, "Energy Transitions from the Cradle to the Grave: A Meta-Theoretical Framework Integrating Responsible Innovation, Social Practices, and Energy Justice," *Energy Research & Social Science* 75 (May 2021): 1–16.

11. William F. Lamb, Giulio Mattioli, Sebastian Levi, J. Timmons Roberts, Stuart Capstick, Felix Creutzig, Jan C. Minx, Finn Müller-Hansen, Tara Culhane, and Julia K. Steinberger, "Discourses of Climate Delay," *Global Sustainability* 3, no. e17 (2020): 1–5.

12. Li et al., "The Clean Energy Claims of BP" Gregory Trencher, Mathieu Blondeel, and Jusen Asuka, "Do All Roads Lead to Paris?," *Climatic Change* 176 (2023): 83; Yutong Si, Dipa Desai, Diana Bozhilova, Sheila Puffer, and Jennie C. Stephens, "Fossil Fuel Companies' Climate Communication Strategies: Industry Messaging on Renewables and Natural Gas," *Energy Research & Social Science* 98 (2023): 103028.

13. Emily L. Williams, Syndney A. Bartone, Emma K. Swanson, and Leah C. Stokes, "The American Electric Utility Industry's Role in Promoting Climate Denial, Doubt, and Delay," *Environmental Research Letters* 17, no. 9 (2022): 094026.

14. NAACP, "Fossil Fueled Foolery: An Illustrated Primer on the Top 10 Manipulation Tactics of the Fossil Fuel Industry," 2019, https://live-naacp-site.pantheonsite.io/wp-content/uploads/2019/04/Fossil-Fueled-Foolery-An-Illustrated-Primer-on-the-Top-10-Manipulation-Tactics-of-the-Fossil-Fuel-Industry-FINAL-1.pdf.

15. Laura Kuhl, Jennie C. Stephens, Carlos Arriaga Serrano, Marla Perez-Lugo, Cecilio Ortiz-Garcia, and Ryan Ellis., "Fossil Fuel Interests in Puerto Rico: Perceptions of Incumbent Power and Discourses of Delay," *Energy Research & Social Science* 111 (2024): 103467.

16. Noel Healy and John Barry, "Politicizing Energy Justice and Energy System Transitions: Fossil Fuel Divestment and a 'Just Transition,'" *Energy Policy* 108 (2017): 451–459.

17. Jennifer Jacquet, "Guilt and Shame in U.S. Climate Change Communication," *Oxford Research Encyclopedia of Climate Science*, 2017, https://oxfordre.com/climatescience/view/10.1093/acrefore/9780190228620.001.0001/acrefore-9780190228620-e-575.

18. Geoffrey Supran and Naomi Oreskes, "Assessing ExxonMobil's Climate Change Communications (1977–2014)," *Environmental Research Letters* 12, no. 8 (2017): 084019.

19. Li et al., "The Clean Energy Claims of BP."

20. S. Hiltner, E. Eaton, N. Healy, A. Scerri, J. C. Stephens, and G. Supran, "Fossil Fuel Industry Influence in Higher Education: A Review and a Research Agenda," *WIREs Climate Change* 15, no. 6 (2024): e904, https://doi.org/10.1002/wcc.904.

21. Wim Carton, Inge-Meret Hougaard, Nils Markusson, and Jens Friis Lund, "Is Carbon Removal Delaying Emission Reductions?," *WIREs Climate Change* 14, no. 3 (2023): e826.

22. Holly Jean Buck, *Ending Fossil Fuels: Why Net Zero is Not Enough* (Verso Books, 2021).

23. Isak Stoddard, Kevin Anderson, Stuart Capstick, Wim Carton, Joanna Depledge, Keri Facer, Clair Gough, Frederic Hache, Claire Hoolohan, Martin Hultman, Niclas Hällström, Sivan Kartha, Sonja Klinsky, Magdalena Kuchler, Eva Lövbrand, Naghmeh Nasiritousi, Peter Newell, Glen P. Peters, Youba Sokona, Andy Stirling, Matthew Stilwell, Clive L. Spash, and Mariama Williams, "Three Decades of Climate Mitigation: Why Haven't We Bent the Global Emissions Curve?," *Annual Review of Environment and Resources* 46, no. 1 (2021): 653–689.

24. Paul Lachapelle, Patrick Belmont, Marco Grasso, Roslynn McCann, Dawn H. Gouge, Jerri Husch, Cheryl de Boer, Daniela Molzbichler, and Sarah Klain, "Academic Capture in the Anthropocene: A Framework to Assess Climate Action in Higher Education," *Climatic Change* 177, no. 3 (2024): 40.

25. Climate Social Science Network, https://cssn.org (accessed January 13, 2024).

26. Sophia Hiltner, Emily Eaton, Noel Healy, Andy Scerri, Jennie C. Stephens, and Geoffrey Supran, "Fossil Fuel Industry Influence in Higher Education: A Review and a Research Agenda," *WIREs Climate Change* 15, no. 6 (2024): e904.

27. "Input Brazil," Climate Policy Initiative, https://www.climatepolicyinitiative.org/pt-br/programas/input-brasil (accessed September 29, 2023).

28. Dyani Lewis, "Censored: Australian Scientists Say Suppression of Environment Research Is Getting Worse," *Nature* (2020), https://doi:10.1038/d41586-020-02669-8.

29. Oliver Milman, "The Silenced: Meet the Climate Whistleblowers Muzzled by Trump," *The Guardian*, September 17, 2019, https://www.theguardian.com/environment/2019/sep/17/whistleblowers-scientists-climate-crisis-trump-administration.

30. "Attacks on Science," Union of Concerned Scientists, March 30, 2023, https://www.ucsusa.org/resources/attacks-on-science.

31. Silencing Science Tracker, Columbia Law School, https://climate.law.columbia.edu/Silencing-Science-Tracker (accessed January 13, 2024).

32. EnviroDataGov Project, https://envirodatagov.org (accessed January 13, 2024).

33. Riley E. Dunlap and Aaron M. McCright, "Organized Climate Change Denial," in *The Oxford Handbook of Climate Change and Society*, ed. John S. Dryzek, Richard B. Norgaard, and David Schlosberg (Oxford University Press, 2011), 144–160.

34. Jacquet,"Guilt and Shame."

35. Banking on Climate Chaos 2023, https://www.bankingonclimatechaos.org (accessed January 13, 2024).

36. "We Won't Be Tricked: How the Fossill Fuel Industry Is Using the Dangerous 'Abatement' Distraction to Stay in Business," Oil Change International, November 30, 2023, https://priceofoil.org/2023/11/30/we-wont-be-tricked/.

37. David J. Hess and Kaelee Belletto, "Knowledge Conflicts: The Strategic Use and Effects of Expertise in Social Movements," *Sociological Inquiry* (2022), https://doi:10.1111/soin.12508.

38. "Backgrounder: Japan's Fossil Finance Threatens to Derail the Energy Transition in Asia and Globally," Oil Change International, November 2023, priceofoil.org/content/uploads/2023/11/Japan-fossil-finance-backgrounder.pdf.

39. Climate Science Rapid Response Team, https://www.climaterapidresponse.org (accessed April 1, 2025).

40. ClimaInfo, https://climainfo.org.br (accessed January 13, 2024).

41. "About CO," Climate Observatory, https://www.oc.eco.br/en (accessed September 30, 2023).

42. "About," Serrapilheira Institute, https://serrapilheira.org/en (accessed September 30, 2023).

43. Graham Dixon, Jay Hmielowski, and Yanni Ma, "Improving Climate Change Acceptance Among U.S. Conservatives Through Value-Based Message Targeting," *Science Communication* 39, no. 4 (2017): 520–534; Julian Matthews, "Maintaining a Politicised Climate of Opinion? Examining How Political Framing and Journalistic Logic Combine to Shape Speaking Opportunities in UK Elite Newspaper Reporting of Climate Change," *Public Understanding of Science* 26, no. 4 (2017): 467–480; Hong Tien Vu, Yuchen Liu, and Duc Vinh Tran, "Nationalizing a Global Phenomenon: A Study of How the Press in 45 Countries and Territories Portrays Climate Change," *Global Environmental Change* 58 (July 2019): 101942, https://doi:10.1016/j.gloenvcha.2019.101942;Nic Badullovich, Will J. Grant, and Rebecca M. Colvin, "Framing Climate Change for Effective Communication: A Systematic Map," *Environmental Research Letters* 15, no. 12 (2020): 123002, https://doi:10.1088/1748-9326/aba4c7.

44. Dominik A. Stecula and Eric Merkley, "Framing Climate Change: Economics, Ideology, and Uncertainty in American News Media Content from 1988 to 2014," *Frontiers in Communication* 4 (2019): 1–15; Sam Rowan, "Extreme Weather and Climate Policy," *Environmental Politics* 32, no. 4 (2023): 684–707.

45. Ibid.

46. Candis Callison, "Learning from Indigenous Journalists Covering Climate Change," *NiemanLab*, January 2019, https://www.niemanlab.org/2019/01/learn-from-indigenous-journalists-on-covering-climate-change.

47. Jessica L. Bolin and Lawrence C. Hamilton, "The News You Choose: News Media Preferences Amplify Views on Climate Change," Environmental Politics 27, no. 3 (2018): 455–476; John Cook, Stephan Lewandowsky, and Ullrich K. H. Ecker, "Neutralizing MisinformationTthrough Inoculation: Exposing Misleading Argumentation Techniques Reduces Their Influence," *PLOS One* 12, no. 5 (May 2017): e0175799, https://doi:10.1371/journal.pone.0175799.

48. Michael E. Mann, *The New Climate War: The Fight to Take Back Our Planet* (Public Affairs, 2021).

49. Rowan, "Extreme Weather and Climate Policy."

50. Kevin D. McCartney and Garry Gray, "Big Oil U: Canadian Media Coverage of Corporate Obstructionism and Institutional Corruption at the University of Calgary," *Canadian Journal of Sociology* 43, no. 4 (Winter 2018): 1–22.

51. Climate Nexus, climatenexus.org (accessed January 13, 2024).

52. DeSmog, https://www.desmog.com (accessed January 13, 2024).

53. Skeptical Science, https://skepticalscience.com (accessed January 13, 2024).

54. Maxine Joselow, "Exclusive: GM, Ford Knew About Climate Change 50 Years Ago," *E&E News*, 2020, https://www.eenews.net/articles/exclusive-gm-ford-knew-about-climate-change-50-years-ago/.

55. Cook et al., "Neutralizing Misinformation Through Inoculation"; Vu et al., "Nationalizing a Global Phenomenon."

56. Eisha Maharasingam-Shah and Pierre Vaux, *Climate Lockdown and the Culture Wars: How COVID-19 Sparked a New Narrative Against Climate Action* (Institute for Strategic Dialogue, 2021).

57. Wendy Hui Kyong Chun, "Beyond Verification: Algorithmic Authenticity and Polarizing Trust," *SSRC Items* (2021), https://items.ssrc.org/beyond-disinformation/beyond-verification-algorithmic-authenticity-and-polarizing-

trust; Thomas Thurnell-Read, Michael Skey, and Marie Heřmanová, *Introduction: Cultures of Authenticity* (Emerald Publishing, 2022).

58. Robert W. Gehl and Sean T. Lawson, *Social Engineering: How Crowdmasters, Phreaks, Hackers, and Trolls Created a New Form of Manipulative Communication* (MIT Press, 2022).

59. Kathie M. d'I. Treen, Hywel T. P. Williams, and Saffron J. O'Neill, "Online Misinformation About Climate Change," *Wiley Interdisciplinary Reviews: Climate Change* 11, no. 5 (2020): e665, https://doi:10.1002/wcc.665; Heidi Tworek and Sonja Solomun, "Beyond Technology: The Role of Information Interference in Climate and Election Obstruction," The Centre for Media, Technology and Democracy, 2023, https://www.mediatechdemocracy.com/all-work/beyond-technology-the-role-of-information-interference-in-climate-and-election-obstruction.

60. Conscious Advertising Network, https://www.consciousadnetwork.com (accessed January 13, 2024).

61. CAAD, https://caad.info (accessed September 30, 2023).

62. Climate Feedback, https://climatefeedback.org (accessed January 13, 2024).

63. Fergus Green and Declan Kuch, "Counting Carbon or Counting Coal? Anchoring Climate Governance in Fossil Fuel–Based Accountability Frameworks," *Global Environmental Politics* 22, no. 4 (2022): 48–69.

64. Jacquet, "Guilt and Shame."

65. Jennifer Jacquet, *Is Shame Necessary? New Uses for an Old Tool* (Knopf, 2015).

66. Chapter 12 of this volume.

67. Yadin, "Regulatory Shaming and the Problem of Corporate Climate Obstruction."

68. Sharon Yadin, *Fighting Climate Change Through Shaming* (Cambridge University Press, 2023), pp 40–42.

69. Yadin, *Fighting Climate Change Through Shaming*, 36; Jennifer Jacquet, "Shame," in *Fueling Culture: 101 Words for Energy and Environment*, ed. Imre Szeman, Jennifer Wenzel, and Patricia Yaeger (Fordham University Press, 2017), 311.

70. Elisa Aaltola, "Defensive Over Climate Change? Climate Shame as a Method of Moral Cultivation," *Journal of Agricultural and Environmental Ethics* 34, no. 1 (February 2021): 1–23.

71. Yadin, *Fighting Climate Change Through Shaming*, 68–70.

72. Mark A. Cohen and W. Kip Viscusi, "The Role of Information Disclosure in Climate Mitigation Policy," *Climate Change Economics* 3, no. 4 (2012): 1250020, https://doi:10.1142/S2010007812500200.

73. Doron Teichman and Eyal Zamir, "Exponential Growth Bias and the Law: Why Do We Save Too Little, Borrow Too Much, and Fail to React on Time to Deadly Pandemics and Climate Change?," *Vanderbilt Law Review* 75, no. 5 (2022): 1345–1400.

74. Jacquet, "Guilt and Shame."

75. Ibid.

76. "Koch Industries: Secretly Funding the Climate Denial Machine," Greenpeace, https://www.greenpeace.org/usa/fighting-climate-chaos/climate-deniers/koch-industries (accessed September 29, 2023).

77. Polluterwatch, http://polluterwatch.org (accessed September 29, 2023).

78. Greenpeace, "Koch Industries."

79. Ibid.

80. Ibid.

81. "Hashtag in Twitter: Exxonknew," https://twitter.com/hashtag/ExxonKnew?src=hashtag_click (accessed January 13, 2024).

82. "Exxon: The Road Not Taken," Inside Climate News, https://insideclimatenews. org/project/exxon-the-road-not-taken (accessed April 14, 2025).

83. Geoff Brady, "Exxon Climate Predictions Were Accurate Decades Ago. Still, It Sowed Doubt," KTOO, January 12, 2023, https://www.ktoo.org/2023/01/12/ exxon-climate-predictions-were-accurate-decades-ago-still-it-sowed-doubt.

84. Mychaylo Prystupa, "ExxonMobil Put on Mock Trial for Climate Crimes by Bill McKibben," *National Observer*, December 6, 2015, https://www. nationalobserver.com/2015/12/06/news/exxonmobil-put-mock-trial-climate-crimes-bill-mckibben.

85. Energy & Policy Institute, https://energyandpolicy.org/about (accessed September 29, 2023).

86. "Southern Company Knew: How a 'Clean Coal' Utility was Warned about Climate Change Risks Years Before It Funded Climate Disinformation, 1964–2022," Energy & Policy Institute, https://energyandpolicy.org/reports/southern-company-knew-climate-change/#southern-company-knew-executive-summary (accessed September 29, 2023).

87. InfluenceMap, https://influencemap.org (accessed September 26, 2023).

88. "The Law Firm Climate Change Scorecard," Law Students for Climate Accountability, https://www.ls4ca.org/scorecard (accessed September 29, 2023).

89. Ibid.

90. Climate Action Against Disinformation, "Climate of Misinformation: Ranking Big Tech," 2023, https://caad.info/wp-content/uploads/2023/09/Climate-of-Misinformation.pdf.

91. Ibid.

92. Climate Action Network, "Fossil of the Day at COP26," Climate Action Network, 2021, https://climatenetwork.org/cop26/fossil-of-the-day-at-cop26/

93. Climate Action Network, "Fossil of the Day at COP26."

94. Oliver Milman, "Australia Named 'Colossal Fossil' of Cop26 for 'Appalling Performance,'" *The Guardian*. November 13, 2021.

95. Chiara Giorgetti, "From Rio to Kyoto: A Study of the Involvement of Non-Governmental Organizations in the Negotiations on Climate Change," *NYU Environmental Law Journal* 7 (1999): 201.

96. Sugandha Srivastav and Ryan Rafaty, "Political Strategies to Overcome Climate Policy Obstructionism," *Perspectives on Politics* (2022): 1–11, 6.

97. C. Downie, "Business Actors, Political Resistance, and Strategies for Policymakers," *Energy Policy* 108 (2017): 583–592.

98. Adam Morton, "Tesla Accuses Australian Car Lobby Group of Making 'False Claims' About Labor's Vehicle Emissions Plan," *The Guardian*, March 4, 2024.

99. Joshua Posaner, "Germany Aims to Swerve European Parliament in Car Engine Deal," *Politico*, March 17, 2023, https://www.politico.eu/article/germany-emissions-transport-aims-to-swerve-parliament-in-car-engine-deal.

100. Climate Group, "Businesses Call for a Swift Adoption of the Regulation on CO2 Emission Standards for Cars and Vans," March 20, 2023, https://www.theclimate group.org/sitesdefault/files/2023-03/Letter%20calling%20for%20adoption%20 of%20Regulation%20on%20CO2%20emission%20for%20cars%20and%20vans %20Final%20Clean%2020.03.2023.pdf.

101. Joshua Posaner, "EU Ministers Pass 2035 Car Engine Ban Law," *Politico*, March 28, 2023, https://www.politico.eu/article/eu-ministers-pass-2035-car-engine-ban-law.

102. Robert J. Brulle, "The Climate Lobby: A Sectoral Analysis of Lobbying Spending on Climate Change in the USA, 2000 to 2016," *Climatic Change* 149, no. 3

(2018): 289–303; Anne Therese Gullberg, "Equal Access, Unequal Voice: Business and NGO Lobbying on EU Climate Policy," in *Global Corruption Report: Climate Change*, ed. Transparency International (Routledge, 2011), 41.

103. Karin Bäckstrand, Jonathan W. Kuyper, Björn-Ola Linnér and Eva Lövbrand, "Non-State Actors in Global Climate Governance: From Copenhagen to Paris and Beyond," *Environmental Politics* 26, no. 4 (2017): 561–579.

104. Natural Resources Defense Council, "With the Inflation Reduction Act, Congress Takes a Monumental Step in the Climate Fight," August 16, 2022, https://www.nrdc.org/stories/inflation-reduction-act-congress-takes-monumental-step-climate-fight.

105. U.S. Chamber of Commerce, "Coalition Letter on Reconciliation," November 18, 2021, https://www.uschamber.com/workforce/coalition-letter-on-reconciliation.

106. Natural Resources Defense Council, "NRDC Analysis: Robust Jobs and Clean Energy Growth Spurred by Build Back Better Agenda," February 9, 2022, https://www.nrdc.org/press-releases/nrdc-analysis-robust-jobs-and-clean-energy-growth-spurred-build-back-better-agenda.

107. ClimateAction100, "2023 Proxy Season: An Introduction to Climate Lobbying," https://www.climateaction100.org/news/2023-proxy-season-an-introduction-to-climate-lobbying (accessed January 15, 2024).

108. "Position on Conflicts-of-Interest and Polluting Industry Obstruction of Climate Policy in the UNFCCC Process," Climate Action Network, April 2020, https://climatenetwork.org/wp-content/uploads/2020/11/can_position_on_conflicts_of_interest-2.pdf.

109. Karen C. Seto, Steven J. Davis, Ronald B. Mitchell, Eleanor C. Stokes, Gregory Unruh, and Diana Ürge-Vorsatz, "Carbon Lock-In: Types, Causes, and Policy Implications," *Annual Review of Environment and Resources* 41 (2016): 425–452.

110. Louise Michelle Fitzgerald, "Tracing the Development of Anti-Fossil Fuel Norms: Insights from the Republic of Ireland," *Climate Policy* 23, no. 9 (2023): 1101–1114; Fergus Green, "Anti-Fossil Fuel Norms," *Climatic Change* 150, no. 1–2 (2018): 103–116.

111. Mathieu Blondeel, "Taking Away a 'Social Licence': Neo-Gramscian Perspectives on an International Fossil Fuel Divestment Norm," *Global Transitions*, 1 (2019): 200–209.

112. Gay W. Seidman, "Divestment Dynamics: Mobilizing, Shaming, and Changing the Rules," *Social Research* 82, no. 4 (2015): 1015–1037.

113. Mathieu Blondeel, Jeff Colgan, and Thijs Van de Graaf, "What Drives Norm Success? Evidence from Anti-Fossil Fuel Campaigns," *Global Environmental Politics* 19, no. 4 (2019): 63–84.

114. Georgia Piggot, "The Influence of Social Movements on Policies That Constrain Fossil Fuel Supply," *Climate Policy* 18, no. 7 (2018): 942–954; Fergus Green, "The Logic of Fossil Fuel Bans," *Nature Climate Change* 8, no. 6 (2018): 449–451.

115. Magdalena Budziszewska and Zuzanna Głód, "These Are the Very Small Things That Lead Us to That Goal: Youth Climate Strike Organizers Talk about Activism Empowering and Taxing Experiences," *Sustainability* 13 (2021): 11119, https://doi:10.3390/su131911119; Maria L. Bright and Chris Eames, "From Apathy Through Anxiety to Action: Emotions as Motivators for Youth Climate Strike Leaders," *Australian Journal of Environmental Education* 38 (2022): 13–25.

116. "Actions," Fridays for Future, https://fridaysforfuture.org/what-we-do/actions (accessed September 29, 2023).

117. "Barnard College Endowment to Divest From Climate Change Deniers," https://www.bloomberg.com/news/articles/2017-03-04/barnard-college-endowment-

to-divest-from-climate-change-deniers?embedded-checkout=true (accessed August 29, 2024).

118. Healy and Barry, "Politicizing Energy Justice and Energy System Transitions."

119. Leehi Yona and Alex Lenferna, "Fossil Fuel Divestment Movement within Universities," in *Environment, Climate Change, and International Relations*, ed. Gustavo Sosa-Nunez and Ed Atkins (E-International Relations, 2016), 190–205.

120. Julie M. Ayling, "A Contest for Legitimacy: The Divestment Movement and the Fossil Fuel Industry," *Law & Policy* 39, no. 4 (2017): 349–371; Neil Gunningham, "Review Essay: Divestment, Nonstate Governance, and Climate Change," *Law & Policy* 39, no. 4 (2017): 309–324; Noel Healy and Jessica Debski, "Fossil Fuel Divestment: Implications for the Future of Sustainability Discourse and Action within Higher Education," *Local Environment*, 22 (2017): 699–742; Luis E. Hestres and Jill E. Hopke, "Fossil Fuel Divestment: Theories of Change, Goals, and Strategies of a Growing Climate Movement," *Environmental Politics* 29, no. 3 (2020): 371–389.

121. UnKoch My Campus, https://www.unkochmycampus.org (accessed January 13, 2024).

122. Jasmine Banks, "UnKoching: No Masters, No [Corporate] Gods," OnlySky Media, March 14, 2023, https://onlysky.media/jbanks/unkoching-no-masters-no-corporate-gods.

123. Paul Lachapelle, Patrick Belmont, Marco Grasso, Roslynn McCann, Dawn H. Gouge, Jerri Husch, Cheryl Boer, Daniela Molzbichler, and Sarah Klain, "Academic Capture in the Anthropocene: A Framework to Assess Climate Cction in Higher Education," *Climatic Change* 177, no. 40 (2024), https://doi.org/10.1007/s10584-024-03696-4.

124. Jessica Grady-Benson and Brinda Sarathy, "Fossil Fuel Divestment in US Higher Education: Student-Led Organising for Climate Justice," *Local Environment* 21, no. 6 (2016): 661–681.

125. Jennie C. Stephens, Peter C. Frumhoff, and Leehi Yona, "The Role of College and University Faculty in the Fossil Fuel Divestment Movement," *Elementa: Science of the Anthropocene* 6 (2018): 41.

126. Eve Bratman, Kate Brunette, Deirdre C. Shelly, and Simon Nicholson, "Justice is the Goal: Divestment as Climate Change Resistance," *Journal of Environmental Studies and Sciences* 6 (2016): 677–690; Healy and Debski, "Fossil Fuel Divestment."

127. "Enhancing Gifts and Grants Golicies," https://president.brown.edu/president/enhancing-gifts-and-grants-policies (accessed August 29, 2024).

128. "Grace on Fossil Fuel Industry Ties," https://www.cam.ac.uk/notices/grace-on-fossil-fuel-industry-ties (accessed August 29, 2024).

129. "Accountable Allies: The Undue Influence of Fossil Fuel Money in Academia," Data for Progress, March 1, 2023, https://www.dataforprogress.org/memos/accountable-allies-the-undue-influence-of-fossil-fuel-money-in-academia.

130. Sierra Club, "America, Let's Move Beyond Coal and Gas," Sierra Club, 2023, https://coal.sierraclub.org.

131. Cristiana Losekann, "The Politics of People Affected by Extractivism in Latin America," *Revista Brasileira de Ciência Política* 20 (2016): 121–164; Oilwatch, "Oilwatch Latin America Statement: The Climate Debate is Not About CO2 Molecules!," October 4, 2021, https://www.oilwatch.org/2021/10/04/2039.

132. Michael L. Cepek, *Life in Oil: Cofán Survival in the Petroleum Fields of Amazonia* (University of Texas Press, 2018).

133. Terence Garrett, "Proposed Liquefied Natural Gas (LNG) Terminals in the Rio Grande Valley of Texas—Citizen Group Participation Versus Natural Gas Corporations," *Journal of Global Responsibility* 9, no. 1 (2018): 58–72.

134. D. J. Hess, Y. R. Kim, and K. Belletto, "How Do Coalitions Stop Pipelines? Conditions That Affect Strategic Action Mobilizations and Their Outcomes," *Energy Research & Social Science* 98 (2023): 102914;George Hoberg, *The Resistance Dilemma: Place-Based Movements and the Climate Crisis* (MIT Press, 2021).

135. Laurence L. Delina, "Coal Development and Its Discontents: Modes, Strategies, and Tactics of a Localized, Yet Networked, Anti-Coal Mobilisation in Central Philippines," *The Extractive Industries and Society* 9 (March 2022): 101043, https://doi:10.1016/j.exis.2022.101043.

136. Ibid.

137. Noel Healy, Jennie C. Stephens, and Stephanie A. Malin, "Embodied Energy Injustices: Unveiling and Politicizing the Transboundary Harms of Fossil Fuel Extractivism and Fossil Fuel Supply Chains," *Energy Research & Social Science* 48 (2019): 219–234.

138. Peter Newell, Harro van Asselt, and Freddie Daley, "Building a Fossil Fuel Non-Proliferation Treaty: Key Elements," *Earth System Governance* 14 (2022): 100159, https://doi:10.1016/j.esg.2022.100159.

139. Fergus Green, "Fossil Free Zones: A Proposal," *Climate Policy* 22, nos. 9–10 (2022): 1356–1362.

140. Damian Carrington, "Fossil Fuel Recruiters Banned from Three More UK Universities," *The Guardian*, December 1, 2022, https://www.theguardian.com/environment/2022/dec/01/fossil-fuel-recruiters-banned-from-three-more-uk-universities.

141. Clean Creatives, https://cleancreatives.org/smoke-and-mirrors (accessed April 1, 2025).

142. Gerard Toal, *Oceans Rise and Empires Fall: Why Geopolitics Hastens the Climate Catastrophe* (Oxford University Press, 2024).

INDEX

Tables, figures, notes, and boxes are indicated by *t*, *f*, *n*, and *b* following the page number

For the benefit of digital users, indexed terms that span two pages (e.g., 52–53) may, on occasion, appear on only one of those pages.

animal agriculture industry's use
    of, 105, 119
  coal industry's use of, 8–10, 75
  confronting obstruction of, 16, 174,
    371–372, 380
  disinformation spread
    through, 162–163, 167, 175
  government investigations
    on, 351–352
  legislation and regulation proposed
    for, 105, 119
  litigation and, 333–335, 345, 347
  misinformation spread through, 10,
    16, 30, 166–167, 333–335, 345, 353
  oil and gas industry's use of, 8, 30, 33,
    39, 162, 166–167, 351
  PR firms' use of, 9–10, 33
  research directions for, 175–176
  utilities industry's use of, 75
Africa. *See also* South Africa
  climate-induced violent extremism
    in, 151
  dangers of data collecting in, 260
  development rhetoric in, 214, 221,
    228
  oil and gas lock-in in, 221, 286
  think tanks in, 228
  unions in, 20
agribusiness. *See* agriculture; animal
    agriculture industry
agriculture. *See also* animal agriculture
    industry
  adaptation obstruction in, 316
  carbon pollution from, 165
  mitigation efforts for, 255–256, 285
  population growth prompting increase
    in, 218
  protests by farmers against
    decarbonization of, 282
  research directions for, 295–296
Al Jaber, Sultan, 34–35, 291
Alliance of Small Island States
    (AOSIS), 15, 225–226, 288, 294
Almeida Azevedo, José Carlos
    de, 199–200*b*
Amazon, the
  deforestation in, 14, 107, 227, 255,
    279, 309
  dismantling of protection measures
    for, 218, 227, 281–282

exploitation in exchange for jobs and
    investments in, 254
  infrastructure projects in, 224–225
  misuse of funding protection of, 228
  social carbon credits exploited in, 344
  sunsetting industry campaigns
    in, 378–379
American Clean Energy and Security
    Act (Waxman-Markey bill) (2009)
    (US), 41–42, 75
American Coal Council, 66*t*, 67–68
American Farm Bureau Federation
    (AFBF), 9, 103–104
American Petroleum Institute (API)
  advertising used to mislead by, 30,
    41–42, 351
  Climate Action Framework of, 39
  consultant groups used by, 40–42
  early knowledge of climate change
    by, 35, 346
  economic scare tactics used by, 40–42
  Energy Citizens initiative of, 41–42
  government hearings on, 351
  key role in climate obstruction of, 2, 8,
    32–33
  litigation against, 14, 335
  net-zero emissions pledges of, 30, 45
  research funded by, 19, 35, 346
  solutionism strategy used by, 45
*Amorema and Amoretgrap v. Sustainable
    Carbon* (2021), 344
Andes Libre (think tank), 9, 224
animal agriculture industry
  academic research and, 112–113, 118,
    119, 166–167
  advertising used by, 105, 119
  CAFOs in, 114*b*, 114–115*b*
  campaigns of, 104–105, 115
  coalition building and management
    by, 116–118
  common narratives of obstruction
    in, 108–109
  confronting obstruction by, 119–120
  Dutch nitrogen crisis and, 117*b*
  emissions role of, 98–101
  essentiality/good for the environment
    claims used by, 109–110
  legislation and regulation
    and, 101–104, 110–111, 119
  litigation and, 119

research directions for, 154–155
strategy mobility use by, 143
think tanks and, 2–4, 141–146, 171,
192–193, 202, 274, 279
"wise use" roots in, 140–141
conspiracy theories, 7, 147–151,
169–170, 187, 196–197, 203
consultant firms, 14, 19, 39–42, 113,
170
contrarians, 2–4, 37, 118, 192–193
CORSIA scheme, 70, 76
COVID-19, 70, 163, 309–310

D
Danish Crown, 14, 19, 98, 112–113, 119
da Silva, Luiz Inácio Lula, 254, 281–282
DCI Group, 9–10, 67–68
delay tactics. *See* climate delay
Democratic Party (US), 41–42, 189,
190–193, 244, 247
denialism. *See* climate denial
DeSmog (news platform), 17, 119, 121,
167, 174, 293–294, 371
Dieselgate (2015), 74, 84
disinformation
academic research exposure of, 174
accountability for, 172
advertising's spread of, 162–163, 167,
175
calls for universal definition of, 172
confronting obstruction
and, 368–372, 374, 375–376
definition of, 4–5
digital platforms' spread of, 16,
164–167
media coverage's spread of, 163–165,
168–172, 174–175
misinformation distinguished
from, 4–5, 185–186
rapid attribution of, 174
rebuttal of, 16, 171–172, 375, 376
regulation of, 340
think tanks' spread of, 9
tracking of, 170–172
typology of publics susceptible
to, 194–195
utilities industry and, 78–79
Dublin Declaration of Scientists on
the Societal Role of Livestock
(2022), 113
Dutch nitrogen crisis, 117*b*

E
Edelman (PR firm), 9–10, 18, 41–42,
67–68, 167
Edison Electric Institute (EEI), 68–69,
78–79
electric vehicles (EVs), 8, 18, 67, 74, 82,
241–242, 280–281, 375
elite cues, 188, 191–192, 202
emissions. *See* animal agriculture
industry; coal industry; Kyoto
Protocol (1997); net-zero emissions
pledges; oil and gas industry; Paris
Agreement (2015); transportation
industry; utilities industry
Energy Citizens (initiative), 8, 41–42
England. *See* United Kingdom (UK)
Eni (Italian oil major), 36, 348
Environmental Foundation Limited
(EFL), 230
environmental justice, 195, 229–230,
369
environmental movements, 40, 141,
143–144
ESG (environmental, social, and
governance) ratings, 173, 339,
340–341
Europe. *See also* European Union (EU);
France; Germany; Netherlands;
Russia; United Kingdom (UK)
cap-and-trade programs in, 47–48
climate denial in, 38, 145–146
far right in, 150
fossil fuel industry in, 35–38, 42,
47–48, 278*t*
industry lobbying in, 42
litigation in, 348
misinformation in, 256
non-state actor obstruction in, 291
online media networks in, 164–165
public opinion in, 255–256
subnational governments in, 255–259
transportation industry in, 74
unions in, 20
European Automobile Manufacturers'
Association, 8–9, 69
European Union (EU)
advertising restrictions in, 105
animal agriculture industry
in, 100–102, 104–105
ban on gas vehicles in, 75
Climate Law of, 282

coal industry in, 10, 65–67, 83,
219–220
emissions of, 277*t*
just transition in, 83
lobbying in, 219–220
NGOs in, 229
political suppression of activism
in, 76–77
PSNs in, 229
right to develop claims in, 81–82
industries. *See* animal agriculture industry; coal industry; military industry;
oil and gas industry; transportation
industry; utilities industry
Inflation Reduction Act (IRA) (2022)
(US), 16, 39, 46, 280, 342–343, 352,
376
InfluenceMap (NGO), 19, 70, 374
influencers, 17, 147–148, 164–165, 167,
168–169, 174–175
information and influence campaigns.
See campaigns
Inside Climate News, 2–4, 17, 44, 103,
374
Institute for Agriculture and Trade Policy
(IATP), 99–100
Institute of Economic Affairs
(IEA), 142–143, 222–223
Instituto Liberdade (Brazil), 9, 224, 228
Intergovernmental Panel on Climate
Change (IPCC). *See also* Conference of the Parties (COP); Kyoto
Protocol (1997); Paris Agreement
(2015); United Nations Framework
Convention on Climate Change
(UNFCCC)
adaptation obstruction at, 309,
311–312
assessment reports released by, 1–2,
161, 283, 311–312
consensus-based decision-making
at, 9, 275
discrediting attempts against, 31, 69,
148, 284–285, 292
establishment and goals of, 36
healthy and sustainable diets
emphasized by, 99–100
historical emissions structural factor
for, 276
leaked draft of AR6 WGIII report
of, 133 n.155
limitations on effectiveness of, 11–12

loss and damage claims and, 216,
288–289
material endowments structural factor
for, 276
media coverage emphasized by, 166
mitigation emphasized by, 311–312
overview of, 272, 294–296
procedural obstruction of, 283–284
role of, 11
scientific findings and, 284–285
substantial obstruction of, 284–290
UNFCCC's relationship to, 312
vested interests as structural factor
at, 279
International Air Transport Association
(IATA), 8, 63–64, 70, 76
International Chamber of Commerce
(ICC), 47, 291–292
International Civil Aviation Organization
(UN ICAO), 74–76, 84
International Climate Change
Partnership (ICCP), 291–292
International Maritime Organization
(IMO), 69–70, 74–75, 82–84, 86
international obstruction. *See also*
Intergovernmental Panel on
Climate Change (IPCC); Kyoto
Protocol (1997); Paris Agreement
(2015); United Nations Framework
Convention on Climate Change
(UNFCCC)
academic research and, 295–296
adaptation in, 287–288
campaigns of, 279, 292, 294
CBDR-RC principle in, 289–290
confronting of, 294
conservative leaders
facilitating, 276–277
domestic politics and, 277–279
equity claims in, 289–290
fossil fuel industry and, 279–280,
291–294
historical emissions and
responsibilities in, 276, 277*t*
lobbying and, 274, 279, 295–296
loss and damage claims in, 288–289
major countries in, 279–282
making sense of, 270–273
material endowments, values, and
ideas in, 276–277
mitigation with energy focus
and, 285–287

international obstruction (*Continued*)
NGOS and, 274, 290–292, 294
non-state actor strategies
for, 290–294
overview of, 270–273, 294–296
procedural form of, 283–284
research directions for, 272–273
scientific findings and, 284–285
state actor strategies for, 283–290
structural and domestic factors
in, 276–279
substantial form of, 284–290
think tanks and, 274, 279
vested interests and, 277–279
International Petroleum Industry Environmental Conservation Association
(IPIECA), 47, 291–292
investigations. *See* academic research;
government investigations and
hearings; non-governmental
organizations (NGOs)
IPCC. *See* Intergovernmental Panel on
Climate Change (IPCC)
Iran
carbon tax opposed by, 285–286
coal as key industry in, 11
fossil fuel industry in, 11, 32, 278*t*
international obstruction of, 285–286
Irish animal agriculture industry, 9–10,
115, 118

J

Japan
coal promoted by, 10, 222
emissions of, 277*t*
fossil fuel industry in, 278*t*, 369
international obstruction by, 10
nationalism in, 75
transportation industry in, 74, 79
utilities industry in, 68–69, 80–81
JBS (food company)
animal agriculture industry and, 120*b*
environmental rating of, 118
litigation against, 14, 119, 120*b*, 335
net-zero emissions pledge
accountability for, 14, 119,
120*b*
size and importance of, 100
journalists, 17, 32, 164, 167–171, 174,
175, 192–193. *See also* confronting

obstruction; *DeSmog* (news platform); *Inside Climate News*; media
coverage
just transition, 11, 20, 80, 85–86, 214,
257

K

Khanna, Ro, 351–352
Koch, Charles, 16, 39–40, 145, 378
Koch, David, 16, 145, 378
Koch Industries, 37–40, 42, 145
Kyoto Protocol (1997)
climate policy reactions to, 41,
191–192
conference establishing, 38, 47–48,
285–286
as first quantified obligation to reduce
emissions, 273–274
mitigation emphasized over adaptation
by, 312
Paris Agreement as indirect recognition
of, 274
rejections of, 38, 273–274, 277, 280,
289, 292
undermining of, 38, 103, 274,
283–284

L

labor movements, 20, 220
labor unions. *See* unions
Latin America. *See also* Amazon, the;
Brazil; Chile; Colombia; Mexico;
Peru; Venezuela
agribusiness in, 218
climate denialism in, 148
conservatism in, 148, 228
deforestation in, 255
fossil fuel industry in, 253–255
main features of subnational
government level in, 251–252
online media networks in, 164–165
other organized interests in, 255
subnational governments
in, 251–255, 259–260
unions in, 20
Laudato Si (Francis), 146
La Via Campesina, 119–120
Law Students for Climate
Accountability, 17–18, 374
lawsuits. *See* litigation

government investigations and, 30
legislation and regulation and, 338
litigation and, 335
oil and gas industry and, 6, 29–33, 35,
44–46
New Zealand
animal agriculture industry
in, 100–102, 105–106
attempt to weaken methane accounting
rules by, 11
emissions of, 100–102
non-governmental organizations (NGOs)
adaptation obstruction and, 308,
317–318
advertising challenged by, 16
BINGOs, 118, 290–293
confronting obstruction by, 367–370,
373–376
conservatism and, 141–142
disinformation rebutted by, 16, 375
international obstruction and, 274,
290–292, 294
litigation by, 14, 119
lobbying and, 14, 16, 375–376
media coverage monitoring
by, 370–372
misinformation weakened by, 16
PR firms and, 165–166
research by, 16, 367–370
role of, 16
strategies of, 14, 16
think tanks and, 142
non-state actors. *See* consultant firms;
front groups; non-governmental
organizations (NGOs); public
relations (PR) firms; think tanks;
trade associations
North America. *See* Canada; Mexico;
United States

O

Obama, Barack
attempt at first comprehensive climate
legislation under, 41–42
Clean Power Plan of, 67–68, 75, 77
coal industry lobbying during
presidency of, 75
NDC pledge under, 280
racial spillover effect and, 195
Ocasio-Cortez, Alexandria, 192
Oceania. *See* Australia; New Zealand

OECD. *See* Organization for Economic
Co-operation and Development
(OECD)
oil and gas industry
academic research and, 30–31, 34–37
advertising by, 8, 30, 33, 39, 162,
166–167, 351
attribution of responsibility for climate
obstruction to, 5, 32
campaigns of, 37–40, 43
cap-and-trade proposals for, 41–42,
47–48
carbon tax opposed by, 40
CCS and, 45–46, 48
climate denial by, 34–43
consultants and experts used
by, 40–41
co-opting the policy arena by, 43–48
early knowledge of climate change
by, 2–4, 36, 37
economic scare tactics used by, 39–43
emissions role of, 29–33
government investigations on, 15,
29–30, 32–33, 45, 351–352
growth plans despite climate rhetoric
of, 30, 38, 45–47
internal documents in, 19, 30, 35, 37,
44, 48, 336
legislation and regulation and, 36–48
litigation and, 14, 44–45, 50, 336
lobbying and, 18, 39–43, 46, 293
net-zero emissions pledges and, 6,
29–33, 35, 44–46
overview of, 29–33, 49–50
as part of the solution to climate
change, 43–48
pivoting to denial in, 36–39
PR firms used by, 33
public acceptance of climate change
reality by, 38
research directions for, 50
role of, 7–8, 31
scientific uncertainty amplified
by, 34–35
solutionism strategy used by, 43–48
strategies for obstruction used
by, 34–48
subnational governments
and, 246–248, 253–255, 258–259
tactical delay used by, 39–43